HISTORY OF ENTOMOLOGY

HISTORY OF ENTOMOLOGY

RAY F. SMITH, *Editor*
University of California

THOMAS E. MITTLER, *Editor*
University of California

CARROLL N. SMITH, *Editor*
U.S. Department of Agriculture, Ret.

1973

PUBLISHED BY
ANNUAL REVIEWS INC.
IN COOPERATION WITH THE
ENTOMOLOGICAL SOCIETY OF AMERICA

ANNUAL REVIEWS INC.
4139 EL CAMINO WAY, PALO ALTO, CALIFORNIA 94306

ANNUAL REVIEWS INC.
PALO ALTO, CALIFORNIA, U.S.A.

Library of Congress Catalog Card Number: 73–76435

ISBN 0–8243–2101–7

Assistant Editor	Jean E. McComish
Indexers	Mary Glass, Leigh Dowling
Subject Indexer	Frances M. Hall

PRINTED AND BOUND IN THE UNITED STATES OF AMERICA

BY GEORGE BANTA COMPANY, INC.

PREFACE

The production of this supplementary volume to the *Annual Review of Entomology* was far longer and more arduous than the typical annual issues. It had its origin in the 1966 meeting of the ARE Editorial Committee in Portland, Oregon. In response to a suggestion from Edward A. Steinhaus, the Committee discussed at length the merits of a volume devoted entirely to the history of entomology. At that point there was agreement that Annual Reviews should undertake such a volume and that it should emphasize "the personalities of past scientists who contributed to the development of entomological ideas and principles". The Editors of ARE were charged with developing an outline for the volume. In the following year, Annual Reviews Inc. agreed to publish the history volume with the Entomological Society of America underwriting the first thousand copies. Subsequent discussions of the Editorial Committee with the able assistance of S. L. Tuxen modified the original emphasis as perusal of this volume will reveal. A final format was agreed upon in 1968 and invitations went to prospective authors in 1969. In all these preliminary preparations the Editorial Committee is indebted to the valuable assistance of Miss Beryl V. Daniel, Senior Assistant Editor.

When an author or a small group of authors working together write a book, the book starts as an outline and ends as a complete book. But when an editorial committee outlines a book, and then seeks 25 to 30 authors for the individual chapters, there is no assurance that the resulting book will be complete. In fact, it is almost certain not to be complete. In the first place, certain specialized items not obviously necessary to a particular chapter might be included by some potential authors but omitted by others. In the second place, some authors will stop their treatment of a subject at a certain point in time whereas others will bring their treatments up to the present. And, third, some having agreed to write a particular chapter find because of health, the pressure of other duties, or some other reason that they cannot produce their chapter on schedule. All of these difficulties plagued preparation of the present volume. Perhaps the most obvious deficiency in the history is the absence of a chapter on insecticides. Steps are being taken to fill some of the more important lacunae by the inclusion of omitted subjects in forthcoming volumes of the *Annual Review of Entomology*.

We hope you will enjoy reading this history of man's growing awareness of the intricate relationship between insects and the other components of his environment. The account begins with the recognition of the importance of insects, the more obvious aspects, as depicted in the art, customs, and religions of the earliest human cultures. It follows the first steps taken to utilize the benefits derivable from insects and to minimize the damage they may cause. It follows the growth of knowledge about insect development, the recognition of the identities of the myriad species, and the continuing refinement and precision of scientific information on insect morphology, behavior, physiology, ecology, and evolution. It culminates in the history of recent advances in man's effort to apply this information to his coexistence with insects in a common environment. The story is replete with famous names from all ages and continents of leaders in the development of our understanding of the group often called our most important rivals.

The Editorial Committee takes this occasion to thank the authors who labored so diligently and searched so many widely scattered sources to bring this history together. Our thanks also go to our efficient Assistant Editor, Miss Jean E. McComish, for the care and effort she devoted to the preparation of this volume, and to the George Banta Company for their customary skill in its publication.

THE EDITORIAL COMMITTEE

CONTENTS

EARLY ENTOMOLOGY IN EAST ASIA[1]

MASAYASU KONISHI AND YOSIAKI ITÔ[2]
Hokko Chemical Industry Co. Ltd. and National Institute of Agricultural Sciences
Tokyo, Japan

INTRODUCTION

There are two attitudes among Westerners on the history of the sciences in the Far East, especially in China. The major one is the ignorance of Chinese contributions in the development of the Western culture. Many authors on the history of science consider that "nearly everything of value in Chinese science came from the West" (36). But Europe could not conquer the world without paper, printing, gunpowder, and magnetic compasses for shipping, all of which came from China (36–38, 40, 61, 62). The minor one is blind admiration of old Chinese science, especially Chinese medicine. A new step for the development of precise understanding of Eastern science has been taken by Joseph Needham in his uncompleted article "Science and Civilisation in China" (1954–) (36–38). He recognized not only that the Chinese had the priority in a great many fields of early science and technology but also that the basic philosophy of Chinese science was completely different from that of the West.

In contrast with the Western atomic or particle viewpoint on the natural world which enhanced the European's development of experimental and analytical sciences, Chinese scientific philosophy can be characterized by its organismic or field viewpoint (36, 37, 62). For the ancient Chinese scientists, a supra-organismic interrelationship among things is the basic character of the world. This organismic philosophy was, according to Needham (36, 37), considered the reason why the Chinese could develop 1. a unique system of medicine where all the organs in a human body are interrelated, 2. a deep insight on the magnetism which allowed the Chinese to invent the magnetic compass in 119 B.C., and 3. recognition of the cause

[1] Romanization of Chinese names and the translation of special terms and titles of books are made, as a rule, according to Needham's (1954) usage (36–38). Needham's romanization system is a modification of the Wade system and different from the new one adopted at the General Meeting of the People's Delegation, the People's Republic of China in 1965. Here we used the former because the change of spelling from a classic one to a new one is very difficult. Japanese names are romanized following the kunrei-system (Ministry of Education) unless the person in question used a special system. The names of persons living in pre-Meiji periods were arranged following the order of surname–middle name or conjunction–given name. The pen name was given in parenthesis. Thus Kaibara-Atunobu (Ekken) is equivalent to A. Kaibara (Ekken Kaibara).

[2] Present address: Okinawa Prefectural Agricultural Experiment Station, Naha, Japan.

of tides in the San Kuo (three kingdoms) period (221–265 A.D.), that is, an approach to the concept of gravitation. The organismic philosophy could also enhance a remarkable development of an idea of the food-web in natural communities in ancient China (see later section). Thus, if we want to understand the development of Chinese science, we must discard our atomic viewpoint. We have not, however, enough pages for a full discussion of this attractive concept here.

The following review covers the development of entomology in China and Japan before the modern sciences had started in these countries. We cannot discuss in this review early entomology in Korea and Indochina; but we would like to point out that although these nations were under the influence of China, they developed their own culture. Thus the Hanoi National University was established in 1070, 17 years after Bologna and 50 years before Paris (17) and printing with copper-made types was first carried out in Korea in the early thirteenth century, some two hundred years before Gutenberg.

EARLY ENTOMOLOGY IN CHINA

In his extensive review on the history of entomology, Bodenheimer (5) recognized four epochs, of which the first was "Die orientalische Urzeit." This epoch started from "Der chinesische Kurturkreis." Thus the origin of Chinese entomology was considered the oldest in the world. Although the Chinese invented sericulture in 4700 B.C. (7), there was a large gap in the historical study of early entomology in China, because Western entomologists (e.g. 14, 22, 43) neglected this. Chou's review (7) on the history of early Chinese entomology may be the first one to fill this gap, although some publications are incorrectly dated in this book.

Chinese entomology had developed along three routes: 1. the use of insects for purposes other than drugs, 2. pest control, and 3. the use of insects for medicinal use (that is, a part of *Pên Tshao,* herbals or pharmacopoeia, which developed in the later period in the studies on the natural history of insects). The following summary is made (according to this order) based primarily, unless otherwise stated, on Chou's article (7).

Useful Insects

Silkworms.—Archaeological evidence (fossil cocoons in stone age remains) shows that the cultivation of the silkworm, *Bombyx mori,* was begun before 4700 B.C. Sericulture occupied an important part of peasant life in China between 4000–3000 B.C., and the culture of mulberry plants and the indoor rearing of the silkworm were started in 1200 B.C. (for details on the history of sericulture, see Yokoyama in this volume).

Honey bee.—Rearing of the oriental honey bee, *Apis indica,* was carried out in the fifth century to obtain honey, and by the sixth century the beeswax was also utilized. Apiculture had so developed by the thirteenth century that it became a major occupation in some agricultural villages. So Beck's words that "... China is the native land of the sugar cane, and for this reason bees were rarely cultivated ..." (4) are to be doubted.

Scale insects.—According to ancient Chinese records, the culture of wax insects dates back to the middle of the thirteenth century; about that time Chinese candles, made of pela wax, were produced from *Ericerus pela* (9). Pela wax was also used as a drug. Lac, produced from *Laccifer lacca,* was also used in the seventh century as a drug and dye, but this was not of Chinese origin but was brought from India (9).

Galls.—The Chinese utilized insect galls to create drugs, dyes, and tanning agents. The major type of gall was that produced by *Melaphis chinensis* on *Rhus javanica* plants (66). By the end of the sixteenth century it was recorded that galls were produced by insects (29, 45, 50).

> In the fifth [May] and sixth [June] moons there are small ant-like insects which eat its [*Rhus javanica*] sap and when old bear their eggs [larvae] and form small balls on the leaves. . . . The shell [of galls] is hard and brittle and empty within except for a small insect. . . . The natives in the hills collect them before the frosts set in, steam them to kill the insects, and sell them. (45)

It must be emphasized that Redi (46), who first denied experimentally the spontaneous generation of flesh-eating flies in 1668, believed that the insects in galls occurred spontaneously. The Chinese discovered that the insects found in galls grew from young laid by adults a hundred years earlier than Vallisnieri's observation in Europe (51). In this respect, Porter-Smith's (44) statement on the Chinese ignorance of morbid character of galls is considered to be a misinterpretation.

Pest Control

The development of the agricultural sciences in China dates back a remarkably long time. It must be noted that ancient agricultural literature not only described the cultural practice of crop plants with the method of fertilization, but also the pioneering use of metals as insecticides. This fact corresponds to a feature of ancient Chinese pharmacology; they were not unfamiliar with the use of metals as a drug for man (36).

For example, *Shen Nung Pên Tshao Ching* (Pharmacopoeia of the Heavenly Husbandman) (2), which was believed to be written about 100 to 200 A.D., mentioned that mercury and arsenic could be used to control the body louse.[3] The use of naturally occurring arsenic sulfide to control lice and

[3] Chinese and Japanese books published before the Civil Revolution (1911) and Meiji Restoration (1868), respectively, are cited only in the text. *Italics* show original Chinese or Japanese titles. 'Books' which were not printed but were spread by hand-written copies were also cited with the year of the completion of the original text.

the use of *Veratrum* (Liliaceous plant) for killing insect pests affecting man was also described. Fumigation by burning toxic plants to kill insect pests dates back to ca 1200 B.C. (*Chou Li:* Record of the Rites of Chou, ca 200 B.C. to 9 A.D.). The use of pig oil to control parasites of sheep and lime to control crop pests were described in Chia Ssu-Hsieh's (528–549) *Chhi Min Yao Shu* (Important Arts for the People's Welfare). Sulfur (powder) and copper (verdigris) were used to kill lice in the seventh century (Su Kung, 659; *Thang Pên Tshao:* Pharmacopoeia of the Thang Dynasty) and in the tenth century (7), respectively. According to Li's (1596) *Pên Tshao Kang Mu* (The Great Pharmacopoeia) (29), Ko Hung, a great alchemist of the fourth century, recorded that white arsenic, obtained as a byproduct from copper smelting, was applied to the roots of rice plants during replanting to protect them from insect pests (38). Studies along this line developed the use of cinnabar (HgS) and copper alum to control head lice (7), green vitriol (iron sulfate) to control fly larvae (7), and sulfur powder to fill the hole made by flowering tree borers [Hsü Kuang-Chhi (1639); *Nung Chên Chhüan Shu:* Complete Treatise on Agriculture], all before the tenth century. Oil was used in 1185 to control rice-plant insect pests (49).

The Chinese were the first to use natural enemies to control insect pests (36). Nests of an ant, *Oecophylla smaragdina,* were sold near Canton in the third century to use for control of citrus insect pests such as *Tesseratoma papillosa* (Chi Han, ca 300; *Nan Fang Tshao Mu Chuang:* Records of the Plants and Trees of the Southern Regions).

The idea of the food web, which was so important in the work of Charles Darwin (1859), was first recorded in China in the third century.

> A factor which increases the abundance of a certain bird will indirectly benefit a population of aphids because of the thinning effect which it will have on the coccinellid beetles which eat the aphids but are themselves eaten by the bird. (37).

According to Needham, this "Chinese principle of control" may be based on an organismic philosophy or five elements theory (37). [For the uniqueness of the five elements theory, see Needham, 36.] Thus an important bud of modern ecology had sprouted in China 1500 years ago.

China developed an outstanding system of bureaucracy, which was then introduced into Japan during the Tokugawa Shogunate. At first this would be a factor which could enhance the development of applied science (36), but it became, in reality, a factor that suppressed further development.

The government of the Shang Kingdom (ca 1520 B.C. to 1030 B.C.) appointed antilocust officers and used a bonfire to catch insect pests in 1200 B.C. Yao Chhung, a minister of the Thang Dynasty (650 to 721 A.D.), organized nationwide counter measures for forecasting locust plagues (37). The first law to control insect pests was written in China as "laws of anti-locust" in 1182. A leaflet for locust control was published by the government in the beginning of the thirteenth century (Anon. 1208; *Puhuang Thu*:

Illustration of Methods to Collect Locusts). Chou (7) cited long lists of known outbreaks of the locust, *Locusta migratoria manilensis,* (from 707 B.C. to 1642), lepidopterous stalk borers (from 718 B.C. to 1635), and the army-worm, *Pseudaletia separata* (from 275 A.D. to 1724). Governments of ancient China also carried out detailed astronomical observations, which allowed present radioastronomers to use ancient Chinese records for their study; for example, the number of sun spots visible was recorded before 100 A.D. Such a record was used by Ma (30) to carry out an extensive statistical analysis of more than 1000 years of records of locust plagues (Figure 1). Ma could conclude that there was no correlation between locust outbreaks in China and the sun-spot cycle and that the factor that brought the rise in locust population was the drought.

It is now known that a fluctuation in water conditions at a delta of a large river can cause locust propagation to reach outbreak conditions, mainly by increasing the survival of eggs and young nymphs; the Chinese reached the same conclusion in the seventeenth century. Hsü Kuang-Chhi (1639, loc. cit.) suggested, with numerous technical recommendations to control the locust, that public engineering works to suspend the flood-drought cycle in the delta could bring an end to the locust plagues. Detailed descriptions of the food plants, bionomics, effects of temperature and precipitation on out-breaks, and methods of control were made in Ku Yen's (1878) *Chih Huang Chhüan Fa* (Complete Book of Locust Control).

Pên Tshao and Related Studies

The basic pattern of Chinese medical science was established, according to *Shih Chi* (Historical Record; edited in the Former Han Dynasty ca 90 B.C.), by Pien Chhio in the Later Chou Dynasty (722–221 B.C.). Studies of herbals (*Pên Tshao*) began in this period (23). The first treatise of the herbals is the *Shen Nung Pên Tshao Ching* (loc. cit.) (2), believed to be completed in the Later Han Dynasty (ca second century).[4] Twenty-nine kinds of insects (including 21 kinds of true Insecta) or their products were described in this book as drug materials. Several books on herbals were published thereafter, e.g. Thang Shen-Wei's (1108) *Chêng Lei Pên Tshao* (Reorganized Pharmacopoeia), where natural materials were classified according to their pharmacological effects. A remarkable turning in the classification of natural materials from pharmacological to biological was made in Li Shih-Chen's (1596) *Pên Tshao Kang Mu* (29, 45, 50), which was evaluated as "the greatest scientific achievement of the Ming [dynasty]" (36).

Li classified 1898 kinds of naturally occuring drugs in 16 divisions and 62 subdivisions according to the order of minerals, plants, and animals (42). His descriptions of the various synonyms of organisms, localities, seasons of

[4] The original copy of this article was lost. Three volumes edited by Thao Hung-Ching (500 A.D.) were referred to; another edition was reprinted recently in Taiwan (2).

emergence, and collection methods were an important contribution to natural history (56).

Some people supposed that the classification of organisms in *Pên Tshao Kang Mu* was made under the influence of Western natural history (27, 45, 52). Read wrote: "The classification herein given was established in the Ming dynasty and is undoubtedly of European origin, it follows almost exactly that made by Aristotle."

Uéno (58) and Needham (36) did not agree with this viewpoint. Uéno wrote that he could not find any proof showing the influence of western natural history on the *Pên Tshao.* While Aristotle, in his treatment of insects, made "distinctions between those with wings present and those in which the wings are absent. . . ." (60), Li Shih-Chen did not adopt the wing as a criterion for insect classification.

Some 106 kinds of "insects" (including 73 kinds of true Insecta) were described in Volumes 39–42 of the *Pên Tshao Kang Mu.* They were arranged according to three main classes such as oviparous, spontaneous, and moisture-born insects.[5] Uéno (56) noted that this classification might be based on four births (oviparous, spontaneous, moisture-born, and viviparous) in Buddhism. Thus the criterion of classification was different from that in Aristotle.

Many authors attempted to identify common or scientific names of insects described in *Pên Tshao Kang Mu* (5–8, 44, 45, 66). A table of synonyms is presented here (Table 1) based on Yano (66), Chu & Kao (8), and Chou (7), because all of these authors are entomologists who could read original texts of ***Pên Tshao.***

TABLE 1. Synonymy of insects in *Pên Tshao Kang Mu*

Chinese name[a]	Scientific name
I. Oviparous	
Mi Feng	Hym. *Apis indica* (7, 8, 66), *A. mellifera* (8)
T'u Feng	Hym. *Vespula* (7, 8), *Bombus* (66)
Ta Huang Feng	Hym. *Vespa* (7, 8, 66)
Chu Feng	Hym. Apidae (66), Xylocopidae (8), *Megachile* (7)
Ch'ih Ch'ih Feng	Hym. Psammocharidae (8), *Psammochares* (66), *Pompilus* (7)
Tu Chueh Feng	Hym. Siricidae (8), *Sirex* (67)
Yi Weng	Hym. Trypoxylidae (7, 8), Eumenidae, Psammocharidae & Sphecidae (1)

[5] The phrase translated in the second category *'Hua-Sêng'* (Born from Change) by Read (45) as "produced by metamorphosis" is misleading. Li considered that the insects whose eggs could not be found might be born spontaneously and classified under the category of *'Hua-Sêng'* (56).

TABLE 1.—(*Continued*)

Chinese name[a]	Scientific name
Ch'ung Pai La	Hem. Wax produced by *Ericerus pela* (7, 8, 66)
Tzu Keng	Hem. Lac produced by *Laccifer lacca* (7, 8, 66)
Wu Pei Tzu	Hem. Gall of *Rhus*-tree produced by *Melaphis chinensis* (7, 8, 66)
T'ang Lang	Dicty. Mantidae (7, 8, 66)
Ch'ao Weng	Lep. Cocoon of Cochlidionidae (8, 66), *Cnidocampa flavescens* (7)
Ts'an	Lep. *Bombyx mori* (7, 8, 66)
Shih Ts'an	Larva of Trichoptera (7, 8, 66)
Chiu Hsiang Ch'ung	Hem. *Coridius chinensis* (=*Aspongopus*) (7, 8)
Kou Chi Ch'ung	Lep. Larva on *Lycium* tree (8, 66), *Theretra oldenlandiae* (7)
Huai Hsiang Ch'ung	Lep. *Papilio machaon* (7, 8, 66)
Ch'ing Fu	Hem. Cicadidae (8), *Mogannia* (7)
Chia Tieh	Lepidoptera (7, 8, 66)
Ching Ling	Odonata (7, 8, 66)
Shu Chi	Hem. *Lycorma delicatula* (7, 8), *Fulgora* (66)
Tsao Mao	Col. Scarabaeidae (66)
Pan Mao	Col. *Mylabris* (7, 8, 66)
Yüan Ch'ing	Col. *Lytta* (7, 8, 66)
Ke Shang T'ing Chang	Col. *Epicauta* (7, 8, 66)
Ti Tan	Col. *Meloe* (7, 66), Carabidae (8)
Yi	Hym. Formicidae (7, 8, 66)
Pai Yi	Isoptera (7, 8, 66)
Ch'ing Yao Ch'ung	Col. *Paederus* (7, 66), Mutillidae (66)
Ch'ü	Dip. Larva of Muscidae (7, 66), Cyclorrhapha (8)
Ying	Dip. Muscidae (7, 66), Muscoidea (8)
Kou Ying	Dip. *Hippobosca longipennis* (7, 8, 66)
Pi Shih	Hem. *Cimex* (7), *C. lectularius* (8, 66), *C. rotundatus* (8)
Jen Shih	Anoplura. *Pediculus humanus* (7, 8, 66), *Phthirus pubis* (7, 8, 66)
II. Spontaneous	
Ch' Ts'ao	Col. Larva of Scarabaeidae (7, 8, 66)
Mu Tu Ch'ung	Col. Larva of Cerambycidae (7, 8, 66), Buprestidae (66); Lep. Larva of Cossidae (8, 66), Hepialidae (66)
Sang Tu Ch'ung	Col. Larva of Cerambycidae (8, 66), *Apriona rugicollis* (7, 8, 66)
Liu Tu Ch'ung	Col. Larva of Cerambycidae (8, 66)
T'ao Tu Ch'ung	Lep. *Dichocrocis punctiferalis* (66), *Paranthrene* (8), *Phassus signifer* (7)
Kuei Tu Ch'ung	Col. Larva of Cerambycidae (7, 8)
Tsao Tu Ch'ung	Col. Larva of Cerambycidae (7, 8)
Chu Tu Ch'ung	Col. Cerambycidae (66), *Cyrtotrachelus* (7, 8)
Lu Tu Ch'ung	Lep. Larva of Pyralidae (66)
Ts'ang Er Tu Ch'ung	Col. Larva of Cerambycidae (8, 66)
Ch'a Chu Ch'ung	Larvae of Coleoptera and Lepidoptera injurious to tea-products (8)
Cha Ch'an	Hem. Cicadidae (7, 8, 66)

TABLE 2.—(*Continued*)

Chinese name[a]	Scientific name
Ch'an Hua	Entomogeneous fungus of nymph of Cicadidae: *Cordyceps* (7, 8, 66), *Isaria* (7, 8)
Ch'iang Lang	Col. Coprophagous Scarabaeidae—Coprinae (7, 66); *Scarabaeus Catharsius* (8)
Fou Yu	Col. Coprophagous Scarabaeidae—Coprinae (66), *Aphodius* (8)
T'ien She Ch'ung	Col. Coprophagous Scarabaeidae—Geotrupinae (66)
T'ien Niu	Col. Cerambycidae (7, 8, 66)
Fei Sheng Ch'ung	Col. *Xylotrupes dichotomus* (66), Cerambycidae (8)
Lou Ku	Orth. *Gryllotalpa* (7, 8), *G. africana* (66)
Ying Huo	Col. Lampyridae (7, 8, 66)
Yi Yü	Thysanura. *Lepisma* (8, 66), *Ctenolepisma* (7, 8)
Che Ch'ung	Dictyop. Wingless Blattidae—*Polyphaga* (7, 66), *Eupolyphaga* & *Steleophaga* (8)
Fei Lien	Dictyop. Blattidae—*Periplaneta* (7, 8, 66), *Blattella* & *Blatta* (8)
Hsing Yeh	Col. Carabidae (8), *Carabus* (66); Dictyop. *Blatta orientalis* (7)
Tsao Ma	Orth. *Diestrammena* (7, 8, 66)
T'su Chih	Orth. Gryllidae (7, 8, 66)
Fu Chung	Orth. Locustidae (7, 8, 66)
Chi Ting Ch'ung	Col. Buprestidae (8, 66), *Chrysochroa elegans* (7, 66)
Chin Kuei Tzu	Col. Scarabaeidae—*Anomala* & *Mimela* (7, 66), Chrysomelidae—Cassidinae (8)
T'ien K'e Ch'ung	Col. Carabidae (8)
K'ou T'ou Ch'ung	Col. Elateridae (7, 8, 66)
Mei Tieh	Lep. Nymphalidae (8), *Vanessa* & *Polygonia* (7)
Mu Meng	Dip. Tabanidae (8), *Tabanus* (66)
Fei Meng	Dip. *Tabanus* (7, 8, 66), *Chrysops* (7, 8)
Wen Tzu	Dip. Culicidae (8, 66)
Jui Tzu	Dip. Simuliidae (66)
Chu Shih	Hem. Aphididae (8, 66), Aleyrodidae & Coccidae (8), *Asterolecanium bambusae* (7)
III. Moisture-born	
Ch'u Sou	Dermaptera (8)[b]
Hsi Kuei Ch'ung	Hem. Belostomatidae (8), *Lethocerus* (*Belostoma*) (7)
Shui Min	Hem. Gerridae (7, 8, 66)
Ch'ih Ch'ung	Col. Gyrinidae (7, 8, 66)
Sha No Tzu	Neur. Myrmeleonidae (8, 66)
Chin T'san	Lep. *Tinea* (8)
Tsa La Ch'ung	Col. Silphidae (8)

[a] Romanization of Chinese names in this table follows Read (45), which is basically the same as the Wade system but different from Needham's which was used in the text. The major difference between Read's system and Needham's is the use of an apostrophe in place of 'h', for example: Chhih = Ch'ih.

[b] Yano (66) considered that the Ch'u Sou might not be Dermaptera.

The *Pên Tshao Kang Mu* is the greatest gift to us from the peaceful Chinese Middle Ages, but, by the beginning of the seventeenth century, the position of the government was greatly deteriorating (36). After a long period of internal disturbances, the Chhing (Manchu) dynasty was established in 1644 and was maintained until 1911 when the civil revolution took place. During the Chhing dynasty, the Chinese suffered from long periods of internal disturbances in addition to pressure from Western nations. Thus further development of Chinese classical entomology ceased. The *Pên Tshao,* however, exerted a strong influence in the development of natural history in China and especially in Japan.

EARLY ENTOMOLOGY IN JAPAN

The history of entomology in Japan was reviewed by Esaki (10–13), Higashi (21: a chronological table), Miyake (33), Sasaki (48), Uéno (57), and Yano (67: agricultural entomology). Entomology was also reviewed in Uéno's (56, 58) texts on the history of general zoology. Our review is based mainly on discussions by Esaki and Uéno.

Prehistory to the Establishment of Tokugawa Shogunate (1602)

Ancient Japanese culture was developed under the strong influence of China and Korea. The Japanese learned sericulture from Korea, probably in the third or fourth century (68). The Japanese also learned apiculture from Korea in the seventh century, but this was not widely established before the ninth century (54).

Pên Tshao and Chinese medicine were brought to ancient Japan from China or via Korea. The *Shen Nung Pên Tshao Ching* (or its abridged or revised edition) is believed to have been brought to Japan in the sixth century (41, 42). Sixty kinds of drugs, including the lac (*Laccifer lacca*), galls (*Cynips gallae-tinctoriae*), and beeswax, were kept in Syôsô-in, one of the oldest museums in the world (23, 58). Fukae-no-Sukehito's (918) *Honzô-Wamyô* (Japanese Names of Herbals) and Minamoto-no-Shitagau's (tenth century) *Wamyô Ruizyû-syô* (Abridged Catalogue of Japanese Names) were published; these were Chinese-Japanese dictionaries or pharmacopoeia describing 27 and 59 kinds of insects, respectively.

From this ancient period to the establishment of the Tokugawa Shogunate, the Japanese dealt with insects mainly as subject for literature and ornament. Sericulture and a remarkable richness of insect fauna in Japan (which has more than 230 species of butterflies as compared with 69 in Great Britain) led the Japanese to be more familiar with insects, as compared with people in Western countries (22). Thus, in poems such as *Man-yô-syû* (edited in the eighth century), novels, and diaries we find many names of insects. A notable one is a tale of *Mushi Mezuru Himegimi* (The Lady Who Loved Insects) (1) in the *Tutumi Tyûnagon Monogatari* (Anon., eleventh to twelfth century). In this short story, the life of a princess whose hobby is collecting and rearing insects is described. As her recognition of the metamorphosis of lepidopterous insects was precise, it may be considered that the Japanese had a good knowledge of the natural history of insects in this period.

In the *Azuma-Kagami* (Historical Records of Kamakura Shogunate), we can find a record of the mass migration of butterflies which took place on September 7, 1248 (59). The Japanese used insects for ornamental purposes, for example, a miniature shrine *'Tamamusi-no-zusi'* kept in Hôryû-zi shrine (built in 607) was decked with 9083 elytra taken from a beautiful buprestid beetle, *Chrysochroa fulgidissima* (63). Similar methods of decoration by elytra of the same species can be seen in Korean harnesses of the sixth century which suggests a close tie between Korea and Japan (63).

In 1543, Portuguese castaways brought the gun to Japan. Thereafter, the Japanese began to learn about Western science from Europeans.

Edo Period (1603–1867)

In 1603, Tokugawa-Ieyasu put an end to the period of internal disturbance in Japan and established a new feudal government (the Tokugawa Shogunate) in Edo (Tokyo). The prolonged era of peace in the Edo period favored the development of the study of natural history in Japan.

Pên Tshao Kang Mu and the development of Japanese pharmacopoeia.— The first edition of the *Pên Tshao Kang Mu* arrived in Japan in 1607, beginning a new epoch in the history of Japanese biology which lasted until the Meiji Restoration (58). Many editions of *Pên Tshao Kang Mu* and related literature were printed in Japan during the seventeenth century.

As in the sixteenth century in Europe, where many encyclopedic naturalists such as Gesner, Aldrovandus, and Mouffet were working (51), illustrated encyclopedia of the natural history were published in Japan in the seventeenth and eighteenth centuries; notable ones of which are those by Nakamura-Tekisai (1666) and Terasima-Ryôan (1713). The former's book, *Kinmô Zui* (Illustrated Encyclopedia for Public Education), was the first encyclopedia in Japan which showed many insects in enlarged figures. Ryôan's book, *Wakan Sansai Zue* (Illustrated Encyclopedia of Japanese and Chinese Materials), was the largest encyclopedia published in the Edo period; and the insects were classified in oviparous, spontaneous, and moisture-born groups as in the *Pên Tshao*. Ryôan used a magnifying glass to draw illustrations of the flea, louse, and other small insects.

Among a number of books on herbals published in the Edo period, *Yamato Honzô* [(Natural History of Japan), 16 Vols. Appendices 2 Vols. 1708; Figures 2 Vols. 1715], by Kaibara-Atunobu (Ekken), was considered very important, because Ekken was not a blind follower of the classification system of the *Pên Tshao*. He presented a new system of classification based on his original observations (58). Of 1362 kinds of naturally occurring materials, 772 kinds were common with *Pên Tshao Kang Mu* but 358 kinds were first described in this book. Ekken classified insects under the heading of terrestrial and aquatic. Esaki (10) considered this system of classification as the first ecological classification system in Asia, comparable to similar classification systems by Aldrovandus (3), Jonstonus (25), and Roesel von Rosenhof (47). The European authors, however, made further classifications based on criteria such as presence and absence of wings, the number of legs, and the metamorphosis (10, 60).

Ekken adopted the idea of Hua-Sêng (see the preceding section) in *Pên Tshao* in his book. Although he knew that mosquitoes emerge from wigglers, dragonflies from their larvae, flies (Cyclorrhapha: Diptera) from maggots, and butterflies and moths from caterpillars, he did not observe that these larvae were born from eggs.

Besides *Yamato Honzô,* Ono-Motohiro's (Ranzan) large book on herbals, *Honzô-Kômoku Keimô* (A Textbook of Natural History, 1803–1806), must be mentioned. Ranzan observed that the firefly develops from the egg through aquatic larva to adult, denying the description of *Pên Tshao* that "decaying grass becomes fireflies". Ranzan noted that "the aphids attract ants with their sweet taste and they can reproduce much faster when ants visit them." As Ranzan followed a classic taxonomical system of the *Pên Tshao,* however, the value of his extensive work in the development of Japanese natural history in respect of the systematization of the animal kingdom could not exceed Ekken's contribution (56).

Tyûhu (Illustration of Insects).—A special feature of entomology in the feudal age of Japan may be the development of *Tyûhu. Tyûhu* (Chufu) is a collection of colored paintings of insects with their names, sometimes dates, localities of collection, and notes on habits. The humid climate in Japan and the lack of chemicals to protect specimens from insects and fungi lead amateurs to draw precise pictures of insects. Although many kinds of *Tyûhu* were known, none of them were published.

Some Daimyôs drew (or ordered their artists to draw) paintings of insects as a hobby. Of these, *Tyûhus* by Hosokawa-Sigetaka are extraordinarily scientific (Figure 2). He presented paintings on the metamorphosis of 37 species of insects (16); even the duration of each developmental stage was noted on Hosokawa's *Tyûhu* (28).

The most excellent among the *Tyûhus* is *Sentyûhu* (Illustrations of a Thousand Insects, 1811) by Kurimoto-Masayosi (Tansyû). Miyake (32, 33) wrote that all the paintings in *Sentyûhu* (adults of 268 species and immature stages of several dozen species were illustrated) were quite precise so that he could identify most of the species by their scientific names.

In respect to the *Tyûhu,* the Owari school is worth mentioning. This school, centered in Nagoya (Owari), made quite precise observations on morphology and biology of insects. Since representatives of this school, such as Mizutani-Sukeroku (Hôbun) and Yosida-Takanori (Zyakusô-an), aimed to

FIGURE 2. A colored picture of a sphinx moth, *Theretra japonica,* with two larval stages and pupa (from Hosokawa-Sigekata's *Syôrui Seisya*: Illustrations of Living Insects). From notes, one can read that the larvae and pupa were reared on the Boston ivy, *Parthenocissus tricuspidata,* in the autumn of 1761 and the moth emerged in April 1762.

study insects only as a hobby or for scientific reasons (not from pharmacological and agricultural interests), they can be considered the equals of European naturalists.

Despite much work in this area, neither the comparative morphological study nor the systematic classification of insects developed during the Edo period (10). Lack of a comparative viewpoint was an unfortunate weakness in the study of biology in early Japan. In addition, experiments which were made even by Redi were not conducted by these workers.

Rangaku.—In contrast with China, where Western sciences were introduced by statesmen, development of the *Rangaku* (learning of Western science) in the Edo period was forwarded at first by private citizens. Nagasaki was the only port opened to the West and Holland was the only nation with which the Japanese were in contact under the isolation policy of the

Tokugawa Shogunate. Many citizens visited Nagasaki and attempted to get scientific information there, but they did so under serious surveillance by the government. Thus the Japenese knew of the compound microscope before 1765 (49). Morisima-Tyûryô (1797: *Kômô-zatuwa,* Miscellaneous Notes on Europe) introduced the microscope to the Japanese with the enlarged figures of small insects (56, 58). However, the use of the microscope in biological studies was limited until the Meiji period. This limited use may be due not to a technological factor (the Japanese could reproduce a number of guns in an amazingly short time) but to an organismic philosophy on organisms. For the Japanese (and also for the Chinese) natural subjects must be observed per se. They do not seem to like analyzing a natural subject in isolated parts.

Udagawa-Yôan's *Kontyû-Tûron* (An Introduction to Entomology, 1827) and *Dogaku-Keigen-kô* (A Textbook of Zoology, 1835), both unpublished, were introductions to Western zoology. The Latin names of seven orders of insects by Linné were first introduced in Japan in these books (24, 55, 58).

Among Westerners who visited Japan as members of the Dutch diplomatic establishment, E. Kämpfer, C. P. Thunberg, and P. F. von Siebold are noted for their contribution in collecting Japanese plants and animals and for introducing Japanese natural history to Europe. Kämpfer's description of Japanese insects in his introduction to Japanese history (26) was quoted in Bodenheimer's book (6). Kämpfer noted a basic similarity between Japanese insect fauna and that of Europe (58). The first description of Japanese insects according to the binomial nomenclature system was made by Thunberg (53) (see also Esaki, 10).

Von Siebold's influence on the development of the *Rangaku* was greatest. During his first stay in Japan from 1823 to 1829, he educated many Japanese physicians and naturalists, and, in turn, introduced Japan to European people through his many books. Insect specimens brought by von Siebold into Europe were described by de Haan (18, 19), Felder (15), and others.

Pest control.—The control of agricultural insect pests was an important problem in the Japanese economy, as in China, and there were many descriptions of old methods of pest control (20). Miyashita (34, 35) reviewed old records of pest outbreaks. In Kagawa and Fukuoka Prefectures, outbreaks of planthoppers on rice plants were recorded in 701 and 815 A.D., respectively. In Fukuoka Prefecture, planthopper outbreaks were reported 16 times during the eighteenth century with an average interval of 5.2 years. Even earlier there were many records of planthopper outbreaks. Miyashita (34, 35) writes that in 701 A.D. outbreaks were reported in 15 prefectures.

Fire is considered to have been used at first in connection with prayer for divine help in controlling outbreaks of insect pests (*Musi okuri*—torchlight procession), but during the Middle Ages peasants knew that bonfires could attract and kill insect pests.

One method of pest control widely used in the Edo period was pouring oil on the water surface of rice paddies to control leaf- and planthoppers. The oils were of animal and plant origin; and after 1670, when the remarkable ability of whale oil to kill insects was discovered (31), whale oil was used primarily. The Japanese might have learned of the use of oil from China but whale oil was first used in Japan (20); this method of control became widespread after Okura-Nagatune's *Zoko-roku* (Control of Insect Pests of The Rice Plant, 1826), the first book on applied entomology in Japan, was published (Figure 3).

In a later book (*Zyokô-roku Kôhen;* Further Book of Control of Insect Pests of the Rice Plant, 1884) (Figure 3), Okura-Nagatune described the use of various oils, slaked lime, and bittern as insecticides. Processes such as dusting, spraying, pouring into the soil, fumigation, seed-coating, and dipping were used during the Edo period (20). It can be seen in laws proclaimed by some Daimyôs in the early nineteenth century that the presence of residues of an insecticide or fertilizer (lime) in rice plant tissues and its poisonous effect on fish were already recognized and avoided (20).

Since the Japanese like the chrysanthemum very much, there were many books published in the Edo period on the biology and control of chrysanthemum pests (64, 65). Among them, Simizu-Kanzi's *Kadan Yôgiku-syû* (Cultivation of Garden Chrysanthemums, 1715) is notable because it includes an approach to integrated control. He wrote:

> . . . as the chrysanthemum longicorn beetle (*Phytoecia rufiventris*) emerges from old stubs of chrysanthemum, these stubs must be removed. . . . The attack of the beetle can be avoided by placing water-softened sea-weed (*Eisenia bicyclis*) around the nursery bed.

Thus Simizu recognized the repellent effect of *Eisenia.* Yôzyuken-Unpô's *Kadan Kikka Taizen* (Textbook of Chrysanthemums in Flower beds, 1717) noted also that as the chrysanthemum beetle attacks a weed (*Artemisia vulgaris*) one can arrange this weed around the chrysanthemum bed to attract the beetle. Unpô suggested also that one can avoid injury from the chrysanthemum beetle by mixed-planting of chrysanthemum with a liliaceous plant (*Allium odorum*). The descriptions of the biology of insect pests of flowering plants and methods of controlling them are modern and precise in comparison to those for insect pests of crop plants. The reason might be that the peasants, who were miserably poor in the Edo period, could not study new methods for the control of pests but rich people who liked flowering plants could (10).

A negligible influence of Western science can be seen in the applied entomology of the Edo period, whereas medicine of this period was under strong Western influence. Pest control techniques in the Edo period were, under the influence of China, developed in Japan to a high level. This line of study was continued even after the Meiji Restoration, thereafter scientific education in Japan was completely changed to the European system and modern entomology in Japan started in 1881 in the Komaba Agricultural College in Tokyo under the guidance of C. Sasaki.

After 1897, following a severe outbreak of rice planthoppers throughout Japan, the government recognized the need to study insects and established many agricultural experiment stations. S. Matsumura's *Nippon Kontyûgaku* (The Japanese Entomology), the first Japanese book on modern entomology written by a Japanese with special reference to the classification of Japanese insects, was published in the next year (1898).

DISCUSSION: POSSIBLE CAUSES OF STAGNATION IN SCIENTIFIC DEVELOPMENT IN EAST ASIA

Readers may have understood that biology in East Asia, especially in China, had reached a higher level in the ancient period and early Middle Ages than in Europe. The further development of science and technology, including entomology, however, was retarded in China after the sixteenth century (the middle of the Ming Dynasty). What was the cause of this stagnation? Political and social factors, the deterioration of policy and internal disturbances from the second half of the Ming Dynasty to the end of the Chhing Dynasty, and invasion by foreign countries certainly retarded the development of science (40); but there may be another factor which might be responsible for the fact that the Chinese did not develop modern science in the seventeenth and eighteenth centuries. One of the possible causes of this was the isolation of China from other nations (58). In Europe, more than 20 nations are living together within an area smaller than China. Retardation of scientific studies in a country of Europe could be compensated by scientific development in other countries. Consequently, different ideas could be combined to create a more systematic viewpoint. On the other hand, China was itself a universe in the ancient period, all the needs of people were provided within this area (61). China gave paper and the magnetic compass to Europe and Japan but did not receive anything before the seventeenth century (36). The Chinese developed a unique organismic philosophy but did not develop an analytical point of view or the law of causality They had not developed geometry, formal logic, or mechanics (62, see also 38). Their organismic viewpoint led them to develop unique medical ideas, but retarded the development of comparative morphology, anatomy, and experimental physiology. For entomology, the microscope did not come into wide use in China and Japan. The Chinese and Japanese made detailed observations on insects and formulated many ecological ideas but did not make experiments as Redi (46) did in 1668. The Chinese and Japanese had a Gesner and Mouffet but did not have a Harvey, Malpighi, and Linné.

During the 300 years of the Edo Period, the Tokugawa Shogunate closed the country to Western nations. This isolation policy undoubtedly suppressed the interchange of scientific ideas and techniques. Thus Japanese science

FIGURE 3. Control of planthoppers on rice by a bonfire (extreme bottom-right), whale oil pouring (two peasants with wooden pails), and beating with a cluster of bamboo. A pan above fire at the top right indicates melting the whale oil (After Ôkura-Nagatune's *Zyokô-roku Kôhen,* 1844).

maintained its weak points, which were the same as those of China, until the Meiji Restoration. It must be noted, however, that there might have been some merit in this policy. Without the isolation policy, Japan might have suffered from an invasion by white people.

Although early science in Japan progressed in a path completely different from that of the West, the Japanese developed their cultural background to a high level during this period. The Meiji Government thus could change completely its scientific policy to the European style. Consequently, the amazing rate of the development of science and technology in Japan started.

In this stage, the Chinese and Koreans suffered seriously from European, Japanese, and American imperialism. The development of modern science could not begin until very recently in China and Korea, but the rate of development is now extraordinarily high.

The atomic viewpoint in European scientific philosophy was a prerequisite for the development of modern science. It must be noted, however, that the modern (analytical) science has produced, in turn, a terrible situation on the earth: nuclear war, mental corruption, and pollution. Is a new, systematic viewpoint of scientific philosophy needed now? Is it possible to construct a modern universal science (39) combining viewpoints of the West and East? Future studies on the early history of Asian science must be concentrated on this point.

ACKNOWLEDGMENTS

The authors are greatly indebted to Mr. Hitoshi Hasegawa for his advice and help in searching for rare literature, and to Mr. Noboru Maruyama for his help in the romanization of Chinese names.

LITERATURE CITED[6]

1. 1929. *The Lady Who Loved Insects,* transl. A. Waley. London: Blackmore. 35 pp.
2. 1970. *Shen Nung Pên Tshao Ching.* (Pharmacopoeia of the Heavenly Husbandman), reprint of a copy edited by Wo. Taipei: Chung Hwa. (C)
3. Aldrovandus, U. 1618. *De Animalibus Insectis Libri Septem.* Reprint (Original: 1602). Francofurti: J. Treudel. 299 pp.
4. Beck, B. F. 1938. *Honey and Health: A Nutrimental, Medicinal and Historical Commentary.* New York: McBride. 271 pp.
5. Bodenheimer, F. S. 1928. *Materialien zur Geschichte der Entomologie bis Linné,* Band I. Berlin: Junk. 498 pp.
6. Ibid 1929. Band II. 486 pp.
7. Chou, I. 1957. *Chungkuo Tsaochhi Khunchhung Hsüeh Yenchiu Shih.* (History of Entomological Studies in China before the 20th Century.) Peking: Science Publishing Office. 132 pp. (C).
8. Chu, H. F., Kao, C. S. 1950. (A commentary on the insects mentioned in the Pen Ts'ao Kang Mu.) *Acta Entomol. Sinica,* 1: 234–62. (Ce)
9. Ealand, C. A. 1915. *Insects and Man.* London: Grant Richards. 341 pp.
10. Esaki, T. 1952–1953. (Some topics in the history of entomology in Japan.) *Sin-Kontyû,* 5(8):27–31, (9):21–29, (10):21–24, (11): 21–29, (12):34–37, 6(1):20–25. (J)
11. Esaki, T. 1954. (Seventy-five years of Japanese zoology: entomology.) *Dôbutsugaku Zasshi* 63: 303–9. (J)
12. Esaki, T. 1955–1956. Stories from the history of entomology in Japan since 1868. 1–4.) *Kontyû,* 23:120–24, 182–88, 24:51–64, 111–21. (J)
13. Esaki, T. 1957. (Brief history of entomology in Japan.) *Kontyû,* 25:151–96. (J)
14. Essig, E. O., 1965. *A History of Entomology.* Reprint (Original: 1931). New York: Hafner. 1029 pp.
15. Felder, C., Felder, R. 1862. Observationes de Lepidopteris nonnullis Chinae centralis et Japoniae. *Wiener Entomol. Monats,* 6:22–32, 33–40
16. Fukui, K. 1937. (Studies on Science and Arts by Daimyôs.) *Syolaimyô no Gakuzyutu to Bungei no Kenkyû.* Tokyo: Kôseikaku. 840 pp. (J)
17. Galston, A. W., Signer, E. 1971. Education and science in North Vietnam. *Science,* 174:379–85.
18. de Haan, W. 1835. Mémoire sur les Métamorphoses des Coléoptères. *Nouvelles Annales du Muséum d'Histoire Naturelle,* 4:125–65.
19. de Haan, W. 1842–1844. Bijdragen tot de Kennis der Orthoptera. *Verhandelingen over de Natuurlijke Geschiedenis der Nederlandsche overzeesche Bezittingen,* Lief 16, 18, 19/20, 24. pp. 45–248
20. Hasegawa, H., Konishi, M. 1971. (Progress of insect pest control in Japan before 1900). *Syokubutu-Bôeki* (Plant Protection, Tokyo), 25: 117–22. (J)
21. Higashi, M. 1941. (2600 years of history of Japanese entomology.) *Takarazuka Kontyûkanpô,* No. 5:1–8. (J)
22. Howard, L. O. 1930. *A History of Applied Entomology.* Smithson. Misc. Collections, 84, 564 pp.
23. Huard, P., Wong, M. 1968. *Chinese Medicine.* London: Weidenfeld & Nicolson, cited from Japanese edition (transl. K. Takahasi et al), 1972, Tokyo: Heibonsya. 300 pp.
24. Itô, T. 1917. (Udagawa-Yôan, the founder of the scientific entomology in Japan and his unpublished manuscripts.) *Nawa-Yasusi Si Kanreki-Kinen Kisô Ronbunsyû,* 2:182–86. Gifu: Nawa Entomol. Lab. (J)
25. Jonstonus, J. 1653. *Historiae Nat-*

[6] Symbols C, Ce, J, and Je at the end of Chinese and Japanese literature mean in Chinese, in Chinese with English summary, in Japanese, and in Japanese with English summary, respectively.

uralis de Insectis. Libri III. Francofurti: Haered. Math. Merian. 200 pp.

26. Kämpfer, E. 1906. *The History of Japan,* 3 vols. Reprint (Original: 1727–1728). Glasgow: J. M. Lehose & Sons. 337, 397, 386 pp.
27. Kimura, K. 1942. Honzô (Pharmacopoeia) In *Sina Kagaku Keizaisi* (History of Science and Economy of China), ed. T. Andô, pp. 143–252. Tokyo: Hakuyôsya. (J)
28. Kotake, H. 1913. (Explanation of plate: Picture of insects painted by Hosokawa-Sigetaka.) *Kontyû-Sekai* (Insect World), 17:36–37. (J)
29. Li Shih-Chen 1959. *Pên Tshao Kang Mu* (The Great Pharmacopoeia). Reprint (Original: 1596). Peking. (See also Read, 1941, for English translation). (C)
30. Ma, S. C. 1958. The population dynamics of the oriental migratory locust (*Locusta migratoria manilensis* Meyen) in China. *Acta Entomol. Sinica,* 8:1–40. (Ce)
31. Mine, Y. 1910. (A history of planthopper control with oils poured into paddy fields.) *Kontyû-Sekai* (Insect World), 14:440–42. (J)
32. Miyake, T. 1912. (On the illustrations of a thousand insects by Kurimoto-Zuiken.) *Kontyû-Sekai* (Insect World), 16:28–30. (J)
33. Miyake, T. 1919. *Kontyû-gaku Hanron* (A General Textbook of Entomology). Vol. II. Tokyo: Syôkabô, 672 pp. (see pp. 655–70 for the history of Japanese entomology) (J)
34. Miyashita, K. 1961. (Chronological tables on the outbreaks of insect pests.) *Syokubutu-Bôeki* (Plant Protection, Tokyo), 15:75–81. (J)
35. Miyashita, K. 1963. Outbreaks and population fluctuations of insects, with special reference to agricultural insect pests in Japan. *Bull. Nat. Inst. Agr. Sci., Jap.,* Ser. C. No. 15:99–170
36. Needham, J. 1954. *Science and Civilisation in China. I. Introductory Orientations.* Cambridge: Cambridge Univ. Press. 318 pp.
37. Needham, J. 1956. *Science and Civilisation in China. II. History of Scientific Thought.* Cambridge: Cambridge Univ. Press. 696 pp.
38. Needham, J. 1959. *Science and Civilisation in China. III. Mathematics and the Sciences of the Heavens and the Earth.* Cambridge: Cambridge Univ. Press. 877 pp.
39. Needham, J. 1963. Poverties and triumphs of the Chinese Scientific tradition. In *Scientific Change,* ed A. C. Crombie. London. Reprinted in (40)
40. Needham, J. 1959. *The Grand Titration. Science and Society in East and West.* London: Allen & Unwin. 350 pp.
41. Ogawa, T. 1964. *Igaku no Rekisi* (The History of Medicine). Tokyo: Tyû-ô Kôronsya. 234 pp. (J)
42. Okanishi, T. 1958. (Importation of Chinese pharmacopoeia and its effects on Japanese pharmacology.) In *Meizi-zen Nippon Yakubutugaku-si* (History of Pharmacology in Japan before the Meiji era). II:1–265. Tokyo: Nippon Gakuzyutu Sinkôkai. (J)
43. Osborn, H. 1952. *A Brief History of Entomology.* Columbus: Spahr & Glenn. 303 pp.
44. Porter-Smith, F. 1871. *Materia Medica and Natural History of China.* Shanghai: American Presbyterian. 237 pp.
45. Read, B. E. 1941. *Chinese Materia Medica. Insect Drugs.* Peking Nat. Hist. Bull., Suppl. 213 pp.
46. Redi, F. 1668. *Esperienze Intorno alla Generazione degl' Insetti.* Cited from English translation by M. Bigelow, 1909. Chicago: Open Court. 160 pp.
47. Roesel von Rosenhof, A. J. 1764–1768. *De Natuurlyke Historie der Insecten.* 4 Pts. Amsterdam: C. H. Bohn, H de Wit en H. Gartman. 287 pp.
48. Sasaki, C. 1926. (Development of entomology in Japan.) *Kontyû.* 1:1–5. (J)
49. Shirai, M. 1933. *Honzôgaku Ronkô* (Papers on Natural History). Vol. 1. Tokyo: Syunyôdô. 522 pp. (J)
50. Shirai, M. et al 1930. *Tôtyû Kokuyaku Honzôkômoku,* (Japanese translation of *Pên Tshao Kang Mu* with scientific notes.), Vol. 10. Tokyo: Syunyôdô. 587 pp. (Entomological notes were written by M. Yano, 1930. (J)

51. Singer, C. 1959. *A History of Biology* (Revised Edition) London: Abelard-Schuman. 580 pp.
52. Takahashi, S. 1958. (Medication in China and Its Effect on Japanese Pharmacology.) In *Meizi-zen Nippon Yakubutugaku-si* (History of Pharmacology in Japan before the Meiji Period), 2:268–513. Tokyo: Nippon Gakuzyutu Sinkôkai. (J)
53. Thunberg, K. P. 1781. *Dissertatio Entomologica novas Insectorum Species Sistens,* Pars 1. Uppsala: Edmann. 28 pp.
54. Tokuda, Y. 1913. *Mitubati* (The Honeybee) Sapporo. 629 pp. (J)
55. Uéno, M. 1943. ("Dôgaku Keigenkô" by Udagawa-Yôan.) *Kagakusi Kenkyû* (Research on the History of the Sciences), 8:128–36. (J)
56. Uéno, M. 1948. *Nippon Hakubutugaku-si* (History of Japanese Natural History). Kyoto: Hosino Syoten. 232 pp. (J)
57. Uéno, M. 1959. Japanese entomology in the first half of the nineteenth century. *Kontyû.* 27:4–9
58. Uéno, M. 1960. *Meizi-zen Nippon Seibutugaku-si* Vol. 1. [A History of Japanese Biology (Zoology) before the Meiji era] Tokyo: Nippon Gakuzyutu Sinkôkai. 583 pp. (J)
59. Williams, C. B. 1958. *Insect Migration.* London: Collins. 237 pp.
60. Wilson, H. F., Doner, M. H. 1937. *The Historical Development of Insect Classification.* Univ. Wisconsin. 133 pp.
61. Yabuuchi, K. 1970. *Tyûgoku no Kagaku Bunmei* (Science and Civilization in China). Tokyo: Iwanami Syoten. 221 pp. (J)
62. Yamada, K. 1965. (Tradition of science and technology in China.) *Sisô* (The Thought, Tokyo), 1965:1361–75. (J)
63. Yamada, Y. 1932. *Kodai Bizyutu-Kôgeihin ni Ôyô seraresi 'Tamamusi' ni kansuru Kenykû* (Studies on a buprestid beetle used for art and ornament in the Old Ages of Japan and Korea) Kyoto. 50 pp. (J)
64. Yano, M. 1926. (Progress of knowledge of the chrysanthemum longicorn beetle.) *Kontyû,* 1:57–63 (J)
65. Yano, M. 1927. (Progress of knowledge of the chrysanthemum longicorn beetle. Appendix.) *Kontyû,* 2:63–64. (J)
66. Yano, M. 1930. (Scientific notes) In *Tôtyû Kokuyaku Honzôkômoku* (Japanese translation of Pên Tshao Kang Mu with scientific notes). Vol. 10. Tokyo: Syunyôdô. 587 pp.
67. Yano, M. 1943. *Ôyô-Kontyûgaku-si* A History of Applied Entomology in Japan). In *Nippon Nôgaku Hattatu-si* (A History of Agricultural Sciences in Japan), pp. 335–63. Tokyo: Nôgyô Tosyo Kankôkai. (J)
68. Yasumatsu, K. 1948. *Ari to Zinsei* (Ants and Human Life). Tokyo: Yôyô-Syobô. 189 pp. (J)
69. Yuasa, M. 1950. *Kaisetu Kagaku Bunka-si Nenpyô* (A Chronological Table of Science and Civilization with explanations). Tokyo: Tyûô-Kôronsya. 250 pp. (J)

EARLY ENTOMOLOGY IN THE MIDDLE EAST

Isaac Harpaz
Department of Entomology, Hebrew University
Faculty of Agriculture, Rehovot, Israel

Introduction

The history of a science, as part of historical studies, is bound to work with methods of historical research and is obliged to be competently familiar with the subject—the relevant science itself (in this particular case entomology). A task like this calls for the combination of knowledge and other skills which is beyond the capability of the average individual scholar. Furthermore, a great many of the original texts relating to the history of entomology in Middle Eastern countries in antiquity are unpublished manuscripts, thus being unavailable to the general student. Even when texts in Arabic, Greek, Assyrian, Egyptian, and other languages are published they are usually not available in reliable translations, hence offering only limited use to all those who are not acquainted with these languages. Very often the texts, even printed ones, are not up to the standard of modern philological science. A number of manuscripts must, therefore, be compared in order to reconstitute the original text as far as possible, a work which can only be done with the help of a philologist. Thus, for instance, there are mentioned in the Bible[1] (Leviticus 11:22) a number of Hebrew names of kinds of grasshoppers (*sol'am, hargol,* and *hagav*) which until this day have not yet been unequivocally interpreted and identified with known species of Orthoptera of the Holy Land (cf Montgomery, 1959).

It therefore seems futile to attempt to compile a systematic treatise on the subject, even in the most concise form. Consequently, for the time being, we shall have to be satisfied with a number of factual examples highlighting the impact that insects seem to have had on people's thinking, behavior, and creativity in that part of the globe where some of the greatest civilizations of the human race have flourished in ancient times.

In fact, zoology (let alone entomology) did not at all exist as a methodo-

[1] Nearly all Biblical references mentioned in this review are according to the *New English Bible*, Oxford University Press and Cambridge University Press, Oxford 1970. Only in one instance, as indicated in the relevant citation, the *Revised Standard American Version* was used. For a complete list of Bible references to insects and arthropods the reader should consult Bruce (1958).

logical, written discipline of science, until the days of Aristotle (see Chapter by Morge). Moreover, in Biblical Hebrew, for example, (and probably also in the other Middle Eastern tongues of that time) there is not a word to denote "insect" specifically. The nearest Hebrew term *sheretz* actually embraces all the "teeming creatures," of which insects obviously make up the majority.

In the absence of even a specific term for "insect" one can hardly expect the existence of an organized science of entomology in the sense recognized today. Nevertheless, people of the era under discussion were not in the least ignorant of, or indifferent to, the presence of insects around them. They made and also recorded some extremely interesting observations on insect life. Others attached magic or divine powers to certain kinds of insects, often using them also as motifs in artistic designs. Insect pests of crops and households, as well as verminous ectoparasites of the human body, received their due share of man's attention, while many other species served as either staples or delicacies in people's diets.

Fragments of historical evidence bearing upon the above-mentioned points will be quoted and discussed in the present review. The subject of commodities in human use deriving from insects, such as honey, beeswax, lac, and silk will not be dealt with here as there are special chapters devoted to them in this volume. Regarding the very economic aspect of entomology, dealing with insects as human food in the ancient Middle East, the reader is referred to Bodenheimer's (1951) special book on this topic.

Historically, the period pertaining to this review extends from prehistory to the end of the fifth century A.D., the time when compilation of the Babylonian Talmud (see below) was completed. Geographically, the ancient Middle East in this context comprised the countries bordering on the eastern shores of the Mediterranean, including Egypt and Mesopotamia.

Since the Talmud will be repeatedly referred to as a source of the history of science in the subsequent sections of this chapter, a few informatory sentences should perhaps be added for the benefit of those readers who are not adequately familiar with this work. The Talmud is the authoritative body of Jewish law and tradition developed on the scriptural law after the closing of the Pentateuchal text at about 400 B.C. It incorporates the Hebrew Mishnah (codified at about 200 A.D.) and the Aramaic Gemara represented in one edition of some 12 volumes completed in Babylon in the fifth century A.D. The Talmud is in fact not merely a law book, but rather an encyclopedia covering every phase of human activity. It is undoubtedly a mine of information for the study of religion, history, and civilization not only of the Jews, but of the peoples of the entire Middle East. An English translation of the Talmud was published in 17 volumes by the Soncino Press of London (1935–1952) under the editorship of Rabbi I. Epstein, while a 9-volume German translation was done by Goldschmidt (1925–1936). An early attempt to collate all the zoological references found in the Talmud was undertaken by Lewysohn (1858) in Germany.

Insect Symbols in Royalty, Religion, Art and Folklore

Perhaps the earliest graphic record of an insect species is found in Egyptian hieroglyphic documents dating back to the first dynasty (about 3100 B.C.) suggesting that King Menes, the founder of the dynasty, made the oriental hornet (*Vespa orientalis*) the symbol of the kingdom of Lower Egypt for many years to come (Bodenheimer, 1960, p. 74). The choice of this insect was probably meant to symbolize the spreading of fear before the powerful monarch. The dreadfulness in which hornets were held in people's minds in ancient times is vividly reflected in the Biblical verse (Joshua, 24:12; *Revised Standard American Version*) referring to God's help extended to the Israelite conquerors in their wars against the mighty inhabitants of the Land of Canaan: "And I sent the hornet before you, which drove them out from before you, even the two kings of the Amorites; not with the sword, nor with thy bow."

Another insect that drew people's special attention in the ancient Middle East is the fly (most probably the house fly on account of its infinite and ubiquitous abundance). The hieroglyphic sign of a fly stands for impudence and also for courage. Hence the military decoration for bravery in early Egypt was a fly. Likewise, fly amulets and pendants appear quite often in ancient Mesopotamia and Egypt.

The Philistine city of Ekron in Biblical times had a god by the name of Ba'al Zebub (meaning "lord of the flies" in Hebrew) who was widely believed to possess the magic power of foretelling the course of a disease (II Kings 1:1–18).

The sacred scarabaeus (*Scarabaeus sacer*) is so common throughout Egyptian civilization that it is almost the personification of ancient Egypt. Scarab seals and amulets (Figure 1) wherever found outside Egypt are regarded as reliable indicators of Egyptian influence or domination. This dung beetle was connected with Re the sun god, but the sun was worshiped in Egypt also by other names, such as the deity Kheper with a scarab as its head (Figure 2). The reason for this peculiar connection with the sun is believed to be associated with the beetle's habit of forming a ball of dung and rolling it over the sand into its burrow, where the morsel serves for oviposition and development of the scarab's progeny. The Egyptians were apparently aware of the fact that the beetle lays its eggs in the ball of dung and saw in its life cycle a microcosmos of the cyclical processes of nature, particularly the daily rebirth of the sun. The spherical shape of the dung pellet must have made this parallel with the sun even more striking. Furthermore, the metamorphosis of the beetle, in particular its eclosion from a mummy-shaped motionless pupa, could have been taken as a symbol of the resurrection of the dead in which Egyptians strongly believed. This latter view was challenged by Bodenheimer (1928, p. 33; 1960, p. 80) who argued that the metamorphosis of *Scarabaeus* was not well recognized until Jean-Henri Fabre made his observations late in the nineteenth century. Boden-

1

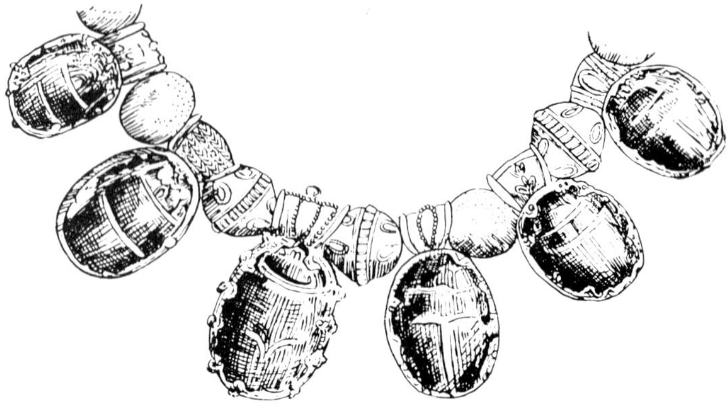

2

5

3

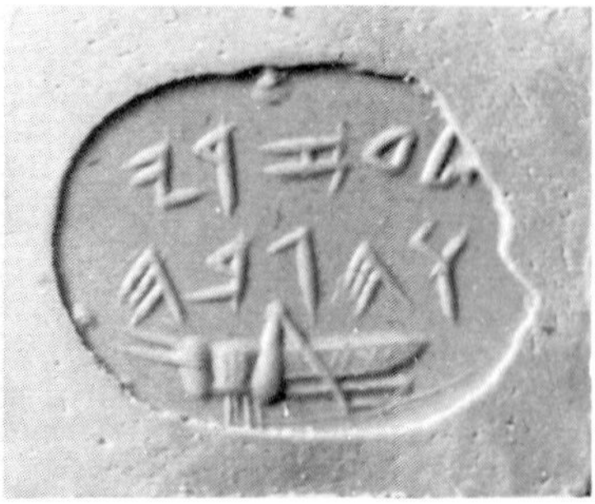

4

FIGURE 1. Necklace with scarab amulets [from Keller (1913)].

FIGURE 2. *Kheper,* the creator sun god, depicted with a scarab head [from Keller (1913)].

FIGURE 3. The deceased with hands raised in adoration standing before a *kheper* beetle on a pedestal [vignette from the *Book of the Dead* (Budge, 1901, p. 146)].

FIGURE 4. Enlarged impression of a seal with a family emblem; the script is the formal cursive Hebrew of the eighth and ninth century B.C., meaning "(belonging) to Azaryaw the locust" [from Avigad (1966)].

FIGURE 5. Hieroglyphic wasp [from Keller (1913)].

heimer based this argument inter alia also on the understanding that the Egyptians regarded the scarabs as unisexual animals, being males only that inseminate the dung ball in order to reproduce. This belief of unisexuality may also have to do with the custom of the Egyptian soldiers wearing scarab rings, as well as with the reason for devoting this insect to the sun god Re, the creator of the gods.

The magic power attached to the scarab is demonstrated by the funerary practice of replacing the heart of the dead person with a scarab prior to mummification. This was taken as an apotropaic measure intended to secure favorable results in the Osirian judgement (also known as the "weighing of the heart") which every noble Egyptian was believed to face after death (Figure 3).

The reverence in which scarabs were held by the Egyptians could also be the reason for their belief that these insects own properties of healing mental (rather than physical) disease. Thus, we find for instance in the Ebers Papyrus, which is a kind of medical compendium of 1550 B.C., the following prescription against all sorts of sorcery: Take a big scarab, cut off its head and wings, boil it, put it in oil and apply as an ointment to the affected person's body. Then cook the head and wings, in snake's fat and give it to the patient to drink (Joachim, 1890).

The significance of figures of flies and scarabaeid beetles depicted on seals, vignettes, or amulets worn as pendants, necklaces, bracelets, and the like have already been mentioned before (Figure 1). An interesting Hebrew seal of the eighth century B.C. bearing the figure of a locust was recently found in Jerusalem (Figure 4). In this case the insect serves as the emblem of the family of the seal's owner, the name of whom is *hgbh,* meaning locust in unvocalized Hebrew (Avigad, 1966). The choice of certain insects as family names seems to be not unusual among Jews in Biblical times. We thus find in the Old Testament names like Hagabah (locust or grasshopper) belonging to a priestly family of temple-servitors (Ezra 2:45; Nehemiah 7:48), or the family name *Par'osh* meaning flea in Hebrew (Ezra 10:25).

The view that people of that time did not display a feeling of disgust towards insects, at least not towards grasshoppers, is further supported by the Talmudic reference (Shabbath 90b) to certain grasshopper species commonly kept as pets for children to play with.

Divination.—Our discussion of the role that insects played in the supernatural would be incomplete without mentioning omens concerning insects. Perhaps the most remarkable document in this respect is a cuneiform text discovered in Mesopotamia and kept in the Kuyunjik Collection of the British Museum, containing some 65 auguries associated with insects (Jastrow, 1912, p. 834). One line in this tablet, for instance, reads as follows in English translation: "When ants [not beetles as Jastrow has misinterpreted it] are seen on the north wall of a house, there will be discord between man and wife."

Morphology, Classification, Ontogeny, and Ethology

Morphology.—Illustrations of insects made in ancient Egypt are fairly accurate in detail and in a few cases the insect depicted can be identified even down to the genus (Figures 1 and 5). However, no records are available from this pre-Aristotelian period suggesting any attempt to formulate even an initial generalization regarding comparative morphology, terminology, or classification of the insect class of animals.

As already mentioned in the introduction, the Hebrews in Biblical times have terminologically included the insects among a wider group of teeming or creeping creatures embracing reptiles, amphibians, mollusks, arthropods, and possibly other invertebrates. The only distinction was made between aquatic and terrestrial kinds, while the latter as a group were further subdivided linguistically into winged and apterous forms. The fact that immature stages of the locust, as well as various phases of its imago, were given specific names, different from the adult type, indicates that they were not adequately familiar with the metamorphosis of Orthoptera. Likewise, the indiscriminate use of the term "worm" (*tola'ath* in Hebrew) to denote nearly everything from a member of the old group *Vermes* through larvae of holometabolous insects and up to adult scale insects of the genus *Kermes,* is also indicative of poor knowledge of animal morphology in its modern sense. On the other hand, however, one should not generalize in this respect since the metamorphosis of calliphorid maggots into adult blowflies was definitely recognized in ancient Egypt as evident from a slip of papyrus found in the mouth of an Egyptian mummy saying: "The maggots will not turn into flies within you" [Gizeh Papyrus, British Museum no. 18026:4:4; cf von Oefele (1901)].

The placing of the insects together with reptiles, amphibia, and possibly also small mammals under one philological taxon (creeping or teeming creatures) has in turn led the Hebrews into an interesting misconception regarding the number of legs an insect has. In keeping with these tetrapodomorphic terms, insects like saltatory Orthoptera are described in the Bible as "teeming winged creatures that go on four legs which have legs jointed above their feet for leaping on the ground" (Leviticus 11:20), in contrast to "every other teeming winged creature that has four legs" (Ibidem 11:23).

It is not improbable, however, that this peculiar leg miscount was based on the ancients' identifying of the insect's forelimbs as hands rather than legs. Hints supporting such a view are found in the writings of an even much later period, viz. the Jerusalem Talmud (Shabbath 1:3), where one scholar is mentioned to have removed a hand and a leg of a louse in order to immobilize it without killing, which is forbidden on the Sabbath.

Taxonomy.—A great collection of cuneiform texts (largely archival), assembled by the Assyrian king Ashurbanipal (669–626 B.C.), was discovered in his royal library at the ruins of Nineveh in Mesopotamia during

FIGURE 6. Left: two fragments from the animal tablet of Har-ra=Hubullu, the oldest book on zoology [from Bodenheimer (1949)]; Right: enamel plate from ancient Ashur (Qal'at Sharqat in northern Iraq of today) representing an Assyrian noble in a locust prayer before the god Ashur [from Bodenheimer (1944)].

the nineteenth century. The collection (now in the British Museum) also includes a series of tablets known as Har-ra=Hubullu (Figure 6) which is in fact a bilingual Sumero-Akkadian lexicographical dictionary in cuneiform script. It was compiled during the ninth century B.C., i.e. at a time when the Sumerian language was no longer spoken. The tablets contain systematically arranged lists of Sumerian names with their current Akkadian translation in the corresponding column (Table 1). The lists originate from Sumerian ones such as were used in the period of Hammurabi (ca. 1792–1750 B.C.), but which had developed from much older lists. Tablets XI-XV of the Har-ra=Hubullu series contain a list of wild and domestic animals of the air, water, and land. They were edited and translated into German by Landsberger in 1934. The names are not arranged by alphabetical order, but according to related groups, all members of one group being characterized by a common prefix. Thus, the prefixes of the Sumerian names, which were written but not pronounced, each indicated a zoological group. Hence, that part of the Har-ra=Hubullu lexicon dealing with animals can quite justifiably be regarded as the oldest book on zoology known to date.

TABLE 1. List of the *buru* group (equivalent to Orthoptera) in the 14th Tablet of Har-ra = Hubullu.[a]

No.	Sumerian name	Translation	Akkadian name	Translation
227	*buru*	locust	*e-ri-bu*	—
228	*buru.sag*	head locust	*si-in-na-ra-bu*	noxious locust
229	*buru.gal*	large locust	*si-in-na-ra-bu*	noxious locust
230	*buru.gal*	large locust	*hi-li-mu*	—
230a	*buru.tur*	small locust	*zi-i-ru*	—
231	*buru.tur.tur*	very small locust	*zir-zir-ru*	—
232	*buru.sahar.ra*	dust locust	*e-rib tur-bu-ti*	locust of the dust
233	*buru.a.ab.ba*	sea locust	*e-rib tam-tim*	—
233a	*buru.id*	river locust	*erib na-a-ri*	—
234	*buru.id.da*	river locust	*ku-li-lum*	dragon fly[b]
235	*buru.gan.na*	field locust	*zi-za-nu*	grasshopper
236	*buru.gan.tir.ra*	forest locust	*zi-za-nu gish-tum*	—
236a	*buru.za.pa.ag*	noisy locust	*sa-si-ru*	cricket
237	*buru.EN.ME.LI*	necromancer	*sha-i-lu*	mantis
238	*buru.EN.ME.LI.a sha(g).ga*	soothsayer of the field	*sha-i-lu eq-lu*	field mantis
239	*buru.ir.gi.lum*	grasshopper	*SHU-lum*	?
240	*buru.ir.gi.zum*	grasshopper	*SHU-zum*	—
241	*buru.sa.a.sa.a*	grasshopper	*si-ig/k-du*	—
242	*buru.ma.su(d).ud.da*	grasshopper	*a-du-dil-lum*	mantis
242a	*buru.DU.HU*	grasshopper	*(X)-ih-tu*	?
242b	*buru.sa.KAL*	grasshopper	*(X)-ti-tu(?)*	?
243	*buru.ha.mum*	noisy grasshopper	*lal-la-ri-tum*	lamenting woman (cricket?)
244	*buru.balag.ga.na*	"harp of the field"	*sar-sa-ri*	cricket

[a] Adapted from Landsberger (1934).

[b] Obviously this is not an orthopteron. According to Bodenheimer (1960) they possibly referred to the large Ephemerid *Palingenia euphratica* Mos., which occurs in huge swarms (reminiscent of locusts) over the rivers of Iraq, and may have therefore been included among the *buru* group.

Insects are rather well represented in the Har-ra=Hubullu list with 121 out of 407 animal names. These 121 insect names are arranged in groups more or less equivalent to the modern taxa Orthoptera (Table 1), Coleoptera?, Odonata, Formicidae, and Diptera/Hymenoptera. The absence of hemipterous and possibly also coleopterous representatives on the list suggests that some tablets are still missing since some of these insects must have been known to the Assyrians.

A very interesting feature of this list of insect species is the placing of pests of crops and stored-products under a separate group, comprising 33 names, regardless of their respective taxonomic affiliation. The Sumerian name of each member of this group is preceded by the determinative *uh* which could be paralleled to the term "bug" in current American slang. It thus appears that the trend to distinguish between economic and general entomology is at least 2900 years old.

The principles of this ancient Sumerian taxonomy were rather simple and resemble those applied by Pliny. They are, however, much less advanced than those developed by Aristotle. Yet, the taxonomic conception, as represented in Har-ra=Hubullu, was maintained for a long time in the Middle East. The zoology of the Talmud was largely based on these concepts apart from some Hellenistic influence. Even some of the medieval Arab writings follow these principles in relation to animal classification. It may very well be that the knowledge of animal life mastered by the ancient Sumerians, Assyrians, and Babylonians was indeed much wider than the underrated impression which we get from the analysis of bare lists of names.

Turning again to Talmud's entomology, the roots of which presumably draw from old Sumerian science, we find there an amazing elaboration on the Orthoptera taxon. While discussing clean and unclean kinds of grasshoppers from a standpoint of Jewish dietary laws, the Talmudical author argues that there are 800 different species of what we now term Orthoptera (Hullin 63b). As he was obviously referring to the orthopterous fauna of the Old World of his time, this should be regarded as an excellent guess for a person living some 1500 years prior to the compilation of Ramme's (1951) faunistic list of Orthoptera of southeast Europe and the Near East comprising 674 species. One may safely assume that in the geographical area under discussion there are at least another 100 species of Orthoptera that are still missing from Ramme's list.

Ontogeny.—The fact that optical magnification aids had not yet been invented has quite understandably led the Talmudical writers to a significant misconception, widely prevalent among early biologists, regarding the ontogenic origin of organisms not clearly visible to the naked eye. Thus, in discussing the kinds of vermin that are permitted to be killed on the Sabbath, the Talmud differentiates between the flea, which, like most animals, propagates by copulation, as opposed to the body louse which is claimed to originate from sweat without mating (Shabbath 107b). This is still somewhat surprising in view of the fact that the pediculine nits were recognized by the Talmudical authors and even termed by them as louse eggs. One should therefore qualify the above statement of the Talmud so as to imply that the nits rather than the actual lice arise from human perspiration. It should, however, be added that in the above-mentioned Har-ra=Hubullu (which has greatly influenced the Talmud's zoology) the nit is listed as a separate species (No. 250 in Landsberger's translation) quite distinct from either the louse or the flea.

From the aforementioned Talmudical discussion regarding the louse's origin we can also see that the age-old concept, or rather misconcept, of spontaneous generation, which was not finally disproved until the middle of the last century, has duly had its use in entomology since the early stages of the discipline's history.

Ethology.—Conversely, however, when referring to larger-bodied insects the ethological observations made by the ancients often amaze us in their deep insight. Statements attributed to King Solomon of Israel (died 934 B.C.) whom the Bible considers the wisest man of his time, read as follows: "Go to the ant, you sluggard, watch her ways and get wisdom. She has no overseer, no governor or ruler; but in summer she prepares her store of food and lays in her supplies at harvest" (Proverbs 6:6–8). "Four things there are which are smallest on earth yet wise beyond the wisest: ants, a people with no strength, yet they prepare their store of food in the summer . . . locusts, which have no king yet they all sally forth in detachments" (Proverbs 30:24, 25, 27).

Social insects with their highly developed instincts and apparent organizational perfection have naturally aroused people's explorational curiosity and admiration since very early stages in human cultural history. Thus, King Solomon's observations on the ant have duly prompted one of the Talmud's scholars (who lived some 1200 years later) to devise a test intended to verify the king's conclusion by experimental evidence. The Talmudical record (Hullin 57b) of this early experiment in insect ecology reads as follows:

> Rabbi Simeon b. Halafta was an experimenter. Why was he called an experimenter? Rabbi Mesharsheya said: It is written, Go to the ant thou sluggard, etc. He (R. Simeon b. Halafta) said: I shall go and find out whether it is true that they have no king. He went out at the summer solstice and spread his coat over an ant hill. When one (ant) came out he marked it, and it immediately entered and informed the others that shadow had fallen, whereupon they all came forth (for ants shun the fierce heat of the sun and in summer only venture forth in the shade or in the dark). He then removed his coat and the sun beat down upon them. Thereupon they set upon this ant and killed it (for having deceived them). He then

said: It is clear that they have no king, for otherwise they would surely have required to obtain royal sanction (for the execution of the delinquent ant).

Considering the facts recorded in this experiment at their face value, we find here a typical anthropomorphic interpretation of the results, an approach which dominated the method of reasoning not only during the period under review but also in many centuries to come.

At any rate, the attempt to gain information about insect life and behavior by experimental means, like the one described above, appears to be a solitary and rather exceptional case. The rule was to try and reach generalized conclusions based on isolated empirical data without any concerted effort to analyze the observed facts or to look for analogies within a complete, comprehensive, and coordinated system. This prevalent approach, coupled with the lack of essential information regarding insect metamorphosis and diapause, has duly led one of the Talmudical writers to assume that all invertebrates (boneless animals, in his words) cannot live for twelve months (another one even puts it at six months) or that no gnat lives a complete day (Hullin 58a–b).

One should not, however, underrate the validity of knowledge obtained solely by empiricism. For instance, treatments recommended by the Talmud against animal venoms, obviously based on empirical experience rather than experimental evidence, may still find their rationale in moden toxicological science. A crushed fly was suggested as an antidote to be applied to the area affected by hornet sting, or a crushed mosquito for a serpent's bite (Shabbath 77b). Since insect hemolymph is now known to contain a number of effective antitoxins, one is inclined to believe that the above remedies rely on more than sheer guesswork.

Locusts and grasshoppers, whether as devastating plagues or as staples in human diet, have understandably drawn the ancients' attention perhaps more than any other insect group. In the same aforementioned chapter of the Talmud dealing with venom antidotes we find the following comment: "Why is the horn [antenna in modern terminology] of a katydid flexible?—Because it dwells on love grass [*Eragrostis* sp., a grass with tough, sharp-edged blades], and if the long and thin horn were rigid it would break off upon brushing against such plants and the katydid would go blind." This indicates clearly that they have correctly identified the antenna as a sensory organ, but ascribed the wrong sense (vision) to it.

In another tractate in the Jerusalem Talmud (Shabbath 13:1) dealing with the permissibility of locust hunting on the Sabbath, the authors differentiate between catching the insects while the night's dew is still present (i.e. early in the morning when they are still inactive) as distinct from the warmer and drier hours of the day when they are much more agile due to the higher temperature. This distinction although made merely for the purpose of determining the extent of the effort required for the capture of the grasshoppers, still demonstrates to us the knowledge which the people had regarding the effects of ambient temperature on the flight activity of poikilothermous animals like insects. In fact a similar observation was made hundreds of years earlier with regard to the same insects by the prophet Nahum (3:17) in the Old Testament, asserting that ". . . your secret agents are like locusts, your commanders like the hoppers which lie dormant in the walls on a cold day; but when the sun rises they scurry off. . . ."

Agricultural Pests and Disease Vectors

The extensive ravages wrought by insect pests upon man's crops and pasture lands since time immemorial have obviously left a profound impression in numerous historical records from all over the ancient Middle East. One of the best known examples in point is the vivid description of a desert locust [*Schistocerca gregaria*] invasion made by the prophet Joel (1:2–20; 2:1–11) in the Old Testament. Referring to the endless swarms of immature hoppers the prophet uses magnificent poetic imagery:

> ...their vanguard a devouring fire, their rearguard leaping flame; before them the land is a garden of Eden, behind them a wasted wilderness . . . bounding over the peaks they advance with the rattle of chariots, like flames of fire burning up the stubble, like a countless host in battle array. Before them nations tremble, every face turning pale. Like warriors they charge, they mount the walls like men at arms, each marching in line, no confusion in their ranks, none jostling his neighbour, none breaking line. . . .

Of the ten plagues, which, according to the Bible, were sent by God to Pharaoh's Egypt at the instigation of Moses and Aaron, three were actually insect plagues, namely lice, flies, and locusts (Exodus 7–11). A very interesting observation concerning the onset of desert locust invasions in the Middle East in general is in fact included in the Pentateuchal narrative relating to this locust plague: ". . . and the Lord sent a wind roaring in from the east all that day and all that night. When morning came, the east wind had brought the locust" (Exodus 10:13). This is perfectly accurate until this very day since only by aid of a strong east wind blowing continually for at least 24 hours (currently known throughout the Middle East as *khamsin*) can the migratory locust swarms reach the invasion areas lying west of the Arabian desert where the swarms are assembled.

There are a number of other agricultural and household pest species mentioned in the Old and New Testaments, e.g. the olive fruit fly *Dacus oleae*, the grape-berry moth *Lobesia botrana* (Deuteronomy 28:39–40), and the clothes moth (*Tineola biselliella*) (Isaiah 50:9; James 5:2), but all without any accompanying information.

More significant, however, is the prevalent attitude of people all over the ancient Middle East towards animal pests in general and noxious insects in particular. These were widely regarded as a kind of divine punishment meted out on the sinful. Hence there is nothing to be done about it except meekly submitting to it in penitence, making prayers, offerings, or other rituals as prescribed by the respective religion. Figure 6, for instance, depicts an Assyrian noble facing the god Ashur either offering a prayer for protection against locusts, or giving thanks for salvation from this plague. Similarly, the method used by Moses and Aaron to stop the three insect plagues mentioned above was also in keeping with the same notion, namely by evincing some miraculous divine intercession. "Pharaoh hastily summoned Moses and Aaron. 'I have sinned against the Lord your God and against you' he said. 'Forgive my sin, I pray, just this once. Intercede with the Lord your God and beg him only to remove this deadly plague from me" (Exodus 10:16–17). The above conversation took place in connection with the locust plague of Egypt, and it is interesting to find further in the same chapter of the Book of Exodus a passage describing the manner by which every locust invasion is naturally liquidated in Egypt, namely by change of the wind into a westerly gale which carries them away and sweeps them into the Red Sea so that not a single locust is left in all the territory of Egypt.Naturally, the Bible attributes this redemptive change of wind to the benevolent act of God in response to Moses' plea. The same procedure is repeated with regard to locust invasions in the Holy Land where the prophet Joel (2:15–20) summons the elders to proclaim a solemn assembly and appoint a day of fast and abstinence. "Let the priests, the ministers of the Lord, stand weeping between the porch and the altar and say, 'Spare thy people, O Lord, thy own people' ". As a result the locusts had their vanguard banished into the eastern sea (the Dead Sea) and their rear into the western sea (the Mediterranean Sea). Likewise, in Talmudical times special fasts were proclaimed and appropriate prayers recited in an attempt to stop outbreaks of hornets, mosquitoes, flies, locusts, etc (Ta'anith 14a).

A similar instance whereby a locust invasion is brought under control by a natural agency while people ascribe it to a supernatural divine intervention is described by the Roman savant Pliny (23–29 A.D.) in his *Historia Naturalis* (10:39). When locusts devastate the fields of the inhabitants of Mt. Cassius in Syria (Jebel el Akra in Turkey of today) Jupiter will respond to the prayers and supplications of the people by sending the migratory Seleucid birds [rose-colored starling (*Pastor roseus*)] which destroy the locust.

People's helplessness in the face of insects' numerical and destructive supremacy has consequently led them to submit to it and in turn learn to live with insects to the maximum tolerable extent. The omnipresence of house flies in human neighborhoods was considered one of the accepted facts of life. Hence the Mishnah, for instance, regards it as one of the ten godly miracles of Jerusalem that not a single fly was seen in the Temple's abattoir (Aboth 5:5). The miraculous significance attached to this unnatural freedom from flies may be even better understood in view of the fact that the Jews at the time of the Talmud were already aware of the role played by flies as disease vectors. Strange as it may sound, however, the specific malady mentioned in the Talmud (Kethuboth 77b) as fly-borne is apparently a venereal disease (*ra'athan* in Hebrew) yet to be identified.

The attitude of choiceless tolerance to the presence of insects in foodstuffs is perhaps best demonstrated by the two following examples cited from the Talmud concerning entomological standards of cleanliness: (*a*) As regards fig marketing the rule was that the buyer is prepared to accept a wormy[2] proportion of not more than one-tenth (Baba Bathra 6b). This is obviously a much higher infestation rate than that tolerated now in most countries. (*b*) If a man finds a hair and a fly in the food cooked by his wife, the fly albeit disgusting is excused since it is not her fault, but the hair is inexcusable and may become grounds for divorce (Gittin 6b).

Physical control measures.—Not in every case were people so submissive in their struggles against insects as described above. The Greek historian Herodotus (fifth century B.C.) recounts in his *History* (2:89) that the Egyptians protect themselves against mosquitoes in two different manners. In the areas above the marsh lands they build towers where they can sleep undisturbed by mosquitoes since these cannot fly up there because of the winds. In the marsh lands, on the other hand, everybody sleeps under a mosquito net which during daytime serves also as a fishing net.

Another instance whereby the ancient Egyptians were known to wage active and direct combat against certain crop pests is cited by Efflatoun Bey (1929) as follows: A certain hieroglyphic papyrus [origin not mentioned], which was identified as a royal decree, contains a reminder to farmers from the person in charge of agricultural administration not to be negligent in checking and killing "the worm" lest it will eat up a large portion of the crop. Efflatoun Bey interprets "the worm" as no other than the present greasy cutworm *Agrotis ypsilon.*

Chemical control.—Much less rational, however, is a prescription found in the Ebers papyrus (see above) against fleas and body lice: One part of date flour and one part water shall be cooked to a volume of two *hennu* (ca 450 ml) jars. Sip up a mouthful while warm and then spray it out onto the part of the body infected by the vermin. Examples of this kind of magic therapy are abundant in the literature of the period under review, but there is no point in quoting any additional ones.

Rational pest control by chemicals (as distinct from magic quackery, or alchemical control like the above-mentioned hieroglyphic prescription) was hardly known in the ancient Middle East. The only exception may be the use of sulfur whether as dust, or as a fumigant by burning it to sulfur dioxide.

It is widely accepted now that the ancient Greeks and Romans utilized sulfur for pest control purposes. However, some historians suggest that this practice had more to do with a ritual designed to enlist some supernatural protection against the pest concerned, than with the proper cognizance of the immediate pesticidal properties of sulfur. This may also help to explain the absence of any clear reference in the Talmud for use of sulfur in crop

[2] The worms in question are the maggots of the Mediterranean fruit fly (*Ceratitis capitata*) and/or the fig fruit fly (*Lonchaea aristella*).

protection in Palestine in spite of the strong influence which the Greek and Roman domination had on the area during that period. It should for that matter be pointed out that the Jews of the era under discussion zealously shunned anything tainted with Hellenistic religious connotations.

On the other hand, however, the Mishnah (Shebi'ith 2:2) in discussing the different practices that are allowed or forbidden during the agricultural sabbatical year, does mention the practice of fumigation with ordinary smoke for the control of "worms" infesting fruit trees.

Biological control.—A most remarkable early reference to biological control of ants is found in the Talmud (Mo'ed Katan 6b) which merits quotation:

> How are they destroyed [the ants]? Rabban Simeon b. Gamliel says: Soil is fetched from one hole and put into another, and they [the ants of the two nests not knowing each other] strangle each other. Rabbi Yemmer b. Shelemia said in the name of Abaye, that is [effective] only if [the ants are] situated on two sides of a river; and that if there is no bridge; and if there is not even a crossing plank; and if there is not even a rope to cross by. How much apart?—up to one parasang [about 4 miles].

Apart from its biological control significance, the above record also proves the peoples' good knowledge of the means by which ant trails cross rivers, as well as of the maximum distances covered by an ant family in establishing filial nests. The latter were presumed to be olfactorily related to the parent colony, therefore unlikely to attack each other.

Chemosterilants.—Even the principle of chemosterilization, the dernier cri of modern economic entomology, was in fact known and practiced in Palestine in Mishnaic times (i.e. at least 1800 years ago), although not exactly in relation to insect pest control but in beekeeping. In laying down the rules applying to the sale of beehives, the *Talmud* (Baba Bathra 80a) states, inter alia, that if a person buys the annual issue of a beehive he takes the first three swarms, and the seller may then emasculate the remaining to prevent them from further breeding and thus induce them entirely to the production of honey. The question of "wherewith does he emasculate?" is specifically asked there, and the answer is given by Rabbi Judah in the name of Samuel—with mustard. One cannot expect the Talmudical sages to understand the exact nature of the effect of mustard glycosides on queenbee fecundity, while the use of the term "emasculate" reveals their vague knowledge of the biology of honeybee reproduction. Yet one of the scholars participating in this discussion is recorded to have expressed the view that the condiment does not act directly, but its sharpness causes the bees to consume a surfeit of honey which in turn deprives them of their reproductive capacity.

LITERATURE CITED

Avigad, N. 1966. A Hebrew seal with a family emblem. *Israel Explor. J.* 16:50–53

Bodenheimer, F. S. 1928. *Materialien zur Geschichte der Entomologie bis Linné. Bd. I.* Berlin: Junk. 498 pp.

Bodenheimer, F. S. 1944. Studies on the ecology and control of the Moroccan locust (*Dociostaurus maroccanus*) in Iraq. *Govt. Iraq Directorate-General Agr.* Bulletin 29. 121 pp.

Bodenheimer, F. S. *Animals in Bible Lands,* Vol. 1. Jerusalem: Bialik Foundation. 350 pp. In Hebrew

Bodenheimer, F. S. 1951. *Insects as Human Food.* The Hague: Junk. 352 pp.

Bodenheimer, F. S. 1960. *Animal and Man in Bible Lands.* Leiden: Brill. 232 pp.

Bruce, W. G. 1958. Bible references to insects and other arthropods. *Bull. Entomol. Soc. Am.* 4:75–78

Budge, E. A. W. 1901. *The Book of the Dead; an English Translation of the Chapters, Hymns, etc.* London: Kegan Paul, Trench, Trübner. 702 pp.

Efflatoun Bey, H. C. 1929. The development of entomological science in Egypt. *Trans. Int. Congr. Entomol., 4th, Ithaca, 1928.* 2:737–42.

Goldschmidt, L. 1925–1936. *Der Babylonische Talmud mit Einschluss der vollstaendigen Misnah, wortgetreu uebersetzt,* 9 Vols. Berlin and Vienna: Harz; The Hague: Nijhoff

Jastrow, M. Jr. 1912. *Die Religion Babyloniens und Assyriens. Bd. II.* Giessen: Töpelmann. 1127 pp.

Joachim, H. 1890. *Papyros Ebers; das älteste Buch über Heilkunde. Aus dem Aegyptischen zum erstenmal vollständig ubersetzt.* Berlin: Reimer. 214 pp.

Keller, O. 1913. *Die antike Tierwelt. Bd. II.* Leipzig: Engelmann. 617 pp. Reprinted 1963, Hildesheim: Ohns Verlagsbuchhandlung

Landsberger, B. 1934. *Die Fauna des alten Mesopotamien nach der 14. Tafel der Serie Har-ra=Hubullu.* Leipzig: Hirzel. 144 pp.

Lewysohn, L. 1858. *Zoologie des Talmuds.* Frankfurt am Main: Baer. 400 pp.

Montgomery, B. E. 1959. Arthropods and ancient man. *Bull. Entomol. Soc. Am.* 5:68–70

Ramme, W. 1951. *Zur Systematik, Faunistik und Biologie der Orthopteren von Südost-Europa und Vorderasien. Mitt. Zool. Mus. Berlin Bd.* 27. Berlin: Akademie Verlag. 431 pp.

von Oefele, F. 1901. Studien über die altägyptische Parasitologie. I. Aussere Parasiten. *Archiv. Parasitol.* 4: 481–530

ENTOMOLOGY IN THE WESTERN WORLD IN ANTIQUITY AND IN MEDIEVAL TIMES[1]

GÜNTER MORGE[2]

Department of Taxonomy of Insects, Institute for Plant Protection, G.D.R. Academy of Agricultural Sciences, Berlin, German Democratic Republic

To understand the origin and the development of entomology in antiquity, modern man must forget all he knows today of the anatomy, biology, ecology, or classification of insects, must ignore all the present equipment for their study, e.g. the optical appliances, and must think in terms of the conception of nature and the supernatural of that time. While the nations of antiquity did not lack the ability to accurately observe nature, they had no scientific interest in insects. Man first began to study insects for purely utilitarian purposes. The question of their usefulness or destructiveness was the starting point of entomology. The early history of entomology from ca 3000 B.C. to 600 B.C. was entirely a period of applied entomology; usefulness was the first link to entomology (bees, silkworms, remedies), damage aroused man to study the life of the insects (locusts, olive fly, vine pests). The total limitation to the utilitarian principle is a characteristic of all the early history of entomology. Anything that was not useful, destructive, or at least annoying was devoid of interest and did not merit any attention.

Through all those centuries bees were kept wherever men lived. Apiculture apparently had no particular home and never fell into neglect. It was described on slabs of stone dating from ca 2000 B.C. Silkworms were reared as early as 3000 B.C., first for guitar strings and fishing lines, but soon after also for silk production.

In the further discussion of the development of entomology we again arrive at a geographical division of our subject which we had limited in this chapter to the western world. In the Old World the development of entomology into a science can be traced by written documents, while for the other parts of the western world we must confine ourselves to interpretations

[1] This chapter is limited geographically to the area west of Suez; the development of entomology in Egypt and the so-called "Arabian epoch of medieval entomology" are not considered here.

[2] Head of the Department of Taxonomy of Insects (former DEI), Institute for Plant Protection of the G.D.R. Academy of Agricultural Sciences, Berlin, Branch Eberswalde (German Democratic Republic). Custodian of the Collections of Natural History and of the Museum of Natural History of the Benedictine Monastery, Admont (Austria).

of pictorial descriptions and traditional customs and to conclusions from the mode of life of the inhabitants until documents of these regions (that are difficult to find and to read) may have been evaluated. From the present state of research it can only be inferred that the people in these regions had a certain knowledge of some insects and their significance, but that no development of an entomology as such is discernible. Not until the ancient Greeks was the study of insects characterized by the delight in observation. The combination of the utilitarian principle with scientific and philosophical problems led to the establishment of biology as a science by Democritus and Aristotle. The latter was the author of the oldest known systematization of insects. The extensive material that he left was the source and the basis of all entomological knowledge until modern times. He compiled many observations on the ecology and especially the reproduction of insects and studied, sometimes very carefully, their morphology and anatomy. Only the natural history of Pliny in Roman antiquity went beyond Aristotle, but in Roman times the utilitarian principle was emphasized again, which amounted to a regression.

While European antiquity is thus characterized by a rise above the purely utilitarian principle and by the establishment of biology as a science, the Middle Ages can hardly claim a place of their own in this development and history. The literature of the early Middle Ages was again based on the utilitarian principle which dominated until it was broken by Michael Scotus and Albertus Magnus.

Greek Antiquity

In European antiquity the first evidence of Greek entomology before Aristotle is to be found in Homer's epics (ca 850 B.C.) in the form of metaphors and similes taken from insect life. For instance, the number of the Achaians is compared to that of fly larvae in dairies, and the impudence and persistence of the horse fly are described. The epics also mention bees, wasps, and the collection of honey. The throng of the ships is compared to the bees, and perhaps even first hints at the keeping of wild bees in clay vessels are given. Fly larvae are referred to in the nineteenth book of the Iliad where Achilles laments that Patroclus' body will soon be destroyed by the maggots in his wounds.

This is also the time of the psyche conceptions of the Mycenaean and archaic Greek eras. The soul is represented as a butterfly. Studies of the origins of the belief in psyche have shown that the original name was not psyche but phalaene (moth). Name and meaning are derived from phallus, and in the old representations the body of the butterfly is emphasized at the expense of the wings. The symbolic development may be explained by the observation of the metamorphosis of the insects with which the Greeks certainly were familiar: the caterpillar represents the living body, the pupa stands for death and the butterfly for the winged resurrected soul which now displays its highest and most beautiful form of life. But it has not yet been ascertained whether metamorphosis really was the model for the spiritualization of the psyche conception.

Other Greek poets besides Homer showed a great interest in the observation of nature. Aesop (sixth century B.C.) with his fables of the ant and the fly may be cited as an example.

If the oldest surviving scientific writings of the Greeks, the origins of which are still largely unknown, marked a definite progress, the principal achievement of ancient Hellas was the establishment of a science that was not limited to the simple observation of nature.

Nordenskiöld (1926) called the era of the Ionian natural philosophers the period of causal science. The use of the causal method was its strength, and the lack of appropriate facts was the weakness by which it failed in the end. The urge to explain the sources of all things led to the assumption that the living beings originated in the mud (Anaximander and Empedocles) or in the earth under the influence of the sunrays (Diogenes of Apollonia). This period culminated in Democritus of Abdera (470 to 370 B.C.). His achievement was the dissection of animals, by which he found the division into blooded animals (Vertebrata) and bloodless animals (Evertebrata). It is still a matter of dispute whether Democritus was a precursor of Aristotle or his source.

The causal period of philosophy was followed by the teleological, with Aristotle as its climax. The founder of this school was Plato (429 to 347 B.C.), Aristotle's teacher, who influenced the history of biology and classification by his conceptual theory and initiated the idea of a scientific entomology. He was the first to give definitions of abstract classifying terms as the species: "Each individuum is only an imperfect reproduction of a perfect eternal conception of the species and genus." One of the precursors and

FIGURE 1. The XVIIth fable of the fly and the ant: ". . . as it happened that a fly and an ant had a quarrel . . ." In *Aesops Fabulae et Vita.* Das zweyt buch, Ulm: Johann Zainer. (ca 1476–1477). Cited from A. Schramm: *Der Bilderschmuck der Frühdrucke,* Bd. 1–23. Leipzig, Stuttgart 1920–1943. Vol. 5, Figure 175.

founders of Greek science was Herodotus, the greatest historian of antiquity, with his critical scientific descriptions of the nations he knew, in which he also mentioned bees and ants. He was the first to follow the principle of preferring observation to theory and to the belief of the narrator. He reported the use of mosquito nets by Egyptian fishermen and pointed out that man can protect himself against mosquitoes at night by sleeping on high towers because the mosquitoes cannot rise to such a height. Pythagoras (ca 450 B.C.) rid a Sicilian town of marsh-fever by drainage, a method which later fell into oblivion.

The absolute climax of the traditional classical biology was Aristotle (384 to 322 B.C.), the disciple of Plato and tutor of Alexander, not a Hellene but a Macedonian. His system of logic predominated for more than fifteen centuries. He may be called the founder of general entomology and of entomology as a science. To him the world owes the first systematization of insects that was ever made.

The flaws in Aristotle's scientific system were inherent in his general system which was characteristically new in its dogmatism and in the deductive method with which he tried to explain all processes and facts from a few general concepts and premises by logical conclusions. This resulted in arbitrary and unprovable assumptions. His theory of categories led him to the principle of teleology: Nature does nothing in vain, every morphological peculiarity has its purpose. As a method this principle of purpose sometimes rendered Aristotle good services, enabling him to recognize connections which otherwise would have remained unknown. Often, however, his endeavor to find a purpose in all events led him astray.

Aristotle comprehended all the knowledge of his time. His zoological writings form only a small part of his work. In spite of all the flaws in his system, Aristotle was not equalled as a zoologist for two thousand years, and for this reason we will discuss his writings at some length. The following works of his are of chief importance for entomology: 1. the *History of the Animals* (general description and biology of the animal kingdom), 2. *On the Parts of the Animals* (comparative anatomy and physiology), and 3. *On the Generation of Animals.*

Aristotle's classification is based on a liberally comparative anatomical and physiological revision of biology. Aristotle named about 500 animals and for some of them additional varieties so that about 600 species could be distinguished in his works. He was the author of the first deductively derived system of insects, his "Entoma" with the most conspicuous characteristic of the notch.

According to him, "Entoma" belong to the bloodless animals and have more than four feet, and some have wings. They are neither bony or fleshy, their body is rigid within and without. He includes the arachnids, myriapods, and worms in his group "Entoma". As systematic subgroups in our sense we can distinguish with certainty only the Coleoptera, the Hymenoptera (Apidae, Vespidae), and the Diptera. The bloodless winged insects (Pterota = winged in contrast to the Ptilota = wingless) are either Coleoptera or Diptera or Tetraptera. Butterflies (Psychae), cicadas, locusts or grasshoppers (Pedetica), and lice appear as uncertain small groups. It is still questionable whether Aristotle had a subsystem comprising all insects known to him, so that conjectures on this matter remain doubtful.

Aristotle classifies the insects firstly according to their wings and secondly according to their mouth-parts:

I. Winged
   1. with elytra
   2. without elytra
      a) with four wings
      b) with two wings
II. Wingless

I. Having teeth and being omnivorous
II. Being without teeth, but having a proboscis
   1. feeding on all saps .... flies
   2. sucking blood only .... mosquitoes
   3. feeding on sweet saps only .... bees

Aristotle has been criticized for ignoring gnats and mosquitoes which must have been as numerous in his time as they are today. But as he emphasized repeatedly, the smallness of the insects made it impossible for him to describe them in detail. So he gave only a general survey of their shape, their development from the larvae and their mode of life. And it must be admitted objectively that there was not much more that he could have seen without optical aids. This example may suffice to characterize the greatness of his achievement for that time. He had also mastered the applied entomology of his time as far as that was possible.

For details of the insects listed by Aristotle and Pliny and by the medieval authors we refer to the excellent tables with comparisons of the different species in Bodenheimer (1929).

Apparently there is in Aristotle no consistent and clear distinction in terms of the different larval stages, often not even of the different instars (egg, larva, pupa) though he had recognized the stages of development as such. Thus with him skolex means worm, but also insect larva. Kampe means caterpillar of a butterfly, but also the larva of *Lampyris* and of the soldier beetle (Cantharidae). Chrysalis means the pupa of a butterfly, but with the "silkworm" of Kos the young caterpillar is called kampe, the adult caterpillar bombylios, and the pupa nekadylos. Skolex and nymphe mean larva and pupa of Coleoptera, Diptera, and Hymenoptera. Kones means nits and larvae of lice, fleas, and bugs.

Aristotle already had a good knowledge of morphology and anatomy and knew of the larval sheddings of the skin. His writings contain remarkable biological and ecological observations on reproduction, feeding, care for eggs and young, and on the production of sounds. In his time the sericulture at Kos and apiculture flourished on a high practical and scientific level, which is also reflected in his writings.

From Aristotle's book on anatomy: Heart between head and abdomen, generally one, in some insects, several. Therefore the latter can live even when cut through. "For nature means always to create only one of a kind, but if that is easy, she forms actually one but potentially several." The mouth parts have different forms; some insects have a proboscis (tongue and lips in one), others have a proboscis-like sense organ between their teeth. The intestines are straight or sinuous, the bigger insects have a maw before them. With the cicadas, mouth and tongue are merged into one organ by which they, like roots, absorb their food from fluids.

> Among the animals, all the insects eat but little, not so much on account of their smallness but because of their coldness (for the warm needs and cooks [= digests] food quickly, the cold is not nutrient), and this is particularly so with the race of cicadas; for they need no food for their bodies but the moisture remaining from the dew, as it is with the ephemerae (these, however, are found around the Pontus), but the latter live only one day, while the former live several, though only a few days.

Insects have more legs because this facilitates their movements, considering their clumsiness and coldness. The coldest among them (the Myriapods) have most legs. Those that have fewer legs have wings instead. Of these, again, those that lead a wandering life have four wings, while the smaller have two wings (flies). The clumsy insects have elytra (e.g. melolonthidans) which "protect the usability" of the wings; owing to their notches they can roll themselves up or contract themselves, which is also a protective function (cantharides turn motionless and rigid at the touch). Some insects have stings, either in front (for feeding) or behind (as a weapon). Bees and wasps carry the sting inside so that it will not be hurt. Scorpions carry the sting outside because they live on the ground. No two-winged insect has its sting behind because they are too weak to sting with their abdomen.

> But it is better, if possible, not to have one and the same organ for different functions; the defensive organ should be very pointed and the tongue-like organ spongy and suitable for sucking up the food. Where it is possible to have two organs for two functions without their hindering each other, nature will not do what a smith does who from thriftiness makes a spit that also serves as a candlestick.

Longer front legs are better for wiping the eyes; longer hind legs are better for flying away or jumping.

With regard to the physiology of the senses, Aristotle attributed to all insects the visual, olfactory, and tasting faculties. They produce sounds by rubbing their legs (locusts and grasshoppers) or by compressed air (singing cicadas, flies). Insects sleep, which is evident from their sitting silent, without humming, throughout the night. Concerning their food Aristotle states that those having mandibles eat many kinds of food while the insects with a proboscis feed on fluids. He also knew the shedding of the skin of the insects, but according to him it happened after birth. "As with the vivipara the pellicle bursts, so it happens with the larval skin of the insects that bear worms." Aristotle considers respiration only as the cooling of the inner fiery principle of life. All bloodless animals cool only from within, i.e. without an exchange with an outer medium.

Aristotle knew four kinds of procreation: a) sexual procreation (the majority of all animals), b) procreation without copulation (most plants, fish, bees), c) procreation by sprouting (some plants and mussels), d) spontaneous procreation or abiogenesis (insects, crustacea, some plants).

The development of insects according to Aristotle begins with the formative stage of the egg, all insects first appearing as worms, larvae, or caterpillars. The first metamorphosis is the quiescent stage of the pupa, which corresponds to the egg stage, and the second metamorphosis or the third stage is finally the winged insect. The development of insects according to Aristotle is as follows:

Worms, larvae, caterpillars (= formative stage of the egg)

↓

Pupa (= egg stage)

↓

Winged adult insect (= third stage, second metamorphosis).

It seems that Aristotle knew the real insect eggs in some cases but did not recognize them as such.

With the insects the females are always bigger than the males. Copulation takes place in all species of animals that have males and females. There are different kinds of fertilization. Insects copulate from behind. The smaller, always the male, mounts the bigger. Unlike other kinds, the female erects its ovipositor, not the male its penis. They hang together for quite some time. Only the cicadas behave differently. They stay paired until the heat and the strength extant in the animal have formed the germ, as the semen does in others. The end of the winter is the season for copulation and parturition.

Animals originating by abiogenesis (from a mixture of fire, earth, water, air, and psychic heat) also have different sexes, but their copulation never produces a creature of their own species but something imperfect (lice produce nits, flies and butterflies produce egg-like worms) which never develops into what its parents are but remains what it is. Insects that are not formed by procreation come from dew fallen on leaves, from putrid mud and manure, from wood, from plants, or develop in the hair of animals, in meat or in animal droppings. Butterflies emerge from caterpillars on radishes and cabbage. Food is taken in and excrements are dropped only in the worm stage, not by pupa and butterfly.

Aristotle gives many examples for the origins of different insects, e.g. the development of the "silkworm": a worm with a kind of horns changes in a first metamorphosis into a caterpillar, then into a bombylios, later into a nekydalos. These changes take place within 6 months. The silk threads (bombykia, the cocoon of this animal) are separated by women, reeled off and spun. It is said that this was first spun in the island of Kos by Pamphile, daughter of Plates.

Mosquitoes (Empis) come from worms (small Ascarides) in the mud of wells. The latter originate in the dirt. The mud first grows white by decomposition, then black, and finally blood-red. In this state small red worms grow out of it which resemble water algae. At first they cling to the ground, then they detach themselves and swim about in the water. After a few days these Ascarides come to the surface of the water, become stationary and harden. The mosquito leaves the split shell and sits still at first, until sun and wind cause it to move.

The common house fly develops in heaped manure, the worms of *Drosophila* in vinegar yeasts. Worms can also develop from old snow, and on Cyprus winged animals spring from a glowing stone in the center of a fire. All animals that stem from worms are first stirred to move by sun and wind. Ichneumon flies kill spiders, carry them into walls, cover them with earth and lay their eggs there, from which young ichneumon flies emerge.

The development of the young takes 3 weeks with the insects that give birth to worms or worm-like young, while it takes 4 weeks with those that lay eggs. With most species the metamorphoses happen on the third or fourth day, which are also the decisive days in the crises of illnesses.

Anthrenus and wasps make combs for their eggs and young, and some bombykia build pointed cases of clay which are coated with a kind of salt. Locusts and grasshoppers lay their eggs in heaps in the earth, forming honeycomb-like clusters. They do this at the end of the summer, and they die soon afterwards. The eggs remain in the earth through the winter, but in the next summer new locusts or grasshoppers emerge from the eggs of the year before. The cicadas are divided into a bigger species and a smaller. They live only in places with trees that do not give much shade, e.g. olive trees. They lay their eggs on fallow ground, in a kind of reed of which vine-props are made, and in the stems of scilla. The young hide in the earth. They appear in masses after rain. The larva of the cicada is most savory before the worms bursts its skin. At the time of the summer solstice they emerge at night, darken in color, harden, and the males begin to sing.

Those insects that live on fluids drawn from flesh (e.g. lice, fleas, bugs) copulate and produce the nits, but these do not develop into anything else. Fleas originate from the lowest degree of putrefaction, e.g. dry manure, bugs from animal moisture that condenses on the outside, and lice from flesh. Aristotle mentions people who died from the louse disease when the humors in their bodies became excessive, as Alkwin the poet and Pherekydes from Syros. People on whom lice are formed more frequently suffer less from headaches.

Moths originate in dusty wool, especially if a spider is included which sucks up the moisture.

The wild fig trees contain in their fruits the so-called psenes. At first it is a small worm, but then it gnaws through the peel and emerges as a gall wasp; it creeps into other unripe figs, pierces them and thus causes them not to fall off; therefore the farmers hang fruits of wild fig trees into their cultivated fig trees and plant wild fig trees around them.

Aristotle also has a chapter on the diseases of bees. Certain animals develop in beehives which destroy the honeycombs. This animal spins a web over them and thus ruins them. (He means the caterpillars of *Galleria mellonella.*) The same damage is caused by a small moth which produces something that is entirely covered by wool (=imagenes of *Galleria mellonella*). The bees are most liable to diseases when there is much mildew in the forest and in dry years.

Even first hints at zoogeographical conditions are to be found in Aristotle: there are two adjacent regions in Milesia of which one is populated by cicadas while the other is not, in Kephalene the river divides the regions with cicadas from those without.

The technical skill of the insects is discussed in the ninth book of the *Natural History*. (This book was not written by Aristotle, its author is unknown.) Of the industrious insects (ants, honey bees, wasps) the wasps are described in detail. Two species are mentioned: One living in the mountains on oaks, having a longer sting, the other living in the earth and being tamer. Two types of wasps are distinguished: queen wasps and worker wasps. In early summer the queen wasps make combs from which wasps' nests develop in which at first worker wasps are formed. When the colony has grown, the queen lays mother eggs. The worker wasps carry food to the queen wasps which do not work. The queen wasps are broader, heavier, and thicker, therefore they can hardly fly. They always stay in the nest. Some of them have stings, others have none, those having stings being bigger and braver. "Many of those which generally have stings seem to lose them with the coming of winter, but we do not know anybody who has witnessed this." If wasps feel threatened they even attack people. They sting chiefly about the eyes.

The next name to be mentioned in the development of entomology as a science is that of Aristotle's disciple Theophrastus who lived from 371 to 286 B.C. Though his main works are devoted to botany they contain a number of valuable entomological observations, e.g. on plant pests, and the statement that the occurrence of diseases and destructive insects is largely influenced by the climate. Theophrastus says little about measures of pest control.

In his chapter on the diseases of trees he remarks, among other things, that wild trees are less prone to diseases that kill them though they may suffer injuries. For the cultivated trees he distinguishes between the diseases that attack all species and those that attack only certain species. All may suffer from "worms", excessive sunlight, or blight, but those with sharp or aromatic saps are attacked less by "worms". Theophrastus also perceived that these diseases differ according to climate, region, and pest, and that they are chiefly caused by injuries of the trees. The examples he gives are the "worms" in figs, the stag beetle *Kerastes,* the destruction of blossoms and leaves of olive trees at Milet by caterpillars, and the attack of "worms" on olives, pears, apples, medlars, and pomegranates. If the "olive worm" only penetrates from outside, the fruit is destroyed, if it lives in the core, it speeds up the ripening. He recognized the increased "invasion of worms" in fallen fruits, the influence of the wind (which makes them fall), and the relation between higher humidity and the stronger attack. Gall wasps on oak trees, fig trees, and vegetables, and caterpillars on vegetables, are mentioned; the fluid excreted by gall wasps is supposed to be the decisive factor in gall production.

In his chapter on wood-worms Theophrastus distinguishes between the wood borer in the sea and the wood-worms on land and recommends a coating of pitch against the latter while he does not yet see a possibility of controlling the former. He discerns rotting as a cause of their attack and describes their boring; and he knows that aromatic or hard kinds of wood, e.g. box-tree and fir, are not attacked. It may be assumed that Theophrastus knew of the scale insect *Ceroplastes rusci* and several Cerambycidae, of leaf-eating and flower-eating caterpillars, of the red spider *Tetranychus telarius,* of the larvae of *Dacus oleae,* and of the caterpillars of *Carpocapsa pomonella,* as well as of the occurrence of certain spiders.

With regard to vegetable diseases he described the controlling effect of rain water on the destructive insects, the invasion of radish by flea beetles and their control by sowing trigonella between the radishes, and the attack of caterpillars and maggots on cabbage and leek and the destruction of the pests by spreading manure.

He divides the cereal diseases into those that are common to all species and those that are specific for one kind. He mentions damage done by caterpillars, flea beetles, "wheat worms" (which attack the roots and the stalks or the ears), and pests of peas, beans, and other leguminous plants. The damage varies according to the climate, the weather, and the origin of the destructive insects (whether grown on the plant or invaders).

Theophrastus gives the first detailed description of caprification (after it had been briefly mentioned by Herodotus). He had studied galls on elms, poplars, and oaks and discusses them at length. He found destructive insects on asphodel, vine, and on medicinal plants. Among the latter, bitter roots were unaffected while sweet roots were attacked.

Though a separate chapter of this volume is devoted to the development of apiology we must now say a few words on this subject in order to maintain the continuity of the history of entomology in antiquity. While the ancient Greeks had only a very incomplete knowledge of apiculture, the statements on this subject in the zoological history of Aristotle (which were obviously not written by himself) clearly reveal a high degree of information on the life in the bee society.

The author reports without comment the various opinions on the origin of the bees: One says that bees do not copulate or lay eggs but gather their brood on the blossoms of trees, another says that only the brood of the drones is gathered from certain plant substances while the bee eggs are produced by the queen bees, and the third says that they do copulate and that the drones are the males and the bees the females.

Later Aristotle regards the three castes of bees not as sexual forms but as different species living together and denies any sexual differentiation within these three species or castes. The king will produce only workers or kings, the workers produce drones, and the drones have no progeny. Aristotle believes in self-fertilization of the bees; their real egg stage was as unknown to him as that of other insects.

He distinguishes four species of bees: the best species is small, roundish, and spotted; a second species is long; a third is called a thief, of dark appearance and with a broad body; the fourth and biggest species is the drone, it has no sting and is lazy. Of the queen bees he holds the opinion that a colony perishes if there are not enough queen bees in it because they are supposed to contribute to the production of bees. On the other hand a colony will also perish, by partition, if its queen bees are too numerous.

About the life and work of the bees it is said that first the honeycombs are made, and then the brood is deposited in them. Later, in the summer and the fall, the honey is gathered. Honey is supposed to drop from the air, especially at the rise of the stars and when a rainbow descends, but never before the rise of the Pleiades. The wax for the combs is prepared from blossoms. The honey thickens after 20 days; before that time it remains liquid. The honey is taken out when the fruit of the wild fig trees appears. The eggs of bees and drones are white, those of queen bees are bright yellow. The former develop first into maggots, then into bees and drones, the latter directly into queen bees without a maggot stage.

Among the peculiarities mentioned in the description of bee life are the following: Their life span is given as 6 to 7 years; it is reported that there are bees in Pontos which prepare honey twice a month; there are said to be combs at Themiskyra that contain hardly any wax but thick honey; bees are mentioned that build combs in the earth which contain honey but no maggots.

The queen bee has a sting but no desire to use it, its tongue is retractable, spongy, and hollow; excessive swarming is harmful, summers with much sunshine yield much honey. The author knew that the flowers that are visited by bees are cup-shaped or tubular; two races of bees are mentioned.

The ninth book of the zoological history of Aristotle, which was not written by himself, contains the detailed report of a beekeeper. It appears that even at that time it was known that normally there is only one queen bee in a hive. Probably there were already professional beekeepers at that time who came from the slave population. They are described as very knowledgeable and efficient. Strong frames were put on base-boards, with the entrance at the bottom. It was known that a good location was one that stayed warm in winter and cool in summer and was near flowing clear water or spring water. Of technical expedients the beekeepers knew the soothing effect of smoke on bees and used it in taking out the honey. They tried to frighten swarming bees by noise and to gather the newly formed swarms into empty hives. Drone traps were used to reduce the number of drones, the formation of swarms was hindered, emergency feeding was practiced, and the enemies and diseases of bees were studied. Some means of controlling the enemies of bees were known. Wax and honey were taken out in due time, hibernation was prepared, and the pasture of the bees was regularly cultivated. Even at that time one swarm brought 4.5 kg to 7 kg honey, and a record yield of 14 kg is quoted.

Finally some biological observations of that time may be mentioned: Cells are built first and in any case for the brood, but for queen bees and drones only when there is enough brood. The building of the combs starts at the top and proceeds without any gap to the bottom. The gathering of the honey is also described: The bees get the wax out of the flowers with their front legs, wipe them off at their intermediate legs and these at the exterior parts of their hind legs. They visit only one kind of flower in each flight, and on their return flight they are visibly burdened with their load. Three or four other bees assist them in shaking off their load in the hive. Having gathered the wax they take care of the brood. Thieves and drones do not work but damage what the bees have built. When they are surprised in this, they are killed, and so are superfluous queen bees. The worker bees are assigned to different tasks. Wasps, swallows, and bee-eaters are the chief enemies of the bees. If bees sting, they die as a rule, because they cannot pull out the sting without their intestines bursting out. Bees are very clean. Colonies in good condition are especially liable to diseases; for instance, they can be attacked by small worms which in their growth cover the whole hive with cobwebs so that the combs are ruined.

A source from the first century A.D. is the work of Dioscorides, a Greek, in which he describes the significance of various groups of insects for pharmacology. His work was the basis of entomological pharmacology throughout the centuries up to early modern times. In the second book of his *Materia Medica* he mentions the following animal remedies. Bed bugs are used against the quartan ague—if the disease is diagnosed early, seven bugs and some beans should be added to the food; they are used without beans against the bite of the asp; their smell cures fainting caused by a spasm of the uterus; added to wine or vinegar they drive leeches out; crushed and administered through the urethra they help against anuresis. Cockroaches when ground with oil or cooked are used against ear-ache. Cicadas when fried are used against bladder complaints. Locusts or grasshoppers are used for fumigation against anuresis of women; dried and taken with wine they are used against scorpion stings. Caterpillars on vegetables when coated with oil are used against the bites of poisonous animals. Beetles containing cantharidin are killed over steam or glowing ashes, preserved, and added to other medicaments for use against leprosy, carcinomas, herpes, and also dropsy. Dyer's coccid of oaks is an astringent for wounds (this scale insect was widely used in Dioscorides' time and probably earlier to produce crimson for dyeing). Gall-apples of elms and oak-apples are mentioned as astringents and desiccatives.

In the book *Toxins and Antitoxins* the symptoms of poisoning by cantharides, pityokampe, and buprestids and the antitoxins are described in detail. Another book, *On Poisonous Animals,* deals with the treatment of people stung by wasps and bees.

Even in Greek antiquity there were two opposing fundamental conceptions of animal psychology. On the one hand Chrysippus (ca 250 B.C.) represented the view of the Stoa: The animal does not act by reason but by the instincts which nature planted in it and which lead it to the useful. Kleanthes studied the mental capacity of the ants. According to Seneca (ca 50 A.D.) animals have no reason but imagination, sensation and instincts. Animal skills, such as the building of the honeycombs, are innate, not learned.

Opposed to this view was that of Plutarch (ca 100 A.D.) who interpreted the behavior of the animals as motivated by reason and understanding. He saw the life of the ants as a mirror of all virtues: friendship, sociability, endurance, courage, moderation, prudence, justice.

Roman Antiquity

The time following the epoch of Aristotle and Dioscorides was marked by a complete loss of interest in entomology. Insects were considered only in so far as they were connected with medical, agricultural, or cultic concerns.

When Greece was enslaved, science suffered a general decline, and Hellas did not regain any importance for the further course of the development of the biological sciences. Rome took possession of this heritage but proved to be unworthy of it. The predominantly sober and practical attitude of the Romans, however, shifted the emphasis to applied entomology like apiculture and the study of agricultural pests. The only name to be mentioned in scientific entomology is that of Pliny. In 77 he wrote the greatest comprehensive work of that time, his *Historia Naturalis,* an encyclopedia formed by a synopsis of "20,000 subjects from 2000 works", but it is highly probable that Pliny had no direct knowledge of the works of Aristotle. His *Historia Naturalis* is a compilation from other works in a particularly successful encyclopedic condensation, but without any philosophical view. He scarcely made any original observations of nature. His work contains the geography, zoology, botany, and mineralogy of his time.

The eleventh book deals chiefly with the insects. Their classification corresponds to that of Aristotle. In his introduction to his chapter on entomology Pliny says:

> We now proceed to describe animals of infinite delicacy which some have thought to be without respiration or even without blood. The life of these

> running or flying creatures shows great variety. Many of them are winged, as the bees, others have winged and wingless forms, as the ants. Still others have neither wings nor legs. They are rightly called insects (notched animals) because of the notches which separate their members at the neck, at the chest and at the abdomen so that their members are connected only by a thin skin.

He adds words of admiration about the indescribable perfection and purposefulness of the insects; nowhere does nature appear as great to him as in her smallest creatures.

Pliny distinguishes himself in various aspects by a marked independence of, for instance, Aristotle, and he supplements Aristotle's entomology by adding its historical dimension. He concedes that insects have respiration and also a sort of blood fluid, and he even counts them as a separate group between the blooded animals and the bloodless. The inner parts of insects are softer than sinews but yet firm. Their "principle of life" is not located in certain organs but distributed throughout the body. Those with the most limbs live longest when torn apart, e.g. the myriapods. Insects can see, touch, taste, and some can also hear.

Pliny also dwells on the life of the bees. He regards them as persevering, creating works, and having a community and leaders. They have a sort of presentiment of the weather. The making of the honeycombs and the division of labor in the society of the bees are described with many details. According to Pliny, honey comes from the air. At certain times the leaves of the trees are wet with honey dew. The bees gather the honey in their honey-stomach and bring it up again through the mouth. In each region there are three kinds of honey: 1. The spring honey or flower honey. It is taken out in May, on the 30th day after the emigration of the swarm. The fifteenth part of the honeycombs must be left in the hive so that the bees will have enough food. 2. The summer honey, which is the best sort but unfortunately often adulterated. 3. The wood honey; it is gathered after the first fall rains when only the heath is still in flower. It is valued least.

Opinions differ on the propagation of the bees. Some assert that the young bees are produced by a suitable combination of flowers, others think that the king bee pairs with bees. It is certain that the bees sit on the eggs as the hens do. A small white worm is hatched, but the king bee does not appear as a worm but directly as the winged insect. The larvae are fed, the brooding time is 45 days. The king bee leaves the hive only in the swarming and then keeps in the center of the swarm. Whether it has a sting or not is not known.

A certain prophetic significance is attributed to the swarms. It is mentioned as an example that bees alighted on Plato's mouth when he was a child and thus announced his later graceful eloquence. When the king bee is caught, the whole swarm follows. It is still doubtful whether the drones form a separate species.

The sting is at the end of abdomen. Some say that the bees must die when they sting, while others hold that they die only if part of their intestines is torn out with the sting. Even horses may be killed by bees. Wasps and hornets (both are related to the bees) are listed as enemies of the bees.

Pliny continues: "The silkworm weaves a thread like the spider. From these threads clothes are made for women that are fond of show. The art of unwinding these threads was invented by a woman, who thus managed to denude a woman by clothing her." Beetles have horny elytra to protect their wings, but no stings. Pliny describes some beetles, e.g. *Lucanus* (with horns divided at the tip which close for the bite); one species shapes big balls of manure with their feet in order to protect their young against the cold of the winter; glow-worms which shine at the sides and at the abdomen; cockroaches which originate in the moist steam of the baths and shun the light.

Furthermore, he gives a general characterization of insects: The wings of all insects are undivided. Mosquitoes and flies have a sting at the oral orifice; with the flies it is blunt and meant not for stinging but for sucking. The legs move at the sides; with the locusts and grasshoppers the longer hind legs bend outwards. Locusts lay their eggs by means of an ovipositor into the earth where they remain through the winter. In wet springs the eggs are spoilt. After dry springs the locusts are all the more numerous, but often the wind drives masses of them into the sea. At times they are a scourge sent by the gods. Then their swarms darken the sun and destroy the crops wherever they go. They even gnaw the doors off the houses. Their home is northern Africa where the laws demand three campaigns against them every year; in Syria even soldiers are sent against them. Pliny reports that the North African and Spanish provinces had to pay a considerable part of their tributes in the form of a scale insect (*Coccus ilicis*) which was used to dye the uniforms of the high-ranking officers. The coccid of Sicily was valued least. The adult animals metamorphosed into small worms.

Apart from these examples taken from the eleventh book of Pliny's *Natural History* other entomological data about galls, medicinal uses, etc are scattered in the other books of his *Historia Naturalis*. Summing up we can conclude that it may be doubtful whether Pliny ranks with Aristotle and Theophrastus but he does belong with the few outstanding naturalists and entomologists of antiquity.

Unlike the Greeks, the Romans left major works on agricultural entomology. They all were called *De re Rustica*. The oldest was written by Cato (235 B.C.), and he was followed by Varro (36 B.C.), Columella (ca 50 A.D.), and Palladius (ca 380 A.D.). Apparently Columella was the most important of these writers; the measures he suggests against grain weevils, for instance, are remarkable for that time: a mixture of clay, chaff, and the fluid of oil-presses is spread on the floor of the granary and allowed to dry before the grain is stored (the same measure is recommended against mice). Another safety measure is the storing of the grain in earth pits. Grain attacked by weevils should not be stirred up; it should either be brought into the open and put next to water vats which suck up the weevils or left untouched because the "corn worm" penetrates only one span deep so that all the grain lying deeper is safe. Against dog fleas he recommends the fluid from the pressing of olives or powdered hellebore mixed with cumin and water or the juice of the snake gourd. Hen's nests must be periodically cleaned to prevent the increase of fleas and other vermin. The fluid from the pressing of olives is also used against house fleas. The following advice is given regarding buprestids and the danger of suffocation is pointed out. The animal that has swallowed a buprestid should be kept in motion and given wine and pulverized raisins or wheat with dry wine of grapes and leek. The palatal vein may be opened to make the animal swallow its own blood.

Columella mentions flea beetles, ants, snails and caterpillars as garden pests. The measures of control which he suggests are primitive and magical: hanging up the heart of an owl in the garden or drawing a circle of ashes or chalk around it. Trees may be protected by a ring of a mixture of tar, butter, and ruddle or a mixture of ruddle and vinegar around their trunks, or by hanging up a *Koracinus* fish in the tree. As a protection against caterpillars the seed should be sprinkled with the sap of houseleek or with the blood of caterpillars, or a woman, ungirded and with flying hair, must run barefoot around the garden; or crayfish must be nailed up in different places in the garden; burning garlic stalks without heads in the garden would help too. To protect vines from caterpillars, the vine-dressers' knives should be rubbed with garlic, or asphalt, or sulfur should be burnt under the vines. You may also gather caterpillars in your neighbor's vineyard, cook them and scatter them in your own.

A very interesting remark, to be found in the fifth book of Palladius, records even at that early time the trapping by light of butterflies that are harmful to bees. Cerambycids were tied to olive trees and other trees to drive away destructive insects by their chirping. Prayers were offered and processions and ceremonies were held in honor of the god of blight and rust of cereals (Robigo), and other rites were held against invasions of pests.

Roman apiculture was probably greatly influenced by its economic development in Greece. Apiculture played a relatively important part in Roman agriculture. Bees were an essential foundation of the economic existence of many a farmer. The study of bees in Roman times, however, was entirely oriented to economic problems and confined to observations and advice that were significant for practical beekeeping. Only Pliny touched upon biological problems. The number of beekeepers must have been considerable; many slaves were engaged in beekeeping, but it was also the occupation of Roman nobles.

No evidence of apiology or apiculture is known from early Roman times, which permits the conclusion that apiculture did not then have the importance it attained in later centuries. The first details are to be found in Varro (36 B.C.), but it is quite obvious that the Romans were not particularly interested in biological and scientific problems. Of the apiaries built by man it is said that they were of different forms and different materials. Those made of bark are recommended. It is advised that the colonies be inspected three times a month, the hives be cleaned and the superfluous queen bees removed. The working and the pasture of the bees are discussed in detail. Beebread, honey, and beeswax are prepared from various plants. A watering-place with clear water for the bees is important. In adverse weather food must be provided, e.g. dumplings of cooked figs and water or honey water or cakes of grapes and figs with grape juice. The harvesting is described, and it is recommended to take out nine tenths of the honey or less.

A literary, not a scientific description of apiculture is given by Virgil (70 to 19 B.C.). In the fourth canto of his rural didactic poem "Georgica" he deals with the bees.

Columella quotes a working calendar for beekeepers by Hyginus: Between the 8th day of Aries and the rise of the Pleiades (which is a period of 48 days) you should make the first inspection of the hives, remove the refuse, blow smoke into the honeycombs and kill the vermin. But before doing this you must observe certain rites, abstain from the enjoyment of love, not get drunk, wash your hands, and avoid malodorous food like garlic or onions. — On May 11th, the colonies begin to gain vigor and grow. From the rise of the Pleiades to the summer solstice they swarm and must be closely watched. From this date to the rise of Sirius, a period of about 30 days, honey and grain are harvested. In the same time you can let the bees reproduce themselves in a killed bull or in the belly of a cow. The hives must be periodically cleaned and the moths removed from the honeycombs. After this period, until the rise of Arcturus, the thyme honey is harvested, during which time the bees must be protected from hornets. At the equinox of Libra (September 19th) honey is harvested for the second time. From the beginning of fall to the setting of the Pleiades (October 28th) the bees gather the honey of the tamarisks and forest trees which they store for the winter. On this they live until the beginning of winter. The hives must be well covered against the winter to keep the bees from freezing, and if necessary additional food must be given.

From places where there are not sufficient flowers and thus sources of food for the bees the colonies should be brought to better localities. The following general habitat conditions are recommended: midday sun, an environment without too much noise and with high walls, surrounded by shrubs, in a valley, and near water. The hives differ according to region and habitat. Most suitable is a foundation of stone, carefully whitewashed. The hives should be easily accessible in front and rear and protected from the rough northern winds by buildings. Narrow entrances keep the cold out.

A long chapter is devoted to the diseases of bees and their cures. The plague is rare; when it occurs, the colonies should be moved to a distant place. The disease to which, among less dangerous ones, the greatest importance is attached is a diarrhea caused by feeding on spurge or elm blossoms, and several cures and magic rites are described.

Pliny the Younger (23 to 79 A.D.) was the only Roman to study the biology of the bees more intensively; he commented on the pasture for bees and on the practical management of the stocks. Later Palladius (fourth century A.D.) wrote a manual of agriculture which closely followed Columella and contained a working calendar for apiculture that was of importance for practical beekeeping and held in esteem up to modern times. Finally Aelian's observations (though they are not scientific) on the life of the bees and on apiculture in his animal stories may be mentioned.

With regard to silk production in general and the rearing of silkworms in antiquity it must be stated first that the genuine mulberry silkworm became known only at a later date. It is not identical with the Greek "silkworm" at Kos (*Pachypasa otus* Drury). Genuine silk was imported into the occident from China only at the beginning of the Christian era. A Roman delegation at the time of Marcus Aurelius (166 A.D.) had spread fabulous news of silk production. In ancient Rome all-silk clothes were an unusual luxury which was at times prohibited for men. Half-silk clothes, however, were much in use in late Roman times.

While the production of silk from the "silkworm" of Kos had been known for a long time and Pliny had described its rearing and the manufacture of silk, the first living silkworm cocoons were brought to Byzantium by monks in Justinian's reign (ca 550 A.D.). This is reported by Zonaras (ca 1100) in his *Chronikon,* and he says that the real knowledge of these animals, their cocoons, and their rearing dates only from that time.

Among the late Roman authors Aelian (160 to 240 A.D.) devoted 50 chapters of the 17 books of his treatise, *On the Characteristics of the Animals,* to insects. He studied Greek literature and wrote in Greek. His works have no scientific significance; he compiled tales of earlier periods which he selected for their morals or their singularity; 12 chapters deal with bees, 7 with ants, and 4 with wasps.

General Knowledge of the Different Groups of Insects in Antiquity

Before proceeding to consider the development of entomology in the next great era, the medieval times, we may briefly summarize the knowledge of the ancients about the different groups of insects as it appears from the details we related.

Lice received much attention. It was known that they infested especially nomadic peoples. They were feared for the diseases they caused; on the other hand it was supposed that people with lice on their heads suffered less from headaches. It was observed that certain animals were attacked by specific lice only, some kinds more, others less or not at all. A much dreaded disease was phthiriasis (pediculosa passio), especially in consumptives and people suffering from suppuration. Helladius names several well-known people who were supposed to have died of it, among them Sulla, Herod the Great, Democritus, and others. According to Herodotus the Egyptian priests shaved off all hair so that no dirt and no louse might cling to them. Konstantinus Manasse wrote that at the death of a man the parasites leave his body. Symphosius' riddle abount hunting also touches upon lice which pass on from the killed game to man. Lice were used in medieval prescriptions. The Greek word for louse means spoiler or destroyer, the Latin word means little foot.

In one of Aesop's fables it is said of ticks that they are found on foxes. It was said that ticks must not be torn off, which would cause sores, and that dogs can be rid of them by boiled pitch and hog-fat.

According to Aristophanes, Petronius, and other writers, bugs were a veritable scourge in Athens and Italy. The recommended method of control was coating walls and furniture with ox-gall or oil foam or sprinkling them with the juice of the squirting cucumber. Bugs were used in some cures for men and animals. "Bug" was also an abusive name for maliciously sneering people.

Fleas were hated less, they often provoked funny situations. Even the ancients distinguished dog fleas from human fleas and dog lice. The third species they knew were flea beetles. For protection they sprinkled the plants with coriander water.

The cicadas lay their eggs in the soil so that their offspring seem to emerge directly from the earth. Thus the people of Attica regarded them as a symbol of their own autochthony. A clasp in the shape of a cicada was a favorite ornament for the hair in Athens, even for men. A metal spiral of this hair cicada served to hold a braid or lock together. Much older than the Attican national symbolism was the oriental connection of the cicada with the sun god, though it did not play any important part in religious symbolism; but it seemed ideally suited for idyllic poetry. Cicadas playing flutes or pipes are depicted on Greek gems. A Pompeian mosaic shows a cicada as the driver of a coach to which a parrot is harnessed. Cicadas were used as toys and were found in children's graves. They were held in cages made of rushes, and people enjoyed their singing. The grub (mother or larva) of the cicada was regarded as savory until the bursting of the outer skin, also regarded as savory are the males before pairing and the females with the white eggs. It was known that only the males of the cicadas make music by means of an apparatus (not too clearly described) in the middle of the body (not by rubbing the wings, as the crickets do) and that in the hot season they start chirping in the fourth hour and are loudest at noon. They were supposed to like sunshine and isolated trees, especially olive trees. The swallows were thought to be their chief enemies. Homer compared the eloquence of the old men of Troy to the chirping of cicadas. In medicine the cicadas were used as a remedy for bladder troubles.

Most of the knowledge of wasps was derived from Aristotle's observations. Twenty-seven stings of wasps or hornets were considered fatal, a number derived from the ancient superstition about the magic numbers three and nine. Recommended remedies for wasps' stings were rue, mallow, balmmint, wild thyme, origan, marsh-mallow leaves (either boiled or as a poultice), unmixed wine, salt, and vinegar. Whoever had survived a scropion's sting was regarded as permanently immune against wasps. A hornet was credited with being able to put a bull to flight. Aesop relates the suicide of a snake on whose head a wasp had settled. The snake threw itself under a wheel and was run over together with its tormentor. Vinedressers and beekeepers strove to exterminate the wasps. The owl, woodpecker, and fox were considered the chief destroyers of wasps. The fox was said to put its tail into a wasps' nest, pull it out full of wasps, slay the wasps, and get the honey. Dreaming of wasps foreboded evil, e.g. death in battle. Proverbs warned against stirring a wasps' nest. Wasps were represented in Egyptian and Greek art. The Greek name for wasp is related with the word for constricting, the Latin name with floating and flickering and is meant to denote the eternal restlessness of a swarm of wasps. Often wasps, hornets, and *Anthrenus* were confused. Later, however, the Romans consistently distinguished between wasps and hornets. *Thendredon* was classified as vespine, and so was *Ichneumon;* the latter, however, was counted among the digger wasps.

The ancient descriptions of bombylios and nekydalos clearly differ from the small Chinese silkworm. Fabulous reports of the silkworm appeared in antiquity, but it was brought to Byzantium only in Justinian's reign.

The common house fly, the blow fly and the gad fly are mentioned in the *Iliad* and figure in metaphors and similes. Pliny, Aelian, and others describe how a drowned fly may be revived by ashes and sunbeams. Flies may live even with their heads cut off. The fly is the symbol of importunity, greediness, curiosity, worthlessness, and impudence. It appears in fables, proverbs, metaphors, and similes; it is a term of abuse in Homer, Aristophanes, and Plautus; it was the sign for impertinence in the Egyptian hieroglyphs. Flies are also characterized as demonic beings and related to death (maggots in the process of decomposition). The demon of decomposition, Eurynomos, was depicted as a carrion vulture or a carrion fly. In Hungary witches were supposed to appear as flies, in Germany the wicked god Loki took the form of a fly to slip through a keyhole. The demons of diseases and the ghosts of death visited men in the shapes of flies. As a demonic animal the fly served as a protection against the evil eye. Beelzebub is the fly god and the lord of vermin. The expulsion of the flies was credited to Zeus, Heracles, or Apollo. Sacrifices, prayers, and vows were offered against the flies, and they were driven away with fans which sometimes were very splendid, e.g. made of peacock's feathers. Domitianus stabbed flies with his own hands, using a sharp style. Farmers fumigated with origan, black caraway, and copper sulfate, the walls were smeared with coriander seed and oil and sprinkled with a decoction of elder. The spiders were regarded as the most serious enemies of the flies. The usefulness of the flies was rated very low: they were food for the birds and were supposed to carry away refuse. The Roman consul Mucianus always had a living fly in a linen cloth about him as a protection against blear-eye. The behavior of the flies was interpreted to predict the weather: when they bit eagerly and viciously, a storm was expected. According to Cassiodar the professional seekers of springs believed that one of the surest signs of an underground spring was masses of the very smallest flies flying about in one place. Phidias is said to have made a statuette of a fly. Myrmekides mentions a miniature sculpture showing a fly. A fly is found on a gold medallion and as a cameo of amethyst. On a carnelian, now in Berlin, a fly is seen flying against a flame.

The harmless common house flies and blow flies were distinguished from the gad flies. In Italy the latter were a nuisance though a tolerable one, but in Egypt they were particularly bad. They figure in the biblical report of the exodus from Egypt. Herodotus relates that the people in Egypt slept on high towers or in tents made of fishing-nets to protect themselves against the mosquitoes. From the Augustan period these Egyptian mosquito nets, called conopeum or conopium, were exported to the occident. The buzzing of the fly was supposed by the Greeks to be produced by an opening in its abdomen. Aristophanes says: "So the behind of the empides is a trumpet." The following remedies for mosquitoes are mentioned: vinegar, hemp, onion, burnt shells, etc.

About the day-flies Aristotle gives these details: At the time of the summer solstice the river Hypanis at the Bosporus carries down a sort of skins, bigger than grapes; out of these come winged animals with four legs. They live and fly about, but at noon they grow very weak and die, because they live only one day they are called day-flies. Aelian assumes that they are born of wine: When the vessel is opened, the day-flies swarm out, see the light of the world, and die. Nature gives them life but takes it away soon after so that they may not feel their own misery nor see that of others.

In the Cyrenaica they had laws that enjoined the extermination of locusts (eggs, young, and adult animals). The violation of this duty was punished as severely as desertion. Also on the island of Lemnos each citizen had to deliver a certain tribute in locusts. Other remedies were prayers, sacrifices, processions round the fields, and conjurations. The Christian times knew special saints: Theodosius in the fifth and sixth centuries, also Stephanus, Gregory, Seraphinus, Theodorus in Galatia, and others. Even excommunication and exorcism were employed against these diabolical insects. Aristotle distinguished two species of locusts. The frequent representation of these insects on gems obviously served magical and phophylactic purposes. The cricket is mentioned by Pliny and appears also in connection with medicine. The mantis is described by Theocritus as the singer sitting in the reeds. Evil will befall any animal it looks at.

Medieval Times

So far we do not know of any original scientific research in the first centuries A.D. In the historical development which marked the end of antiquity its knowledge was also lost. After the partition of the Roman Empire in 395 the East Roman Empire of Byzantium survived for a thousand years. Its librarians guarded the collected knowledge of antiquity and preserved and copied the books. These were the roots of the important era of Arabian science and of the Renaissance of the fifteenth century in Italy. As far as entomology was concerned this meant chiefly the biological works of Aristotle.

The migration of the nations put an end to the West Roman Empire. The knowledge of the classical authors, of Greek and Latin and especially of the scientific traditions was gradually lost. This was not the Church's fault; it was the Church that stimulated efforts to preserve the cultural heritage of antiquity. An extraordinarily significant event in this respect was the founding of the Benedictine Order in 529. Monastic schools took the places of the imperial provincial schools. Unfortunately this happened too late to save the fundamental works of Aristotle and Pliny for the new era. They were replaced by compendia and epitomes. The mystic mentality of that time, dogmatism and scholasticism impeded scientific research. In the sixth and seventh centuries the Irish monasteries did much to preserve the works of the classical authors and to cultivate the knowledge of Greek and Latin. In the Irish monasteries the movement was born that grew into the Carolingian Renaissance in the Frank Empire in the ninth century. Isidorus and especially Rhabanus Maurus were the most important scientific writers of that time.

The popular zoological book of an unknown author dominated the scene throughout the Middle Ages: the *Physiologus*. It was the most widely distributed work on zoology of its time, and it marked the transition from antiquity to the Middle Ages. Numerous versions in Greek, Latin, Ethiopian, Armenian, Syrian, Arabic, and in almost all Germanic and Romance languages have come down to us. In 496 the *Physiologus* was put on the index of heretical writings, and it was not removed from the index until the time of Pope Gregory the Great (ca 600), but then it was even included in the list of useful books.

The *Physiologus* had its origin in the first Christian centuries with their mysticism, a time in which even the sciences were influenced by symbolism. The fly was the symbol of the devil and of sorrows, the ant was the symbol of the provident worker, the bee of virginity and wisdom, the scarabaeus of the sinner. Christ was the locust or the grasshopper, which also stood for demons and for pride. The moth symbolized the temptations of the flesh, the worm was the symbol of Christ, of desire, and other things. The *Physiologus* does not contain any original observation or study of nature. Of its 63 chapters, 56 are concerned with animals; the insects among them are dung beetle, bee, wasp, and ant. Not all of the later versions of the *Physiologus* include all the chapters and all the insects mentioned here, but the ant appears in almost all editions.

The fifty-second chapter deals with the dung beetle which is supposed to develop from dung in the month of flowers and to live in dung and in stench. It is said to form its eggs of dung and to warm them until its offspring develop in the centers of the eggs and live with it in the same stench. The dung beetle is declared to be a heretic, sullied by the stench of heresy, and the balls of dung are explained as evil thoughts and heresies.

The ant is described in the fifty-fifth chapter and credited with three qualities: 1. Ants do not rob each other of grain, but each gathers for itself. This trait is quoted with an allegory from the Bible, comparing the ants with the prudent virgins and the foolish virgins. 2. The ant bites the kernels in two before storing them in the earth so that they cannot germinate and it will not starve. Ants sense the weather in advance and act accordingly: when rains threaten and winter approaches they carry their food inside, and before fine weather they carry it out. This, too, is explained with words from the Old Testament, but unintelligibly. 3. Ants know wheat from barley by the smell of the stem, before creeping up. They gather only wheat. Barley is food for animals and comparable to the teachings of the heterodox. Man should abstain from barley and take only wheat which represents the true faith in Christ.

The third chapter of the *Physiologus* is devoted to the "ant lion"; however, it does not describe this insect but a fabulous animal with the front of a lion and the rear of an ant, descended from an herbivorous mother and a carnivorous father. So the ant lion could not eat meat on account of its mother and could not eat plants on account of its father and consequently had to perish. This, too, is followed by a comparison. Every man is said to have likewise a double soul and to be inconstant in all his actions. But he must not walk in two paths and equivocate in prayer.

The most important work for the sciences, next to the *Physiologus,* was, throughout the middle ages, the book *Origines sive Etymologiae,* written in Latin by Isidorus, Bishop of Sevilla. Almost all the works of the following centuries were based on it. It is an etymological compendium covering all fields of knowledge; for each field the subjects are arranged in alphabetical order. It was probably compiled from excerpts of other compendia, scholia, etc. It adds nothing to the knowledge of entomology, and its contents are poor in comparison to the standard that had been reached in antiquity.

In the twelfth book, the fifth chapter, "de vermibus," and the eighth chapter, "de minutis volatilibus," are devoted to the insects, of which only 27 are mentioned; they formed the whole basis of the knowledge of insects nearly to the end of medieval times.

Of the worms it is said that they are animals which originate in wood, meat, earth, etc without fertilization; only the scorpion is supposed to develop from eggs. Cantharis (*Lytta vesicatoria*), an earthworm, causes blisters on human skin. Eruca, the caterpillar, and teredo, the wood borer, are mentioned as well as the louse, a skinworm, properly called little foot, and the fleas, named thus because they feed only on dust.

Worms develop in rotting meat, moths in clothes, caterpillars in cabbage,

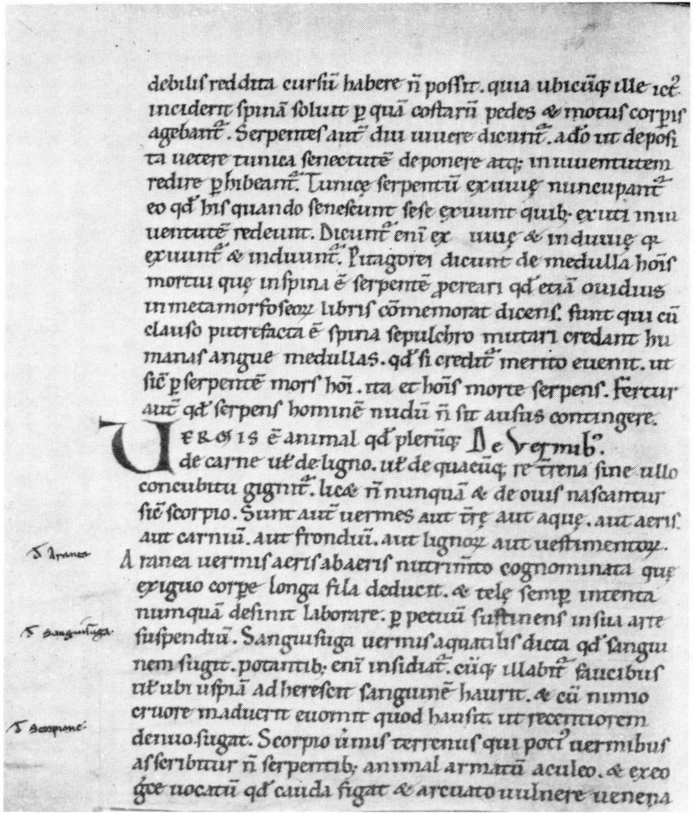

debilis reddita cursu̅ habere n̄ possit. quia ubicu̅q; ille ict9
inciderit spina̅ soluit p qua̅ costaru̅ pedes & motus corporis
agebant᷑. Serpentes aut̄ diu uiuere dicunt᷑. adeo ut deposi
ta uetere tunica senectute̅ deponere atq; in iuuentutem
redire phibeant᷑. Tunicę serpentu̅ exuuię nuncupant᷑
eo qđ his quando senescunt sese exuunt quib; exuti in iu
uentute̅ redeunt. Dicunt᷑ eni̅ ex uuię & induuię q̄
exuunt᷑ & induunt᷑. Pitagorei dicunt de medulla hoi̅s
mortui quę in spina ē serpente̅ pcreari qđ etia̅ ouidius
in metamorfoseoꝝ libris co̅memorat dicens. sunt qui cu̅
clauso putrefacta ē spina sepulchro mutari credant hu
manas angue medullas. qđ si credit᷑ merito euenit. ut
sic p serpente̅ mors hoi̅. ita et hoi̅s morte serpens. Fertur
aut̄ qđ serpens homine̅ nudu̅ n̄ sit ausus contingere.

De Vermib9.

VERMIS ē animal qđ pleru̅q;
de carne ul' de ligno. ul' de quacu̅q; re terrena sine ullo
concubitu gignit᷑. licet n̄ nunqua̅ & de ouis nascantur
sic̄ scorpio. Sunt aut̄ uermes aut t̄re aut aque. aut aeris.
aut carniu̅. aut frondiu̅. aut lignoꝝ aut uestimentoꝝ.

Aranea. Aranea uermis aeris ab aeris nutrimto cognominata quę
exiguo corpe longa fila deducit. & telę semp intenta
numqua̅ desinit laborare. p petuu̅ sustinens in sua arte
Sanguisuga. suspendiu̅. Sanguisuga uermis aquatilis dicta qđ sangui
nem sugit. potantib; eni̅ insidiat᷑. cu̅q; illabit᷑ faucibus
ul' ubi usp ia̅ adheserit sanguine̅ haurit. & cu̅ nimio
Scorpione. cruore maduerit euomit quod hausit. ut recentiorem
denuo sugat. Scorpio u̅mis terrenus qui poti9 uermibus
asscribitur n̄ serpentib; animal armatu̅ aculeo. & ex eo
gręce uocatu̅ qđ cauda figat & arcuato uulnere uenena

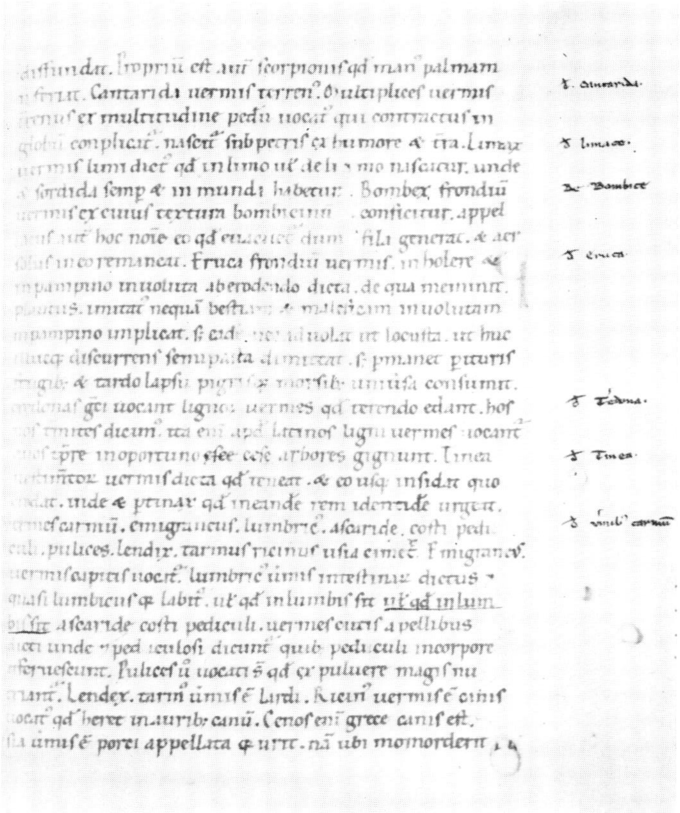

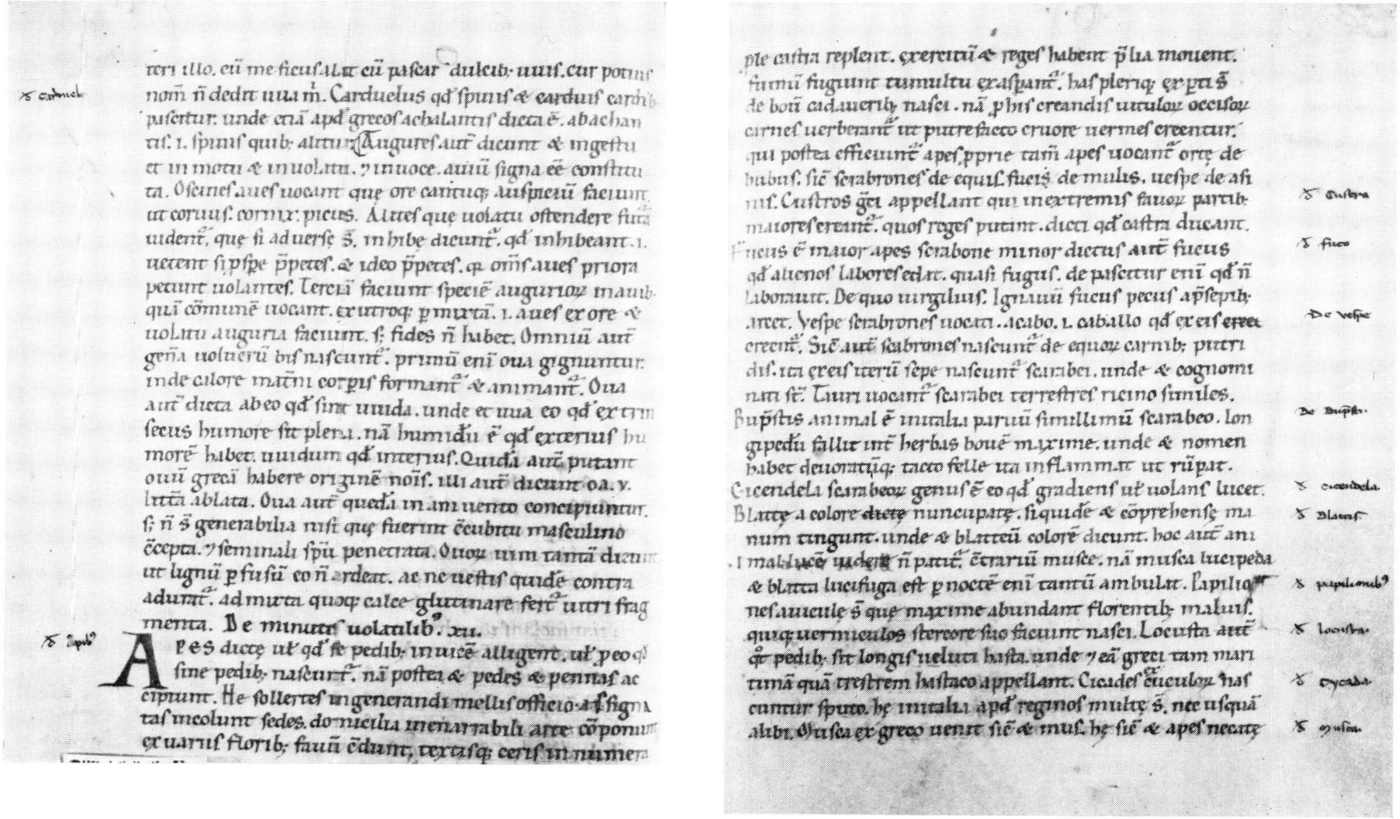

FIGURE 2. Fol. 38v–39r and 48v–49r of Isidor Hispalensis: Originum sive Entymologiarum libri posterius. Parchment manuscript in the library of the monastery at Admont, no. 278, 168 pp., twelfth century. Made at Admont, white leather cover was restored in 1955.

the wood borer in wood, and Tarmus (the larva of *Dermestes*) in bacon. The worms move by stretching their bodies and pulling them up. Of the "minutis volatilibus," the honey bee is described with king, workers, and drones. Bees come out of dead cattle, hornets of horses, drones and bumble bees of mules, and wasps of asses.

The work deals with the following beetles: Thaurus as an earth-beetle, the buprestes of Pliny, and cicendula the glow-worm. Platta is mentioned and papiliones, the butterflies from whose droppings small worms are supposed to develop, locusta, the grasshopper, the cicadas which are said to originate in the saliva of the cuckoo, the blood-sucking culex (probably *Tabanus*), spinifes (= Culicidae), the "third plague of the Egyptians", and bibiones (= *Drosophila*) which are thought to develop in wine.

Several compendia date from the eighth and ninth centuries, the time of the Carolingian Renaissance, as *De visione Naturae* by Johannes Scotus Erigena, *Natura Rerum* by Bede and *De Universo* by Rhabanus Maurus. We have a detailed knowledge only of the work of Rhabanus Maurus (776 to 856), abbot at Fulda and later Bishop of Mainz. It follows largely that of Isidorus but proves its superiority by the much better descriptions of the insects dealt with in the eighth book which is devoted to the animal kingdom.

The interpretation of the species mentioned by Rhabanus shows that he combined the ant lion, the ant, and the cricket under the heading of "meat worms"; the section "vermes" contains louse, flea, *?Dermestes,* bug, *Meloe* or *Lytta, Bombyx mori* as a caterpillar, wood-boring insect larvae, and the clothes moth. The "minuta volatilia" include "apis" with king, worker, drone, hornet, and wasp; the "scarabaei" are *Geotrupes,* stag beetle, *?Meloe,* and the glow-worm. They are followed by cockroaches, butterflies, and cicadas (probably *Aprophora spumaria*). The heading "Muscae" combines flies, gad flies and probably *Hippobosca,* and "Locusta" includes locusts, grasshoppers, *Tabanus,* and *Drosophila.*

The descriptions of the species are much more comprehensive and detailed than those in earlier compendia, especially with regard to the mode of life. They show a considerably better knowledge. As examples we quote the descriptions of "formicaleon" and "eruca": "The ant lion is a small animal exceedingly hostile to ants. It hides in the dust and kills the ants that carry provisions. It is rightly called formicaleon because it is a lion to ants, if only an ant to all other animals."

"Eruca is a leaf worm which rolls up in leaves and tendrils. It does not come as the locust does, hurrying from one place to another and leaving things half eaten, but it stays on the doomed plants and eats them up slowly and sluggishly but entirely. Plautus: malefica involuta!"

An illuminated manuscript of the fables of Phaedrus made in the monastery of St. Martial near Limoges dates from the eleventh century. The drawings, which are sometimes primitive but show the characteristics of the animals they describe, probably follow models of the fourth or fifth centuries.

There is an interesting commentary on the Talmud by Rabbi Shlomo Jizchaki (Rashi) who was born in the Champagne and lived from 1030 to 1105. He tried to elucidate dark passages in the Talmud and included descriptions of the insects mentioned in it without adding anything materially new. Obviously he had a certain talent for the observation of nature, which is evident from his remarks on apiculture and on the mode of life of mosquitoes (he seems to mean *Drosophila* and the larvae of Culicidae) and of the "worms of meat" (probably *Hypoderma*).

A pharmacological book not connected with the science of antiquity but based on the popular tradition was written by a Benedictine abbess, Saint Hildegardis (1099 to 1179). The fourth book of this work deals with the animals, which are divided into Pisces, Volatilia, and Animalia. Nine insects are listed as remedies: Bee, fly, mosquito, cicada, bumble bee, wasp, and glow-worm among the flying animals, and ant and flea among the terrestrial animals. Even the subcutaneous injection of living ants as a remedy for neurasthenia is recommended in this work.

Of the useful insects, bee, silkworm, and the dyer's coccid played important parts in medieval times. As it was mentioned above, shortly after 500 two monks brought the cocoons of the genuine silkworm (*Bombyx mori*) from China to Constantinople at the risk of their lives. The rearing of silkworms spread slowly in the occident. The Arabs brought it to Spain, and it reached Sicily in 1130, the Italian mainland in the fifteenth century, and southern France not before 1470.

Probably *Margarodes polonicus* was the dyer's insect "kermes" that was used even in the medieval monetary system: Many monasteries demanded from their peasants a tribute of a certain amount of kermes or the payment of an equivalent sum of money.

We have shown that the zoological literature of the early Middle Ages in Europe, being limited to simple compendia, reflected a decline of the knowledge in this field as compared to the high level attained in antiquity. This is contrasted by the development of the sciences by the Arabs from the eighth to the fifteenth century which is called the Arabian epoch in the history of the sciences. This was the first Aristotelian renaissance; in this epoch the Arabs made the classical sources accessible again and thus gave decisive impulses for further development. It was only because this Arabian epoch (and naturally also its entomology) coincided with the beginning of the period dominated by scholasticism that it failed to achieve a real renaissance of the sciences; for the mentality of those five centuries (about the time from 1000 to 1500) was characterized by an alienation from nature which, to say the least, made fundamental progress difficult.

It was the supreme principle of scholasticism to combine knowledge with faith; methodologically it was, in spite of its weaknesses, a triumph of exact and abstract thinking over the traditional mysticism, and thus it broke the ground for the later development of the sciences. One of the most outstanding representatives of scholasticism was Thomas Aquinas, a disciple

FIGURE 3. Of the punishment. In Hans von Vintler, *Die Blumen der Tugend*. Bl. r. Illuminated manuscript, Upper Germany, fifteenth century (latter half). In Forschungsbibliothek Gotha (GDR), Sign.: Chart. A 594.

FIGURE 4. Of the wrath. In Hans von Vintler, *Die Blumen der Tugend*. Bl. v. Illuminated manuscript, Upper Germany, fifteenth century (latter half). In Forschungsbibliothek Gotha (GDR), Sign.: Chart. A 594.

of Albertus Magnus. He formulated his view of the world by dividing existence into three realms: nature, grace, and salvation. Of these, he supposed nature to be accessible also to heathens, and in this realm the writings of a heathen like Aristotle were as important to him as those of any Christian or even of any theologian. Grace and salvation were in his opinion attainable only by Christians and were ruled by theology.

At the end of the twelfth century Spain was the country where the Christian occident came into contact with the heritage of antiquity, revived by the Arabs. Toledo, the capital of the kingdom of Castile and the seat of Arabian science under King Alfonso VIII (1158 to 1214), became the center of the "translation of science" and the bridge to Europe. For the great scholars of that time the study at Toledo, the university of Arabism, was the necessary foundation of their work and their dignity. Physicians were especially greatly interested in the sciences. The most important translator of Arabian sources for sciences and medicine at Toledo was the Scotsman Michael Scotus.

Without describing the Arabian epoch in detail and without discussing its importance for the Arabian world we may here honor its achievement in preserving and transmitting the scientific knowledge of antiquity to a Europe that was passing through centuries of cultural decay. That it was limited in its own and proper sphere of the exploration of the animal kingdom by the strictness of Koran and Islam is immaterial for its importance for Europe. While it was at first content with the exclusive tradition of the scientific knowledge of antiquity, the later works of the Arabian epoch contain in addition a great number of original observations, though these are mainly related to practical purposes. As the zoological research of the Arabian epoch was chiefly concentrated on the orient and was important for the world west of Egypt only in the connection mentioned above, we may here confine ourselves to summarizing some of its biological studies which were concerned with the migratory locust, the dung beetle, the silkworm, the house fly, the origin and the swarming of the flies, and other subjects.

The revived interest in Aristotle's scientific findings caused dissensions within the Church. The most pronounced opposition to Aristotle's writings and their propagation came from the university of Paris and from three French councils in the thirteenth century, and the Parisian faculty succeeded in having them banned; but these struggles against Aristotle were locally limited. About 1260 Wilhelm von Mörbeke even made a direct translation of Aristotle's zoological writings from Greek into Latin. Albertus Magnus later won due respect for the scientific thoughts of Aristotle; he not only obtained the sanction of the Church but even its support.

The most important works of the thirteenth century were written by authors who belonged to the Dominican Order: *Liber de Natura Rerum* by Thomas Cantipratanus, *De Animalibus* by Albertus Magnus, and *Speculum Naturale* by Vincentius Bellovacensis.

Thomas Cantipratanus (born in 1201, died between 1263 and 1293) studied in Cologne under Albertus Magnus and later in Paris. He wrote several hagiographies and two scientific works on the basis of the early medieval books noted above. Unlike Albertus Magnus he hardly added anything of his own. He was a symbolist of nature who generally failed to make observations of nature, but his works were important as scientific compendia and as models for later translations. Like other symbolists of nature he was moved by the bee society to make comparisons with the human community and especially with the monastic regulations, which is apparent in his work *Bonum Universale de Apibus.*

His book *Liber de Natura Rerum* which, according to his own words, he laboriously compiled in 15 years (1233 to 1248) from several other works, was widely known from medieval times to the beginning of the modern age. In its content it resembles the works of Isidorus of Sevilla and Albertus Magnus. The animals are treated in the fourth to ninth of its 19 books, and the ninth book deals with the worms which include amphibians and insects. Again, in these descriptions the comparisons with man and his communities, already noted for the bee society, appear. So he says of the ant lion that it may be likened to those idlers that do not allow the workers what they have earned by their hard toil. He also explains the medicinal use of *Lytta vesicatoria* for protection against fleas.

The greatest of the three thirteenth century encyclopedists mentioned above was the Dominican of noble descent, Albert von Bollstädt (1193 to 1280), called Albertus Magnus. He was a teacher in many German monasteries, obtained a Master's degree in Paris and figured in public life; he often mediated in political conflicts between towns and the Church. He wrote 34 quarto volumes, which is a singular achievement in itself. As stated before, his great merits were the restoration of the authority of Aristotle and his success in making science largely independent of theology. As late as 1251 a synod prohibited the reading of any of Aristotle's works. In fighting for their recognition Albertus Magnus lived under the constant threat of being charged with heresy. It was not before 1366 that the Pope not only recognized Aristotle's works but made their study a condition for obtaining the *venia legendi.*

Though the scientific works of Albertus Magnus are paraphrases of Aristotle and his contemporaries, he yet made his own observations, and his botanical and chemical knowledge exceeds that of Aristotle by far. Entomology benefitted by his original summary of his anatomical and physiological knowledge. With this work he was for a long time the outstanding figure in the exploration of nature, in spite of occasional errors in his observations.

He wrote his main zoological work *De Animalibus* between 1255 and 1270. It contains 26 books, as he added 7 books of his own to the 19 books of Aristotle. In the last five of these books he gives a survey of the species he knew in the form of an index with a short description of each species, which indicates what was known about the animal kingdom in his time.

This survey is characterized by objective and sober observation which flags unfortunately with the insects. Of all the species he mentions, about 450, only 33 are insects, and they are listed without differentiation together with the amphibians, molluscans, and worms under the heading of "small bloodless animals". Only the bee is discussed in detail, true to the Aristotelian model.

In his description of the construction of the articulates he mentions rings (=body segments) and the numbers of legs and wings, but as with Aristotle his view is teleological and causative. In the general discussion of development he divides the articulates basically into those with a complete development (by fertilization) and others with an incomplete development (by parthenogenesis). Thus the caterpillars on the cabbage plants, the larvae of the clothes moth, the wood-borers and the insects attacking stored grain are said to develop without fertilization. The caterpillar (worm stage) is followed by an egg-like pupa stage, from which the imago emerges. In the special part, however, Albertus Magnus sometimes abandons his fundamental division into complete and incomplete fertilization and assumes again that many insects develop from dirt, etc.

There is a chapter written by himself (not taken from Aristotle) on the intelligence and the perfection of animals which also has a section on insects; he regards them as very perfect. The criteria of perfection are the development of the senses, which are fully in evidence in the insects (sense of touch poor, smell excellent, taste in the proboscis, hearing not found in anatomical dissection but demonstrable in experiments, eyes hard so that they will not be injured), and of the locomotive organs (in completely ringed animals more than two wings and more than four legs).

Physiological remarks form the introduction to the index of species. It is stated that the bloodless animals have another body fluid instead of blood. They are coldblooded, their bodies are segmented into rings, and the intestine extends through the whole body. The periodical changes of the skin are due to the surrounding heat and to an innate quality. Only caterpillars and locusts eat plants, almost all others live on saps only, flies on animal fluids. It is important for these animals to maintain their inner moisture content. Those that have a surplus of it build houses from it or spin silk. Unlike the spiders with their anal secretion, the caterpillars spin their silk with their mouths.

The chapter on internal anatomy shows that Albertus Magnus identified the abdominal cord of the articulates and knew that it extends from the brain through the whole length of the body. He also discusses the anatomy of the bee.

The 33 insects listed by Albertus Magnus are among the 49 animals in the last book of *De Animalibus*. A few examples of his descriptions:

> Cinifes are mosquitoes (Culicidae), flying worms with long legs. They pierce the human skin with a small proboscis. They originate in moisture and are frequent near water. They have a predilection for men and animals that sweat and therefore they are found so much on sleeping persons. In humid countries

the beds must be covered with special nets to protect the sleepers against their bites.

"Cantarides" = green blister beetles (*Lytta vesicatoria*) develop in masses on alder and ash from the moisture on the leaves and gnaw the leaves like caterpillars. During the day they fly about, but at night they can be gathered in clusters, which the physicians do in August. They are steeped in vinegar and used in medicine.

"Formica", the ant, is a small insect which increases in size and intelligence even in its old age. Some begin to fly in their old age. They stay on their roads and keep good order. They store provisions against the winter. Wet kernels are dried to keep them from rotting. The ants sense the weather. If a powder of sulphur and origan is sprinkled on their nests they leave them immediately. With their bite they squirt out a poisonous fluid that causes blisters.

That Albertus Magnus recognized the "ant eggs" as pupae is evident from the following remark: They first produce eggs which develop into white worms wrapped in small pellicles. These are carried to the surface of the heap into the sunshine, and then the ants emerge from them.

"Formicaleon [=larva of *Myrmeleon formicarius*] is not first an ant as some say. It is of a tick-like shape, hides in the sand, and catches ants looking for food. As it does not gather provisions in summer it is said to rob the provisions of the ants in winter."

Albertus Magnus asserts that the fly has two wings and eight legs.

Fleas originate from moist warm sand when it suddenly comes into touch with the warm bodies of animals. They suck blood with their proboscis so that the skin swells in these spots. They have long saltatorial legs besides six legs for walking. As they are very small they jump very quickly. They suck so much blood that they drop it continually as a blackish, dry secretion. Their eggs are lentiform. Always a small male and a big female are found together. The fleas born in March and April die in May. In that month there are very few fleas or none at all. If born later they live until winter when they are particularly obnoxious. As a protection against fleas it is recommended to spray the houses with decoctions of colocynth or rubus.

Lice love heat; they develop in the dirty pores of men; gluttons are preferred. Lice are particularly numerous on birds of prey.

The third encyclopedist to be noted for this period is Vincentius Bellovacensis. He created the most voluminous of these works which on account of its very size never became so widely known among his contemporaries. In spite of the bulk of his work he is the least important of these three Dominicans and the one who added the fewest original observations. Apart from the early fathers and the annotators of the Bible the authors whom he quotes most often are, among others, Aristotle, Pliny, Palladius, Dioscurus, Physiologus, Isidorus of Sevilla, and Thomas Cantipratanus, the latter with his *Liber de Natura Rerum.*

As a Dominican and the tutor of the sons of King Louis (the Saint) of France, he compiled for the king an encyclopedia called *Speculum Maius.*

Within the frame work of this he finished (ca 1250) the *Speculum Naturale* which consists of 33 books and is entirely devoted to the natural sciences. Books 17 to 23 deal with the animals. In his encyclopedia Vincentius Bellovacensis used all the quotations he knew; only Albertus Magnus does not appear, while Thomas Cantipratanus' *Liber de Natura Rerum* is quoted in full.

In the twenty-first book "Of reptiles and worms" 34 species of insects are mentioned, but only two are new, one fabulous animal and one animal called "simultas", a worm in the heads of rams, by which he possibly means the larva of *Oestrus ovis.*

Of the works written by these three encyclopedists the *Liber de Natura Rerum* by Thomas Cantipratanus was the one that had the earliest and widest distribution because of its translation from Latin (the scholar's language) into the vernacular and the many more or less complete manuscripts produced. Thus it spread the knowledge of nature existing at the time among the people. The works of Albertus Magnus were printed or made accessible only much later and then were badly mutilated. The oldest translation of the *Liber de Natura Rerum* was a Dutch translation in verse by Jacob von Maerlant without any addition of original observations. It was made between 1265 and 1269, and its author thought that the original was by Albertus Magnus.

More important was a German translation by Conrad von Megenberg (1309 to 1374). He translated the *Liber de Natura Rerum* with a personal touch and a distinct moralizing twist as *Buch der Natur,* again in the assumption that it was by Albertus Magnus, and it has been reprinted in various forms up to the present. In the sixth book of the history of animals only 20 insects are dealt with (as worms) but, compared with other parts of the translation, without any change from the original. The oldest printed editions contain as their only illustration a plate of woodcuts showing some insects.

About the same time when the encyclopedias of the three Dominicans were written, the Franciscan monk Bartholomaeus Anglicus compiled a small compendium in which he also included a few insects. He counts among the "birds" (Aves): bee, mosquito, cicada, and locust; among the "land animals" (Animalia): silkworm, caterpillar, ant, ant lion, bumble bee, cricket, glow-worm, louse, flea, moths, wood-worms, and worms (including maggots of flies, etc).

Bartholomaeus says about the butterflies that they are small birds which are frequent on apples and hatch worms which originate in their vile-smelling droppings. The caterpillars grow up into butterflies, and from their droppings, which stick to the leaves, new caterpillars develop.

The flea is a small worm which plagues people terribly. It feeds on dust. It is very light and flees from danger by hopping, not by running. It grows slowly and is absent in the cold season. It is very fast in summer. The flea is first white but immediately turns black and thirsts for blood. Its sharp bite does not even spare royalty. It is most painful before rain. As a protection against this pest, wormwood, the leaves of the wild fig-tree, and colocynth are recommended.

The grasshoppers take their name from their legs (crooked and folded) which are as long as a lance-shaft. They have no king and yet migrate in well-ordered processions. In the place of a tail they have a sting. The south wind creates them and stimulates them to migrate, the north wind kills them.

To prevent damage by moths, Bartholomaeus Anglicus advises to put bay leaves or needles of cedar or cypress between clothes or books.

Petrus Candidus Decembrus (1399 to 1477) was a humanist who served at the courts of several Italian princes as a secretary. About 1460 he wrote a codex of animals in complete accordance with his models: Thomas Cantipratanus and Albertus Magnus, and in parts probably also using Pliny and Vincentius Bellovacensis. Long after the death of Petrus Candidus, in the sixteenth century, each page of his codex of animals was illustrated by an unknown artist with wonderful water colors on a 65 mm high margin, an entirely original achievement of this anonymous painter. (This work is now a particular treasure of the Library of the Vatican in Rome.) These excellent representations of insects, all drawn after nature, are among the oldest entomological water colors. The codex comprises five books, the fourth treating the snakes and worms with 35 species of insects.

The last great work of the Middle Ages related to entomology and belonging to the period covered by this chapter is *Ruralium Commodorum Libri XII* written between 1304 and 1309 by Petrus de Crescentii in Italy. Petrus was a political official in Italian towns, travelled widely throughout Italy for thirty years, and then retired to the country to write this book which, by virtue of its lucidity, its transparent style, and its objective comparison of contemporary practice and old traditions, became the European manual of agriculture for about 300 years. His classical models are mainly Palladius, Columella, and others; unlike other authors of his time, he cites them all as the sources of his quotations. Petrus de Crescentii's book was probably the only original work of that epoch besides that of Albertus Magnus.

Of special interest with regard to the level of knowledge at that time is the theory advanced in Petrus' work that the "procreation" of the worms in fruit or trees is due to physiological causes as surplus or lack of sap in the tree, etc; the scientific importance of this idea was recognized only centuries later.

Besides a detailed survey of apiology Petrus' work contains remarks on plant pests. They refer to the protection of granaries against beetles, mice, and other pests (the grain should be covered not with straw but with the leaves of cedar or cultivated olive tree), to the protection of vineyards against caterpillars (which should be picked off and crushed or burned), and against ants and worms in fruit trees in general (the trees should be ringed with pitch or similar material so that the ants cannot crawl up). Swellings at the trunk should be cleft, and trees that bear little or poor fruit should be given better soil; excessive moisture should be drained. One chapter is devoted to gardens and their cultivation. Here the control of ants by sprinkling origan and sulfur is recommended. As a protection against caterpillars he suggests steeping the seed in the juice of houseleek or the blood of caterpillars or picking off the caterpillars and killing them; against pests in general he recommends putting each row of plants between trigonella.

Though apiculture is the subject of another chapter of this history of entomology and thus need not be discussed here in detail, in connection with the work of Petrus de Crescentii a brief reference to its development in the Middle Ages is due. Already the early medieval Teutonic laws recognized artificial apiculture besides the collection of wild honey. While the Slavs in their densely wooded countries utilized the natural advantages for the cultivation of wild bees and developed the hollowed trunk as the artificial apiary, the hive made of straw is characteristic of the Teutonic nations. Our knowledge of early medieval apiculture in general is derived from the contemporary laws, in particular: the laws of the Visigoths (ca 650), of the Lombards (about the same time), of the Bavarians and the Burgundians (recorded ca 470). Charlemagne was greatly interested in apiculture. In his instructions for the management of the royal lands special bee-masters are mentioned. Besides the honey the production of wax increased in importance by the great demands of the monasteries and convents. Wax was even sometimes substituted for money.

A special line of development of medieval apiculture was the cultivation of wild bees by cooperatives organized like guilds and endowed with extraordinary privileges. The keepers of wild bees at Nuremberg were mentioned in documents even before 1000. This branch of apiculture gained great importance in the twelfth to fourteenth centuries. The cooperatives cultivating only wild bees developed into estates which combined this craft with an extensive cultivation of domestic bees and even had their own jurisdiction.

Finally mention should be made of a work consisting of 20 books and called *Geoponika* which also has a chapter on apiculture. It is a Greek compilation on agriculture, but it is not a uniform work; it originated in different centuries (from the second to the tenth) and was put together by a Byzantine ca 950. It became known in Europe only near the end of the Middle Ages. It deals with the various branches of agriculture, and the transition to animal husbandry (including bees) is formed by a section on pests and their control. The chapters on bees are headed: "Of bees and how their brood grows in a cow," "When the honey should be taken out of the hives," "Remedies for the bewitching of beehives," etc.

The decline of Roman apiology as it is evident from the primitive descriptions in *Geoponika* was followed near the end of the Middle Ages by a scientific treatment in the ninth book, which is devoted to apiculture, of

Petrus de Crescentii's fundamental work on agriculture. Besides drawing up a calendar for beekeeping he wrote about beehives and suitable locations for them, about the origin of bees (partly from their kind and partly from dead cows), about their life, their keeping and maintenance, their injuries and the remedies for them, about the gathering of honey and wax, and about the general usefulness of bees.

Even in ancient times bees were used for purposes of defense. In the Roman civil war Virgil took refuge with his valuables among his beehives and was safe from marauders who were chased away by the stings of the angry bees. By the same means Lucullus was forced to raise the siege of Themiskra. The castle Güllingen in Hungary was saved from an Austrian siege in 1289 as on several other occasions by beehives thrown from its walls.

Other Sources for the Study of the History of Entomology in Antiquity and Medieval Times

The historical study of entomology in antiquity and medieval times must not neglect the careful records in the old chronicles especially of the years of locusts but also of extraordinary invasions of caterpillars or other pests, though the locusts on account of their conspicuousness and the immense damage they did were most widely noticed. The great migrations of locusts in those times were made by four species with different modes of life and distribution. The most detailed reports about the appearance of locusts come from Germany, where between 591 and 1477 11 years of locusts were recorded in the Tyrol, 25 in the Rhine districts, 9 in Bavaria, 1 in Silesia, and 3 in Saxony. Local chronicles and other documents give as the earliest years of locusts 232 for Italy, 873–874 for France, 1062 for Hungary, and 1084–1086 for Poland and Russia. In the fourteenth century, for instance, 5 years of locusts were recorded in Hungary, 16 in Germany, 4 in France, 6 in Italy, and 2 in Eastern Europe.

The kermes scale insects had been used for the production of crimson in the Mediterranean countries from time immemorial and had been praised by Dioscurus and Pliny as a remedy for wounds. In medieval times the crimson dyers at Marseille, Montpellier, Genoa, Venice, and in other towns formed their own guilds whose statutes have in some cases been preserved. In the Middle Ages the kermes must have been traded in great quantities, for this scale insect played an important part in commerce. It came mainly from Provence.

The representation of insects in illustrations began in the Middle Ages with clumsy attempts like the drawings added to Aelian's fables and single illustrations on the margins of prayer-books, etc. Good representations of natural objects did not appear before the fifteenth century, the oldest technique being that of water color drawing. With the inclusion of plants and animals in book illumination the interest in these illustrations increased. At first they were intended to ornament the books, but later water color

FIGURE 5. The second fable of the eagle and a hornet: "An eagle flew swiftly in pursuit of a hare. But when the hare was quite without help, it met by chance a hornet. . . ." In *Aesops Fabulae et Vita,* Neuw geteutschet fabel Rimicy. Ulm: Johann Zainer (ca 1476–1477). Cited from A. Schramm, *Der Bilderschmuck der Frühdrucke,* Bd. 1–23. Leipzig. Stuttgart 1920–1943. Vol. 5, Figure 327.

FIGURE 6. Two pottery vessels from Pachacamac and Marque, Peru. Two persons examining the soles of their feet showing holes from which sandfleas have supposedly been removed. Kept in the American Museum of Natural History, New York. Courtesy of the American Museum of Natural History, cited from Hoeppli, 1969.

miniatures appeared also in scientific manuscripts, e.g. in the wonderful *Codex Vaticanus* of Petrus Candidus Decembrus mentioned above.

However, it must be generally kept in mind when studying the antique and medieval representations of insects that the fact that a present plant pest is shown does not prove by itself that it was recognized as such at that time nor is the existence of such a species established, unless this assumption is confirmed by a relevant text.

The public erection of figures of insects should be mentioned in this connection. A few examples from the geographic area under consideration may be given. Pisistratus (ca 500 B.C.) had an iron locust put up on the Acropolis of Athens which was supposed to lessen the effect of the evil eye. Virgil is said to have fastened an iron locust to a tree at Naples and thus to have saved the city from this plague. Locusts and other insects were depicted in Greek and Roman times on gems, signet rings, and coins. Sometimes these representations had a magic or prophylactic character, sometimes they were simply ornaments or amulets. But in some cases they also show the misery caused by the locust invasions.

Entomology in the Non-European Western World

It will be difficult to establish traditions in the history of entomology for the antique and medieval periods in Ethiopian Africa, in North and South America, and in Australia, for even the great voyages of discovery at the beginning of the modern age had for a long time no influence on the development of entomology. But it must be pointed out that even the ancient Mayan manuscripts contain some indications of representations of insects, especially bees. In one description of their culture it is said that the Mayans used full, live wasps' nests as missiles against their enemies, and that they cultivated bees. Much information on America and Ethiopian Africa may still be buried in Spanish and Portuguese manuscripts.

The American Indians ate the seventeen-year locust (*Tibicen septendecim*) as a delicacy, and they regarded it as an omen for pestilence and war. The Hottentots are reported to have worshiped the praying mantis. The Aztecs even before the conquest of Mexico by the Spaniards produced the scarlet dye from the cochineal insect (*Coccus cacti*). Old documents and drawings confirm that even the tribute of certain regions was partly paid in cochineal grains.

Insects are also mentioned in studies made to determine which diseases existed in the western hemisphere before Columbus and thus are not imports from Europe (Hoeppli, 1969). A further investigation of this problem will certainly reveal new source material for entomology. In this connection researchers found eggs of lice in the hair of a Peruvian mummy of the time about 200 B.C. Lice were also discovered in the scalps of three prehistoric Indians in the southwest United States. Apart from bones and mummies inscriptions also document the existence of diseases in ancient times.

Some hieroglyphs in the temples of Denderah and Edfu possibly refer to malaria. In 1373 an Arab described Mari Djata's death of sleeping sickness. Ancient popular belief connected mosquitoes with malaria, the tsetse fly with sleeping sickness, etc. Sailors, traders, and adventurers spread reports of diseases in those times. Early works of art show patients and their pathological changes. Herodotus (500 to 424 B.C.) described how Egyptian fishermen hung nets about their beds to protect themselves against mosquitoes. Hippocrates mentions several forms of malaria. An account of the African sleeping-sickness was given by Ibn Khaldūn and Al-Qalgashandī before 1406. They stated that this disease attacked the inhabitants of certain countries and was known there. The connections between its germ and its carrier, the tsetse fly, are said to have been suspected in West Africa for a long time.

The description of the dermal leishmaniasis, which is spread by the *Phlebotomus* sand flies, dates from a later time. But vessels have been found in Peru from pre-Columbian times (the pre-Inca period) on which this disease is depicted. The Somali have the same word "kaneo" for mosquitoes and for malaria. The Africans of Usambara call mosquitoes and malaria "mbu". A connection between flies and diseases had been known from very early times: Among the arthropod designs on pottery made by the Pueblo people in the Mimbres Valley, New Mexico (ca 1200 A.D.), is a swarm of mosquitoes.

In the same way that these studies on the dates of the first appearances of certain diseases reveal the knowledge of insects which the people had at that time, some further details might surely be gathered from old books of popular remedies. Though the therapeutics mentioned in them and the references to insects to be derived from these mostly go back to oral traditions and the time of their origin is difficult to determine, they might yet furnish more information about such entomologically and scientifically poor periods as the Middle Ages or about similarly "poor" geographical regions.

Ticks were mentioned in early historical times. In China, for instance, children were given baked cattle ticks as a remedy or prophylactic for chicken pox. Aristotle, Cato, Varro, Columella, Pliny, Dioscorides, and others described ticks. Pliny thought that they had no anus. Several authors pointed out that some animals were often attacked by ticks while others were not, e.g. mules and asses were never attacked. The oldest representation of an animal suffering from ticks is the fragment of the head of a hyena-like animal with three swellings at the inside of its ear. It has been known for a long time in Africa that tick bites generally cause fever and nausea. Spanish conquerors reported that the Indians killed ticks by fire.

Sandfleas (chigoe; *Tunga penetrans*) have been known for a long time on the American continent. There are clay vessels in Peru of the pre-Inca period (400 to 900 A.D.), so-called "huacos", with human figures showing their infested feet. Sandfleas in Africa were mentioned first in 1324 in the report of a pilgrimage to Mecca in which the sultan of Mali took part; its progress was impeded by a disease of the feet, and it is highly probable that this was an infection by sandfleas. That fleas were known in South America in antiquity is evident from a pre-Columbian representation in terracotta and stone that has survived.

FIGURE 7. Two fleas. Above: Flea in terracotta. Height 20 cm, Western Mexico, ca 200–800 A.D. Below: Flea in white stone. Length 50 cm, Mexico, ca 1200–1500 A.D. Courtesy of the Museo Nacional de Anthropologia e Historia, Mexico D. F., cited from Hoeppli, 1969.

Lice must have been known from very early times. A louse plays a part in the famous mythological and historical book of the Maya-Quiche called Popol-Vuh which a Christian Indian named Diego Reynoso put in writing about 1530 in the Indian language but in Roman letters. The book was found in the convent of St. Thomas of Chichicastenango; in the second part it deals with the twins Hunahpú and Ixbalamqué, who fight the forces of "evil". There Popol-Vuh tells the story of Ixmucané, a kind of goddess in the shape of an old woman who sends an important message by a louse to her grandsons.

A mochica huaco in the Museo Nacional de Antropologia y Arquelogia in Lima shows a woman delousing herself. In Montezuma's palace the con-

FIGURE 8. Delousing. Woman with big lice on her garment. They supposedly dropped from her hair while she was delousing herself, Mochica pottery, fourth century A.D., northern coast of Peru. Measurements: 227 mm x 107 mm. Courtesy of the Museo Nacional de Antropologia y Arqueologia, no. 1/2862 (802). Lima, Peru; cited from Hoeppli, 1969.

querors found sacks which they thought were filled with gold. The sacks, however, contained lice, apparently from payments of tribute. Montezuma employed old people who were unfit for any other work to make the rounds of the houses and delouse the inhabitants. The Incas defeated the Urus who lived near Lake Titicaca and imposed on them a yearly tribute of lice.

There are also representations of insects in African art, e.g. bronze objects from eastern Nigeria which are ornamented with crickets, flies, and several kinds of beetles. From early American art the picture of a swarm of mosquitoes on a piece of Mimbres pottery has survived.

Essig (1931) describes the use of insects by some Californian Indians who lived on a particularly low level of civilization. To them, locusts as well as many other small animals were delicacies. Insects served them not only as food but also in their religious customs and as remedies. Lice were eaten by several tribes. With the Cupeño Indians the word "nauwilot" (body louse) was also the name of a clan. Fleas played parts in myths and legends.

FIGURE 9. Tick, greatly simplified, terracotta. Nok culture, length 13 cm. Museum of Jos, Nigeria. Reproduced by permission of the Director of Antiquities of the Federal Republic of Nigeria; cited from Hoeppli, 1969.

A blow fly song was sung in the days of the Shasta Indians to increase their power of scent and the ability to frighten game. The purpose of the song is sufficient to show that the Indians were familiar with the habits of these insects. That flies were troublesome is shown in a girl's ceremony held by the Luiseño Indians in which each girl had her head covered with an openwork basket to keep the flies off. An interesting paper by Böning (1971) deals with the meso-American illustrations of grasshoppers and more literature is cited there.

It is surprising that mosquitoes are mentioned rarely in the early reports. Coccidae were used to produce wax and rubber, and *Cerococcus quercus* served the Indians as an eraser. Ants and the therapeutic effect of their bite were well known. In some regions, also, fly larvae were eaten, especially *Holorusia rubiginosa,* several species of *Tipula, Ephydra hians,* and one species of *Atherix.* The sweetish juice of the secretions of aphids and coccids was called Indian honey.

These few examples from countries outside the Old World demonstrate that some entomological knowledge certainly existed there in the ancient times. Possibly a study of nearly untouched sources, e.g. the Portuguese and Spanish manuscripts mentioned above which have hardly been noticed, will supplement the knowledge obtained in countries which in classical antiquity were the birthplaces of entomology.

LITERATURE CITED

Aelianus, C. 1839–1841. *Tiergeschichten.* Transl. F. von Jacobs, Vol. 2. Stuttgart

Aristoteles. 1860. *5 Bücher von der Zeugung und Entwicklung der Tiere,* Transl. and ed. von Aubert, Wimmer. Leipzig

Aristoteles. 1868. *Historia Animalium,* Transl. and ed. von Aubert, Wimmer, Vol. 2. Leipzig

Balme, D. M. 1970. Aristotle and the beginnings of zoology. *J. Soc. Bibl. Nat. Hist. London* 5(4):272–85

Bessler, S. 1885. *Geschichte der Bienenzucht.* Ludwigsburg

Bodenheimer, F. S. 1928. *Materialien zur Geschichte der Entomologie bis Linne.* Bd. 1. X. Berlin: Junk. 498 pp.; 1929. Bd. 2 VI, Berlin: Junk. 486 pp.

Böning, K. 1971. *Heuschreckendarstellungen aus dem Altertum und ihre Bedeutung für die Geschichte des Pflanzenschutzes.* Anz. Schädlingskde u. Pflanzenschutz. (Berlin, Hamburg) 44:21–31

Böning, K. 1971. *Mesoamerikanische Heuschreckendarstellungen.* Anz. Schädlingskde u. Pflanzenschutz. (Berlin, Hamburg) 44:185–89

Burmeister, H. C. C. 1832. *Handbuch der Entomologie,* Vol. 1. Berlin

Carus, J. V. von. 1872. *Geschichte der Zoologie.* München

Dannemann, F. 1921. *Plinius und seine Naturgeschichte in ihrer Bedeutung für die Gegenwart.* Jena

Dingler, M. 1938, 1939. *Das Insekt in der Darstellung.* Ausstellung der Bayerischen Staatsbibliothek München. München. 43 pp.

Eiselt, J. N. 1836. *Geschichte, Systematik und Literatur der Insektenkunde von den ältesten Zeiten bis auf die Gegenwart,* VIII. Hartmann. Leipzig, 255 pp.

Essig, E. O. 1931. *A History of Entomology.* New York: Macmillan. 1029 pp.

Geisenheyner, L. 1911. Über die Physika der heiligen Hildegard von Bingen und die in ihr enthaltene älteste Naturgeschichte des Nahegaues. *Ber. Versamml. bot.-zool. Ver. Rheinland-Westfalen, Bonn,* pp. 49–72

Herodot. *Geschichten*

Hoeppli, R. 1969. *Parasitic Diseases in Africa and the Western Hemisphere.* Early documentation and transmission by the slave trade. Verl. f. Recht u. Gesellschaft. Basel. (Acta Tropica. Suppl. 10) XII, 240 pp.

Homer. *Ilias und Odyssee*

Horn, W. 1926. Über die Geschichte der ältesten Entomologie und den Einfluss des Christentums in seinen ersten Jahrhunderten. Uschmann. Weimar. *Aus: II. Internationaler Entomologen-Kongress, Zürich, Juli 1925.* Bd. 2, pp. 38–52

Kaufmann, A. 1899. *Thomas von Cantimpré.* Köln

Keller, O. 1913. *Die antike Tierwelt.* Bd. 2. XV. Leipzig: Engelmann 617 pp.

Killermann, S. 1914. *Das Tierbuch des Petrus Candidus.* Zool. Annalen, VI: 113–22

Kirby, W., Spence, W. 1826. *Introduction to Entomology,* Vol. 4. London

Klek, J. 1926. *Die Bienenkunde des Altertums.* IV. Archiv f. Bienenkunde,, VIII:41

Langenberg, H. 1890–1891. *Aus der Zoologie des Albertus Magnus.* Programm Realschule Elberfeld, 40 pp.

Lewes, G. H. 1865. *Aristoteles.* Transl. J. V. von Carus. Leipzig

Lissner, I. 1964. *Rätselhafte Kulturen.* Wien: Deutsche Buch-Geneinschaft. 337 pp.

Maerlant, J. von. 1878. *Der Naturen Bloeme.* Ausgabe von Verwyijs. Groningen

Manitius, K. 1923. Naturwissenschaftliches in der Geschichtsschreibung der Karolingerzeit. *Archiv fur Geschichte der Medizin* 15:68–77

Marlatt, C. L. 1898. A brief historical survey of the science of entomology. *Proc. Entomol. Soc. Washington* IV

Meyer, J. B. 1855. *Aristoteles' Tierkunde*. Berlin
Nordenskiöld, E. 1926. *Die Geschichte der Biologie*, Jena
1902. *Des Pedanios Dioskorides aus Anazarbos Arzneimittellehre in 5 Büchern*. Übersetzt und mit Erklärungen versehen. Stuttgart: J. Berendes
Pelster, F. 1920. *Kritische Studien zum Leben und zu den Schriften Alberts des Grossen*. Freiburg: Herder
Pfeiffer, F. 1861. *Das Buch der Natur von Konrad von Megenberg*. Stuttgart
Phädrus. *Äsopische Fabeln*
Pritzl, J. 1920. *Das ehemalige Zeidelgericht zu Feucht*. Arch. f. Bienenkunde, II:310–33
Prosser, J. 1915. *Geschichte der Bienenzucht in Oesterreich*. Wien
Raschke, W. 1898. *Die Zoologie in Konrad von Megenberg's Buch der Natur I*. Programm Realgymnasium. Annaberg
Schulz, H. 1897. *Das Buch der Natur von Conrad von Megenberg*. Greifswald
Stadler, H. 1906. Albertus Magnus, Thomas von Cantimpré und Vincenz von Beauvais. *Natur und Kultur*. 4:86–90
Stadler, H. 1909. Albertus Magnus von Köln als Naturforscher und das Kölner Autogramm seiner Tiergeschichte. *Verhandlungen der Gesellschaft deutscher Naturforscher und Ärzte zu Köln, 1908*. 1:29–37. Leipzig
Stadler, H. 1913. Irrtümer des Albertus Magnus bei Benutzung des Aristoteles. *Arch. f. d. Gesch. d. Naturwiss. und d. Technik*. 6:387-93
Stadler, H. 1917–1921. *Albertus Magnus: De Animalibus Libri XXVI*. Nach der Kölner Urschrift, Vol. 2. Münster i. Westf., 1664 pp.
Steier, A. 1913. *Aristoteles und Plinius*. Würzburg
Steier, A. 1913. *Der Tiersbestand in der Naturgeschichte des Plinius*. Würzburg
Strunz, F. 1926. *Albertus Magnus*. Wien und Leipzig
Theen, H. 1898. Die Biene im Kriegsdienst. *Illustrierte Zeitschrift für Entomologie*, 3:6–9
1822. *Theophrast's Naturgeschichte der Gewächse*. Übersetzt und erläutert von K. Sprengel, Vol. 2. Altona
Wasmann, E. 1925. Zur Vollendung der Neuherausgabe der Tiergeschichte Alberts des Grossen. *Stimmen der Zeit* (Freiburg: Herder). Vol. 106, Jg. 54:79–80

THE EARLY NATURALISTS AND ANATOMISTS DURING THE RENAISSANCE AND SEVENTEENTH CENTURY

MAX BEIER
Naturhistorisches Museum, Vienna

The second half of the fifteenth century was dominated by the influence of the medieval scholastics and their aftereffects. Whereas all kinds of art and the technic-mathematical sciences flourished, there was no equivalent development of entomology, as in most of the other biological sciences. This development was delayed by the prevalent mysticism and the all-controlling doctrinal dogma of the Church. Freedom of opinion and original research could not rise above the absolutely deductive-speculative, sensitivity and religious-oriented thinking, which was predominately directed towards the "next world".

The intellect of this period is significantly illustrated by the frequent secular and ecclesiastical court trials of different pests like locusts, cockchafers, and caterpillars, which were especially prevalent in Portugal, Spain, the south of France, Switzerland, and many other European countries. At first, as in Switzerland (ca 1450), they took a contumacious course without any legal proceedings. However, by 1478 a method of prosecution was adopted, the justice appointed a counsel for the defense for the indicted pests. At the end of the century this method changed into a limited mandated process. The first part of this process always finished with an extradition order for the defendants, who had to abandon the country in a fixed time. If the pests remained past this term they were threatened with malediction or excommunication. In the second process, in which only episcopal judgements were recognized, the threatened ecclesiastical punishment was pronounced in a solemn manner and often some of the pests were brought before the court and executed. There was much sophistry employed in these processes and the famous lawyer Bartholomaeus Chasseneux (1480–1542), and later Felix Malleolus (1582) and Gaspard Bailly (1668), showed this fully in their publications. But we cannot reject the suspicion that even in later times the secular and ecclesiastical justices of high rank passed their sentences with the knowledge of the short duration of the plague of pests, thus giving the impression, that these judgements caused their disappearance.

Perhaps it was this reason which for so long preserved the process of prosecuting insects, despite different warnings in later times on account of prejudicial reaction in faith and religion. At the end of the seventeenth century excommunication was forbidden, although it nevertheless took place; the last ecclesiastical judgement was made in Europe in 1733.

It is understandable that with such mentalities and ideologies the biological sciences could not succeed and thus were not cultivated. With the invention of the printing press (ca 1450), the possibility developed of raising great numbers of educated people. However, printed biological literature, previously exclusively cultivated in monasteries and passed on by handwriting, still remained in the future, discouraging any impulse for individual observation and original research. Following this and beginning in Italy there was a period of humanism, in which Aristotle with his science of nature, and Plinius were the only authoritative source of study recognized by the church, and only after much resistance. When the Byzantine Empire fell with the conquest of Constantinople by the Turks in 1453, a period of emigration began. Many scientists emigrated to Italy; one of whom was Theodorus Gaza who left his home in Thessaloniki. His first publication in Venice, a Latin translation from the papers of Aristotle (1476–1498), followed (1497) an edition in the Greek original. The translation of the books of Plinius into Italian by Christoforo Landino (1473) made ancient literature more available to a greater social class. Periods following this have seen comments and interpretations about these books without the addition of any fundamental new insights. On the contrary, misunderstandings, errors, mistakes, misinterpretations, and depreciations of the insights were debasing the knowledge acquired by Aristotle and Plinius, particularly in the domain of entomology.

In this period, for medicinal purposes, insects are particularly mentioned in medical books; as in *Ortus Sanitatis* by the physician of the town of Frankfurt, Johannes Wonneke von Caub (Cuba), which was first published in Latin in 1480, followed by several editions in German (1485 in Mayence, 1493 in Frankfurt on Main, 1520 in Lübeck). In the first great edition of this book, (Mayence, 1491) are found the oldest drawings of insects (lice, fleas, ants, gnats, caterpillars, and others), reproductions of very primitive and clumsy wood engravings, in respect to their natural forms with many imperfections which barely reflect the original. The text in its compilation misses all scientific character, is absolutely uncritical, and, therefore, equal in character to the illustrations. The divers insects are arranged alphabetically between "Animalia" and "Reptilia" on the one hand, and "Aves" and "Volantia" on the other. No distinction is made between crickets and cicadas, which were already known to Aristotle and Plinius. It was followed by a series of further medical books, dealing with a very limited number of insects in a similarly superficial manner. The wood engravings in the text are partially painted by hand but they are still unrecognizable. To this series belongs the medical compendium *Opus de re Medicina* by Paulus Aeginetus (1534), the *Prosopeia Animalium* by J. Ursinus (Vienna, 1541) which was written in latin poetry, the *Naturaliae Historiae Opus Novum* by Adam Lonicer (Egenolph publishers, Frankfurt, 1551) and the *Commentary of Dioscorides* by the Italian Pietro Andrea Matthioli (1548). This author represents an important adherent of the Dioscurmedicines. His work appeared in many editions in both Latin and Italian. Many men of medicine were observers of insects. Among them the English medic Edward Wotton is to be particularly mentioned. His book *De Differentiis Animalium Libri Decem* (Paris, 1552) does not contain any new names or observations, but he summarizes all the knowledge of Aristotle and that of the other scientists of antiquity, with remarkable objectivity in view of his period. Wotton, therefore, is the first precursor of the second Aristotelian Renaissance in the history of zoology. Above all, his merit lies in the reference to primordial sources and his rejection of the worthless scholastic compendium of literature. This book, nevertheless, did not run to a second edition, therefore Wotton's influence on the further development of entomology remained slight. Guilelmus Rondeletius on his own observations gave the description of some larvae of insects in his great history of the nature of water-dwelling animals. He was in 1555, professor of medicine in Montpellier. Ferrante Imperato, a medic in Naples, also brings to his *History of Nature* (1599) rather more natural illustrations of insects. On the other hand, the two northern Italian doctors, Hieronimus Cardanus (principal work: *Offenbarung der Natur;* Basel, 1559) and Julius Caesar Scaliger (*Exotericarum Exercitationum;* Lutetiae, 1557), were both absolutely bound up in humanism and wasted their substance in indulging in useless polemics.

Authors of this time not belonging to the medical profession, thought in another fashion; an example of such an author is Archbishop Olaus Magnus (Olaf Ster), a Swede, who escaped to Rome as a consequence of the Reformation. His work there published in 1555 was titled *Historia de Gentibus Septentrionalibus,* concerned the local fauna of the Scandinavian animals, and presents an ability to observe, coupled with an interesting writing style and figures. Yet many of his other conceptions are bordering on fairyland, such as his descriptions of flies, crane flies, bees, and ants. Assessments of an ecological way of thinking are pointed out in the small book *De Ortu et Causis Subterraneorum* (Basel, 1546 and 1558) by Georg Agricola (Georg Bauer), a mineralogist and chemist, who also mentions different soil-dwelling insects. The *Questiones Physicae* by Johann Thomas Freigius from Basel contains practical demonstrations of apiculture and honey production as well as a didactic poem on the rearing of silkworms. But these works remain inferior in their forms of scientific description. Among numerous other authors the Italians Baptiste Porta and Antonio

Mozaldo, as well as the Englishman Lupton (1595), gave instructions for controlling injurious and pain-giving insects. But these books are rather mixed up with mysticism and absurdity, as it was popular reading at that time. Finally the poem "Floh Hatz" by Johann Finhart (1573), must be set in the domain of literature despite mentioning in its contents numerous bits of advice against plagues of fleas. *Des Flohes Zank und Straus gegen die stolze Laus* (1610) and many other little works which are not worth mentioning, are mostly written in the form of dialogues in Latin, Spanish, French, Italian, and German.

Beside the more or less scientific publications, the theologically oriented literature occupied a large space. Above all, it was appropriate to place at the preachers' disposal impressive comparisons and symbols, but this also had no more than a passing relation to real observation or knowledge of nature. The big folio volume titled *Reductoria Moralia* by the priest Petrus Berchovius (Paris, 1521) is one example of scientific absurdity in the manner in which insects, besides terrestrial animals and birds, are mentioned. Its presentation with moralizing and motif-less references to Biblical passages shows the negative scientific mind. At that time, the most popular form of dialogue writing is exemplified by the *Universae Naturae Theatrum* by Joannus Bodinus (1546), the *Physica Christiana* by Lambertus Daneus (2nd edition; Genova, 1579) built on the history of the creation, and the *Dierum canicularium* by Bishop Simon Majolus (1600), which is textually rather pleasing, and though mixed with many errors, is not so importunately moralizing. The collection of ethics, *Symbolorum et Emblematum Centuriae Tres,* by Joachim Camerarius (2nd edition, 1605) contains pretty, but scientifically insignificant engravings in copper of insects and other natural objects, yet in its text it is scarcely different from the above-mentioned books.

From this beginning developed a Biblical-zoological literature, which set about to deal with all animals mentioned in the Bible. This suggested, on the one hand, partly new and rather clear data, sometimes not absolutely strange to nature, to be used by the preacher for his metaphorical comparisons; examples are the *Biblische Bilderbuch* by H. H. Frey (1595) and the *Animalium Historia Sacra* by the professor of theology Wolfgang Franz (Wittenberg, 1612). The latter work was printed in many editions both in the Netherlands and Germany. On the other hand, that form of literature tries to fix exactly by name, all the animals mentioned in the Bible. This resulted in the *Hierozoikon* by Samuel Borchartus (London, 1663; Frankfurt, 1675; Leipzig, 1795–1799); a great display in historical, philological, and literary–historical erudition but absolutely worthless zoologically. Nevertheless, this form of literature endured for a rather long time. Its range extended from the *Arce Noa* by the priest S. J. A. Kircher (1675) with the fauna of the legendary Noah's ark up to the *Kupferbibel* by Johann

Jakob Scheuchzer (Augsburg and Ulm, 1731–1735), a lavish donation to science of more than 500 worthless copper plates, in folio size. He copied, without consideration to the facts, previously known forms of extremely bizarre exotic insects to illustrate his quotations from the Bible.

The scholastic mentality had already overcome the past, so it was possible to maintain many serious works of natural science and independent investigation. In this, their contributions to development gave the philosophers of nature, Telesio, Patrici, and Campanella, the chance to be the first critics of the blind confidence of authority in Aristotle, in respect to natural science. The deciding factor for this was the alteration in the faculties of the universities. Namely in former times a "lector simplicium" of the medical faculty had to teach as a subsidiary subject, pharmacology, including the animals (also insects) used as remedies in conformity with Dioskur and Galen. In addition, an "Ostensor simplicium", whose task was to inform the students of the practical knowledge of the natural objects, was appointed. In Italy in 1561 these "semplici" were ultimately promoted to professorships. The next two centuries saw a close connection between the science of medicine and natural history, because physicians (Malpighi, Redi, Vallisnieri, and others) made important contributions to the further development of entomology.

The first of these professors, and the most significant for entomology, was the philosopher and physician Ulysse Aldrovandi (1522–1605) in Bologna. He was in his scientific orientation a savant eclectic, but conforming to his time, very much subjected to the influence of Aristotle and the scholastics. In spite of this, with his abilities as an observer, he exposed facts which he had determined by his own research. As a pioneer of pure natural research he was by far the most outstanding among the compilers of his time and remained independent of the fixed aims of theology and medicine, which became the essence of the Renaissance. During his 50 years of study, he produced several hundred volumes of manuscripts and excerpts, a voluminous collection now found for the most part in the Museo Aldrovandi in Bologna. The big folio-volume *De Animalibus Insectis libri VII* was first published in 1602, then in other editions in 1618 and in 1623 in Frankfurt on Main and later in Bologna (1638). This was the first work of literature in the world dealing with insects, thus finally establishing entomology, and especially systematical entomology as a science. Objects from this collection were taken as illustrations; these recognizable, well-cut wood engravings, though rather primitive, were numerous. They contained the first known dichotomic key for determination of the higher groups, which are all morphologically defined. The arachnids, centipedes, worms, Echinodermata, and even the naked snails were among these groups. A special chapter was devoted to the morphology of the insect body. Metamorphosis and reproduction had been described excellently and supplemented with notes on rearing and biological notations; with this physiological problems like breathing, smelling and the senses of taste and touch are mentioned. Very extensively described are the honeycomb-making Hymenoptera and especially the honey bee. The abundance of species in this book, mainly relating to the fauna of northern Italy, is astonishingly high; for instance 81 different species of Lepidoptera are delineated. In this volume the total literature up to his day was used by Aldrovandi almost to exhaustion.

The English doctor Thomas Mouffet (1550–1599 or 1604) wrote a similar book independently from Aldrovandi, also in Latin; the *Insectorum sive Minorum Animalium Theatrum* was published many years after his death by Theodore Mayern (1634) in London. In this book the copious unpublished notes by Konrad Gesner and Edward Wotton are utilized; the volume was completed by the English natural scientist Thomas Penn. This manuscript was finally purchased by Thomas Mouffet. Despite the added essential enrichment of scientific knowledge and new treatments, the book still retains its inhomogeneous character. With the exclusion of the echinoderms the system remains strictly formal. Although his correlation was known to Mouffet, based on the presence or absence of wings, he separated widely by his descriptions, for instance, the butterflies from their caterpillars. In other respects also, Mouffet, in his less systematical talent and accuracy of observation, is not to be compared to Aldrovandi. The British fauna gives the chief species, but also many continental and tropical species are illustrated and described. Terms used by Mouffet were frequently, at a later date, utilized by Linné for the names of his genera. The illustrations were mostly set on plates, consisting of 500 wood engravings of different quality, several of which were surprisingly exact. Mouffet had also profound knowledge of literature, but not as all-embracing as that of Aldrovandi.

Suggested from the works of these two great scientists, a series of more comprehensive books came out in the next decade, representing almost exclusively compilations on their models, without fundamentally new contributions. Among these were the *Theatrum Animalium* by the Silesian doctor John Johnson (1653) and the insect volume *Historiae Naturalis de Insectis libri III,* which was preceded in 1633 by the little pocket edition entitled *Thaumatographia Naturalis,* which was circulated in many copies and contained a remarkable contribution on the silkworm by Andreas Libavius. Johnson's main work depends absolutely upon Aldrovandi and Mouffet, but is more concisely written and with clearer arrangements, though without evidence of critical selection, so that several species are mentioned two or three times in different places. The illustrations are nearly exclusively taken from the publications of the two cited authors, but for the first time the illustrations are done on copper print. The good quality of the new technique, indeed, does not come into play because the plates were engraved, as in the old wood blocks but not with regard to observations of insects as prototypes. Johann Sperling in his *Zoologia Physica* (Lipsiae, 1661), presents a compendium of lectures, about the vast number of known species of insects: 40 species of beetles, 50 species of caterpillars, 70 species of flies, and more than 100 species of butterflies; he considers also the segmentation as a specific sign of insects. No observable progress is noted in the *Onomasticon Zoicon* by Walter Charleton (Oxoniae, 1668 and 1677) and the *Regnum Animale* by Emanuel König (Frankfurt, 1682).

Faunistic publications of interest in this period deal with the fauna of a smaller area. Often resulting from observations through several decades, they are not in general burdened with literary knowledge showing, therefore, in many cases a close connection with nature. *Description of Britain* by William Harrison (1577 and 1586) is their precursor and the sparse notices on insects are mixed with old superstition. *Historia Fontis et Balnei Admirabilis Bollensis Liber Quartus* by Johann Bauhin (Montis Beligardi, 1598) is more remarkable because it contains the author's own observations. The Medicus of Möpelgard, J. Bauhin, is concerned in this book with the fauna of the neighboring spa of Bolles. Written by the same author, moreover, is a brief study on venomous insects or those with a poisonous sting, published in 1583 in Montbelliart with other contributions. Caspar Schwenckfeld, the doctor from Hirschberg, describes in alphabetical order 80 species of insects in the Silesian local fauna *Theriotropheum Silesiae* (Lignicii, 1603), a considerable number for that time. This book includes plenty of observations, proving the author to be an excellent scientist. He not only described the morphology, but he also obtained impressions of the inner anatomy of insects by dissections; a detailed index of synonyms in Latin and German completes the valuable scientific work. The description of the water fauna in the surroundings of Strassburg represents, without doubt, the high point; completed for the most part in 1666, it was written by Leonhard Baldner, a simple fisherman and autodidact. This work never saw print, but a few copies of the manuscripts are preserved; among these the most complete and most beautiful is the Casseler manuscript. This magnificent work contains very fine writing and is provided with excellent water-colors, describing with their biology 26 species of insects. Great talent is shown in description, based on long intensive observations of nature. Familiarity with the biology of the species is also indicated in the *Historia Naturalis Helvetiae Curiosa* by Johann Jakob Wagner (Figuri, 1680), in which only a few insects are mentioned. In comparison with these works are the *Scotia Illustrata sive Prodromus Historiae Naturalis* by Robert Sibbald (Edinburgh, 1684) and *The Natural Historiae of Stafford Shire* by Robert Ploth (Oxford, 1686) (in an entomological sense, of little value).

Notable on account of its supplement is the *Historia Naturalis Curiosa Regni Poloniae* by Gabriel Rzaczynski (Sandomiriae, 1721) which contains descriptions of the massings of insects.

Two philosophers gave natural science an impulse, which finally lead it to new directions. The English empiricist Francis Bacon of Verulam (1561–1626) holds claim in his work *Sylva Sylvarum sive Historia Naturalis,* published posthumously at first in English (London, 1627) and later in Latin (Amsterdam, 1661), to any research that could be based inductively on empirical data and experiments. But he himself in the domain of entomology did not follow his postulates; on the contrary, he held further to divers and erroneous ideas, based on the old precept that insects had their origin in decay or other substances. René Descartes, named Cartesius (1596–1650), who stated that "doubt is father of knowledge", also believed still that insects arose from spontaneous generation. He exerted a deciding influence on the physiologists of his century by his machine theory of life, in accordance with the great progress of the physical sciences of that time, which resulted in the founding of the school of iatromechanical tendency.

The Englishman William Harvey (1578–1657) is to be considered the first of these physiologists. He is the discoverer of blood circulation in animals, including insects (*Exercitatio Anatomica de Motu Cordis et Sanguinis in Animalibus;* Francoforti a.m., 1628), is the creator of the ideas of metamorphosis and epigenesis (*Exercitationes de Generatione Animalium;* Amstelodami, 1651), but held in his imagination only a vague idea of the egg, in which he included the pupa. He was to be compared to Daniel Sennert in his *Hypomnemata Physica* (Frankfurt, 1636) and Jakob Wolff *Dissertatio Zoologica* (Leipzig, 1669) as still a disciple of the theory of abiogenesis of insects. In the first comprehensive textbook of physiology *De motu Animalium* (Leyden, 1680–1681), insects are also mentioned throughout. In this respect the Italian Giovanni Alphonse Borelli (1608–1679) proves himself to be a typical iatromechanic.

Unexpected possibilities opened for entomology with the invention of the microscope by the Dutchman Janssen (1599). Foundations of academies and scientific societies in Florence, Rome, London, Paris, and Schweinfurt, encouraged more people in entomological research. Of the entomological books the oldest based on microscopical studies is the *Apiarium* by sovereign Frederico Cesi (1625) with beautiful illustrations and precisely drawn plates by the Roman doctor Francesco Stelluti. They represent the external morphology of the body of the honey bee and its parts, slightly enlarged. The illustrations by the French doctor Petrus Borellus in the *Observationum Microscopiarum centuria* (1656) seem to be clumsy in comparison with those of the *Apiarium.* In writing the *Micrographia* (London, 1667) Robert Hooke (1635–1730) too, made use of the already improved microscope. In addition, this book is illustrated with impressions and drawings of the habits of various small insects; in its content the book is not arranged coherently. Many biological observations and speculations are inserted in the descriptions. Despite this, and even as a pioneer work in microscopy, the scientific worth of the book is of no great significance. The elaboration, interestingly and brilliantly written, shows the work to be that of a highly educated amateur, as in fact Hooke was in reality a professor of geometry in London. *Micrographia Nova* by Johann Franciscus Grindel (Nuremberg, 1687) is textually, and in respect to illustrations, much worse, although it was edited 20 years later.

Francesco Redi (1626–1698), the Italian physician and poet, also made use of the microscope in his work and brought out, as a result, very accurate drawings; his special importance being, however, the foundation of experimental biology in entomology. He faced admirably and impartially all problems and examined the traditional opinions for correctness. In exact experiments, he held only to his perceptions and pointed out numerous errors of Aristotle; he demonstrated that insects do not come from decay but develop without exception from eggs, which are deposited by the female on the convenient substratum chosen by their sense of smell. *Osservazioni Interno agli Viventi che si Trovano Negli Animali Vivent* (Firenze, 1684), *Opusculorum pars prior sive Experimenta circa Generationem Insectorum* (Amsterdam, 1686), and *Esperienze Intorno alla Generazione degli Insetti* (Firenze, 1668) are his most important works, the last one published in its fifth edition in 1688. Redi was an excellent experimenter, though he did not remain uncontradicted. The priest S. J. Athanasius Kircher (1618–1680) with his *Arce Noe* (Amsterdam, 1675) tried to disprove Redi on the basis of absolutely insufficient and perhaps partly adulterated experiments (*Physiologia Kircheriana;* Amsterdam, 1680) to save, once more, the theory of abiogenesis of insects. In this effort he was supported by his disciple Filipo Bonani (1638–1725) (*Observationes circa Viventia, quae in rebus non Viventibus Reperiuntur;* Roma, 1691). But despite their efforts the era of the theory of abiogenesis was now definitely over.

After the first tentative attempts extended to insects by Marco Aurelio Severino (*Zootomia Democritaea;* Nuremberg, 1645) the comparative knowledge of anatomy of insects had undergone a rapid extension by the works of three outstanding scientists. The first was by the Italian physician Marcello Malpighi (1628–1694); an exact observer and striving to be sure of his aim, he separated the animal anatomy from that of medicine, thus creating an independant field of study and giving ecouragement for future possibilities for their development. Of sheer ingenuity, and of timeless validity is his richly illustrated monograph of the silk spinner *De Bombyce,* which was written at the instigation of the British Royal Society and was printed in London in 1669. Morphology and anatomy in all stages of development are well described in his work and again and again the book supplies us with pointers to the condition of other insects. Malpighi was the discoverer of numerous organs, among them the Malpighian tubules. He also interpreted correctly the function of most of the organs he discovered. Furthermore, his excellent studies directed to galls were the basis for the future scientific study of galls; he proved their origin with different insects. That a row of later books are based on his research is not surprising, among them are *De Anima Brutorum* by Thomas Willis (Amsterdam, 1672), *Anatomia Animalium* by G. Blasius (Amsterdam, 1681), and *Amphitheatrum zootomicum* by M. B. Valentini (Frankfurt, 1720 and 1742). The papers of Malpighi were collected in the *Opera Omnia* (London, 1687) and *Opera Postuma* (London, 1697).

The second of these famous anatomists, a religious fanatic, was the Dutchman Jan Swammerdam (1637–1685) who was absolutely devoted to the natural sciences and especially that of entomology. He stood in astonished admiration for the magnitude of God in the multiplicity of creation. His earliest years he occupied with the study of insects; he had already published (1669) in the *Historia Insectorum Generalis* (Utrecht) his first comprehensive book, later (Utrecht, 1685) edited in French and Latin. His principal work the *Bybel der Natuure* was brought out by Boerhave and Gaubius in Dutch and Latin (Leyden, 1737–1738) long after his death and then translated several times (Leipzig, 1752, London, 1758, Dijon, 1758). Swammerdam had developed an exceedingly refined technique of preparation, having made for himself, for experimental purposes, numerous instruments. With these he succeeded in the anatomical dissection and description of even the smallest insects, for instance, lice. He described for the first time the venom gland and the apparatus of the sting of the honey bee, and discovered the ovary of the queen bee, and the male reproductive organs of the drone. By his research on insects of various orders he put their anatomy on a large comparative basis making certain Harvey's unformulated idea of the egg. Of his biological studies those concerning bees, ants, dragonflies, and ephemerids are especially to be emphasized. He recognized as parasites the Hymenoptera and Diptera hatching from caterpillars and pupae of butterflies. He described the molting of various insects and concerned himself about the function of the facet-eye. With accord to his extensive studies of metamorphosis Swammerdam created finally a system, in which he classified the insects as ametabolic, hemimetabolic, and holometabolic; although he did not, as yet, use these terms. This classification is essentially maintained up to this date. Swammerdam was, therefore, a pioneer and leader in many fields of entomology, so we cannot overestimate his significance.

The third of these anatomists was the Dutchman Antony van Leeuwenhoek (1632–1723), a self-taught man who devoted himself with enthusiasm to the natural sciences, his favorite occupation, throughout his life. He had a number of microscopes, some with more than one lense, whereby enlargements up to 270 times were obtained; thereby making visible objects and structures till then unknown because of minuteness. By this means he discovered the spermatozoa of insects and other animals, details of the histology of the muscles and their transverse strips, and the compound structure of numerous fibers of the nervus opticus of insects. He demonstrated on the basis of the dissected cornea of a beetle, whose facets he computed to be in excess of 3000, the mosaic sight structure of insects. His ingenious talent for observation and reliable conclusions made it possible for him to detect the parthenogenesis and the vivipary (ovovivipary) of aphids. He has in his paper about the "Hemelt" (*Tipula paludosa*) correctly recognized, for instance, the proterandry and the limiting circumstances of population dynamics. Leeuwenhoek's papers are summarized in his 4 volumes of *Arcana Naturae Detectae Ope Microscopiorum* (Delphis Batavorum; 1695, 1697, 1719, 1722), which were published in several editions, among them German (Delft, 1696) and English (London, 1698) versions.

John Ray (1628–1705) with his friend Francis Willughby, who died before his full potential was realized, was devoted to, besides botany, the taxonomy of insects; basing his work mainly on the work of Swammerdam. Ray's system of insects is grounded on morphological and biological facts, consequently it is divided into different categories. For the first time the species is understood as a systematical unity and defined as a collective idea for all specimens that come from the same progenitors. Every species is exactly differential-diagnostically described. Thereby, his main work *Historia Insectorum,* published posthumously in 1710 in London, is of special significance to the history of entomological taxonomy.

Later Antonio Vallisnieri (1661–1730) was opposed to Ray, in an ecological system of insects described in *Nuove Idea d'una Divisione Generale degl'Insetti* (1713), which by no means in its expression recognized the unity in nature and natural relationship. But nonetheless his studies are excellent in connection with the cattle warble fly, sheep nasal bot fly, and the armed horse bot fly (1715), as well as the web-spinning sawfly. Of the last he investigated particularly the ovipositor, observing the development of the embryo through the transparent eggshell. In the beginning of the eighteenth century, he was the leading entomologist of Italy.

Finally, several more or less amateur investigators must be remembered, who, by self-sacrificing study of nature and especially in the rearing of many species, contributed greatly to our knowledge of species and to the enlightenment of the biologies of holometabolic insects; although they did not settle difficult problems. Objects of their study were delineated in a notably perfect and artistic manner so that an aesthetic pleasure is gained in looking through their books. The first voyages of discovery find their outcome in these works; their entomological results stemming from the Dutch overseas colonies. Of these works the first is from the Dutch painter Jean Goedart (1620–1668) which, he says, originate from his own observations and are not taken from any authority. He studied the development and the metamorphosis of 140 species, but otherwise he scarcely surpasses in his scientific views, those of Aldrovandi. His compatriot Johann de Ney (1617–1678) edited his *Metamorphosis Naturalis* (1662–1668) in three volumes though he was not in any way entomologically distinguished. It was composed rather of a series of incoherent single observations. A later edition followed which was better arranged with regard to the subject matter and was illustrated with some new copper plate engravings and was published by Martin Lister (1638–1711). He himself wrote a catalogue of the insects of England, clarified the biology of ichneumon flies and thus contributed very much to the propagation of interest in entomology in England. Inspired by Goedart, also, the Dutch physician Stephen Blankaart devoted himself to entomology and had summarized his observations in his book *Schouburg der Rupsen, Wormen, Maden* (Amsterdam, 1688; in German in Leipzig, 1690), which contains besides other matters a fine study on the cuckoo-spit.

Exceedingly beautiful pictures, also true to life, were created by a German woman, Maria Sibylla (Gräffin) Merian (1647–1717), who had published at her own cost, books with wonderful plates. Her watercolors, done by her own hand, were engraved on beautifully colored copperplates in many copies. Furthermore she presents in her two volume work *Der Raupen wunderbare Verwandlung* (Nürnberg, 1679 and 1683) the metamorphosis of 150 European insects, each of them on their host-plant; the text was written from her own observations. In the years 1699–1701 she even undertook a voyage to Surinam to study and portray in drawings the then unknown development of the insects of this country. This was the first voyage of exploration, whose aim was purely entomological. She summarized the result of this voyage in the *Metamorphosis Insectorum Surinamensium* (Amsterdam, 1705). This book contains more than 100 insects portrayed, again with their host-plant, on 72 plates in a beautifully executed manner characteristic of Maria Sibylla Merian. This is the first and for a long time to come only large illustrated work concerning exotic insects with their stages of development. After her death a complete edition of her books in French was augmented by an addition of 34 new tables of only minor entomological importance.

The *Beschreibung von allerley Insekten in Teutschland* (Berlin 1720–1738) by the pedagogue Johann Leonhard Frisch (1666–1743), published in 13 parts in German, contains an abundance of excellent observations, in addition to about 300 endemic insects and their stages of development. The copper plate engravings of this book do not equal, of course, those by Maria Sibylla Merian. Attention was given by Frisch especially to the injurious insects, emphasizing that their control could not be possible without knowledge of their biology. *A Natural History of English Insects* (London, 1720) by Eleazer Albin, an English painter, was illustrated by himself and brings to an end the work of the seventeenth century, although much of his work, containing moderately good pictures, was published as late as 1736. What, however, remains noteworthy is that Albin based his research on all stages of butterflies, reared mostly by himself.

LITERATURE CITED

Albin, E. 1720. *A Natural History of English Insects*. London
Aldrovandi, U. 1602. *De Animalibus Insectis Libri VII*. Bologna
Bacon von Verulam, F. 1661. *Sylva Sylvarum sive Historia Naturalis in 10 Centurias Distributa*. Amstelodami (Amsterdam)
Baldner, L. 1666. *Vogel-, Fisch-, und Thierbuch*. Manuscript Library, Strassburg
Bauhinus, J. 1598. *Historiae Fontis et Balnei Admirabilis Bollensis Liber Quartus*. Montis Beligardi
Berchovius, P. 1521. *Reductoria Moralis*. Paris
Blankaart, S. 1688. *Schouburg der Rupsen, Wormsen, Maden*. Amsterdam
Bochartus, S. 1663. *Hierozoicon sive Bipartitum Opus de Animalibus*. London
Bodenheimer, F. S. 1928–1929. *Materialien zur Geschichte der Entomologie bis Linné*. 2 vols. Berlin
Bodinus, J. 1546. *Universae Naturae Theatrum*
Borelli, G. A. 1680–1681. *De Motu Animalium*. Leyden
Borellus, P. 1656. *Observationum Microscopicarum Centuria. Hagae Comitis.*
Camerarius, J. 1605. *Symbolorum et Emblematum Centuriae Tres.*
Caub (Cuba), Wonnecke, J. v. 1491. *Ortus Sanitatis*. Mainz
Cesi, F., Stelluti, F. 1625. *Apiarium*. Manuscript in Library of Academia dei Lincei. Rome
Charleton, W. 1668. *Onomasticon Zoicon. Oxoniae.*
Daneus, L. 1579. *Physica Christiana. Genf.*
Franz, W. 1612. *Animalium Historia Sacra*. Wittenberg
Freigius, J. T. 1579. *Quaestiones Physicae*. Basel
Frisch, J. L. 1720–1738. *Beschreybung von Allerley Insecten in Teutschland*. Berlin
Goedart, J. 1662–1667. *Metamorphosis Naturalis.*
Griendelius, J. F. 1687. *Micrographia Nova. Norimbergae.*
Harvey, W. 1651. *Exercitationes de Generatione Animalium*. Amsterdam
Harvey, W. 1628. *Exercitatio Anatomica de Motu Cordis et Sanguinis in Animalibus*. Frankfurt a. M.
Hooke, R. 1667. *Micrographia*. London
Johnston, J. 1653. *Historiae Naturalis de Insectis Libri III.*
Kircher, A. 1675. *Arca Noe*. Amsterdam
Magnus, O. 1555. *Historia de Gentibus Septentrionalibus*. Rome
Majolus, S. 1600. *Dierum Canicularium Tomi VII.*
Malpighi, M. 1687. *Opera Omnia*. London
Malpighi, M. 1697. *Opera Postuma*. London
Matthioli, P. A. 1604. *Dei Discorsi Nelli sei Libri di Pedacio Dioscoride Anazarbeo Delle Materia Medicinale*. Venice
Merian, M. S. 1679, 1733. *Der Raupen Wunderbare Verwandlung*. Nürnberg
Merian, M. S. 1705. *Metamorphosis Insectorum Surinamensium*. Amsterdam
Mouffet, T. 1634. *Insectorum sive*

Minorum Animalium Theatrum. London

Ray, J. 1710. *Historia Insectorum.* London

Redi, F. 1668. *Esperienze Intorno alla Generazione degli Insetti.* Florence

Redi, F. 1686. *Opusculorum pars Prior sive Experimenta circa Generationem Insectorum.* Amsterdam

Scheuchzer, J. J. 1731–1735. *Kupferbibel.* Augsburg and Ulm 8 vols.

Schwenckfeld, C. 1603. *Theriotropheum Silesiae in quo Animalium vis, Natura et Usus Sex Libris Perstringitur.* Lignicii

Severino, M. A. 1645. *Zootomia Democritaea.* Norimbergo

Sperling, J. 1661. *Zoologia Physica.* Lipsiae

Swammerdam, J. 1669. *Historia Insectorum Generalis.* Utrecht

Swammerdam, J. 1737–1738. *Bybel der Natuure.* Leyden

Vallisnieri, A. 1713. *Nuove Idea d'una Divisione Generale degl'Insetti.*

van Leeuwenhoek, A. 1695–1722. *Arcana Naturae Detectae Ope Microscopiorum Delphis Batavorum.* Vol. 4

Wagner, J. J. 1680. *Historia Naturalis Helvetiae Curiosa.* Figuri

Wottoni, E. Oxoniensis 1552. *De Differentiis Animalium Libri Decem.* Lutetiae Parisiorum

ENTOMOLOGY SYSTEMATIZES AND DESCRIBES: 1700–1815

S. L. Tuxen
Universitetets Zoologiskie Museum
Copenhagen, Denmark

> *Dès qu'un Dessinateur se contente de n'experimer qu'en gros ce qu'il voit, le faux s'y mêle avec le vrai, & defigure le tout; aussi me suis-je constamment interdit cette licence.*
>
> Lyonet (1762, p. XI)

It is fascinating to observe how in different times, even in different centuries, at an always accelerating speed, science has sought different ways of organizing itself; and how much science has been achieved without any organizing at all. This holds true for science in general, and it holds true for entomology in particular.

Aristotle taught entomology to his pupils in his open schools in the peristyles of Athens. In the Middle Ages monasteries were the centers where science and even entomology were studied, though mostly on a literary basis and they were probably not taught to the novices. At the end of the Middle Ages universities began to spring up, but though university means all-round knowledge, it was not until the sixteenth and seventeenth centuries that natural history began to play a role in them, small as it was. Till the beginning or even middle of the nineteenth century, however, universities often had more of a retarding than a promoting effect, and scientists, therefore, felt the need for other centra to their activity. Thus the academies originated, beginning in 1603 with Accademia dei Lincei in Rome and in the same century with the French, British, and German academies, though the eighteenth century was the real time for opening of academies.

But even academies, though publishing "letters" and later transactions, were too cumbersome, and so the nineteenth century saw the birth and growth of many natural history societies concerned with entomology, again beginning in France, Great Britain, and Germany. These national societies published journals in which the scientists of their country could publish their results; but the need of an international forum was felt by the turn of the century, and thus the international congresses which mark the first half of the twentieth century originated. These congresses, however, are now over-worked for the constantly accelerating science, and so the latter half of the twentieth century will see them replaced by symposia on smaller delimited topics with fewer participants. Thus new centra, which at first took several centuries to develop, are now following each other within less than a century, and who knows how fast they will replace each other in the twenty-first century.

In any event the eighteenth century is the century of the academies and many of its entomologists advertised their membership in academies more than their affiliation to universities, on the title pages of their books. They sent letters to the academies and the academies formulated problems for solution. It is relevant to point out that Linné was one of the founders and the first president of the Swedish academy, while he was at the same time professor of medicine at the university!

The "letters" they wrote were on the anatomy and biology of insects, and as the age just before Linné has been called the Age of Bionomics, of "life-histories" or "life-descriptions", it is natural to begin here with the biologists and observers. Yet the limit will not be set as before and after Linné, but the development of the different disciplines will be viewed as an entity during the whole eighteenth century. Lacordaire (1838), who took it from Kirby and Spence, distinguishes between "la période de Swammerdam (–1735), de Linné (1735–1775), de Fabricius (1775–1798) et de Latreille (1798–1815)". This system, too, is artificial, though it gives some clue to the development of entomological systematics.

But entomology is much more than systematics. It is actually part of the whole cultural pattern of the Age, and so it is not surprising, but still interesting, to note that the pivot switches over from Italy, the Netherlands and partly England in the seventeenth century to France, Germany, and the northern countries in the eighteenth century. It is the century of Voltaire and Rousseau, of David Hume, of Herder, Lessing, and Kant; humanistics flourishing above all in Germany and natural history in France. So it is no wonder that it is the century of Réaumur, Bonnet, Linné, O. Fr. Müller, Fabricius, Lamarck, and Latreille; just as the seventeenth was the century of Malpighi, Redi, Valisnieri, Goedart, Swammerdam, and Leeuwenhoek.

Observers: Biology and Anatomy

Madame Merian had made her drawings, beautiful and correct as they were, from what she saw and in a few words explained them; she even stayed two years with the Indians in Surinam to further her aims ("ç'a été une espèce de phenomène, de voir une dame traverser les mers pour aller peindre [les insectes] de l'Amerique" says Réaumur, who went nowhere. Goedart had described and made primitive drawings of the metamorphoses of a number of insects, and he added, naively, what he had been told as well as curious reflections on the intelligence of the insects. Frisch described what he saw, especially the metamorphoses of the insects which took the fancy of all entomologists at that time, but his drawings were bad and he did not enter further into the problems. Leeuwenhoek used his microscopes on everything which came near him and was the only one of them all who tried to get to the bottom of problems.

Such was the state during the first decades of the eighteenth century and such were the predecessors of Réaumur. To him all this was not enough, and he knew it. To read the first 50 pages of his *Mémoires* is to read an instruction to students of our day.

Réaumur (Réné Antoine Ferchauld, Seigneur de Réaumur, des Alpes et de la Bermondière) (Figure 1) was born in 1683 in La Rochelle (Poitou); he studied law and later, in Paris, mathematics and natural history. He was elected a member of the French Academy in 1708 and devoted his whole

FIGURE 1. René Antoine Réaumur (1683–1756). The foremost entomological observer in the eighteenth century. Courtesy of Musée d'Historie Naturelle, Paris.

life to various studies until his death in 1756. He is perhaps best known for his temperature scale and for his porcelain; he was a chemist as well, but he made natural history studies of the most varying objects. Still his most important work is the *Mémoires pour servir à l'histoire des insectes* in six volumes from 1734–1742. In his preface to the first volume he tells us his intentions. He is not very pleased with lists of forms with different colors or different distributions of the same color; let them remain confused! What is necessary is to study and describe the life history and "industries" of as many insects as possible, and not only "pour des biens réels", but just out of curiosity. Insects may be useful, yielding silk, ripening figs, producing galls for ink, etc, and they may be noxious to grain storage, but studies made from pure curiosity may result in useful discoveries. The observations should be made with the utmost care; the more unexpected the observations the more important to check them. Hence the purpose of *Mémoires* was to stimulate attention. Never trust what you hear, look for yourself. The best observations in earlier times were made by those who, perhaps happily, were ignorant of the classical languages. It is important also to write clearly, to "plaire"; it demands a great patience to read only twenty of Ray's descriptions. It is also necessary to make drawings: Réaumur could not draw himself, so others made them. For these drawings he frequently used a woman whose modesty, however, prevented him from giving her name. In these drawings, too, it is necessary to show the exterior as well as the anatomical details as carefully as possible; for in deciding to which "classe" or "genre" an insect belongs, the exterior characters are essential.

It is evident from this working program that Réaumur could not accept generatio aequivoca since it could not be proved; also he would not assign intelligence to insects, since renewed and improved experiments generally disprove it, but he would not regard them as true machines either as Leibniz did, for example. Who would deny that the Almighty can create machines who can multiply, work, and think; and who would deny that He can also create and place intelligence where He wills?

In using these principles it is not astounding that Réaumur succeeded in making his many *Mémoires* small monographs. To take just an example: Volume 5, Mémoire IV of *Mémoires* deals with three species of cicadas, "les grandes cigales", "les cigales de moyenne grandeur", and "les tettigonies". The exterior is described in detail, the mouthparts with a clear understanding of their use, the sound-producing organs with even such details as the small tendon from the tymbal muscle to the tymbals, the ovipositor, the process of egg laying, the nymphs in detail, and the development. It is here he makes the pun about those having written about cicadas "depuis Aristote, ou plus exactement d'après lui". The function of the sound-producing apparatus is discovered in studies of living and dead specimens and in experiments made to reproduce their song artificially on dead animals. Everything appears with careful drawings, legends, and detailed explanations of the tables; a recent account of the function is best illustrated by his plates.

There is a jump even from Swammerdam to Réaumur, no less than from Ray to Linné, and so it is no wonder that his work gave inspiration to successors as did his sympathetic personality.

His closest successor in time did not have his scientific mind, but he had the same ability to observe and, moreover, he could depict what he saw; this was the German miniature painter August Johann Rösel who later took over his old nobility title von Rosenhof (1705–1759). He learned painting from his father's brother and later (1726–1728) at the Danish court, but finally he settled down in Nürnberg as a painter and engraver. He had no education as a naturalist, but an indefatigable interest especially in insects; so he ventured in 1740 to give out illuminated plates to subscribers, two every month with text, which were later combined in his *Insecten-Belustigung* I–IV 1746–1761 (Vol. 4 and a later edition given out by his son-in-law C. F. C. Kleemann).

His drawings which he engraved himself are among the most beautiful of the century and meticulously accurate: others then illuminated his printed engravings. Furthermore his tales on the life of insects and the way he observed them are most charming. Read for example his description of the *Necrophorus* beetles (curiously enough placed at the beginning of the posthumous fourth volume); how he first saw them move a dead mouse "as if it was walking itself" and later depicted their whole development as well as their ovaries, and even the tarsi of the parasitic mites covering them. He also made lenses and microscopes for his observations.

Karl (Carolus, Charles) de Geer (1720–1778) was Swedish, but of Dutch origin. He studied biology in Utrecht and was so impressed by Réaumur's work that he continued observations in the same way and even published them under the same title, *Mémoires pour servir à l'histoire des insectes,* I–VII 1752–1778. He never met Réaumur, but was even induced by him to publish this "journal of observations" in French, because the reader would be indulgent with an author writing in a foreign language. Also he points out in his modest preface that the truth very often is first found after many errors. It was the observations which had his interest, not the literature; he even writes that he did not find it necessary to learn Latin for studying insects. When, nevertheless, many species carry his name as author it is because he actually later gave Latin diagnoses though in pre-Linnean style, which A. J. Retzius (1742–1821) tabulated in Linnean shape in *Caroli de Geer genera et species insectorum* (1783).

De Geer made the drawings himself and though they are not particularly artistic they show a great many structures not known before and are accompanied by new observations. To give examples, he was the first to describe living cochineal coccids and his drawings of Collembola, which nobody had seen till he described them (in 1740 and 1743), are astonishingly exact, as were his observations. He saw their eggs, realized that the insects were active in the winter, that the young resemble the adult, but that they nevertheless molt, since he saw the exuvia. He even saw the collophore and supposed it to gather humidity to help the insect to adhere to smooth surfaces. De Geer was also a social pioneer on his estates and became Lord Chamberlain and Baron at the Swedish Court.

The French-Swiss Charles Bonnet (1720–1793), born the same year as de Geer, was another contemporary inspired by Réaumur's books and correspondence with him, but he could only enjoy this inspiration for a few years. By the age of 25 his sight was so destroyed by hard microscopy that for the rest of his life he could barely read or write. He studied jurisprudence in Geneva, but made his biological observations on caterpillar respiration, on the freshwater polyp, on cestodes, etc, simultaneously. In 1745 he published his *Traité d'insectologie,* "insectology" termed in conformity with conchyliology, etc "since no special name exists for the science on insects". But insects meant all animals which are not vertebrates or mollusks, and the entomologically most important part of this treatise is that on the parthenogenesis and heterogony of aphids. Leeuwenhoek had already observed parthenogenesis, but Bonnet proved it convincingly in well-cogitated experiments carried out on different species during longer periods. In his preface he considers the use of knowing that some insects propagate without copulation, and he gives many reasons: from a better understanding of the laws of nature and the structure of animals on the whole, to usefulness in medicine and to his "scale of life" (Stufenleiter) theory. "One truth explains another, especially in natural history", he said. From his 25th year to his death Bonnet lived for his philosophy and the popularization of his science; *Contemplation de la nature* 1–2, 1764–17645, and *Oeuvres d'histoire naturelle et de philosophie* 1–8, 1779–1783 were his later works.

Among this spiritual circle around Réaumur must also be reckoned Lyonet, although his interest concentrated on the anatomy of insects. (Figure 2). Pieter Lyonet (or Lyonnet) (1707–1789) was a Dutch lawyer who was impressed by the idea of finding God's perfection in nature and so at the age of 35 he translated Lesser's (1692–1754) *Insecto-Theologie* (1738) into French. We know from his preface (1762) that he intended to publish a "Recueil historique" on entomology, but he realized that his discoveries were made also by others and so he abandoned this idea to help his country with his knowledge as a lawyer. This gave him so much disappointment that he returned to entomology, and so, he tells us, the ignoble feeling of annoyance made him an entomologist! And with juristical perseverance he concentrated on one species, *Cossus ligniperda;* he described and depicted its anatomy in a way which could hardly be improved on today. *Traité anatomique de la chenille qui ronge le bois de saule* appearing in 1762, had more than 600 pages and 18 plates, all engraved by himself. From Bonnet's preface to his *Insectologie* we know that Lyonet was among the best engravers and he points out the precision with which he depicts and engraves the different sorts of tissue, etc. It is, in fact, astounding how much he could see in his larvae and how well he could keep the muscles, nerves, tracheae, etc clear of each other in the drawings, particularly since he only killed eight or nine larvae for the whole work. As many as 1647 muscles are accounted for, as well as the finest branching of the nerves and tracheae, so that 1804 tracheal branches are counted! Exact observation was what interested him; he had so little taste for theories that he often wished they could be banned from all science! Of physiology he had little understanding; he maintained, for instance, that the larvae did not respire and that the dorsal vessel was for producing nerve substance. His intention was to make similar works on the pupa and imago, but he never did; some drawings were published posthumously by W. de Haan in 1832 (in *Mém. Mus. d'Hist. Nat. Paris*).

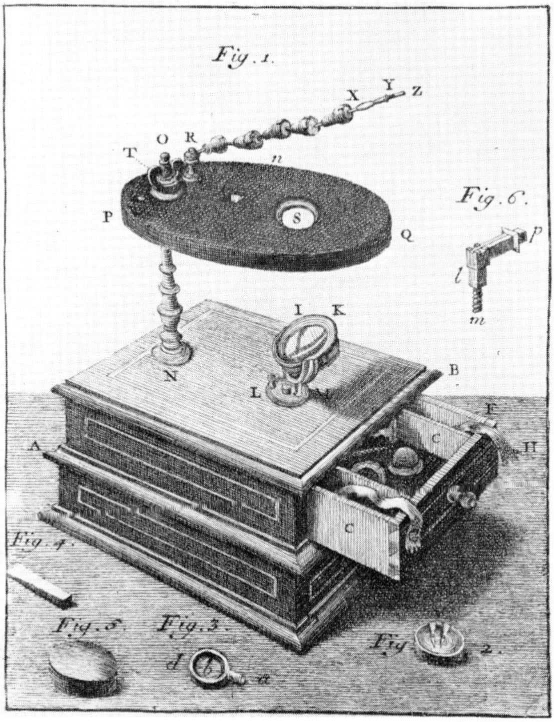

FIGURE 2. The dissecting tools of Lyonet (1707–1789). After Lyonet 1762.

Just as Malpighi stood out in the seventeenth century, so Lyonet stood out in the eighteenth, and it was not until well past the middle of the nineteenth century that insect anatomy of the same standard was carried out. Still some parts of the internal anatomy and especially the morphology were already being analyzed in the eighteenth century; beginning (if we disregard the many anatomical details in the works of Réaumur and de Geer) with the Dane Thomas Jacobæus, who in 1708 wrote a dissertation on the eyes of insects, *De oculis insectorum,* which was a purely scholastic dissertation. This was followed in 1813 by Marcel de Serres' (1780–1862) *Mémoires sur les yeux composés es les yeux lisses des insectes et sur la manière dont ces deux espèces d'yeux concourent à la vision,* which, besides describing the eyes, also attempts a physiological explanation of vision. This problem was, however, taken up in a broader sense a few years later by the German Johs. Müller (1801–1858) in his well known *Zur vergleichenden Physiologie des Gesichtssinnes* (1826), in which he promotes the famous mosaic theory, until recently regarded as the final solution of the problem.

On the antennae and their use Job Baster (1711–1775) in Holland wrote: *Over het gebruik der sprieten by de Insecten* (1770) and the German-Dane M. C. G. Lehmann (1775–1856) who was born in Holstein and died as a high official in Copenhagen, wrote two dissertations *De antennis insectorum* in 1799 and 1800. He concluded that to ascribe to the antennae an olfactory function would be vain, an auditory function untrue, and a gustatory one ridiculous. Perhaps they are organs to defend the head since they react on air currents? It was not till the observations by Alexandre Lefebvre (1798–1868) that their olfactory function was demonstrated (*Remarques sur le sentiment olfactif des antennes,* 1838–1839) and still later before all their other functions were known. Lehmann wrote in 1798 also a paper *De sensibus externis animalium exsanguium insectorum scilicet ac vermium* on which many of the sense-physiological speculations by Kirby & Spence in the fourth volume (1826) of their *Introduction to entomology* are based. Time was not ripe for sense physiology; conclusions were based on analogical sites of the organs in man and insects (the antennae are analogous to the ears of vertebrates, since they occur in the same number, are placed on the head, and especially because no other organ can be regarded as ear, Kirby & Spence loc cit).

General anatomy was topical for several papers after 1800. C. F. Posselt (?–?) published on the anatomy ("Splanchnologie", i.e. intestine and genitalia) of some beetle larvae (*Beyträge sur Anatomie der Insekten,* 1804), but his work was never continued. K. A. Ramdohr (?–?) wrote on *Carabus monilis* (1807) and later on the honey bee, and H. M. Gaede (1796–1834) on different insects (1815–1823). In 1821 (*Physiologische Bemerkungen über die sogenannten Gallgefässe der Insecten*) Gaede ascribed an absorbent function to the Malpighian tubes which the Swiss J. R. Rengger (1795–1832), who later travelled for eight years in Paraguay, had shown four years earlier (*Physiologische Untersuchungen über die tierische Haushaltung der Insecten,* 1817) to be excretory organs (comparing their excreta with the feces of birds!). Another Swiss, J. J. Hegetschweiler (?–?) in 1820 gave a description of the genital system of several orders of insects, *Dissertatio de insectorum genitalibus,* and the well-known German systematist G. A. W. Herrich-Schäffer (1799–1874) wrote the dissertation *De generatione insectorum partibusque ei inservientibus* in 1821. These smaller, more or less comparative descriptions were followed in 1828 by a study of the same quality as that of Lyonet, the author of which was the Frenchman, H. E. Straus-Durckheim (1790–1865). His *Considérations générales sur l'anatomie comparée des animaux articulés, auxquelles on a joint l'anatomie descriptive du hanneton vulgaire* (1828) included 500 pages and 10 double plates and is simultaneously a scrupulous examination of one species, the cockchafer, and the first true work on comparative anatomy in insects.

Réaumur had already made investigations on respiration in insects, the existence of which Lyonet denied. The Dutchman J. F. Martinet (1734–1795) opened a mainly scholastic dissertation (though containing some experiments) *De respiratione insectorum* (1753), with the words "Omne quod vivit respirat" and found respiration to take place "through stigmata, skin, anus and mouth". Much later J. F. L. Hausmann (1783–1859) with *De animalium exsanguium respiratione* (1803) and F. L. A. W. Sorg (1773–1827) with *Disquisitiones physiologicae circa respirationem insectorum et vermium* (1805) made experiments on respiration, and Kurt Sprengel (1766–1833) with *Commentarius de partibus quibus insecta spiritus ducunt* (1815) described the spiracles; but still time was not ripe for a true physiology.

The field of comparative morphology, however, had ripened a little earlier. After Fabricius had stressed in 1775 the importance of the mouthparts for his system, it was natural that specialists began to investigate these organs, and here and there in books we find drawings of the mouthparts of different insects. But it was left for Savigny to present their comparative morphology. Marie-Jules-César Lelorgne de Savigny (1777–1851), born in Provence, went as naturalist (Cuvier having declined) with Napoleon to Egypt in 1798 and began in 1802 to identify the material he had brought home. He felt forced to compare the mouthparts of insects in order to be able to compare the orders of the insects (i.e. Arthropoda). "Which parts of the proboscis of a fly are also found in the mouth of a wasp, a spider? Or the crab? The entomologists created new genera every day and the most important foundations of this building didn't exist. What nobody had attempted, I ventured to do, though it was doubtless above my capacity" he writes in the preface to the *Mémoires sur les animaux sans vertèbre. 1. fascicule: Théorie des organes de la bouche des Crustacés et des Insectes* (1816). Far from being above his capacity, he actually shaped the scheme for the mouthparts, sucking or biting, and the comparison of their different parts, with which and on which we are still working. And he did it in the form of tabulations with figures, comparing also what Fabricius, Latreille, and Cuvier had seen and named. As so many of the earlier entomologists, he lost his sight from too intense microscopy.

As Fabricius based his system on the mouthparts so did Linné on the wings and it is natural that thereafter the wing venation was studied to distinguish groups within the orders. Moses Harris (1731–1785), a London engraver and miniature painter whose *The Aurelian or natural history of English insects namely Moths and Butterflies* (1766) with illuminated coppers is very rare, was the first to use the venation in *An essay wherein are considered the tendons and membranes of the wings of butterflies; first as useful in describing the situation of their spots or markings; secondly of great assistance in discovering their different genera etc.* (1767). In 1802 William Kirby (1759–1850), an English clergyman, used it in his *Monographia Apum Angliae,* but the true comparison of wing venation is due to the Swiss physician and naturalist Louis Jurine (1751–1819) who shaped it in his *Nouvelle methode de classer les Hymenoptères et les Diptères* (1807). It was, however, not till almost a hundred years later that a useful comparative system was shaped (Comstock & Needham).

More success followed the attempt at comparing the thorax sclerites made by Jean-Victor Audouin (1797–1841), a French physician, who gave entomological lectures at the Musée d'Histoire Naturelle and followed Latreille as a professor in 1833. He published many smaller papers on the morphology of insects and a *Resumé d'Entomologie* in 1828–1829, but his outstanding work is the *Recherches anatomique sur le thorax des animaux articulés et celui des insectes hexapodes en particulier* (1820, printed 1824) in which the sclerites of the insect thorax were described and analyzed in a way which Crampton and Snodgrass were the first, thereafter, to improve upon. He later turned to applied entomology, noxious insects, and the silkworm, and published *Histoire des insectes nuisible à la vigne* in six parts 1840–1842.

In 1820 another important comparative paper appeared: *Essai sur le vol des insectes* by J. Chabrier (?–?) who calls himself "Ancien Officier superieur". He seems to have worked quite independently, though building upon the work by Latreille, and he gives not only a beautiful description of the thorax and its muscles as well as the articular sclerites in all groups of winged insects, but also an attempt at a physical explanation of the flight of insects.

Still, the father of comparative anatomy is the great Cuvier (1769–1832), of whom Kirby and Spence in their *Entomology* (IV:472) say that he is "himself a host" and who in his universality is second only to Humboldt. Although his first publications are small entomological papers, the entomological part of his *Règne Animal* in both editions (1816 and 1829) is by Latreille. His own entomological system, based on comparisons, appeared in *Leçons d'Anatomie comparée* 1799–1805, but is actually inferior to that of Latreille's (1796) *Précis des caractères génériques des insectes disposés dans un ordre naturel,* or Latreille's *Histoire naturelle, générale et particulière des crustacés et des insectes* (1802–1805). Both works, however, belong to systematics more than to comparative anatomy. Cuvier and Latreille were close friends and both were friends of Fabricius.

The meticulous observation of details as exemplified in the motto at the beginning of this chapter is characteristic of the eighteenth century, be it biology or anatomy, but the comparative and corollary views on anatomy, morphology, and physiology belong to the germinating and ripening nineteenth century. How far they got was compiled and augmented by personal reflections by the Englishmen William Kirby (1759–1850) and William Spence (1783–1860) in *Introduction to entomology* in four inspiring volumes 1815–1826; as well as by the French-Belgian coleopterist Jean Théodore Lacordaire (1801–1870) in his *Introduction à l'entomologie* 1–2 (1834–1838) and the German H. Burmeister (1807–1892) in his *Handbuch der Entomologie* 1 (1832).

A few important biologists from this time (ca 1800) must also be mentioned. In Switzerland, Franz Huber (1750–1831), though blind, made outstanding observations on the life of the honey bee aided by his son and his valet; they told him what they had seen and he asked them to look further for this and that. The result was the book *Nouvelles observations sur les abeilles* (1792), a classic on bee-life. The son Jean Pierre Huber (1777–1840) concentrated on the life of ants in *Recherches sur les moeurs des fourmis indigènes* (1810) and wrote smaller papers on several insect groups. Another extremely important activity in the life of the bee was discovered by the German Chr. Konrad Sprengel (1750–1816), namely their part in the pollination of plants. His observations were published in a book with the characteristic title *Das entdeckte Geheimniss der Natur im Bau und in der Befruchtung der Blumen* (1793).

Systematizing

The main achievement in eighteenth century entomology, however, is the systematizing; out of chaos came order. This is primarily due to Linné, or Linnaeus as he is known in English-speaking countries (Figure 3). Carl Linné was born in 1707 at Råshult in Småland, Sweden and died at Uppsala in 1778. His father was a poor clergyman, but from the beginning of his studies (medicine) he always found protectors and promotors who realized his genius. In 1732 he made a journey to Lapland, with subsequent journeys to less remote parts of Sweden, resulting in important surveys of their nature. From 1735–1738 he studied in Leiden where he published first editions of two important works, *Systema naturae* in 1735 and *Fundamenta botanica* in 1736. He became physician to the Queen of Sweden in 1738, was one of the founders of the Swedish Academy of Science in 1739, was professor

FIGURE 3. Carl von Linné (1707–1778). The founder of true systematics. After Tullberg: *Linnéporträtt* (1907).

in Uppsala by 1741, and in 1758 he bought Hamarby in the neighborhood of Uppsala. A most charming picture of his lectures in Uppsala and Hamarby is given by Fabricius (1780) who studied with him from 1762–1764.

Linné was not the first to try to bring order to the army of insects, of which a horrified author well into the nineteenth century says that more than 20,000 species are known! Aldrovandi built his system in 1602 on the wing- and leg-morphology after having distinguished between land- and fresh-water-insects. Mouffet used the same characters in 1634 though in another arrangement. Swammerdam, on the other hand, in 1669 based the system on the kinds of metamorphosis, dividing the insects into those (*a*) without metamorphosis, (*b*) with a metamorphosis in which the wings develop little by little, (*c*) with a metamorphosis including a pupal stage, and (*d*) with a metamorphosis within the original larval skin [puparia]. The second group contains the hemimetabolous insects, to use the modern word, although Carus (1872:446) calls it "die logische Ungeheuerlichkeit einer halben Verwandlung". Valisnieri, in 1713, divided the insects in four groups: those living in plants, in water, in earth or hard substance, and in or on other animals or meat. "If we want to arrange the insects in a small number of classes we can hardly find a better grouping than that of Valisnieri or that of Swammerdam" says Réaumur, "but they are not without inconveniences."

Finally Ray in 1705 and 1710 (posthumous) used and elaborated Swammerdam's system, dividing the insects into those with metamorphosis and those without; the first ones with or without pupae; within those with pupae he distinguished between pupation outside and within the larval skin. Next in order come morphological and ecological characters. This, of course, is a more logical system than any of the earlier, but Ray did more than that, he gave a definition of the concept of the species, namely as offspring from the same parents or identical parents, and to some degree also of the genus, namely groups of species which resemble each other (*Historia Plantarum* 1686 cap. XX and XXVI).[1]

Even though Carus (1872:432) regrets the "Unbeugsamkeit" of this concept, it was an enormous step towards a useful system, towards the system of Linné and Linné knew it himself. In the first edition of *Systema Naturae* (1735), he writes (Regnum animale § 2): "Paucissimi vero sunt, qui Zoologiam in Genera & Species secundum leges Systematicas redigere tentarunt, si Nobiliss. Willughbejum & Clariss. Rajum excipiamus" (extremely few have tried to arrange zoology in genera and species, if we except Willughby and Ray). It is in the same edition that the first sentence runs as follows: "Si opera Dei intueamur, omnibus satis superque patet, viventia singula ex ovo propagari, omneque ovum producere sobolem parenti simillimam. Hinc nullæ species novæ hodienum producuntur" (if we regard the works of God, it is evident to everybody, that every living being originates from egg and every egg produces offspring like the parents. Thus no new species are produced today). And § 15: "Lapides crescunt. Vegetabilia crescunt & vivunt. Animalia crescunt, vivunt & sentium. Hinc limites inter hæcce Regna constituta sunt" (stones grow, plants grow and live, animals grow, live and feel. Thus the limits between these kingdoms are given).

These phrases show Linné's clear thought and his wonderful, lapidary way of expressing it. They also show his view on nature, of which I shall speak later; here we shall deal with his systematizing.

It is often said that Linné's system was an artificial one, based on just one character. This is only partly true. Already in the first edition of *Systema Naturae* he says: "In Tetrapodologia Ordines Animalium à Dentibus; in Ornithologia à figura Rostri; In Entomologia ab Antennis & Alis &c. inprimis desumsi" (I have chosen above all . . .), but this "&c" is not without importance. It was the only way to make order in chaos: to choose a character and try to arrange the beings accordingly. Linné's predecessors, even Ray, had used ecological, biological, and several morphological characters at random. Even though the rigid use of one or a few characters resulted in

[1] Dr. Anne Fox Maule found these references.

arrangements, which we now know are wrong, it was the only way to overcome the obstacle; and this was done in one stroke with the first edition of *Systema Naturae* in 1735. Here classes, orders, genera, and species are used consistently as never before. Many emendations are then made in those following editions that Linné published himself: 2nd edition 1740, 6th edition 1748, 10th edition 1758, and 12th edition 1766–1767 (see preface to 10th edition).

As is well known the 10th edition (1758) is the starting point of zoological nomenclature, Clerck's *Svenska Spindlar* from 1757 being the only exception. But it is not the first work with binomial nomenclature. This accomplishment, to denote a species by a generic and a specific name, is an idea very far from just "a natural way to refer to things" in Latin, as Usinger puts it in his excellent paper from 1964 and it took some time for Linné to arrive at it. Even in the 6th edition from 1748 only genera and, as examples, some species with one or several adjectives are given, and though *Fauna Svecica* from 1746 is the first work with descriptions of every species, these are usually given in several words (one to twelve). The stroke of genius was to make it consistently one word and that is done universally for the first time in *Pan Svecicus* (1749) for plants and in *Museum S. R. Maj. Adolphi Frederici Regis* (1754) for animals.

In using the wings as the primary character, Linné in 1735 distinguished between four orders of "insects": Coleoptera, Angioptera (later Gymnaptera), Hemiptera, and Aptera. Insects then comprised all arthropods, but not worms, etc which even Ray had included. From *Fauna Svecica* (1746) and onwards he extends the system to seven orders: Coleoptera, Hemiptera, Lepidoptera, Neuroptera, Hymenoptera, Diptera, and Aptera, of which the last one is very heterogenous. In the 12th edition, however, the Aptera is divided in sections where what we now call myriapods, crustaceans, and arachnids form a section each, different from true insects.

Linné wrote many other papers, besides the great systematic works, very often in the shape of dissertations read by his pupils under "præside" of Linné; these are collected in his *Amoenitates academicae* 1–7 (1749–1769) and 8–10 (1787–1790), which contain several entomological papers. One of them, *Fundamenta entomologica* from 1767 (prop. A. J. Bladh), is an introduction to entomology where e.g. the word palpi is used for the first time as well as stemmata for the dorsal ocelli. Others contain reflections on the larval stages, on exotic species and the danger of introducing foreign insects, on protective resemblance, and quite especially on useful and noxious insects. (For Linné's concept of the species see section on Philosophy.)

Linné's *Systema* and perhaps quite especially *Fauna Svecica* had an enormous inspiring effect; new descriptions and more or less local lists followed each other. Still at first it was not generally accepted. The most important opponent at that time, though now completely forgotten, was a German, J. Th. Klein (1685–1759). He was a lawyer in Danzig and his idea was to make a system by means of which it was possible to identify all animals on external characters. So he was opposed to using teeth as characters in mammals because it was too dangerous to open their mouths! Linné never answered his attacks, but it is well to remember him when we now accuse Linné of making an "artificial system based on only one character". Klein's system shows us what an artificial system really is, combining beaver and frog because their hind toes are united, worms and snakes as naked animals without legs, and so on—and this is very far from Linné's system! It is not without reason when Aurivillius (1909) calls Linné's animal system "ein natürliches System" in contrast to his sexual system of plants.

Already in the 2nd edition of *Systema* Linné proposed a study of the mouthparts ("Character essentialis genericus insectorum ab ore primario desumendus est"). This thread was picked up by his Danish pupil J. C. Fabricius (1745–1808) (Figure 4). He was the son of a physician and he studied for two years with Linné in Uppsala, an experience for which Fabricius was grateful all his life and often referred to in writing. After some years' travel abroad he was appointed professor in Copenhagen (1770), then in Kiel (1775) where he stayed till his death. He had not, however, the promotors and protectors of a Linné and received so little encouragement in Kiel, that though he was beloved by his students, he could not make Kiel the center of entomology as Linné had made Uppsala the center of natural history. Instead he travelled every year all over Europe, consulted the entomological collections, and discussed with colleagues, especially in Paris and London. The result was descriptions of more than 10,000 species, whereas Linné had described well over 2,000. But he did more than that, he determined a new basis for the system.

In 1778 he published his *Philosophia Entomologica,* which is patterned on Linné's *Philosophia Botanica* (1751) and is actually the world's first textbook in entomology. It is comprised of the following sections: Literature ("Bibliotheca"), Morphology, Mouthparts, Metamorphosis, Sex, Systematics ("Dispositio"), Nomenclature, Distinguishing characters, Description ("Adumbrationes"), Ecology and Biology ("Oeconomia"), and Applied Entomology ("Usus"). "Bibliotheca" is a history of Entomology since Gesner. "Nomenclature" and the following sections contain advice on how to describe new species; "Dispositio", however, is the most important of the sections since it contains his view on systematic problems. It is characteristic that a whole section is devoted to the mouthparts, since his system is built on them: "Dispositionem insectorum artificialem a solis instrumentis cibariis desumsimus" (we have chosen an artificial system of insects based solely on the mouthparts) (VI. 10). Earlier authors, even Linné, used now this, now that character, but to Fabricius it seemed that only the mouthparts exhibit sufficient and constant characters. It is very important, however,

FIGURE 4. J. C. Fabricius (1745–1808). The foremost entomologist of the eighteenth century. Engraving from 1797.

that he realizes that this produces an artificial system and that he distinguishes between artificial and natural systems: "Dispositio insectorum est artificialis, quae classes et ordines, vel naturalis, quae genera, species et varietatis docet" (artificial with respect to classes and orders, natural with respect to genera, species, and varieties) (VI.2). [Varieties are the curse ("flagellum") of science, he says in VI.5.] But he also knows that natural classes exist ("Naturales existere insectorum classes vix dubitandum. Suadent ratio, delecta, observata"), but the time has not come for elaborating them, as long as we only are novices in science ("quum tyrones adhuc scientiae simus") (VI. 7). With "classes" he means our orders, and with "orders" our families which, however, he never named. The arrangement of the genera shows their relationship (VI.14). The whole book contains many valuable reflections and should be translated.

It was logical, therefore, that Fabricius began with a book on his system (*Systema entomologiae*, 1775), then one on the genera (*Genera insectorum*, 1776) and then followed his famous works with species descriptions: *Species insectorum* 1–2 (1781), *Mantissa insectorum* 1–2 (1787), *Entomologia systematica* 1–4 and *Supplementum* 1792–1799, and finally a book on each order: *Systema eleutheratorum* 1–2 (1801), *Systema rhyngotorum* (1803), *Systema piezatorum* (1804), *Systema antliatorum* (1805), and the never finished *Systema glossatorum* (1807). In accordance with his new fundamental characters he changed the Linnean names into Eleutherata (Coleoptera), Rhyngota (Hemiptera), Piezata (Hymenoptera), Antliata (Diptera), and Glossata (Lepidoptera). In addition he made other "classes": Ulonata (Orthoptera), Synistata (Neuroptera), Odonata, and, among noninsects, Mitosata (Myriopoda), Unogatoa (Arachnida), Polygonata, Kleistognatha, and Exochnata (all Crustacea). His enormous importance lies in his descriptions, his grouping into genera, and his systematizing based on the mouthparts of which he says in a Danish paper (1790) that it is reasonable that they mark the most natural genera, since they must be built up according to the nourishment of every insect and their whole biology is dependent on their nourishment.

A more influential opponent to Linné's system than Klein was Buffon (1707–1788) whose name actually was Georges Louis Leclerc. As with Bonnet his sight was weak so he could not make personal investigations. His thesis was that to put nature into a system like that of Linné was to impose constraint on it and draw attention from what was most essential: the whole world as an entity and everything as part of and connected with the whole. "In nature only individuals are found, genera, orders, classes exist only in our imagination" (Nordenskiöld 1921:144). To show this he published his *Histoire naturelle* from 1749 and onward, which was written in beautiful language and had an enormous influence. His view on the geographical distribution of animals was epoch-making. Linné never answered either Buffon or Klein.

Whether Linné's (and later Fabricius') system was natural or artificial was a question which absorbed their contemporaries as well as later scientists. Linné's system was natural, although in a rigid scheme, as it combined groups which have later proved related. Fabricius' system, which combined related groups even better, was called artificial by himself. Already four years after the "10th edition" the question was touched upon by the French physician E. L. Geoffroy (1727–1810). In his *Histoire abrégée des insectes qui se trouvent aux environs de Paris* (1762) (new edition "an VII", 1799) he extends Linné's system, i.e. by introducing the number of tarsal joints

FIGURE 5. P. A. Latreille (1762–1833). The founder of a "natural" system. After his *Genera Crustaceorum et Insectorum* 1(1806).

in Coleoptera as a taxonomic character. He knows his system is artificial, but "the more you can introduce of combinations and characters in an artificial method, the less you are remote from a natural order" (preface p. XVII); curiously enough his nomenclature is not binomial.

J. A. Scopoli (1723–1788), who was first a physician in Idria and later a professor in Pavia, expressed his doubts clearly a few years after the appearance of Fabricius' *Systema,* namely in his *Introductio ad historiam naturalem* (1777); not the mouthparts alone, the wings alone, the antennae alone should form the basis for the system but the whole structure ("structura totius") (p. 401). Also the French coleopterist A. G. Olivier (1756–1814) expressed his doubts concerning the Fabrician system, i.e. in his "Introduction" to the *Encyclopédie méthodique* Vol. 4 (1789), partly because it was so difficult to see the mouthparts, partly because Fabricius had separated groups which, to Olivier, had almost identical mouthparts and combined some with very divergent ones.

The scientist, however, who most of all should promote the insect system and who, in combining Linné's and Fabricius' systems, arrived at a more "natural" one based on real relationship, was Fabricius' good friend Latreille

(Figure 5). Although this is apparent in his book from 1796, *Précis des caractères générique des insectes,* he will be dealt with under the following period. He also introduced the concept "family", though Fabricius already had this in mind under his concept "order". Latreille's aim was to "rendre à l'entomologie cette aimable simplicité qu'elle a eue dans les temps de Linnaeus, de Geoffroy et des premières productions de Fabricius" (*Regne animal* IV p. VII).

Also the mild and modest Lamarck (1744–1829) gave a "natural" system in his *Système des animaux sans vertèbre* (1801) built on mouthparts, wings, metamorphosis, etc. He was also the first who removed crustaceans and arachnids (the latter he designated) from "insects". Though Fabricius did not think the time ripe for a natural system of "classes" (orders) and wrote that a system built on more than one set of characters would lead to chaos, it is quite improbable that since Fabricius visited the museum where Lamarck was chief and Latreille assistant every year for months, the three of them should not have discussed their systems in an unbiased way. All three were sympathetic and nonfanatical spirits. Latreille changed his system several times, as Burmeister (1832:673) puts it: "to please everybody" but, actually, no doubt, because he thought it thus improved. It should be borne in mind, however, that only Lamarck had a modern phylogenetical point of view in mind (see e.g. 1801:409).

Descriptions and Local Faunas

Artificial or natural, Linné's system had an enormously encouraging effect. A great many books appeared in the latter part of the eighteenth century which only dealt with new descriptions of insects, on the whole or of a special group (though often with titles such as *Histoire naturelle des insectes*), or which consisted of local faunas. The treatment of special groups will be found under the following period, but some selected faunas or descriptions will be mentioned here.

On the divide stands the German J. C. Schaeffer (1718–1790) who was a clergyman in Regensburg. In his books *Elementa Entomologica* (1766) and *Abhandlungen von Insecten* 1–3 (1764–1779) he praises Linné's systematization, but he does not use any Linnean names for the insects which he depicts in beautiful colored plates and describes in a Réaumurian way. He gives the first picture of a gynandromorphous moth.

Though J. H. Sulzer (1735–1813), a Swiss physician, calls his books *Die Kennzeichen der Insecten nach Anleitung des Ritters Carl Linnaeus* (1761) and *Abgekürzte Geschichte der Insecten nach dem Linneischen System* (1776), they are merely descriptions in a pre-Linnean manner.

Later in the century the iconographies were given with true systematic names of the species. Mention should be made of two works comprising all insect groups: *Genera insectorum Linnei et Fabricii iconibus illustrata*

(1789) by the Swiss J. J. Roemer (1761–1819) and especially *Illustratio Iconographica Insectorum, quae in Musaeis parisinis observavit et in lucem edidit Joh. Christ. Fabricius* (1799–1804) by A. J. Coquebert de Montbret (1753–1825); the latter can give clues to the Fabrician type specimens. Also important are the dissertations (1781 et seq) with new descriptions by the Swede C. P. Thunberg (1743–1828) who was a pupil of Linné and his successor in Uppsala. He described mostly Swedish insects, but also many from his visits to South Africa, Indonesia, and Japan.

Typical of what may be called collector's descriptions is the *Illustrations of natural history . . . of exotic insects* 1–3 (1770–1782) by the English jeweller D. Drury (1725–1803) whose collection was sold in 1805 for a fortune. The illuminated coppers are wonderful.

The local faunas began beautifully with Linné's *Fauna Svecica* (1746) (2nd ed., 1761) and multiplied after the 10th edition. Here might be mentioned *Insecta Musei Græcensis* (from Graz in Austria) (1761) by Nic. Poda (1723–1798); *Entomologia carniolica* (from Krain) (1763) by Scopoli; *Verzeichnis der ihm bekannten Schweizerischen Insecten* (1775) by J. C. Füessly (1743–1786); *Zoologiæ Danicæ prodromus* (1776) by O. Fr. Müller (1730–1784), who by 1764 had published his *Fauna insectorum Fridrichsdalina* and later several excellent biological observations; *Fauna Groenlandica* (1780) by Otho Fabricius (1744–1822), a missionary in Greenland in 1768–1774; *Enumeratio insectorum Austriae indigenorum* (1781) by F. von Schrank (1747–1835), who also wrote *Fauna Boica* (Bavaria) 1798–1804; *Entomologia parisiensis* (1785) by A. F. Fourcroy (1755–1809); *Forsøg til en Islandsk Naturhistorie* (1786) by N. Mohr (1742–1790); *Fauna Etrusca*, 1–2 (1790) and *Mantissa insectorum exhibens species nuper in Struria collectas* 1–2 (1792–1794) by P. Rossi (1738–1804); and *The natural history of British Insects* (1792–1813) by Edw. Donovan (1768–1837). Rossi was the world's first professor in Entomology, in Pisa 1801–1804.

Many of Linné's pupils went abroad and sent him insects from their new domiciles; insects were also sent from distant countries to other collectors such as Olivier, Fabricius, etc. Collecting expeditions were sent out, as for instance the Danish "Arabian Expedition" 1761–1767, which is famous for its tragic course. Entomologists also took part in expeditions, among them Banks and Solander in Cook's first expedition 1768–1771 to the tropics for instance. In this way great materials came to the museums and were used for descriptions. Thus Donovan described insects from his collection in *An epitome of the Insects of Asia, New-Holland, New Zealand, New Guinea, Otaheita, and other islands in the Indian, Southern, and Pacific Ocean* (1805), to mention just an example. From his own travels Palisot de Beauvois (1752–1820) described insects from Africa and America in *Insectes recueillis en Afrique et en Amérique . . . pendant les années 1781–1797* (1805–1821).

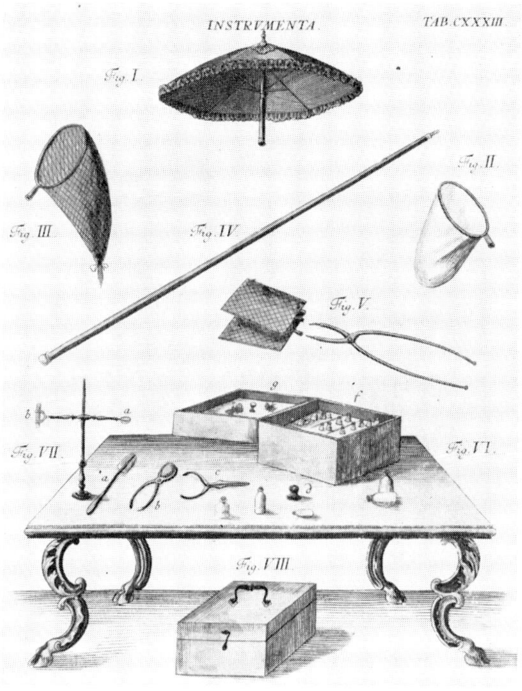

FIGURE 6. Collecting tools of the eighteenth century. After J. C. Schaeffer (1718–1790): *Elementa entomologia* (1766).

Most important were, however, the travels made by P. S. Pallas (1741–1811). Born in Berlin, where in 1760 he wrote his important dissertation on endoparasitic insects *De insectis viventibus intra viventia,* he became dissatisfied and went to St. Petersburg in 1767 and was sent by Katharina II; on collecting and exploring travels all through the great Russian Empire from whence he did not return to Germany till 1810. He intended to work up his enormous collection of animals himself; some descriptions appeared as *Icones insectorum praesertim Rossiae Sibiriaeque peculiarium* 1–4 1781–1806; his *Insecta Rossica* was never published. His views on animal relationships, on Linné's system (concerning "Vermes"), on fossils, on the influence of the climate, and on domestication of animals, were far ahead of his time.

PHILOSOPHY

In his preface to *Abhandlungen* Schäffer tells us that Bonnet called the eighteenth century "le siècle des observateurs". So it was, but it was also "le siècle des philosophes" and some of our entomologists were involved in this philosophy.

I have already mentioned that Réaumur argues against Descartes and Leibniz on the question of whether the insects have a soul or are pure machines, without, however, arriving at a result. Réaumur was an observer, not a philosopher; but his spiritual pupil Bonnet was, and became more and more so. In the preface to his *Insectologie* (1745) he puts forward the idea that all beings could be arranged in a row; his "idée d'une échelle des êtres naturels" was what the Germans later called "die Stufenleitertheorie". The insects are at the bottom of the animal ladder, beginning with coelenterates, the whole series culminating, via monkey and orangutan, in man. The more numerous our observations the smaller the invervals between the steps. Fabricius later (1781) changed the concept of a chain into that of a net in which all beings are woven together. It all goes back to Leibniz's "natura non facit saltus", no discontinuity in nature. Olivier says in his *Introduction* (1789:10): "best would it be if we could place all species one after the other in an uninterrupted chain, but then we should know all species in the whole world, and that far we will never get". Pallas (1766), on the other hand, was against the metaphor of a chain and gave instead that of a tree, the branches only coming together at their "roots", actually the first pedigree, but without genetical background.

Linné's philosophizing took another direction, he shaped his Systema which was a practical one and conquered all metaphors. It was built upon the concept of a species which did not change ("Hinc nullæ species novæ hodienum producuntur", 1735) and of which there exists as many as created by God ("in principio creatæ", 1736). But even to theologians trouble arose with the increasing number of known species: how could they have been in the ark of Noah? And so they turned to the possibility of hybridization as a source of new species. Linné himself had got new "species or varieties" out of crossing plants.

Fabricius in 1778 coined his well-known phrase "Species tot numeramus, quot diversae formes constantes existunt hodie" (we count so many species as constant forms existing today). This word "today" (hodie) against Linné's "ab principio" (from the beginning) has made people believe that Fabricius had a more modern species concept than Linné and many of them regard him as a "pre-Darwinist." But Linné uses himself the word "hodiene" (hodierne) when he says "no new species are produced today", so actually no controversy in thought is expressed. Nevertheless Fabricius, in general, expressed thoughts (German language works, 1778, 1782, 1804) which seem quite evolutionistic and which made a Danish scientist (Henriksen, 1922) suppose that he had influenced Lamarck. Helveg Jespersen (1946) is certainly right in discarding such suppositions. Still, as already said, Lamarck and Fabricius may have discussed the problems, and both were influenced by Buffon, in whose *Histoire Naturelle* many of Fabricius' thoughts, and even the same examples, are found.

Buffon cautiously tried to alter the view on the Animal Kingdom and the whole of nature from a static to a changeable concept, but it was Lamarck, who, though a child of the eighteenth century (Hodge, 1971), first directly expressed an evolutionary thought, though without using the word (*Systéme des animaux san vertèbre,* 1801:11–18, 408–11; and later in his *Philosophie zoologique,* 1809). His philosophy had not the success as that of Darwin and was met with general indifference. Lacordaire (1838:660) spends four lines on him, Burmeister (1832:673) only one, and it is characteristic when Kirby & Spence (IV 1826:462) mention him as "Lamarck, whose merits as a zoologist, except in one point [his views on Creation, see III:349] are of the highest order".

LITERATURE CITED

Aurivillius, Chr. 1909. *Carl von Linné als Entomolog*. Jena. 43 pp.

Baccetti, Baccio 1962. Pietro Rossi. Naturalista toscana del 1700. *Frustula Entomologica* 5, 3:1–30

Bodenheimer, F. S. 1928–1929. *Materialien zur Geschichte der Entomologie bis Linné*. I–II. Berlin

Burmeister, Hermann 1832. *Handbuch der Entomologie*. I. Berlin

Carus, J. Victor 1872. *Geschichte der Zoologie*. München

Eiselt, J. N. 1836. *Geschichte, Systematik und Literatur der Insektenkunde*. Leipzig

Fabricius, J. C. 1780. *Einige nähere Umstände aus dem Leben des Ritters von Linné*. Deutsches Museum. 1: 431–41, 2:39–48

Hagen, H. A. 1862–1863. *Bibliotheca Entomologica* I–II. Leipzig

Henriksen, Kai L. 1922–1937. Dansk Entomologis Historie. *Ent. Medd. XV, Kbh.*

Hodge, M. J. S. 1971. Lamarck's science of living bodies. *Brit. J. Hist. Sci.* 5:323–52

Jespersen, P. Helveg, 1948. Linnés artsbegreb. *Sv. Linné-sällsk, årsskr*. 31: 45–56

Kirby, Will., Spence, W. 1826. *An Introduction to Entomology*. IV:419–73

Kleemann, C. F. C. 1761. *Ausführliche und zuverlässige Nachricht von dem Leben, Schriften und Werken des verstorbenen Miniaturmahlers und scharfsichtigen Naturforschers August Johann Rösels von Rosenhof. Rösel von Rosenhof: Insecten-Belustigung* IV:1–48

Lacordaire, J. Th. 1838. *Introduction à l'entomologie*. Paris II:619–81

Linné, Carl von 1823. *Egenhändige anteckningar om sig sielf*. Stockholm

Löwegren, Yngve 1952. *Naturaliekabinett i Sverige under 1700-talet*. Uppsala and Stockholm

Marlatt, C. L. 1898. A brief historical survey of the science of entomology. *Proc. Entomol. Soc. Wash*. 4:83–120

Nordenskiöld, Erik 1921. *Biologiens historia*. II. Stockholm

Tuxen, S. L. 1967. The entomologist J. C. Fabricius. *Ann. Rev. Entomol.* 12:1–14

Usinger, Robert L. 1964. The role of Linnaeus in the advancement of Entomology. *Ann. Rev. Entomol.* 9: 1–16

SYSTEMATICS SPECIALIZES BETWEEN FABRICIUS AND DARWIN: 1800–1859

CARL H. LINDROTH
Zoological Institute, University of Lund
Sweden

"In der Beschränkung
zeigt sich erst der Meister"
(Goethe, "Natur und Kunst")

When, in the year 1788, the young J. W. Meigen first decided to devote himself to the study of Diptera, his starting point was of course the descriptions of Linné. His frustrations were movingly described in some autobiographical notes (*Stett. ent. Zeit.*, 7, 1846:70; translated from German): "Like most students I was at that time under the illusion that such a great scientist as Linné necessarily had known and described in his works all living beings existing in the whole world. It did not occur to my mind that actually an immense amount of knowledge was still wanting."

During his lifetime Linné described less than 3000 species of insects. We estimate now that this means between 0.2 and 0.3% of the actually existing fauna. In spite of this, a thorough knowledge of all Linnean species would be a respectable achievement and no one has ever tried it.

The capacity of the human brain will always be disputed. How many species or similar entities is the taxonomic specialist really able to know; not only the name, but the concept as such with at least the essence of its structural properties?

Experience tells us that this capacity is limited under all circumstances. Notably at advanced age it is easily observed that attempts to enter new fields by incorporating into our minds new taxonomic groups or faunas inevitably lead to corresponding loss of previous knowledge.

Years ago, when visiting one of America's most famous museums, I was introduced by a friend to a prominent mollusk specialist, Mr. X. He showed very little interest in the procedure and in an attempt to excuse him, my friend confessed afterwards: "You see, Mr. X is not fond of making new acquaintances, because every time he has to learn the name of a new person, he has to drop one of a mollusk." This is the proper attitude of a devoted taxonomist.

Specialization is thus the necessary sacrifice on the altar of too much knowledge. How it should best be executed, is a matter of personal inclination, or of sheer chance. It is interesting and still useful to observe how the

FIGURE 1. J. O. Westwood (1805–1893). The last entomological polyhistor. [From *Entomol. Mon. Mag.* (2) 4, 1893]

problems of specialization arose among the entomologists of early post-Linnean time, and how they were solved.

The Last Polyhistors

Of course, some few failed to realize that a choice was at all necessary and still stuck to the idea of being able to grasp it all, at least as taxonomists, and these were the latter day polyhistors, either men of extraordinary capacity or men with a deficiency in self-criticism and accuracy.

The most notorious among the latter was no doubt Francis Walker (1809–1874). A more devastating obituary than that devoted to him by an anonymous author (*Ent. m. Mag.* 11, 1874:140–141) has never been written: "More than twenty years too late for his scientific reputation, and after having done an amount of injury to entomology almost inconceivable in its immensity, Francis Walker has passed from among us."

Walker worked in practically all orders of insects and described ca. 20,000 new species (a world record!), mostly in the catalogues of the British Museum, which paid him 1 shilling for each new species and 1 £ for each new genus.[1] "He may be said to have become a mere describing machine" (loc cit).

[1] For other prominent species and subspecies producers, see section on Coleoptera, to which should be added Edward Meyrick (1854–1938) in England (ca. 15,000 Lepidoptera) and Charles P. Alexander in the US who is still active (more than 10,000 Diptera).

The last entomological polyhistor, in the good sense of the word, was J. O. Westwood (1805–1893), Hope Professor in Oxford (Figure 1). He worked in all insect orders and also on exotic material. His entomological production was enormous and he was an active expert in archaeology as well. His most important single work was *An Introduction to the Modern Classification of Insects,* (1839–1840); two volumes of more than 1000 pages, in which he followed what could be termed a combination of the principles of Latreille with those of MacLeay (see below). To the end he remained a faithful anti-Darwinist, which may have saved him a good deal of valuable time, as did, perhaps, his reported lack of a sense of humor (a good laugh at the futility of your own efforts may stop orderly thinking).

The obituary of Westwood, written by McLachlan (*Ent. m. Mag.* 29, 1893:49–51), gives the following summing-up: "There probably never has existed, and in the present state of science, there can never again exist, one who had so much general knowledge, both from personal investigation and a study of the works of others; one who was less of a specialist in the modern acceptation of the term."

In North America, where the scientific study of insects started more than half a century later than in Europe, the father of entomology was Thomas Say (1787–1834). In this branch of science he was the Linné of the New World, and a specialist in birds and mollusks at the same time. His descriptions were good for their time and often of a comparative nature, which makes an interpretation of his species easier. This is important because his collection was almost entirely destroyed. Say's achievement is the more remarkable since, in every respect, he was a complete autodidact. After Say, all North American entomologists have specialized, one way or the other.

Japan, in spite of its present high standards in entomology, was very late in entering the international arena of our science, not until the Meiji Era, from the 1860s, when the country opened cultural relations with the Occident. It is no wonder, therefore, that the last real polyhistor in entomology was a Japanese, S. Matsumura (1872–1960). His first paper was written in 1895; since 1926 they have mostly appeared in the series *Insecta Matsumurana.*

A still existing remnant of the entomological polyhistor is the perfect excursion leader, who produces a Latin name for each winged creature, hoping that it will immediately escape.

Gross Taxonomy

The leading entomologists of the eighteenth century, notably Linné and Fabricius, were species describers and system builders at the same time. As the number of known insects rapidly increased towards the end of the century, the need for a satisfactory general system became urgent, and this task was so heavy that it consumed most of the efforts of the scientists involved. Gross taxonomy thus developed into a kind of specialization.

The leading man, above any comparison, was the Frenchman P. A. Latreille (1762–1833). He had been active as a specialist on the specific level, mainly in Aculeate Hymenoptera, but soon felt the urge to work on higher categories and in 1796 published a first outline of a new insect system, under the title *Précis des caractères génériques des insectes disposés dans un ordre naturel.*

Both Linné and Fabricius had used a single set of characters as fundament of their insect systems: Linné in the wings, Fabricius in the mouthparts. One of the first entomologists to oppose such biased methods, and thus a forerunner of Latreille, was the Italian J. A. Scopoli (1723–1788). Initially, in the *Entomologia Carniolica* (1763), he followed Linné, but in his *Introductio ad Historiam Naturalem* (1777:401) expressed the following heretic views:

> Classes & Genera naturalia non sola instrumenta cibaria, non solae alae, nec solae antennae constituunt, sed structura totius, ac cuiusque vel minimi discriminis diligentissima observatio. (Natural classes and genera are based not only on the mouth-parts, the wings or the antennae, but on careful observation of the entire structure, even of the smallest differences.)

These words could serve as a motto for the work of Latreille.

The system of Latreille, just because it paid attention to morphological characters from all parts of the insect body and thus, to some extent, was a compromise between the Linnean and the Fabrician systems, was termed eclectic (e.g. in Kirby & Spence, IV, 1826:154). (Since the word has a slightly depreciatory tinge, the term integrated system would perhaps have been preferable.)

Latreille developed his system in several successive works and under considerable changes (1802–1805, 1806–1809, 1810), the last version, in which he finally restricted the name Insecta to the Hexapod Arthropods, being first published in 1825 and then in volumes 4–5 of Cuvier's *Règne Animal,* (2nd edition, 1829). Following Linné he divided the Insecta into Orders (by Latreille originally named Classes) but invented the important Family concept, and also the Tribus, as intermediate categories above the Genus. A. M. C. Duméril (1774–1860), in his *Considérations Générales sur la Classe des Insectes* (1823), presented a system very similar to that of Latreille and tried, in vain, to maintain his priority, to the Family concept, which gave rise to an animated discussion in the Académie des Sciences in Paris, as late as 1860 (Nussac, 1907:163).

It is also important to point out that Latreille conceived of and published the first formation of his principles (1796) before the two famous works on general zoological systematics of this period appeared in France: the *Leçons d'Anatomie Comparée,* (1–5, 1799–1805; by G. C. L. D. Cuvier, 1769–1832) and the *Système des Animaux sans Vertèbres* (1801; by J. B. P. A. deM. Lamarck, 1744–1829). However, the latter influenced him considerably when, in his later works, Latreille changed and improved his system of the arthropods.

Latreille met Fabricius several times and they apparently had ardent disputes about the general principles of insect taxonomy. Fabricius, in the preface of his *Supplementum Entomologiae Systematicae* (1798), warns against the use of basic characters other than those of the mouthparts: "Mixta semper chaos praebent" (if mixed, chaos is produced), and Kirby (*Introduction to Entomology,* IV:455) is certainly right in regarding this phrase as aimed at Latreille. On the other hand, Fabricius named Latreille "princeps Entomologiae" (Nussac, 1907:156). Schönherr (*Genera et Species Curculionidum,* 1, 1833:IX) used the still more profuse expression "entomologorum nostri temporis omnium princeps" (the foremost entomologist of our time).

The taxonomic principles of Latreille have greatly influenced the insect systems used by later entomologists. This was the case also for those who pronounced a rather negative attitude to much of Latreille's work. The most important were W. Kirby (1759–1850) in England and H. C. C. Burmeister (1807–1892) in Germany.

Kirby was a clergyman, who wrote several theological works but found sufficient time for his entomological hobby to become "the father of entomology in England" (Essig, 1931). He was a specialist mainly in Aculeate Hymenoptera, his first major work being *Monographia Apum Angliae* (1802), and in Coleoptera; he also discovered the order Strepsiptera. His outstanding achievement was *An Introduction to Entomology,* (I–IV, 1st ed., 1815–1826), written with W. Spence (1783–1860). This was a general entomology but in the last volume Kirby presented a history of entomology and his own system of the higher units, in which he was highly influenced by W. S. MacLeay (1792–1865). This man, in his *Horae Entomologicae, or Essays on the Annulose Animals* (1819–1821), had produced an original but quite artificial system of the entire animal kingdom, though with special stress on the arthropods, in which all higher categories were divided into groups of five, each forming a circle. The sympathy with which Kirby regarded this Quinary system (III, 1826:12 et seq; IV, 1826:465 et seq) was probably due to the fact that it suggested the orderly hand of the Creator.

Burmeister presented his insect system at the age of 22 in his doctoral thesis, *De Insectorum Systemate Naturali* (1829), and followed it later in his famous *Handbuch der Entomologie,* (I–V, 1832–1855). His attitude towards the Latreille system is pronouncedly negative and Burmeister himself introduced the different types of metamorphosis as the leading principle (I, 1832: 683 et seq). He estimated (loc cit:642) the total number of known insects as about 80,000 species and the original plan was apparently to describe them all. However, after the first volume of general entomology, in the four to follow he managed to treat only the Hemiptera, the smaller orders, and the Lamellicornia among the Coleoptera. His move to South America interrupted the work.

Not until the idea of evolution had been accepted by the taxonomists, could the gross taxonomy of insects be arranged according to a natural system entitled to the name. Therefore this development lies beyond the scope of this chapter. But it is difficult to avoid mentioning that the *Campodea* hypothesis, so important for the later discussions of insect phylogeny, was created in 1869 by the great Austrian F. M. Brauer.

Special Taxonomy

Systematic specialization is by no means a phenomenon of the post-Linnean period alone. Personal inclination may have also induced older naturalists to concentrate upon certain limited subjects. The Swede C. A. Clerck (1710–1765) became fascinated by spiders and published a book with colored plates of the Swedish species, *Svenska Spindlar,* in 1757. This work later received the unique honor of being declared valid in zoological nomenclature although it preceded Linné's *Systema Naturae,* (10th ed., 1758), by one year.

Otherwise it is obvious, before as well as during the nineteenth century, how often specialization was forced upon an author against his original plans for his work, simply because he eventually realized that the entire field was beyond his capacity. Optimistic titles, like *Fauna* or *Insecta Suecica* (G. Paykull, L. Gyllenhal), *Fauna Austriae* (C. Duftschmid), *Insecta Fennica* (C. R. Sahlberg), *Histoire Naturelle des Insectes* (J. T. Lacordaire, a.o.), *Deutschlands Insecten* (J. Sturm), *Entomologie, ou Histoire Naturelles des Insectes* (G. A. Olivier), *Entomologia Britannica* (T. Marsham), often ended up in a treatment of only one insect order, usually the Coleoptera.

The period of true specialization, which may be defined as the time after Fabricius, is here treated by orders of insects.

Coleoptera

The number of taxonomic specialists in Coleoptera is overwhelmingly rich since the beginning of the nineteenth century. The Lepidoptera may have been the favorite object of collectors, but in part from esthetic or other somewhat irrelevant reasons. Coleoptera, generally speaking, were easier to mount and describe, with less infraspecific variation, as a rule.

The first entomologists who devoted the greater part of their interest to Coleoptera appeared in the two last decades of the eighteenth century. J. F. W. Herbst (1743–1807), a German priest, in the period 1783–1806, followed up the *Natur-System Aller Bekannten In- und Ausländischen Insecten,* started by C. G. Jablonsky (see below), in 21 volumes, strictly following the system of Linné. Lepidoptera were included but the majority of species were beetles.

Next to follow were G.-A. Olivier (1756–1814) in France, G. W. F. Panzer (1755–1821) in Germany, and G. von Paykull (1757–1826) in Sweden.

The *Entomologie, ou Histoire Naturelle des Insectes* by Olivier appeared in 8 volumes, including 2 with plates, 1789–1808, but, in spite of its name, treated Coleoptera only. The first volumes preceded Latreille in time and the system used was a combination of those of Linné and Fabricius.

Panzer started his series *Faunae Insectorum Germanicae Initia* in 1793.

It consisted of separate issues, each containing 24 plates with accompanying description of each species, with beetles as dominating subjects. The work was continued by the same author until about 1810, then taken over by the lepidopterist G. A. W. Herrich-Schaeffer.

Paykull was an ardent collector of all kinds of zoological objects but his publications were almost exclusively in coleopterology. He revised (1789–1792) several families of Swedish beetles and summarized his knowledge in a *Fauna Suecica* (1798–1800). A geographic extension of his interests appeared in the *Monographia Histeroidum* (1811).

Two other Scandinavian coleopterists directly followed Paykull. The *Insecta Suecica,* (1–4, 1808–27) by L. Gyllenhal (1752–1840) meant great progress, both in accurate descriptions of species and on the generic level, where the Latreille system was followed.

C. R. Sahlberg (1779–1860) in his *Insecta Fennica* (1817–1839), published as a series of dissertations, closely followed Gyllenhal and was the first to give an account of the beetle fauna of Finland. Through the joint efforts of Paykull, Gyllenhal, and Sahlberg, the beetles of this part of Europe were probably better known at that time than in any comparable area.

The Sahlberg family provides the unique case of four generations devoted not only to entomology in general, but quite particularly to the study of Coleoptera: after C. R. Sahlberg, R. F. Sahlberg (1811–1874), J. R. Sahlberg (1845–1920), and U. Sahlberg (family name changed to Saalas; 1882–1969).

Around the turn of the century extensive works on local faunas, with a first description of a large number of species, were also produced in German-speaking countries. Most noteworthy, besides Herbst and Panzer (above), are J. C. W. Illiger (1775–1815) with *Verzeichniss der Käfer Preussens* (1798) and C. Duftschmid (data unknown) with *Fauna Austriae* (1805–1828), which later had a successor in the work by L. Redtenbacher (1814–1876), *Fauna Austriaca* (1849). Finally, J. Sturm (1775–1848), an engraver who made the plates for Panzer's above-mentioned work, also wrote a book of his own, *Deutschlands Insecten, Käfer* (1805–1857). In Switzerland, O. Heer (1809–1893), famous for his work on fossils (see below), published the important *Die Käfer der Schweiz* (1837).

The first British beetle fauna was by T. Marsham (?–1819), namely *Entomologia Britannica* (1802) which treated Coleoptera only and therefore was also published under the title *Coleoptera Britannica.* It contained descriptions of a large number of new species. Also J. F. Stephens (1792–1852), who worked on all insect orders, paid special attention to the beetles and, in 1839, produced *A manual of British Coleoptera.*

The great name in coleopterology during the early nineteenth century was P. F. M. A. Dejean (1780–1845), with a career quite exceptional among entomologists; a general who rose to the title "pair de France" and served as Napoleon's first aide de camp in the battle of Waterloo.

Dejean (Figure 2) was an ardent collector and eventually, in part by

FIG. 2. P. F. M. A. Dejean (1780–1845). The first great coleopterist.
[From *Ann. Soc. Entomol. Fr.* (2) 3, 1845]

purchase and exchange, brought together the largest private insect collection of his time. He lost no opportunity to fetch rare specimens, which is amply evident from what was told about his somewhat unmilitary behavior during raging war (Boisduval, *Ann. Soc. ent. Fr.* (2)3, 1845:502–503) (translated):

> Before the battle of Alcanizas, which Dejean won after a long-contested fight, taking a great number of prisoners, when the enemy had just appeared and he was prepared to give the signal of attack, Dejean, at the border of a brook caught sight of a *Cebrio ustulatus (nomen nudum)* on a flower. He immediately dismounted, pinned the insect, applied it to the inside of his helmet which, for this purpose, was always supplied with pieces of cork, and started the battle. After this, Dejean's helmet was terribly maltreated from cartouche fire; but, fortunately, he refound his precious *Cebrio* intact on its piece of cork.

Dejean soon specialized entirely on Coleoptera. In 1802, he started a series of catalogues of the species preserved in his collection, the final edition of which (1837) contained more than 22,000 names. These lists were generally used by other entomologists as models for arrangement of collections; but it is important to state that Dejean's names of new species in the catalogues are *nomina nuda*.

Together with Latreille (1822–1824), and later, in a new edition (1829–

1838) with the lepidopterist Boisduval, who was Dejean's private curator, and C. Aubé, Dejean commenced an *Iconographie* of the Coleoptera of Europe, but this remained fragmentary.

The great enterprise was the *Spécies Général des Coléoptères* (1825–1838). Dejean's original plan was to describe, in this work, the world fauna of Coleoptera, but this was soon changed to the more limited aim of the species of his own collection. Twenty volumes were foreseen but, as it turned out, only six appeared; the last of which, containing the *Hydrocanthares,* was written by C. Aubé (1802–1869). Dejean was the author of the five other volumes, entirely devoted to the Carabidae (s.l.); they contained close to 3000 pages in all, a masterpiece in descriptive entomology.

Dejean was a persistent opponent of the priority principle in nomenclature and expressed his standpoint very clearly in the preface to the first volume of the *Spécies* (translated): "I have made it a rule always to preserve the name most generally used, and not the oldest one; because it seems to me that general usage should always be followed and that it is harmful to change what has already been established". Dejean acted accordingly and often introduced in litteris names, mostly given by himself in his catalogues, to replace those already published by other authors. A commanding general was not likely to allow simple privates to act up!

The descriptive branch of coleopterology flourished greatly in France in the period after Dejean. The most prominent representative was J. T. Lacordaire (1801–1870), who worked also on exotic material collected during his travels in South America. His main work, *Histoire Naturelle des Insectes. Genera des Coléoptères* (1854–1876), was continued by F. Chapuis (1824–1879) after his death. In his diagnoses of tribes and genera in this book "he has excelled all other entomological writers" (H. W. Bates, *Proc. ent. Soc. London,* 1870).

In the scope of production, though not in quality, Lacordaire was surpassed by two contemporary French coleopterists, E. Mulsant (1797–1880) and L. M. H. Fairmaire (1820–1906). The former, together with C. Rey (1817–1895), wrote the famous work *Histoire Naturelle des Coléoptères de France* (1839–1885), a giant in 38 parts, amounting to more than 12,000 pages!

Fairmaire had an enormous production, mainly small papers containing more or less isolated descriptions of new species from all over the world. Thus he was a worthy predecessor of the man, likewise French, who set the world record in Coleoptera, M. Pic (1866–1957), with more than 20,000 new species or varieties of Coleoptera; the Austrian Edmund Reitter (1845–1920) with ca 10,000 and the American T. L. Casey (1857–1925) with a modest 9,600 new names follows next.

The Russian entomologists were somewhat late in starting. They had an immense area to explore within the boundaries of their own country, to which most of them devoted their entire efforts. This made them more or less isolated from their colleagues in western Europe. On the other hand, they were remarkably early in specializing and Coleoptera became a favorite group.

The first prominent Russian entomologist, though born in Germany, was G. Fischer von Waldheim (1771–1853). He published on several insect orders but his main work was *Entomographia Imperii Russici,* with more than 1500 pages, printed 1820–1851, (volumes 1–3 mainly consisted of Coleoptera, volume 4 treated Orthoptera, and volume 5, on Lepidoptera, was written together with E. v. Eversmann).

F. A. von Gebler (1782–1850) specialized almost exclusively on Coleoptera, notably belonging to the Siberian fauna.

C. G. Mannerheim (1804–1854) reached the high position of Governor of Finland, which at that time belonged to Russia. He was exclusively a coleopterist but not restricted geographically, as his work included exotic insects, though his main field was the fauna of the Russian Empire including the possessions in North America. These possessions had been visited frequently by Russian collectors who went as far south as Fort Ross in northern California, but Mannerheim was the one who worked up the collections of Coleoptera (*Bull. Soc. Nat. Moscou,* 16, 1843; 19, 1846; 26, 1853) and his name stands as author for a considerable number of North American species.

An example of taxonomic but not geographic specialization was M. de Chaudoir (1816–1881) who, in a long series of important revisions, treated the world fauna of carabid beetles. He followed the example of Dejean in clearness and precision of his descriptions and made valuable contributions to the definition of higher taxonomic units within the family. In 1859, he purchased the Dejean collection and so had an immense working material of Carabidae at his disposal.

The same year that Chaudoir published his first entomological paper (1835), an entirely opposite type of personality made his entrance on the public entomological scene in Russia, V. von Motschulsky (1810–1871). A quite amusing though perhaps slightly exaggerated description of his career has been written by W. Horn (*Ent. Mitteil.* 16, 1927:1–6, 93–98). Motschulsky, who restricted himself almost entirely to the Coleoptera, made a good impression upon foreign entomologists by his earlier works, such as *Die Coleopterologischen Verhältnisse und die Käfer Russlands* (1845) and *Insectes de la Sibérie* (1845). In spite of his limitation to one single order of insects, his interests seemed versatile enough, covering natural history, zoogeography, bibliography, and so forth. But his restless mind was soon apparent. Not only was he perpetually on the move through most of Europe and Asian Russia, as well as North America, but his entomological work was suffering under a similar lack of stability. The number of genera and species created by Motschulsky is impressive but the superficial and careless way in which they were described often prevented a subsequent interpretation of his names. Fortunately, his collection is preserved in almost complete condition at the Moscow University, where lectotypes may be designated.

Several European coleopterists made further specialization in the period considered here. We have already mentioned Dejean and Chaudoir, who restricted themselves to the carabid family; as did J. F. Dawson (1802–1870) in England and J. A. A. H. Putzeys (1809–1882) in Belgium.

L. A. A. Chevrolat (1799–1884), in France, worked on the world fauna and was a general coleopterist from the beginning, but concentrated more and more on the curculionid family, both in his publications and as the proprietor of a vast collection. The Swede C. J. Schönherr (1772–1848) was in the same domain with his famous *Genera and Species Curculionidum* (1833–1845), in which most of the descriptions of new species were written by L. Gyllenhal and, particularly, by C. H. Boheman (1796–1868), also known for his *Monographia Cassididarum* (1850–1862). The German E. F. Germar (1786–1853), with an extensive production in Coleoptera and Hemiptera (see below), had a particular preference for the Curculionidae and Elateridae.

The staphylinid family necessarily had its attraction for able taxonomists keen on intricate problems. The first specialist was J. L. C. Gravenhorst with his *Coleoptera Microptera Brunsvicensia* (1802) which was followed by *Monographia Coleopterorum Micropterorum* in 1806. Like the Swede C. G. Thomson, he then turned his interest to the parasitic Hymenoptera; we shall find the names of both under that heading.

The real genius in staphylinid taxonomy, as in that of many other groups of beetles, was the German W. F. Erichson (1809–1849). He may be regarded as a Mozart or Schubert of entomology, dead at the age of forty. Though active in many orders of insects, from many parts of the world, Erichson's main interest was in the Coleoptera. His first extensive work was *Die Käfer der Mark Brandenburg* (1837–1839) which, without being completed, was substituted by the *Naturgeschichte der Insekten Deutschlands*. In spite of its title, it treated Coleoptera only and Erichson himself wrote the third part, the first to appear, *Clavicornia* & c. (1845–1848); his early death prevented him from treating the group that was his real specialty, the Staphylinidae, which was later written by G. Kraatz. Other authors continued the *Naturgeschichte* but it was never finished.

Erichson has been named perhaps the greatest entomologist of his time ("grösster Insektenforscher seiner Zeit"; Horn & Kahle, 1935–1937). This judgment is no doubt based on his masterpiece, *Genera et Species Staphylinorum* (1839–1840), a work of almost 1000 pages entirely written in Latin. This was an attempt to treat all known species of the family, including many described as new; it was certainly not a compilation but the construction of an entirely new system based on previously neglected characters. Fundamentally, this system, termed by Ganglbauer (*Die Käfer von Mitteleuropa*, II, 1895: 11), as one of genius ("genial entworfen"), has been followed by subsequent specialists: G. Kraatz (1831–1909) in part 2 of *Naturgeschichte der Insekten Deutschlands* (1856–1858); the Frenchman P. N. J. Jacquelin du Val (1828–1862) in volume 2 of *Genera des Coléoptères d'Europe* (1857–1859); the Swede C. G. Thomson in parts 2 and 9 of *Skandinaviens Coleop-*

tera (1860, 1867); and in North America by J. L. Leconte (below) in his *Classification of the Coleoptera of North America* (1861–1862).

Of these, G. Kraatz, who had an enormous production, was the leading German coleopterist after Erichson. A lasting achievement of his, though not realized until much later, was the foundation in 1904 of a German Entomological Museum, still existing under the name of "Deutsches Entomologisches Institut".

E. C. A. Candèze in Belgium (1827–1898) specialized in the world fauna of the Elateridae; C. J. F. Gillmeister (no data available) in Germany and A. Matthews (1815–1897) in England in the smallest of all beetles, the Ptiliidae, from all parts of the world.

The study of North American beetles in the first half of the nineteenth century was mainly an occupation of foreign entomologists, such as W. Kirby, several of the old Russians, and, for the Carabidae, Dejean. Thomas Say, mentioned earlier, was the first earnest indigenous taxonomist, but he was by no means restricted to Coleoptera.

This, however, was the case with J. L. Leconte (1825–1883), who published his first paper in 1844 (Figure 3). Though a medical doctor by profession, he managed to cover the entire vast field of North American Coleoptera. He described about 4700 new species, very few of which have not remained valid, and revised those previously named. His descriptions and keys are clear and concentrated, often still sufficient for species recognition.

Leconte's accomplishments were equally remarkable in the gross taxonomy of Coleoptera. In 1861–1862 he published a *Classification of the Coleoptera of North America,* of which an enlarged and revised edition, with G. H. Horn as co-author, appeared in 1883. Many of the inventions of this system were accepted by European coleopterists and so included in an international classification of this insect order. Leconte, though restricted in his work to the study of beetles, must be regarded as the greatest North American taxonomist in entomology.

G. H. Horn (1840–1897) was an intimate friend of Leconte's as gracefully described by himself in an obituary (*Science,* 2, 1883:784–86), and the cooperation in science between these two men was of a very unusual character. Horn did not possess the taxonomic genius of Leconte but after the death of his friend pursued their joint work in a most useful manner.

Subsequent North American coleopterists of importance belong to the Darwinian period.

Diptera

The father of dipterology was J. W. Meigen (1763–1845) (Figure 4). He made his living under narrow circumstances as a teacher and part-time communal official, during his last half century in the small German town of Stollberg near Aachen; temporary use was also made of his extraordinary capacity for drawing. Entomology was thus a hobby of his.

Beginning as a general collector, who also published papers on Lepidop-

FIG. 3. J. L. Leconte (1825–1883). The foremost coleopterist of North America. (Courtesy of the Museum of Comparative Zoology, Harvard)

tera, Meigen turned to dipterology at the age of 25. The vast number of valid European species carrying his name testify to his great achievements in descriptive taxonomy. The most important of his works was *Systematische Beschreibung der bekannten Europäischen Zweiflügligen Insecten,* (1–7; 1818–1838), of almost 3000 pages. Meigen's life and thoughts were vividly described in *Stett. Entomol. Z.* (7; 1846).

Meigen's work in gross taxonomy of Diptera was equally important. The system constructed by him was far more natural than that of his Swedish contemporary C. F. Fallén, mainly because he refused to base it on a single group of characters, as did Fabricius on the mouth-parts. Meigen thus followed an eclectic method, as Latreille, but apparently independently. Fabricius visited Meigen in 1802 but could not persuade him to abandon his heretic views.

Two dipterists following Meigen, also Germans, were more specialized, C. R. W. Wiedemann (1770–1840) mostly on exotic forms and J. Winnertz (1800–1896) on European Nematocera.

The development of the nineteenth century dipterology after Meigen has been described in C. R. Osten Sacken's autobiographical book (1903). Due to his mixed personal feelings towards Loew, his opinions are perhaps some-

FIG. 4. J. W. Meigen (1763–1845). The founder of modern dipterology. (Courtesy of Deutshes Entomologisches Institut, Eberswalde, DDR)

what biased but the book makes most interesting reading, and for the most part, is no doubt the expression of fair judgment.

The dominating dipterist from the 1840s and for three decades to come was the German, H. Loew (1807–1879). While Meigen had restricted himself to the European fauna, Loew worked on a worldwide basis, although with a preference for the Palaearctis and North America. His vast capacity is evident from the fact that he described, in an excellent way, more than 4000 species, Nematocera as well as Brachycera. He belonged to the pronouncedly descriptive type of taxonomists, had little interest in bionomics, and was less successful in his treatment of higher categories. "Loew's talent lay in the direction of the particular, the *minute*, and not in that of *generalization*" (Osten Sacken, loc cit:50).

Valuable contributions to both gross and species taxonomy were made by the Frenchmen J. Macquart (?–1855), who also worked on exotic material, and J. B. Robineau-Desvoidy (1799–1857), the Dane R. C. Staeger (1800–1875), the Italian C. Rondani (1808–1879), with his *Dipterologiae Italicae Prodromus* (1856–1862), and above all, the Briton A. H. Haliday (1806–1870), who, according to Osten Sacken (loc cit: 51 et seq), was far superior to Loew on the higher taxonomic levels. Haliday was equally prominent as a specialist in Hymenoptera Parasitica and the judgment of Westwood in his obituary (*Trans. ent. Soc. London,* 1870:XLVII) applies to both fields of activity: "He was our *first* entomologist. His ideas of classification and tabulation were so logical, his latinity so classical, and his knowledge of whatever he touched so masterly, that I fear we shall be long before we look upon his like again."

Prominent specialists in species recognition were J. R. Schiner (1813–1873), with his two faunas on the Diptera of Austria, and the Swede J. W. Zetterstedt (1785–1874). The latter started as a general entomologist, as amply evident from his *Insecta Lapponica* (1838–1840), but then devoted himself entirely to Diptera. His monumental *Diptera Scandinaviae* (1842–1860), in 14 volumes, amounted to more than 6600 pages and included an immense number of new species. The classification used by Zetterstedt, mainly following the *Diptera Sueciae* (1814–1825) of his teacher C. F. Fallén (1764–1830), was rather old fashioned and artificial, as admitted by Zetterstedt himself (*Dipt. Scand.,* I:VIII).

The outstanding monograph of the European Oestridae (1863) by F. M. Brauer, treated as a neuropterist below, should also be mentioned. He was the first to introduce the division of Diptera into Orthorrhapha and Cyclorrhapha.

Finally, Osten Sacken (1828–1906), a Russian diplomat who spent part of his life in the United States, was especially interested in bionomics and gross taxonomy of Diptera. His system eventually became strongly influenced by Darwinian principles and therefore reaches beyond the scope of this chapter.

Lepidoptera

It is rather natural that the early publications devoted entirely or mainly to Lepidoptera consisted of illustrations and thus were pieces of art rather than of science. Such iconographies appeared in pre-Linnean time.

The first work built on Linnean binomial nomenclature was the *Icones Insectorum rariorum* by the above-mentioned Swede, C. A. Clerck, with hand-colored reproductions of Lepidoptera only. Volume one, of two, was printed in 1759; that is, one year after the tenth edition of *Systema Naturae,* and may be regarded as an illustrated appendix of this, facilitating the interpretation of certain Linnean names.

It is striking that the specialists in Lepidoptera devoted themselves to one order of insects earlier than other entomologists.

The first book on the British fauna was *The Aurelian or Natural History of English Insects, namely Moths and Butterflies* by M. Harris (1730–1788), with many colored plates; its first edition was in 1766. Harris was a pioneer in using the wing nervature for systematic purposes. A more modern revision did not appear in Britain until A. H. Haworth's (1767–1833) *Lepidoptera Britannica* (1–4; 1803–ca. 1828).

On the continent, a strictly local fauna was the first work to contain a considerable number of new species, *Systematisches Verzeichniss der Schmetterlinge der Wiener Gegend* (1776) by I. Schiffermüller (1727–1809), in cooperation with M. Denis. The following year, the German E. J. C. Esper (1742–1810) published the first of the five volumes of his series, *Die Europäischen Schmetterlinge,* finished in 1794. This was a great achievement, a combination of iconography and descriptive text. It was soon followed by two other voluminous works from the same country: *Natursystem aller Bekannten In- und Ausländischen Insecten,* of which two volumes on Lepidoptera (1783–1784) were by C. G. Jablonsky (1756–1787), the following nine by F. W. Herbst (see under Coleoptera); then, in 1788–1794, *Naturgeschichte der Europäischen Schmetterlinge,* by M. B. Borkhausen (1760–1806), a huge work in five volumes with more than 2000 pages of descriptive text.

The most important life work in Lepidoptera started before the end of the eighteenth century was that of J. Hübner (1761–1826), designer at a cotton factory in Augsburg. His *Sammlung Europäischer Schmetterlinge* was commenced in 1796 and after his death continued by Herrich-Schaeffer. It consisted of hand-colored plates with accompanying text, which, unfortunately, was often incomplete though sufficient for making "Hb." one of the most frequent author designations among European Lepidoptera. Hübner later (1806–1824) added a *Sammlung Exotischer Schmetterlinge,* volumes 1–2 (continued by C. Geyer), and several smaller works. He has been named "the first great world lepidopterist" (Essig, 1931).

A contemporary of Hübner was F. Ochsenheimer (1767–1822), who with H. Edwards in the US shared the profession, somewhat unusual in the entomological context, of actor ("Hofschauspieler" in Vienna). *Die Schmetterlinge von Europa* (1807–1835) appeared in ten volumes, of which the last six were written by F. Treitschke. The work has importance also for the taxonomy on the generic level.

One of the really great names in lepidopterology was the German G. A. W. Herrich-Schaeffer (1799–1874). His profession as a medical doctor did not prevent him from fulfilling an entomological life work of almost unbelievable dimensions. In his twenties he took over the interrupted publication of Panzer's *Faunae Insectorum Germaniae Initia,* and for this purpose learned the art of engraving. Together with the insect painter C. Geyer, he also continued Hübner's great work, *Sammlung Europäischer Schmetterlinge.* As a supplement of this, his own most important contribution, *Systematische Bearbeitung der Schmetterlinge von Europa,* in six volumes (1843–1856), appeared, also in cooperation with C. Geyer as illustrator. This meant not only great progress in the description of species and genera, but also the presentation of a new system for the Lepidoptera, including the Micros, mainly based on the wing nervature, at which an attempt had earlier been made by M. Harris in England (see above). In his later years, Herrich-Schaeffer published extensively also on exotic material.

The first real French fauna was written by P. A. J. Duponchel (1774–1846), in cooperation with J. B. Godart, namely *Histoire Naturelle des Lépidoptères, ou Papillons de la France* (1–11; 1821–1838), with descriptions of many new species. The leading French lepidopterist of this period was J. B. A. D. de Boisduval (1799–1879). A medical doctor by profession, he was an amateur entomologist of great general knowledge, in his youth a private curator of the collection of Dejean, with whom he cooperated in the above mentioned *Iconographie* of Coleoptera, but also a prominent botanist who produced an extensive book on horticultural entomology. His first interest however, was in the Lepidoptera of the world and he published a long series of first class revisions and monographs. Following Dejean's example in the Coleoptera, he started a *Spécies Général des Lépidoptères,* supposedly to treat all known species, but only two volumes (1836, 1874) were written by himself, while A. Guenée was the author of the noctuid sections. According to the expert judgment of C. Oberthür (*Ann. Soc. ent. Fr.* (5)10, 1880:129–38): "Boisduval had a more prominent position in our science than any other lepidopterist had won in the opinion of his contemporaries" (translated).

The leading German lepidopterist in the second half of the century was O. Staudinger (1830–1900), who had an enormous production of usually small descriptive papers (from 1854), but also published several useful catalogues and epitomes, the first being *Catalog der Lepidopteren Europas* (with M. Wocke, 1861).

In Sweden, H. D. J. Wallengren (1823–1894) revised the Scandinavian fauna, both Rhopalocera (1853) and Heterocera (1863–1885).

The first Russian, though German-born, who mainly specialized in Lepidoptera was E. von Eversmann (1794–1860). *Les Noctuélites de la Russie* (1855–1859) was a valuable contribution with description of many new species. He was also the co-author of volume 5 in Fischer's *Entomographia Imperii Russici* (see above, Coleoptera). E. Ménétries (1802–1861), who also published on exotics, contributed particularly to the knowledge of the Siberian fauna.

In North America, the Lepidoptera, like other insect orders, were first studied by foreigners. Of considerable historical interest is *The Natural History of the Rarer Lepidoptera of Georgia* (1777), with more than 100 colored plates, which was written by two Englishmen, J. Abbott (ca. 1760–ca. 1810) and J. E. Smith (1759–1828). Then followed several works of Boisduval, one in cooperation with J. E. Leconte (father of J. L. Leconte): *Histoire Générale et Iconographie des Lépidoptères et des Chenilles de l'Amerique Septentrionale* (1829–1834).

There was little achieved by native American lepidopterists, though a *Synopsis of the Lepidoptera of North America* by J. G. Morris (1803–1895) appeared in 1862, and, from 1868 onwards, *The Butterflies of North America* in three volumes, by W. H. Edwards (1822–1909). The revisions by A. S. Packard (1839–1905) did not commence until the 1870s.

Further specialization on a certain part of the insect order was not common among the early lepidopterists. But three men should be mentioned, two Germans and one Englishman, who laid the foundation for understanding the most delicate and, in part, the most difficult of all insects, the so-called Microlepidoptera.

The first was P. C. Zeller (1808–1883), a teacher by profession. From 1838 onwards be published a long series of revisions of genera and higher groups, mostly among the Micros. His interests were not limited to pure taxonomy but very much directed towards observations of the biology, including behavior, of his favorites. Also geographically he eventually extended his work to cover exotic faunas, in particular that of North America. His descriptions were so excellent and important for other specialists that Stainton in Britain was said to have learned German for the sole purpose of being able to read them properly. A most amusing compliment, as good as any, bestowed upon the ability of Zeller, was the statement of Osten Sacken (1903:138): "In Zeller, the lepidopterist, the ideal of a perfect dipterologist was lost". He was pronounced the greatest lepidopterist of his time, even of the entire nineteenth century (H. Frey, *Stett. ent. Zeit.*, 44, 1883: 406 et seq).

H. von Heinemann (1812–1871) was a customs officer in Braunschweig. His literary production was not extensive but included one big work, *Die Schmetterlinge Deutschlands und der Schweiz* (1859–1877), of which the second volume, treating the Microlepidoptera, was especially important.

H. T. Stainton (1822–1892) published his first paper in 1845 and then developed an admirable production on Lepidoptera, mainly Micros. Most of his work was restricted to the British fauna, the main works being *Insecta Britannica. Lepidoptera: Tineina* (1854), and *Manual of British Butterflies and Moths* (1–2; 1857–1859). Together with Zeller, and in part with his compatriot J. W. Douglas and the Swiss H. Frey, he started the great enterprise, *The Natural History of the Tineina,* which appeared in 13 volumes, containing about 2000 pages, 1855–1873. It was written in four languages: English, French, German, and Latin!

Stainton was extremely meticulous and self-critical. "He goes so far as to recommend that no species should be described upon less than twenty to thirty specimens, and advocated an amount of self-denial in such matters, which I imagine hardly any of us are prepared to put in practice" (F. D. Godman, *Trans. ent. Soc. London,* 1892:XLVIII). This is an admirable, truly biological approach to taxonomy.

It is remarkable that the gross taxonomy of Lepidoptera was so long neglected. The division into butterflies (Rhopalocera) and moths (Heterocera), or, mostly in German-speaking countries, into Macro- and Microlepidoptera, apparently was regarded as sufficient. The reason may partly have been that the dense vestiture of scales and hairs in Lepidoptera conceals so many important structural characters, not visible except by partly destroying the precious specimens. A sound classification of the Lepidoptera, based on phylogenetic relationships, was not attempted until in the days of A. S. Packard (1839–1905), T. A. Chapman (1842–1921), J. H. C. Comstock (1849–1931), H. G. Dyar (1866–1929), and J. W. Tutt (1858–1911); that is, not before the 1890s, well after the end of our period.

Hymenoptera

In temperate regions this is the largest of all insect orders and many groups are extremely difficult from a taxonomic point of view. The three main conventional suborders, Symphyta (Phytophaga), Parasitica, and Aculeata, or whatever names are used, to a great extent offer taxonomic and biological problems of their own, which explains why many specialists restricted themselves to one of them only.

General hymenopterists are exceptions. The oldest was the Swiss L. Jurine (1751–1819), with only few entomological papers, one of which, however, *Nouvelle Méthode de Classer les Hyménoptères et les Diptères* (1807), was important for the gross taxonomy of Hymenoptera. As a curiosity, Jurine used the family concept as subordinate to the genus.

C. G. Thomson, treated below, was active in all groups of Hymenoptera; so were the early hymenopterists of North America, E. T. Cresson (1838–1926) ("the greatest American general hymenopterist", Essig, 1931), and L. Provancher (1820–1892) in Canada, who also published on most other insect orders, mostly in the *Naturaliste Canadien.*

The Hymenoptera Phytophaga (Symphyta) were revised early by the famous director of the Berlin Museum, J. C. F. Klug (1775–1856), with *Die Blattwespen nach ihren Gattungen und Arten Zusammengestellt* (1808–1818) as his most comprehensive work. He was, however, a most all-embracing entomologist, the advantage of which, even to a specialist, was nicely expressed in the obituary by Gerstaecker (*Stett. ent. Zeit.*, 17, 1856: 225; translated): "Klug provides the best proof of the truth, only too little admitted by many contemporary scientists, that the activities within a *special* branch only then may be of real importance if supported by broad *general* knowledge."

An important contribution to the Symphyta was the *Monographia Tenthredinetarum* (1823) by the Frenchman A. L. M. Lepeletier (1770–1845). Later, he, together with A. Brullé, produced a huge general work of more than 2500 pages, *Histoire Naturelle des Insectes. Hyménoptères* (1–4, 1836–1846), on the entire order.

The German T. Hartig (1805–1880) was a professional forest entomologist but he made taxonomic studies on aphids, as well as on cynipid and phytophagous Hymenoptera; his main achievement in the latter field being *Die Familien der Blattwespen und Holzwespen* (1837).

The Aculeata, notably the social groups of course, have always attracted attention, but usually from other than taxonomic points of view. Actually, their systematics are often very complicated and difficult to grasp, for instance in many genera of ants.

The great Latreille had a particular weakness for the Aculeata and his first two papers (1792) were on the mutillids. The ants were, however, his favorites in his youth and the most important result of these studies was *Histoire naturelle des Fourmis* (1802).

The Austrian G. L. Mayr (1830–1908) was active also on other Hymenoptera, notably the Cynipidae and wrote a book entitled *Die mitteleuropäischen Eichengallen* (1870–1871); but his main preoccupation was with the ants. Besides many smaller papers, he produced two important works, *Formicina Austriaca* (1855) and *Die Europäischen Formiciden* (1861), and also revised extensive exotic collections. He was the first to tackle the difficult problems of identifying ants in Baltic amber, in his *Die Ameisen im Baltischen Bernstein* (1868).

More modest contributions to the taxonomy of European ants were made by W. Nylander (1822–1899) in Finland between 1846 and 1856.

As mentioned above, W. Kirby monographed the British Apidae in 1802. The Swede A. G. Dahlbom (1806–1859) started a comprehensive work under the title *Hymenoptera Europaea Praecipue Borealia,* but his early death prevented him from completing more than two volumes (1843–1854), treating the Sphecidae, Pompilidae, and Chrysididae. Earlier he had dealt with Scandinavian bumble bees (1832) and other Apidae.

H. L. F. de Saussure (1829–1905) was a leading Swiss orthopterist but he started with Hymenoptera. He was an all-around naturalist with personal experience from field work in the tropics and he was also active in ethnography and archaeology. A prominent work of his young days was a revision of social wasps, *Monographie des Guêpes Sociales* (1853–1858).

The first Russian scientist to specialize in the Aculeata was F. Morawitz (1827–1896), who concentrated upon the fauna of his own country and had his first paper printed in 1864.

As usual, the start was later in North America, but the polyhistor A. S. Packard (1839–1905) published *The Humble Bees of New England* as early as 1866; later he also published many papers on other groups of Hymenoptera.

The first entomologist to devote his efforts entirely to Hymenoptera Parasitica was the German C. G. Nees von Esenbeck (1776–1858), professor of botany. Starting in 1811, he produced six publications on the Ichneumonidae, two of which were extensive monographs. He worked at first in cooperation with his famous compatriot J. L. C. Gravenhorst (1777–1857). This, as mentioned above, created for himself a name, through the investigation of staphylinid beetles, but then, from 1814, he turned almost exclusively to the Ichneumonidae. His magnum opus, *Ichneumonologia Europaea* (1829), appeared in three parts, almost 3000 pages altogether, and is fundamental for all subsequent research in this family of wasps.

In Ireland, the great Haliday, whom we have met as a dipterist, developed into a first class specialist also in parasitic Hymenoptera, with emphasis on previously neglected groups, such as braconids, chalcids, and proctotrupids.

His most comprehensive work was *An Essay on the Classification of the Parasitic Hymenoptera of Britain* (1833–1838). A contemporary of his was C. Wesmael (1798–1872), in Belgium, whose *Monographie des Braconides de Belgique* (1835–1838) was a most important work. J. T. C. Ratzeburg (1801–1871) in Germany was a forest entomologist; his most extensive publication on Parasitic Hymenoptera had an applied approach, as evident from its title, *Die Icheumonen der Forstinsecten* (1–3; 1844–1852).

From the middle of the nineteenth century, two Swedes were the dominating students of parasitic Hymenoptera. A. E. Holmgren (1829–1888), an applied entomologist by profession, had a strong interest in purely taxonomic work as well, where he almost exclusively concentrated upon the ichneumonid family. From 1855 onwards he carried out many important revisions, the two most comprehensive published under the titles *Monographia Tryphonidum Sueciae* (1855) and *Ichneumonologia Suecica* (1–3; 1864–1889).

The leading man in hymenopterology, covering all groups, was C. G. Thomson (1824–1899) (Figure 5). At the same time he commanded the entire Scandinavian fauna of Coleoptera, which he monographed. He was just as interested in the description of species as in the definition of higher systematic categories, which he established in a very independent manner, often making use of previously unnoticed characters. Thomson is generally considered the foremost of Scandinavian taxonomists in entomology.

His publications in coleopterology began in 1851 and led to the monograph *Skandinaviens Coleoptera* (1–10; 1859–1869) which was more than 3500 pages; written as typical of him, in a combination of Latin and Swedish.

Thomson's first paper on Hymenoptera was published in 1857 and consisted of the first part of a revision of the Swedish proctotrupids (finished in 1861). His first extensive work in this field, *Skandinaviens Hymenoptera,* volumes 1–5, appeared in the period 1871–1878; in it the Phytophaga, the Aculeata (except the ants), and the chalcids were monographed. The revisions to follow covered the remaining Parasitica (except the Mymaridae) and were almost entirely published in an independent series edited by the author himself, *Opuscula Entomologica* (1–22; 1869–1897), amounting to almost 2500 pages. In the year of the last issue failing eyesight stopped further work.

Thomson to a very great extent used material collected by himself, mostly in the province of Scania in southernmost Sweden. He was an excellent field entomologist. On his excursions he used to bring a flute (still preserved at the entomology department in Lund) which he used with professional skill, and, after the capture of a particularly fine specimen, he would sit down in the grass to play a joyful tune. Science and arts may sometimes join in fruitful combination.

Hemiptera

The first entomologist devoting himself almost entirely to Hemiptera was the Dutchman C. Stoll (?–1795), with his iconography *Natuurlyke*

FIG. 5. C. G. Thomson (1824–1899). Greatest specailist in Coleoptera and Hymenoptera in Scandinavia. (Courtesy of Zoological Institute, University of Lund)

en naar't Leeven Naauwkeurig Gekleurde Afbeeldungen en Beschryvingen der Cicaden en Wanzen (1–12, 1780–1788). A similar work, his only publication, was by the German J. F. Wolff (1778–1806), *Icones Cimicum Descriptionibus Illustratae* (1–5; 1800–1811); it was entirely devoted to Heteroptera.

Two other men, less specialized, whom we have mentioned above under other insect orders, Diptera and Coleoptera, respectively, gave valuable contributions to the taxonomy of Hemiptera; these were the Swede C. F. Fallén and the German E. F. Germar, the latter notably on the Cicadina.

An exclusive hemipterist, with one single very important revision of the Heteroptera, was C. W. Hahn (?–1836), who in 1831 started the series *Die Wanzenartigen Insecten;* after his death it was continued by the indefatigable G. A. W. Herrich-Schaeffer, the lepidopterist.

The famous dipterist H. Loew had an interest also in the Hemiptera, notably in the Psyllina, on which he published many important papers. The German J. H. Kaltenbach (1807–1876) was the first great specialist in the Aphidina. His *Monographie der Familie der Pflanzenläuse* appeared in 1843; later (1856–1869), under the name of *Die Deutschen Phytophagen aus der Klasse der Insekten,* he produced a large work on plant-feeding insects of all orders, arranged according to their host plants.

The most important hemipterists around the middle of the nineteenth century were less restricted; at least they studied both Heteroptera and Cicadina; F. A. Kolenati is mentioned below under Neuroptera. The Frenchman J. G. A. Serville (1775–1858), occupied also with studies of Orthoptera and Coleoptera, published a *Histoire Naturelle des Insectes Hémiptères* (1843) with the collaboration of C. J. B. Amyot (1799–1866). F. X. Fieber (1807–1872), from what is now Czechoslovakia, produced his important *Fauna Austriaca, Die Europäischen Hemiptera* (1860–1861), containing Heteroptera only, and, as a posthumous work, *Les Cicadinae d'Europe* (Paris, 1875–1879). The Frenchman V. A. Signoret (1816–1889) worked on all groups of Hemiptera, except the Aphidina, and his revisions of the Coccina were of special importance.

The Swede C. Stål (1833–1878), except for a monograph of North American chrysomelid beetles, studied Hemiptera almost exclusively, with his first paper appearing in 1853. He was a pioneer in gross taxonomy of the Heteroptera of the world and, upon his death, was named "the most distinguished hemipterologist of the present time" (*Ent. m. Mag.*, 15, 1878:94–96).

Subsequent European hemipterists, though starting to publish earlier, made their main contributions in the Darwinian era. Thus, in the period 1866–1879 appeared the *Histoire Naturelle des Punaises de France* (dealing with Heteroptera) by the coleopterists E. Mulsant and C. Rey, mentioned above, in six volumes of more than 900 pages. C. von Flor (?–1883), in Livland (Esthonia), published *Die Rhynchoten Livlands* (1860–1861), a work of almost 1500 pages which is especially important for the study of Cicadina. Finally, J. W. Douglas (1814–1905), in Britain, with a preceding large production in other insect orders, in his later years turned to Hemiptera and, together with J. Scott, published *The British Hemiptera. 1. Heteroptera* (1865; Homoptera not published). The same authors compiled a catalogue of British Hemiptera, in 1876.

The North American hemipterists were late in starting and only two should be mentioned here, both with previous work on other insects. P. R. Uhler (1835–1913) presented his first paper on Hemiptera in 1861, which was to be followed by a long series of descriptions and revisions of both Heteroptera and Homoptera, including exotics. He was named "America's greatest hemipterist" (Essig, 1931).

The main work of C. Thomas (1825–1910) was on grasshoppers but above all he was the pioneer in taxonomy of the North American Aphidina (*Ann. Rep. State Ent. Illin., 3rd*, 1879).

Small Orders

Several small orders are here brought under one heading, mainly because the early authors treated two complexes of what are now considered several distinct orders under the collective names Orthoptera and Neuroptera.

Specialists on Apterygotan insects, primarily Collembola and Thysanura, were scarce. Abbé Bourlet in France, for whom no biographical data are available, published a few papers on Collembola in the short period 1839–1842. In the same country, H. Nicolet (?–1872) made more comprehensive revisions of both Collembola and Thysanura, his main work being *Essai sur une Classification des Insectes Aptères de l'Ordre des Thysanoures* (1847). Later, though still in the middle of the century, came the works of the Englishman J. W. Lubbock (1834–1913), who had broad interests in anatomy, embryology, ethology, etc of insects, but also, from 1862 onwards, made advances in the taxonomy of the Apterygota. Most important is the *Monograph of the Collembola and Thysanura* (1873).

In the study of parasitic lice (Anoplura and Mallophaga) little was accomplished in the early nineteenth century. The German C. L. Nitzsch (1782–1837) published several small papers on the subject, from 1818, but the main portions of his observations and descriptions were not printed until long after his death by C. G. Giebel (1851, 1874). In Britain, H. Denny (1803–1871) started with a fine monograph of the beetle families Pselaphidae and Scydmaenidae (1825), then turned to the Anoplura (including Mallophaga) and published *Monographia Anoplurorum Britanniae* (almost 300 pages) in 1842. The Dutchman E. Piaget (1817–1910) wrote two valuable revisions, but not until after 1864.

The Orthoptera in the old sense, including the now recognized orders Blattodea, Mantodea, Saltatoria, Phasmida, and Dermaptera, sometimes even Odonata and Thysanoptera, were usually treated together by the same specialist. Only one of these orders, the true Orthoptera or Saltatoria, is widely distributed, rich in species, and of utmost economic importance, and the related groups (orders) were regarded more or less as appendices of the Saltatoria.

Several of the entomologists treated under other headings in this chapter have made valuable contributions to the knowledge of orthopteroid insects. Especially important works of such authors are J. G. A. Serville's *Histoire Naturelle des Insectes Orthoptères* (1839), the second volume of H. C. C. Burmeister's *Handbuch der Entomologie* (1839), G. Fischer von Waldheim's fourth volume of *Entomographia imperii russici* (1846–1849), also under the name of *Orthoptera Imperii Russici,* and H. L. F. de Saussure's *Orthoptera nova Americana* (1859–1862), *Mélanges Orthoptèrologiques* (1–6, 1863–1878), especially important for the Gryllodea, and several of his revisions of Blattodes.

A more pronounced specialist in Orthoptera was the German L. H. Fischer (1817–1886), with his *Orthoptera Europaea* (1853), and, later, K. Brunner von Wattenwyl (1823–1914) in Austria, a world authority on these insects, whose activity did not start until 1865. The first important American specialist in the Saltatoria was S. H. Scudder (see below, fossil insects), who began publishing his valuable revisions in 1859.

No one specialized entirely in Thysanoptera during this period but the famous A. H. Haliday, in 1836, published *An Epitome of the British Genera in the Order Thysanoptera.*

The Neuroptera in its old sense was an extremely composite group, including all orders with aquatic larvae, irrespective of the type of metamorphosis, and even the Isoptera, the Embioptera, etc. In *The Zoological Miscellany,* (3; 1817), the above-mentioned W. E. Leach removed the Trichoptera, but he was not generally followed in this.

Some entomologists worked within the whole complex covered by the name Neuroptera. The first of importance was the Frenchman J. P. Rambur (1801–1870) who, besides his main occupation with the Lepidoptera, published a large work of more than 500 pages on the *Histoire Naturelle des Insectes, Neuroptères* (1842). The German W. G. Schneider (1814–1889) treated the true Neuroptera, that is, Rhaphididae and Chrysopidae, in condensed monographs (1843 and 1851, respectively). F. Kolenati (1813–1864), from present-day Czechoslovakia, was a versatile entomologist but his *main* field was the Trichoptera, as manifested in his *Genera et species Trichopterorum* (1–2; 1848–1859).

The first monograph of the Odonata *Libellulinae Europaeae* (1840), by T. de Charpentier (1779–1847) appeared in Germany. The author was also an important specialist in Orthoptera.

"The pioneer of modern neuropterology in its broad sense" (McLachlan, *Ent. m. Mag.,* 30, 1894:12–20) was H. A. Hagen (1817–1893), born in Germany but from 1867 curator of insects at the Museum of Comparative Zoology, Harvard. He has secured the gratitude of the whole entomological world by his unique Bibliotheca Entomologica (1–2; Leipzig, 1862–1863), covering the entire field of entomological literature up to the years of publication.

Hagen had an enormous entomological production, commencing in 1839 with a paper on Odonata, and he kept rather faithfully to the groups of insects at that time lumped as Order Neuroptera. His working material was mostly from Europe and North America but he wrote a *Monographie der Termiten* (1855–1860) and several revisions of the world fauna of the Odonata, mostly in cooperation with M. E. de Sélys-Longchamps (1813–1900), a Belgian nobleman with high offices in politics, who devoted his spare time to ornithology and entomology, in the latter field almost entirely concentrating upon the Odonata.

In Britain, the dominating neuropterist was R. McLachlan (1837–1904). He started as a botanist, then turned to Lepidoptera and finally to the Neuroptera in the widest sense. Initiated by Hagen, he became the pioneer of the world fauna of Trichoptera, where he introduced the study of genital characters. His greatest work was *A Monographic Revision and Synopsis of the Trichoptera of the European Fauna* (1874–1884).

A scientist with extremely broad interests was F. M. Brauer (1832–1904), for good reasons named the greatest entomologist of Austria. He was a general biologist by inclination, with a special interest in the metamorphosis of insects, but at the same time a first class taxonomist, with Neuroptera s. str., Odonata, and Diptera as specialties. His first paper (1850) described observations on the biology of Chrysopidae. In 1876 appeared *Die Neuropteren Europas insbesondere Oesterreichs,* but he also revised exotic material of the same group, as well as of Odonata.

Geographical Specialization

A young person starting his entomological activities will necessarily gain his experience from a limited geographical area, the fauna of which he tries to explore as completely as possible. Some entomologists restrict themselves to goals of this kind for their entire lives or extend the investigated area only slightly, so as to cover a province, county, and so on. An advantage is that the study of a small area gives the opportunity of collecting and learning a greater diversity of insect groups.

The efforts of these local entomologists should not be belittled. Field notes and faunal lists ensuing are often important contributions to ecological knowledge and provide an indispensable fundament for zoogeography. Without all these devoted amateurs no reliable maps of insect distribution could have been drawn.

The restriction is, however, sometimes driven in absurdum. Horion (1941:25) tells a charming story of a German coleopterist who collected exclusively the beetles of his own district (Gau). On an excursion he arrived at a big stone marking the boundary of his Gau, on which several interesting beetles had settled. He most carefully collected only specimens on his side of the stone, whereas the others, as foreigners, were strictly avoided. One rare beetle happened to be seated exactly on the corner of the stone and therefore brought him into a state of perplexity. It turned out, however, that the head—with the brain!—was located on the right side; and so the insect was found worthy of meeting its heroic death for science.

Even serious first class taxonomists sometimes are entirely disinterested in the fauna of foreign countries. Westwood [*Tr. ent. Soc. London,* (2)2, 1853: 47] describes Stephens' attitude in this respect, and it is obviously meant as praise: "The earnestness with which he studied British entomology, induced him to reject from his cabinet every specimen not indigenous, barely tolerating foreign examples of rare or doubtful species, which he kept in a detached cabinet". The drawback of this way of national thinking is evident from the fact that a proportionally large number of Stephens' species names have later proved to be synonyms of earlier continental ones. But Stephens' importance for the development of entomology in Britain should not be underrated. His *Illustrations of British Entomology* (1828–1846), among other things, was a most valuable complement to Kirby and Spence (1815–1826).

A famous example of a man who, for most of his life, concentrated his efforts on the fauna, almost exclusively beetles, of a restricted geographical area, was T. V. Wollaston (1822–1878). He published several papers on British beetles but then, at the age of 25, was obliged to stay on Madeira because of poor health, and he became fascinated by its peculiar fauna. In years to follow he visited all groups of the Makaronese Islands and, from 1854 through 1867, published a series of fundamental revisions of their beetles, and later (1877) a series on those of St. Helena. He collected almost all of his material himself.

Local faunas and catalogues of most insect orders were published during this period from almost all European countries as well as from many states in the US.

The best example of an entomologist who devoted his main efforts to the study of exotic faunas was H. W. Bates (1825–1892). In his early career he spent more than ten years in South America, partly in company with the famous A. R. Wallace; he was a great general biologist, the first to discover the form of mimicry in butterflies that afterwards was named after him. He worked on different insect groups, with his first paper published in 1843. After 1870 he almost exclusively turned to Coleoptera and may be termed the father of coleopterology not only for tropical America, as manifested by his volumes on Carabidae, Longicornia, and Lamellicornia in Godman & Salvin's *Biologia Centrali-Americana* (1881–1890), but also for southern and eastern Asia, notably Japan. The main part of Bates' work, both in time and approach, falls of course in the Darwinian period and thus outside the scope of this chapter.

On the specific level, geographical restrictions involve few disadvantages. However, as soon as genera and higher taxonomic categories are involved, it becomes dangerous to build a system based on the fauna of a limited area. Genera, seemingly well defined, may be joined by transitional species in other regions. On the other hand, a single somewhat aberrant-looking species may prove to be the representative of a large, well-defined group, or a valid genus, if other faunas are studied. The desirable fundament for gross taxonomy (though rarely achieved) is a thorough knowledge of the world fauna of the group.

A specialist with the ambition to produce lasting results and forced to the necessary constraint of restricting his sphere of work should, therefore, limit his material on a taxonomic rather than on a geographical basis. Many of the early nineteenth century entomologists, working in different insect orders, like Boisduval, Candèze, Chaudoir, Dejean, Loew, Sélys-Longchamps, and Stål, already realized this.

Fossil Insects

The study of fossil insects is a specialty of its own. It has usually been executed by specialists on corresponding extant forms, but some entomologists, most of them already named under the pertinent insect order, devoted so much interest to paleoentomolgy that they are worthy of being mentioned separately in that context.

The first of importance was E. F. Germar, mentioned under Coleoptera.

He was a professor of mineralogy and his encounter with fossil insects and plants was therefore inevitable. His numerous papers on these subjects were commenced in 1813.

O. Heer was a Swiss professor of botany but a specialist in Coleoptera as well, about which he published from 1834. His contributions on tertiary fossils, mainly plants, started in 1847; the insect fossils he described included not only Coleoptera but also Blattodea, Hymenoptera, and other orders.

H. Loew, the famous dipterist, had a special interest in amber fossils, though the results were only partially published. According to Osten Sacken (1903), who was a sharp critic of Loew, his treatment of fossil forms was not satisfactory.

We have also mentioned G. L. Mayr as a specialist in Baltic amber ants.

The inevitable prerequisite for successful work on fossils is an intimate knowledge of the corresponding extant fauna. This demand has not always been fulfilled. An example in North America was S. H. Scudder (1837–1911). He was a first class taxonomist on Orthoptera Saltatoria and also on North American butterflies, but he did not at all restrict himself to these groups when describing fossils. In this field, mostly cultivated during his appointment as a paleontologist with the US Geological Survey (1886–1892), he treated almost all groups of fossil insects and described more than 1000 species, from the Paleozoic through the Pleistocene, resulting in the comprehensive book *The Fossil Insects of North America* (1890). A modern taxonomist studying Scudder's specimens very often finds himself in disagreement with his identifications.

Early Stages of Insects

These have of course usually been studied in connection with the adults, notably among hemimetabolous insects. But a few entomologists specialized and published on larvae, mostly of Lepidoptera. An early example is the *Collection Iconographique et Historiques des Chenilles d'Europe* (1832–1843) by J. A. Boisduval, J. P. Rambur, and A. de Graslin, in France. In Germany several so-called Raupenkalender (caterpillar diaries) appeared from time to time.

In the Coleoptera, F. Chapuis (1824–1879) and the above-mentioned E. Candèze published a descriptive *Catalogue des larves des Coléoptères* (1853), but the still unsurpassed name in this field was the Dane J. C. Schiødte (1815–1884).

Schiødte was an extremely all-encompassing scientist, the most versatile in Danish entomology. His first paper was a revision of the carabid genus *Amara* (1836), followed in 1841 by *Genera og Species af Danmarks Eleutherata,* of which only one volume, treating the Adephaga, appeared. This was followed by many other papers in species and gross taxonomy in Coleoptera, but also in other insect orders, in part dealing with exotic material. In everything he touched, Schiødte gave proof of his independence and imagination, and his approach to taxonomy gained considerably from his vast experience in the fields of anatomy, physiology, and applied entomology.

However, Schiødte's most lasting monument to science were his investigations on the early stages of Coleoptera, mostly published under the title *De Metamorphosi Eleutheratorum Observationes* (1861–1883). The descriptions and the engraved plates of larvae belonging to most of the large beetle families are still indispensable. Based on these primary observations, Schiødte attempted to utilize comparative morphology for erecting a larval system of Coleoptera and compared it with that based on characters of the adult beetles. It should be mentioned also that Schiødte is the father of speleology. In his *Specimen Faunae Subterraneae. Bidrag til den Underjordiske Fauna,* from 1849, he not only describes cave insects (and other arthropods), but he arranges them according to their greater or lesser adaptation to the life in caves and attempts an explanation of their origin and development.

Over-Specialization

Restricting oneself to a small taxonomic group or to a limited geographical area involves the temptation to go into too much detail; for instance, to proceed below the species level in naming deviating insect specimens that cannot be regarded as representatives of definable subspecies.

This happened mainly in the first decades of the present century, e.g. in genus *Carabus* and among coccinellid beetles, in Coleoptera, and in the genera *Parnassius* and *Zygaena* among Lepidoptera.

Only one case in the period under consideration should be mentioned, K. W. Letzner (1812–1889), who was described (*Deuts. ent. Z.,* 1890:28–29) as "a local and provincial patriot in the noblest sense of the word" (translated). He published, among many other things, a *Systematische Beschreibung der Laufkäfer Schlesiens* (1847–1852), in which an almost countless number of infraspecific names of Carabidae were given. These cannot be interpreted as of subspecific value, because what Letzner described and named were single characters, and one individual beetle, therefore, often belonged to two or more of these "varieties". Letzner's infraspecific names are thus invalid.

Detailed description of individuals and groups of individuals are of fundamental importance in population genetics. However, they must not be named, because the rules of taxonomic nomenclature in zoology do not accept as valid those that belong to categories lower than subspecies.

Auxiliary Sciences

Of course, entomologists in the early nineteenth century specialized in other branches than taxonomy, such as anatomy, physiology, and ethology, which are not treated here. Some of these investigations were also of great indirect use to taxonomy. An example is the admirable study of the female genitalia in beetles by F. Stein, *Vergleichende Anatomie und Physiologie der*

Insecten. 1. Ueber die Geschlechts-Organe und den Bau des Hinterleibes bei weiblichen Käfern (1847).

Specialized Journals

The earliest serial publications devoted to entomology included private journals, such as those started by J. C. Füessly in 1778, L. G. Scriba in 1790, J. C. W. Illiger in 1801, and E. F. Germar in 1813, as well as those published by the entomological societies in London, Paris, Stettin, etc. These journals were open to contributions from the entire field of entomology. Periodicals specializing in restricted groups were not generally needed until after the end of the period covered by this chapter. Therefore only the two appearing first should be mentioned, both in Germany: the *Coleopterologische Hefte,* edited by E. von Harold in the period 1867–1879 in Munich, and *Iris. Deutsche Entomologische Zeitschrift. Lepidopterologische Hefte* in Dresden, from 1884 to 1911. In the present century, the number of specialized entomological journals is legion, though usually restricted on methodological rather than on taxonomic grounds.

Conclusions

It might have been expected that the breakthrough of the principle of evolution in biological sciences would have caused a fundamental change of goals and methods in taxonomy. This happened to a surprisingly slight extent only. The reason could of course have been that the taxonomists were conservative, to whom description was more important than explanation, and that they therefore refused to apply evolutionary ideas. But this was not always the true explanation. The concept of the species had not become fundamentally changed, at least not in practical use, and in gross taxonomy the natural systems adopted from Latreille onwards were arranged according to affinities. The foundation of such systems was a recognition, though seldom expressed, that relationship is the same as common origin.

Actually, it was many years after 1859, even in gross taxonomy, before a truly phylogenetic system of insects was constructed. As pointed out by A. Handlirsch (1925:13), the systems of the German C. E. A. Gerstaecker (1828–1895), in his *Handbuch der Zoologie* (with J. V. Carus, 1863), and the American J. D. Dana (1813–1895), in *Classification of Animals* (11; 1864), were unsatisfactory from this point of view. The first Darwinian system of insects was that by E. Haeckel (1834–1919) in his *Generelle Morphologie* (1866).

Acknowledgments

I am most indebted to Dr. Bengt-Olof Landin, Lund, who read the manuscript and proposed many valuable changes and additions.

LITERATURE CITED

Abbot, J., Smith, J. E. 1797. *The Natural History of the Rarer Lepidopterous Insects of Georgia.* 1–2. London: Edwards, 229 pp.

Bates, H. W. In *Biologia Centrali-Americana,* ed. F. D. Godman, O. Salvin; I:1, 326 pp. 1881–1884; 11:2, 444 pp. 1886–1890; V, 538 pp. 1879–1886. London: Dulau

Bengtsson, S. 1900. C. G. Thomson. Minnesteckning. *Entomol. Tidskr. Stockholm* 21:1–16

Boheman, C. H. 1850–1862. *Monographia Cassididarum,* 1–4. Stockholm: Norstedt. 2017 pp.

Boisduval, J. B. A. Guenée, A. 1836–1857. Spécies général des Lépidoptères. *Histoire Naturelle des Insectes,* 1:5–10. Paris: Roret. 3695 pp.

Boisduval, J. B. A. 1874. Spécies général des Lepidoptères Hétérocères. *Histoire Naturelle des Insectes.* Paris: Roret. 576 pp.

Boisduval, J. B. A., Leconte, J. E. 1829–1834. *Histoire Générale et Iconographie des Lépidoptères et des Chenilles de l'Amerique Septentrionale.* Paris: Roret. 228 pp.

Boisduval, J. A., Rambur, J. P., de Graslin, A. 1832–1843. *Collection Iconographique et Historique des Chenilles d'Europe, ou Descriptions et Figures des Chenilles d'Europe, avec l'Histoire de leurs Métamorphoses.* Paris: Roret. 42 issues.

Borkhausen, M. B. 1788–1792. *Naturgeschichte der Europäischen Schmetterlinge nach Systematischer Ordnung,* pp. 1–5. Frankfurt: Varrentrapp & Wenner. 2542 pp.

Brauer, F. M. 1863. Monographie der Oestriden. *Zool. Bot. Ges. Wien,* 298 pp.

Brauer, F. M. 1876. Die Neuropteren Europas und insbesondere Oesterreichs mit Rücksicht auf ihre geographische Verbreitung. *Festschr. Zool. Bot. Ges. Wien,* 38 pp.

Burmeister, H. C. C. 1829. *De Insectorum Systemate Naturale.* (Diss., Grunert, Halle) 40 pp.

Burmeister, H. C. C. 1832–1855. *Handbuch der Entomologie,* 1–5. Berlin: Reimer. 4391 pp.

Chapuis, F., Candèze, E. C. A. 1853. Catalogue des larves des Coléoptères connues jusqu'à ce jour avec la description de plusieurs espèces. *Mém. Soc. Sci. Liège,* 8, 313 pp.

de Charpentier, T. 1840. *Libellulinae Europaeae.* Leipzig: Voss. 180 pp.

Clerck, C. A. 1757. *Svenska Spindlar.* Stockholm

Clerck, C. A. 1759–1764. *Icones Insectorum Rariorum cum Nominibus eorum Trivialibus Locisque e C. Linnaei Syst. Nat. Allegatis.* Stockholm

Cuvier, G. C. L. D. 1799–1805. *Leçons d'Anatomie Comparée,* 1–5. Paris: Baudoin

Dahlbom, A. G. 1843–1854. *Hymenoptera Europaea Praecipue Borealia,* 1–2. (Lundberg, Lund) 1008 pp.

Dana, J. D. 1864. The classification of animals based on the principle of cephalization. II. Classification of insects. *Am. J. Sci.* (2)37, 13 pp.

Dejean, P. F. M. A. *Catalogue des Coléoptères,* 5 editions. (Paris: 11 pp., 1802; 136 pp., 1821; 176 pp., 1833; 443 pp., 1836; 517 pp., 1837)

Dejean, P. F. M. A. 1822–1824. *Histoire Naturelle et Iconographie de Coléoptères d'Europe,* 1–3. Paris: Crevot. 198 pp.

Dejean, P. F. M. A. 1825–1838. *Spécies Général des Coléoptères de la Collection de M. le Comte Dejean,* 1–6. Paris: Méquignon-Marvis. 3803 pp.

Dejean, P. F. M. A. 1829–1838. *Iconographie et Histoire Naturelle des Coléoptères d'Europe,* 1–5. Paris: Mequignon-Marvis. 2060 pp.

Denny, H. 1825. *Monographia Pselaphidarum et Scydmaenidarum Britanniae.* Norwich: Wilkin. 80 pp.

Denny, H. 1842. *Monographia Anoplurorum Britanniae, or an Essay on the British Species of Parasitic Insects.* London: Bohn. 287 pp.

Douglas, J. W., Scott, J. 1865. *The British Hemiptera. 1. Hemiptera-Heteroptera.* London: Ray Society. 649 pp.

Douglas, J. W., Scott, H. J. 1876. *A Catalogue of British Hemiptera: Heteroptera and Homoptera (Cicadariae and Phytophthires).* London. 108 pp.

Duftschmid, C. 1805–1825. *Fauna Austriae.* 1–3. Linz. 910 pp.

Duméril, A. M. C. 1823. *Considérations Générales sur la Classe des Insectes.* Paris: Levrault. 272 pp.

Duponchel, P. A. J. 1821–1838. *Histoire*

Naturelle des Lépidoptères de la France, 1–11, with J. B. Godart. Paris: Crevot & Méquignon-Marvis. 5802 pp.
Edwards, W. H. 1868–1897. *The Butterflies of North America,* 1–3. Am. Entomol. Soc., Philadelphia. 151 pp.
Eiselt, J. N. 1836. *Geschichte, Systematik und Literatur der Insectenkunde, von den Ältesten Zeiten bis auf die Gegenwart.* Leipzig: Hartmann. 263 pp.
Erichson, W. F. 1837–1839. *Die Käfer der Mark Brandenburg.* 1–2. Berlin: Morin. 740 pp.
Erichson, W. F. 1839–1840. *Genera et Species Staphylinorum, Insectorum Coleopterorum Familiae.* Berlin: Morin. 954 pp.
Erichson, W. F. 1845–c. 1920. *Naturgeschichte der Insecten Deutschlands.* I. Coleoptera, 1–6, with several coworkers. Berlin: Nicolai. 7412 pp.
Esper, E. J. C. 1777–1807. *Die Europäischen Schmetterlinge in Abbildungen nach der Natur mit Beschreibungen,* 1–5. Erlangen: Walther. 2591 pp.
Essig, E. O. 1931. *A History of Entomology.* New York: Macmillan. 1036 pp.
Eversmann, E. von 1855–1859. Les Noctuélites de la Russie. *Bull. Soc. Nat. Moscou,* 28–30, 32. 706 pp.
Fabricius, J. C. 1798. *Supplementum Entomologiae Systematicae.* Copenhagen: Proft & Stoch. 574 pp.
Fallén, C. F. 1814–1825. *Diptera Sueciae,* 1–2. Lund, diss. 484 pp.
Fieber, F. X. 1860–1861. *Fauna Austriaca. Die Europäischen Hemiptera,* 1–3. Wien: Gerold. 460 pp.
Fieber, F. X. 1875–1879. Les Cicadines d'Europe d'après les originaux et les publications les plus récentes, 1–4. *Rev. Mag. Zool.* 38–42, 566 pp.
Fischer, L. H. 1853. *Orthoptera Europaea.* Leipzig: Engelmann. 474 pp.
Fischer von Waldheim, G. 1820–1851. *Entomographia Imperii Russici; Genera Insectorum Systematice Exposita et Analysi Iconographica Instructa,* 1–5. Moscou. 1520 pp.
Flor, G. von 1860–1861. Die Rhynchoten Livlands in systematischer Folge beschrieben. 1–2. *Arch. Naturk. Livl.,* 3–4. 1472 pp.
Ganglbauer, L. 1892–1904. *Die Käfer von Mitteleuropa. Die Käfer der österreichisch-ungarischen Monarchie, Deutschlands, der Schweiz, sowie des französischen und italiänischen Alpengebietes,* 1–4. Wien: Gerold. 2782 pp.
Gerstaecker, C. E. A., Carus, J. V. 1863. *Handbuch der Zoologie. 2. Wirbellose.* Leipzig: Engelmann. 642 pp.
Giebel, C. G. A. 1851. Anleitung zur Beobachtung der Thierinsecten, aus dem Nachlass des Prof. Nitzsch mitgetheilt. *Jahresber. Nat. Ver. Halle,* 3. pp. 113–35
Giebel, C. G. A. 1874. *Insecta Epizoa. Die auf Säugethieren und Vögeln Schmarotzenden Insekten. Nach Chr. L. Nitzsch's Nachlass bearbeitet.* Leipzig: Wigand. 324 pp.
Gravenhorst, J. L. C. 1806. *Monographia Coleopterorum Micropterorum.* Göttingen: Dieterich. 265 pp.
Gravenhorst, J. L. C. 1829. *Ichneumonologia Europaea,* 1–3. Bratislava. 2916 pp.
Gyllenhal, L. 1808–1827. *Insecta Suecica Descripta. Classis I. Coleoptera sive Eleutherata,* 1–4. Skara: Leverentz. 2767 pp.
Haeckel, E. 1866. *Generelle Morphologie.* Berlin
Hagen, H. A. 1855–1860. Monographie der Termiten. *Linnaea Entomol.* 10, 12, 14; 598 pp.
Hagen, H. A. 1862–1863. *Bibliotheca Entomologica,* 1–2. Leipzig: Engelmann. 1078 pp.
Hahn, C. W. 1831–1853. *Die Wanzenartigen Insecten,* 1–9. Continued by G. A. W. Herrich-Schaeffer. Nürnberg: Zeh. 1885 pp.
Haliday, A. H. 1833–1838. An essay on the classification of the parasitic Hymenoptera of Britain. *Entomol. Mag.* 1–5. 257 pp.
Haliday, A. H. 1836. An epitome of the British genera in the order Thysanoptera. *Entomol. Mag.* 3, 13 pp.
Handlirsch, A. 1905. Friedrich Moritz Brauer. *Verh. Zool. Bot. Ges. Wien,* 55:129–66
Handlirsch, A. 1925. Aus der Geschichte der Entomologie. (Mit besonderer Rücksicht auf die Entwicklung der Systematik.) *C. Schröder: Handbuch der Entomologie.* III. Jena: Fischer. 16 pp.
Harris, M. 1766. *The Aurelian or Natural History of English Insects Namely Moths and Butterflies, together with the Plants on which They Feed.* London. 77 pp.
Hartig, T. 1837. *Die Familien der Blattwespen und Holzwespen, nebst einer*

Allgemeinen Einleitung zur Naturgeschichte der Hymenopteren. Berlin: Haude & Spener. 430 pp.
Haworth, A. H. 1803–1828. *Lepidoptera Britannica,* 1–4. London. 645 pp.
Heer, O. 1838–1842. *Fauna Coleopterorum Helvetica.* Turici: Orellii. 664 pp.
Heinemann, H. 1859–1877. *Die Schmetterlinge Deutschlands und der Schweiz,* 1–2. Braunschweig: Vieweg. 2513 pp.
Herbst, J. F. W. 1783–1806. *Natursystem Aller Bekannten In-und Ausländ ischen Insecten; nach dem System des Ritters Carl von Linné Bearbeitet,* 1:1–10 (1 by Jablonsky); II:1–11 (1–2 by Jablonsky). Berlin: Pauli. 3468 and 2756 pp.
Herrich-Schaeffer, G. A. W. 1843–1856. *Systematische Bearbeitung der Schmetterlinge von Europa, als Text, Revision und Supplement zu J. Huebners Sammlung Europäischer Schmetterlinge,* 1–6. Regensburg: Manz. 1885 pp.
Hofmann, O. 1874. Necrolog. Herr Dr. med. Gottlieb August Herrich-Schäffer. *Corr. Bl. Zool. Miner. Ver. Regensburg,* 28:65–73
Holmgren, A. E. 1855. Monographia Tryphonidum Sueciae. *Act. Reg. Acad. Sc. Holm,* 1. 302 pp.
Holmgren, A. E. 1864–1889. *Ichneumonologia Suecica,* 1–3. Stockholm: Norstedt. 469 pp.
Horion, A. 1941. *Faunistik der Deutschen Käfer,* 1. Krefeld: Goecke. 463 pp.
Horn, W., Kahle, I. 1935–1937. Über entomologische Sammlungen, Entomologen und Entomo-Museologie. *Entomol. Beihefte,* 2–4. Berlin. 342 pp.
Horn, W., Schenkling, S. 1928–1929. *Index Litteraturae Entomologicae,* Berlin. 1447 pp.
Hübner, J. 1796–1841. *Sammlung Europäischer Schmetterlinge,* 1–9. With C. Geyer and G. A. W. Herrich-Schaeffer. Augsburg. Incomplete text
Hübner, J. 1806–1824. *Sammlung Exotischer Schmetterlinge,* 1–2. With C. Geyer and G. A. W. Herrich-Schaeffer. Augsburg
Illiger, J. C. W. 1798. *Kugelann, Verzeichniss der Käfer Preussens.* Halle: Gebauer. 551 pp.
Jacquelin du Val, P. N. C. 1855–1868. *Genera des Coléoptères d'Europe. Manuel Entomologique.* 1–4. With L. Fairmaire. Paris: Migneaux. 1469 pp.
Junk, W. 1912. *Bibliographia Coleopterologica.* Berlin. 162 pp.
Junk, W. 1913. *Bibliographia Lepidopterologica.* Berlin. 168 pp.
Jurine, L. 1807. *Nouvelle Method de Classer les Hyménoptères et les Diptères.* Genève: Paschoud. 319 pp.
Kaltenbach, J. H. 1843. *Monographie der Familie der Pflanzenläuse (Phytophthires).* Aachen: Roschütz. 266 pp.
Kaltenbach, J. H. 1856–1869. Die deutschen Phytophagen aus der Klasse der Insekten, oder Versuch einer Zusammenstellung der auf Deutschlands Pflanzen beobachteten Bewohner und ihrer Feinde. *Verh. Naturh. Ver. Preuss. Rheinl.,* 13–26, 958 pp.
Kirby, W. 1802. *Monographia Apum Angliae.* Ipswich: White. 668 pp.
Kirby, W., Spence, W. 1815–1826. *An Introduction to Entomology, or Elements of the Natural History of Insects,* 1–4. London: Longman. 2430 pp.
Klug, J. C. F. 1808–1818. Die Blattwespen nach ihren Gattungen und Arten zusammengestellt. *Mag. Ges. Naturf. Freunde, Berlin,* 2, 6, 7, 8, 245 pp.
Kolenati, F. A. 1848–1859. *Genera et Species Trichopterorum,* 1–2. Prague: Kronberger. 269 pp.
Kraatz, G. 1856–1858. Staphylinidae. In *Naturgeschichte der Insekten Deutschlands,* ed. W. F. Erichson, 2. Berlin: Nicolai. 1088 pp.
Lacordaire, J. T. 1854–1876. *Histoire Naturelle des Insectes. Genera des Coléoptères,* 1–12 (10–12 by F. Chapuis). Paris: Roret. 6410 pp.
Lamarck, J. B. P. A. deM. 1801. *Système des Animaux sans Vertèbre ou Tableau Général des Classes, des Orders et des Genres des ces Animaux.* Paris: Deterville. 440 pp.
Latreille, P. A. 1796. *Précis des Caractères Génériques des Insectes, Disposés dans un Ordre Naturel.* Bordeaux: Brive. 229 pp.
Latreille, P. A. 1802. *Histoire Naturelle des Fourmis et Recueil de Mémoires et d'Observations sur les Abeilles, les Araignées, les Faucheurs et Autres Insectes.* Paris: Barrois. 445 pp.
Latreille, P. A. 1802–1805. *Histoire Naturelle, Générale et Particulière des Crustacés et des Insectes,* 1–14. Paris: Dufart. 5817 pp.
Latreille, P. A. 1806–1809. *Genera*

Crustaceorum et Insectorum Secundum Ordinem Naturalem in Familias Disposita, Iconibus Exemplisque Plurimus Explicata. Paris. 1257 pp.

Latreille, P. A. 1810. *Considérations Générales sur l'Ordre Naturel des Animaux Composant les Classes des Crustacés, des Arachnides et des Insectes avec un Tableau Méthodique de leurs Genres Disposes en Familles.* Paris: Schoell. 444 pp.

Latreille, P. A. 1825. *Familles Naturelles du Règne Animal.* Paris: Baillière. 570 pp.

Latreille, P. A. 1829. In *Règne Animal,* ed. G. C. L. D. Cuvier, 2. Paris: Deterville. 4–5 ed. 612 pp.

Leach, W. E. 1814–1817. *The Zoological Miscellany,* 1–3. London. 442 pp.

Leconte, J. L. 1861–1862. Classification of the Coleoptera of North America. *Smithson. Misc. Coll.* 3:1–286

Leconte, J. L., Horn, G. H. 1883. Classification of the Coleoptera of North America. Part II. *Smithson. Misc. Coll.* 26:605 pp.

Lepeletier, A. L. M. 1823. *Monographia Tenthredinetarum, Synonymia Extricata.* Paris: Levrault. 176 pp.

Lepeletier, A. L. M. 1836–1846. *Histoire Naturelle des Insectes. Hyménoptères,* 1–4. With A. Brullé. Paris: Roret. 2569 pp.

Letzner, K. W. 1847–1852. Systematische Beschreibung der Laufkäfer Schlesiens. *Z. Entomol. Breslau,* 1–292

Lindroth, C. H. 1957. The Linnaean species of Carabid beetles. *J. Linn. Soc. (Zool.),* 43:325–41

Lubbock, J. W. 1873. *Monograph of the Collembola and Thysanura.* London: The Ray Society. 286 pp.

MacLeay, W. S. 1819–1821. *Horae Entomologicae, or Essays on the Annulose Animals.* London. 524 pp.

Marsham, T. 1802. *Entomologia Britannica, sistens Insecta Britanniae Indigena Secundum Linneum Disposita.* London: White. 579 pp.

Mayr, G. L. 1855. Formicina Austriaca. Beschreibung der bisher im österreichischen Kaiserstaate aufgefundenen Ameisen, nebst Hinzufügung jener in Deutschland, in der Schweiz und in Italien vorkommenden Arten. *Verh. Zool. Bot. Ver. Wien,* 4. 206 pp.

Mayr, G. L. 1861. *Die Europäischen Formiciden. Nach der Analytischen Methode Bearbeitet.* Wien: Gerold. 88 pp.

Mayr, G. L. 1868. Die Ameisen des baltischen Bernsteins. *Beitr. Naturk. Preus.,* Vol. 1. 111 pp.

Mayr, G. L. 1870–1871. Die mitteleuropäischen Eichengallen in Wort und Bild, 1–2. *Jahresber. Rossauer Communal-Oberrealschule,* 9, 10, 77 pp.

McLachlan, R. 1874–1884. *A Monographic Revision and Synopsis of the Trichoptera of the European Fauna.* London: van Voorst. 707 pp.

Meigen, J. W. 1818–1838. *Systematische Beschreibung der Bekannten Europäischen Zweiflügligen Insecten,* 1–7. Aachen and Hamm. 2869 pp.

Morris, J. G. 1862. *Synopsis of the Lepidoptera of North America.* Smithson. Inst., Washington. 385 pp.

Motschulsky, V. von 1845. Die coleopterologischen Verhältnisse und die Käfer Russlands. *Bull. Soc. Nat. Moscou,* 18:2, 1–131

Motschulsky, V. von 1845. Insectes de la Sibérie rapportés d'un voyage fait en 1839–1840. *Mém. Acad. St. Petersbourg,* 13. 289 pp.

Mulsant, E., Rey, C. 1839–1885. *Histoire Naturelle des Coléoptères de France,* 38 issues. Lyon and Paris. 12,644 pp.

Mulsant, E., Rey, C. 1866–1878. Histoire Naturelle des Punaises de France (Scutellérides). *Ann. Soc. Linn. Lyon,* N. S., 12–14, 20, 25; *Mém. Acad. Sci. Lyon,* 18, 935 pp.

Nees con Esenbeck, C. G. 1834. *Hymenopterorum Ichneumonibus Affinium Monographia, Genera Europaea et Species Illustratae,* 1–2. Stuttgart: Cotta. 780 pp.

Nicolet, H. 1847. Essai sur une classification des Insectes aptères de l'ordre des Thysanures. *Ann. Soc. Entomol. Fr.* (2)5, 61 pp.

Nussac, L. de 1907. *Pierre-André Latreille à Brive, de 1762 à 1798.* Paris: Steinheil. 263 pp.

Ochsenheimer, F. 1807–1816. *Die Schmetterlinge von Europa,* 1–4. Leipzig: Fleischer. 1807–1816, 1374 pp. Continued by F. Treitschke, 5–10, 1825–1835

Olivier, E. 1880. *G.-A. Olivier, Membre de l'Institute de France, sa Vie, ses Travaux, ses Voyages.* Paris: Moulins. 98 pp.

Olivier, G.-A. 1789–1808. *Entomolo-*

gie, ou Histoire Naturelle des Insectes, avec leurs Caractères Génériques et Spécifiques, leur Description, leur Synonymie et leur Figure Illuminée. Coléoptères, 1–6. Paris: Baudonin. 3142 pp.

Osten Sacken, C. R. 1903–1904. *Record of my Life and Work in Entomology,* 1–3. Cambridge, Mass. and Heidelberg. 242 pp.

Packard, A. S. 1866. The humble bees of New England and their parasites, etc. *Commun. Essex Inst.,* 4 (1864–1865). 34 pp.

Panzer, G. W. F. 1793–c. 1810. *Faunae Insectorum Germanicae Initia oder Deutschlands Insecten,* 110 issues. Continued by Herrich - Schaeffer. Nürnberg: Felssecker. 2340 pp.

Paykull, G. von 1798–1800. *Fauna Suecica,* 1–3. Uppsala: Edman, 1059 pp.

Paykull, G. von 1811. *Monographia Histeroidum.* Uppsala: Palmblad. 118 pp.

Rambur, J. P. 1842. *Histoire Naturelle des Insectes Neuroptères.* Paris: Roret. 551 pp.

Ratzeburg, J. T. C. 1844–1852. *Die Ichneumonen der Forstinsecten in Forstlicher und Entomologischer Beziehung, als Anhang zur Abbildung und Beschreibung der Forstinsecten,* 1–3. Berlin: Nicolai. 769 pp.

Redtenbacher, L. 1849. *Fauna Austriaca. Die Käfer nach der Analytischen Methode Bearbeitet.* Vienna: Gerold. 910 pp.

Rondani, C. 1856–1862. *Dipterologiae Italicae Prodromus,* 1–5. Parma: Stocchi. 1146 pp.

Sahlberg, C. R. 1817–1839. *Dissertatio Entomologica Insecta Fennica Enumerans,* 1–2. Åbo: Franckel. 807 pp.

Saussure, H. F. de 1852. *Monographie des Gûepes solitaries, ou de la tribue des Euméniens, comprenant la classification et la description de toutes les espèces connues jusqu'à ce jour.* Paris: Masson. 336 pp.

Saussure, H. F. de 1859–1862. Orthoptera nova Americana. *Rev. Mag. Zool.* (2) 11, 13, 14, 74 pp.

Saussure, H. F. de 1863–1878. Mélanges orthoptèrologiques, 1–6. *Mem. Soc. Phys. Hist. Nat. Genève,* 17–25. 1302 pp.

Schiffermüller, I. 1776. *Systematisches Verzeichnis der Schmetterlinge der Wiener Gegend.* With M. Denis. Wien: Beck. 322 pp.

Schiner, J. R. 1854–1858. Diptera Austriaca. Aufzählung aller im Kaiserthum Oesterrich bisher aufgefundenen Zweiflügler, 1–4, *Verh. Zool. Bot. Ver. Wien,* 4–8. 263 pp.

Schiner, J. R. 1860–1862. *Fauna Austriaca. Die Fliegen, Diptera. Nach der Analytischen Methode Bearbeitet.* Wien: Gerold. 754 pp.

Schiødte, J. C. 1841. *Genera og Species af Danmarks Eleutherata,* etc. Copenhagen: Klein. 624 pp.

Schiødte, J. C. 1861–1872. De metamorphosi Eleutheratorum observationes. *Naturh. Tidskr.* (3) 1, 3, 4, 8, 334 pp.

Schneider, W. G. 1843. *Monographia Generis Rhaphidiae Linnaei.* Bratislava: Grass. 99 pp.

Schneider, W. G. 1851. *Symbolae ad Monographiam Generis Chrysopae Leach.* Bratislava: Hirt. 178 pp.

Schönherr, C. J. 1833. *Genera et Species Curculionidum cum Synonymia hujus Familiae,* 1:1. Paris: Roret. 396 pp.

Scopoli, J. A. 1763. *Entomologia Carniolica Exhibens Insecta Carnioliae Indigena et Distributa in Ordines, Genera, Species, Varietatis, Methodo Linneana.* Vienna: Trattner. 456 pp.

Scopoli, J. A. 1777. *Introductio ad Historiam Naturalem, Sistens Genera Lapidum, Plantarum et Animalium,* etc. Prague: Gerle. 506 pp.

Scudder, S. H. 1890. *The Fossil Insects of North America with Notes on Some European Species,* 1–2. New York. 1118 pp.

Serville, J. G. A. 1839. *Histoire Naturelles des Insectes Orthoptères.* Paris: Roret. 776 pp.

Serville, J. G. A. 1843. *Histoire Naturelle des Insectes Hemiptères.* With C. J. B. Amyot. Paris: Roret. 751 pp.

Stainton, H. T. 1854. *Insecta Britannica, Tineina.* London: Lovell Reeve. 313 pp.

Stainton, H. T. 1855–1873. *The Natural History of the Tineina,* 1–13. With P. C. Zeller, J. W. Douglas, and H. Frey. London: van Voorst

Staudinger, O. 1861. *Catalogus Lepidopterorum Europaeorum.* With M. Wocke. Dresden: Burdach. 20 pp.

Stein, F. 1847. *Vergleichende Anatomie und Physiologie der Insecten. I. Monographie. Ueber die Geschlechts-Organe und den Bau des Hinterleibes bei den Weiblichen Käfern.* Berlin:

Duncker & Humblodt. 147 pp.
Stephens, J. F. 1828–1846. *Illustrations of British Entomology, or a Synopsis of Indigenous Insects,* etc. Mandibulata, 1–7; Haustellata, 1–4; Suppl. London. 3353 pp.
Stephens, J. F. 1839. *A Manual of British Coleoptera.* London: Longman. 455 pp.
Stoll, C. 1780–1788. *Natuurlyke en naar't Leeven Naauwkeurig Gekleurde Afbeeldingen en Beschryvingen der Cicaden en Wantzen,* 1–12. Amsterdam: Sepp. 121 + 164 pp.
Sturm, J. 1805–1856. *Deutschlands Insecten, Käfer,* 1–23. Nürnberg. 3490 pp.
Thomson, C. G. 1859–1869. *Skandinaviens Coleoptera,* 1–10. Lund. 3558 pp.
Thomson, C. G. 1869–1897. *Opuscula Entomologica,* 1–22. Lund. 2452 pp.
Thomson, C. G. 1871–1878. *Skandinaviens Hymenoptera,* 1–5. Lund. 1489 pp.
Wallengren, H. D. J. 1853. *Lepidoptera Rhopalocera Scandinaviae Disposita ac Descripta. Skandinaviens Dagfjärilar.* Malmo: Cronholm. 302 pp.
Wallengren, H. D. J. 1863–1885. *Lepidoptera Scandinaviae Heterocera, Disposita et Descripta,* 1–2. Lund: Bülow. 581 pp.
Wesmael, C. 1835–1838. Monographie des Braconides de Belgiques. *Nouv. Mém. Acad. Brux.,* 9–11. 490 pp.
Westwood, J. O. 1839–1840. *An Introduction to the Modern Classification of Insects, Founded on the Natural Habits and Corresponding Organisation of the Different Families,* 1–2. London: Longman. 1072 pp.
Wolff, J. F. 1800–1811. *Icones Cimicum Descriptionibus Illustratae,* 1–5. Erlangen: Palm. 212 pp.
Wollaston, T. V. 1865. *Coleoptera Atlantidum, Being an Enumeration of the Coleopterous Insects of the Madeira, Salvages, and Canaries.* London. 666 pp.
Wollaston, T. V. 1877. *Coleoptera Sanctae-Helenae.* London. 256 pp.
Zetterstedt, J. W. 1838–1840. *Insecta Lapponica Descripta.* Leipzig: Voss. 1145 pp.
Zetterstedt, J. W. 1842–1860. *Diptera Scandinaviae Disposita et Descripta,* 1–14. Lund. 6665 pp.

THE HISTORY OF PALEOENTOMOLOGY

B. B. Rohdendorf
Paleoentomological Institute of the
Academy of Sciences of USSR, Moscow

The history of the study of insects of the geological past reflects a very uneven development in this branch of entomology in respect to its different aspects, that is, to the geological age of the insects under study, to the geography of the localities of their fossil remains, and to the specialized study of certain groups of insects.

The remains of the insects as inclusions in the fossilized transparent resins, in particular in Baltic amber, have been known for a long time and we find records of them in the sources of the sixteenth to eighteenth centuries by Geronimo Lipomani; N. Sendelius, 1742; and M. V. Lomonosov, 1794. However, all these pre-Linnaean indications and descriptions, apart from their insufficient completeness and inevitably superficial descriptions, refer to the insects of one geological age, one geographical area, and give information on quite different groups of the class. Such onesidedness and haphazardness regarding the first information on insects of the geological past are the most typical features of these first data on paleoentomology obtained by the naturalists of the eighteenth century from the most studied part of the earth, namely from Europe, and from a well-known popular mineral, that is, from paleogenous amber of the Baltic area.

A sufficiently greater number of data on paleoentomology appear beginning in the first third of the nineteenth century when the study of fossils of insects begins to rest upon the first success of descriptive entomology which was to result from the works of Linnaeus and his followers in taxonomy and nomenclature of the class of insects. Besides works on Baltic amber there appeared articles on other Cainozoic insects, for instance, on paleogenous insects of France by P. M. de Serres, 1829, and later, on some other more ancient faunas, namely on Mesozoic and Paleozoic ones. Such are first of all works on Jurassic insects of Bavaria by P. L. Van der Linden (1827) and E. F. Germar (1837–1842), and in England by H. E. Strickland (1840) and P. B. Brodie (1842–1849), and on Carboniferous finds in Saxony by Germar (1842, 1844). But all these investigations were characterized by small accidental reports and besides they were all geographically based only on European materials. Other continents remained absolutely unknown in the paleoentomological respect.

Mention should be made of considerable intensification of paleoentomological research in Europe beginning with the second third of the nineteenth century. It found its expression in a greater number of publications and the appearance of special monographs devoted to the whole faunistic complexes. Such are the works by O. Heer (1847–1879) and the C. and L. Heydens (1856–1870) in which fossils of various insects from paleogenous and neogenous localities in Western and Eastern Europe are described. Works on Mesozoic faunas by C. Heyden, and Ch. Giebel (1856–1860) and on Carboniferous insects of Europe by F. Goldenberg (1854) and Giebel in 1850 should be added to the foregoing. In these works, rudiments of systematic specialization of paleoentomological research show for the first time. Descriptions of representatives of separate orders are united in special chapters or parts of monographs.

In 1848, non-European fossils of insects were mentioned in press. Such is the note about the existence of "larva of Neuroptera" in late-Jurassic, or to be more exact in early-Cretaceous deposits of Eastern Siberia (Transbaikalia). These fossils were later described as a remarkable species of mayfly (Ephemeroptera) by Eichwald. The fossils mentioned above turned out to be the first paleoentomological evidence for the whole continent of Asia and at the same time they were the first mesozoic insects for the whole territory of Russia of that time.

During the last third of the nineteenth century paleoentomological research in Europe continued. The number of publications grows and more and more works on paleoentomology carried out by specialist-entomologists appear. Especially many data appear on Cainozoic, Jurassic, and Carboniferous faunas. Mention should be made of investigation on Tertiary insects of France by E. Oustalet (1870–1874), of a great number of works on Baltic amber, e.g. by C. Schaufuss (1888–1896), of some notes on Tertiary Arctic, Spitsbergen, and Greenland by O. Heer (1870–1883), on Upper Jurassic of Bavaria by H. Weynbergh (1869–1874) and P. Oppenheim (1885, 1888), on Lias of Mecklenburg by G. Geinitz (1880–1887), and of Eastern Siberia by F. Brauer, J. Redtenbacher, and L. Ganglbauer (1889); on Carboniferous faunas of Europe by Ch. Brongniart (1885, 1893), J. Kušta (1883), and M. Kliver (1883–1886). New works on insects of Permian faunas of Europe by F. E. Geinitz (1873) and by J. V. Deichmüller (1886), and new data on Cretaceous insects of Australia by Woodward (1880), and on Holocene insects on coleopterous remains of Middle Europe and of the Carpathian area by Lomnicki (1894), have appeared.

In 1867, S. H. Scudder, a paleoentomologist who studied different groups of Cainozoic, Carboniferous, and some other insects of North America and partly of Europe, began to publish his works. Scudder's intensive activity continued up to 1900. His works mark the beginning of paleoentomological research in the New World. In 1885, Scudder made a first review of the achievements in paleoentomology, participating in the well-known manual on paleontology, published by K. Zittel.

Beginning with the twentieth century, an Austrian entomologist, A. Handlirsch, began to publish his works (1904). He tried to compile a monographic work on paleoentomology. That was a well-known two-volume summary *Die Fossilen Insekten* (1906–1908) which covered all the data on the study of insects of the geological past. The publication of this work was undoubtedly a turning point in the development of paleoentomology. Handlirsch gave a description of all known Paleozoic and Mesozoic insects and published a catalogue of numerous descriptions of insects of Cainozoic faunas. It appeared that a greater number of descriptions of insects was made on materials on Tertiary deposits, mainly paleogenous, that is, on Baltic amber of Europe and on a number of other deposits of Europe and North America. Considerably fewer insects were described from Carboniferous deposits of Europe and North America and from Jurassic deposits of Western Europe. The descriptions of the insects of the Permian period were not numerous. This can be accounted for by the fact that Triassic and Cretaceous faunas were very little studied. In addition, geographic unevenness of the study of the whole world remained very sharp. The overwhelming majority of the data on faunas of insects of the past were obtained in Europe and much less in North America. Asia, South America, and Australia gave only separate finds of fossils. No information was obtained from Africa.

Such a review of paleoentomological works made by Handlirsch should be supplemented by the character of the research works which, in the nineteenth century, were usually carried out by scientist-paleontologists who did not connect their research work with the study of recent insects. The majority of comparisons of fossils was carried out on the general level, namely the family, and rarely at the level of separate genera. This inevitably resulted in incompleteness of descriptions and inaccuracy of systematic conclusions. The specialization of research workers was, as a rule, of purely geological character: certain epochs of the past were studied and descriptions of the whole faunistic complexes of insects of some separate localities of remains were made. Zoological specialization, namely, the study of separate groups of insects based on the remains of different geological deposits was rarely carried out. Only accidental descriptions of some, mainly Tertiary fossils, were made by entomologists, specialists in taxonomics of corresponding groups of recent insects. Such works, however, were very few; for example, on Tertiary Diptera by H. Loew (1861), on dragonflies by H. A. Hagen (1854–1858), on caddisflies by F. A. Kolenati (1848), on ants by G. L. Mayr (1867–1868), and some other works. The main part of paleoentomological descriptions was made by nonspecialist paleontologists and, owing to this fact, the greater part of these publications lack validity and the data described demand serious study, for instance, works by F. Meunier in 1892 (1925).

It is necessary to add that the appraisal of the significance of paleontological research work is determined not only by the completeness and level of the descriptions and comparisons, but also by the absolute geological antiquity of fossils. Insects of different epochs of the geological past of the earth differ from modern ones in proportion to their antiquity. This is especially so for the insects of Paleozoic era. These are first of all the faunas of the Carboniferous period; they differ greatly from modern ones by the very different structure of orders, the greater part of which are not now represented in the recent faunas. In some Carboniferous orders which have survived to the modern epoch, the composition of families is quite different. In the Permian period there appear some orders of insects widely represented even at the present time. Much greater changes have taken place in the Mesozoic era when almost the whole order structure of this class became very similar to the modern one. Even some families appear which continue to exist to the present. During the middle of the Cretaceous period, the fauna became very similar to the modern one.

Such character of insects of faunas of the geological past determines the ease of study of their fossils. Really distinctive insects of the Paleozoic era, which differ sharply from modern ones, may be characterized and described much more easily than those which are similar to modern Mesozoic and in particular to paleogenous and neogenous fossils. Plainly speaking, superficial nonexact descriptions of ancient remains, for example, Carboniferous remains of insects made in the nineteenth century, are more informative than descriptions of Mesozoic, and especially of paleogenous insects. These incomplete descriptions of Carboniferous insects give important information on the organization of these original animals. At the same time, the incomplete information on the structure of insects of Baltic amber reveals very little to scientists and often leads to false conclusions about propinquity and even identity of these paleogenous forms with modern ones. Thus, it is obvious that research on the youngest fossils is most difficult and can be carried out only under conditions of complete knowledge of the whole structure of the recent fauna and organization of all representatives of the groups of insects now alive.

The analysis of development of paleoentomology in the nineteenth century makes it possible to understand the significance of further research in this field of knowledge, carried out following Handlirsch's report at the beginning of the present century.

On the one hand, further descriptive works were forthcoming and provided data on new faunas and on separate finds of fossils. Such works were carried out both on relatively well-known faunas of the Carboniferous period, let us say, in England, by H. Bolton, in North America by A. Handlirsch, of the Permian formation in North America by E. H. Sellards, and particularly on various Cainozoic faunas of Europe by R. S. Bagnall, C. T. Brues, T. D. A. Cockerell, F. M. Edwards, and A. C. Kinsey and of North America by C. T. Brues, E. G. Mitchell, G. Ulmer, and H. F. Wickham.

On the other hand, the number of works dealing with certain groups of insects increases considerably. These works carried out by specialist-entomologists became more and more numerous though they were partly accidental. The next important stage in the history of paleoentomology is the development of such a kind which may be called phylogenetic.

During this important change in the science of insects of the past the works by R. Y. Tillyard (1881–1937) and A. B. Martynov (1879–1938) were of primary importance. These scientists may be characterized as men of great learning. Extensive entomological knowledge was natural for both of them and, at the same time, paleoentomology was part of the development of their evolutionary interests.

Beginning in 1916, the Australian entomologist R. Y. Tillyard, who originally studied the biology of larvae of dragonflies, insects of the fauna of Australia and New Zealand, began publishing his works on paleoentomology. His interests were not limited only to Mesozoic and Tertiary insects of Australia. Soon he began to study the rich Permian fauna of Kansas (North America) and Mesozoic insects of England and other countries. As a result of these studies Tillyard drew many new phylogenetic conclusions and various general conclusions on the evolution of many orders of insects.

The Russian zoologist A. B. Martynov began his paleoentomological research when he was already an outstanding scientist. The desire to find paleontological evidence for his morphological general conclusions on the evolution of insects made Martynov begin descriptions of fossils of insects from the territory of the USSR, which had not yet been studied. Martynov's paleoentomological activity (1922–1938) very soon gave important results which he began to publish in 1925. He ascertained that in the USSR there were numerous locations of fossils of insects (more than 50) of various ages, mainly of Permian (in the European part of the USSR) and Jurassic (in Middle Asia) formations. Martynov's discoveries made it possible to describe many very little known faunas and to substantiate numerous new large taxa (among them several orders and a number of families) and, what is especially important, to carry out analysis and to outline a scheme for the systematics and phylogenesis of the whole class of insects.

In the middle of the 1920s, the American entomologist F. M. Carpenter (1926) began to study some groups of insects of Modern and Tertiary faunas of North America (Neuroptera, Raphidioptera, ants), and later Permian, Mesozoic, and Carboniferous insects from various localities, primarily from North America. His research work is very accurate and detailed. Especially important are Carpenter's findings on the study of original collections of fossils studied earlier, that is, his revision of the forms described superficially and not precisely. Such works are of great importance because of the imperfection of the works of the last century. As a result of such revisions, the existing notions about the forms described earlier have greatly changed and their systematic and phylogenetic relations have become more precise.

In spite of the fact that sometimes, even fairly recently, the descriptions of the various whole faunistic complexes were made by one paleontologist and not by specialists in certain orders and groups of insects, the main mass of paleoentomological research came to be carried out by entomologist-specialists and it often assumed a purely phylogenetic character. Mention should be made of some most fruitful paleoentomologists who followed the universality of Handlirsch. Such are, for example, T. D. Cockerell (1906–1928), who described various fossils, chiefly from the paleogenous period in North America; G. M. Zalesski (1931–1955), who studied various insects of the Urals of the Permian period; G. Statz (1937–1950), who investigated insects from one rich paleogenous location in Europe; A. Bode (1953), who described insects from Lias of Western Europe. A great number of new taxa, unfortunately with little reliability, were described by these investigators.

Research work by Tillyard and Carpenter had, without question, great influence on the further development of phylogenetic paleoentomology; first of all in the countries in which they worked. Thus among the studies of Tillyard's Australian followers may be mentioned those by J. W. Evans (1934–1970) on Hemiptera of the Paleozoic and Mesozoic periods, E. F. Riek (1950–1970) on various Mecoptera and some other groups, and C. Davis (1942–1943) on Hemiptera, Psocoptera, and Neuroptera.

Paleoentomological works continued to gain in scope especially in the Soviet Union. After Martynov's death in 1938, the author of this article began to work on Mesozoic Diptera, E. E. Becker-Migdisova on Permian and later Mesozoic and Cainozoic Homoptera and Psocoptera and on Paleozoic cockroaches; O. M. Martynova on Raphidioptera, Neuroptera, and some other groups; and A. G. Sharov on Apterygota and Orthoptera.

Beginning with the 1960s the number of scientist-paleoentomologists has grown in the USSR. Among them, O. A. Tshernova began to work on ancient mayflies; L. N. Pritykina on dragonflies; V. N. Vishniakova on cockroaches and Psocoptera; Ju. A. Popov on Heteroptera; V. V. Zherichin, A. G. Ponomarenko, A. Tichomirova, L. N. Medvedev, and S. M. Iablokov-Khnzorian on various groups of Coleoptera; D. A. Ussatschev in Diptera; D. V. Panfilov on Neuroptera and the paleoecology of insects; and A. P. Rasnitsyn, G. B. Dlussky, and M. A. Kozlov on Hymenoptera.

In North America, namely in the United States, the development of paleoentomology was greatly influenced by Carpenter, who, apart from his active personal investigation, organized collective treatment of fossil faunas chiefly of the Cainozoic and Cretaceous periods. Among these are the numerous investigations on the paleogenous faunas of the western United States, of Florissant and of Creede formation situated in Colorado. A series of works had been carried out also in the paleogenous faunas of Chiapas in Mexico. In recent years, studies on Cretaceous faunas from Canada (Cedar Lake, Manitoba, and Redmond Labrador) have been published. Studies carried out by specialists in the taxonomy of various groups of insects have been on Apterygota (I. W. Folsom), dragonflies (F. C. Fraser), termites, (E. E. Snyder, A. E. Emerson), Rhynchota (E. O. Essig, R. L. Usinger),

Thysanoptera (R. S. Bagnall), Hymenoptera (F. O. Wilson, C. T. Brues, A. C. Kinsey), Diptera (C. P. Alexander, M. T. James, A. L. Melander, F. M. Hull, J. F. McAlpine), among others. It is necessary to mention important research on Cainozoic insects of various orders, carried out mainly on Coleoptera by W. D. Pierce (1944–1963), H. M. A. Rice (1959), and especially by H. F. Wickham (1908–1931). In addition, F. M. Carpenter and E. Richardson (1953–1956) carried out scientific work on insects of the Carboniferous epoch.

By the middle of this century, paleoentomological research was carried out in a number of other countries, mainly in Europe. In France, D. Laurentiaux and his collaborators studied ancient Paleozoic insects, Triassic, and mostly Cainozoic faunas. These faunas were also studied by L. Piton, F. Quievreux, N. Theobald, S. Timon-David, and others. Paleozoic insects were described from the Carboniferous and the Permian of Portugal (and partly from Africa!) by C. Teixeira, from Germany by P. Guthörl, and from Czechoslovakia by J. Kukalová. Numerous are the works on Cainozoic insects of Denmark (O. Heie on aphids), Germany, Austria, Czechoslovakia, Hungary, Rumania, Poland, and of some other countries of Europe.

New data from Asia on Cainozoic insects of Japan are by N. Naora, I. Fujiyama, and others, and from India (by V. P. George). Recently, a work on Cretaceous insects of Lebanon by W. Hennig and D. Schlee and a work on Permain insects of Madagascar by K. Paulian have been published.

South America and Africa still remain little studied except for notes on Chili by P. G. Kuschel, on Uruguay by G. Frenguelli, and on Brazil by others.

Consideration of all the work on paleoentomology, namely on the study of faunas of definite epochs of the geological past, makes it possible to draw brief conclusions. Paleozoic faunas continue to be studied in Western Europe, the USSR, North America, and Australia. The discovery of Devonian winged insects by Rohdendorf was based on a false determination of the remains of uropods of Eumalacostraca (Crustacea). Thus, in that way only representatives of wingless insects (Collembola) were known in the fauna of the Devonian and Pterygota were discovered only in the fauna of the end of early Carboniferous epoch.

The study of the fauna of Carboniferous insects has continued in most regions. Such are the descriptions of Carboniferous insects of Portugal, France, Germany, South Siberia, and North America.

Investigations are numerous on Permian faunas, especially from North America, the European part of the USSR, Czechoslovakia, South Siberia, and Australia.

Triassic insects have been investigated very little. Many research studies were carried out in Australia (R. Y. Wootton), and separate descriptions were made on material from Europe, the South Urals, mid-Asia, Siberia, and South America.

Jurassic insects were mainly studied in Eurasia. First of all, mention should be made of the work on the Lower Jura (Lias) of Germany, Mecklenburg, compiled by A. Handlirsch (1937–1939), and some works on material from England and Siberia. The most rich and diverse material was obtained from mid-Asia. The study of Jurassic fauna of mid-Asia was started by A. B. Martynov (1925) who discovered the rich occurrence of the Upper Jurassic insects in Kazakhstan (Karatau) and of the Lower Jurassic in Fergana (1937). Further investigation considerably expanded the knowledge of Jurassic faunas of this region. Several hundred species of numerous orders of insects were discovered, the knowledge of the composition of the Jurassic fauna of mid-Asia became relatively full. Other works on Jurassic (or early Cretaceous) insects of Eurasia, which were, to a certain extent, accidental, were carried out on the fauna of Mongolia and China. Data on Jurassic faunas of the earth are limited to this information; other continents, among them America, Africa, and Australia have not been studied in this respect up to the present moment.

The Cretaceous fauna remained until recently the least studied. Only some separate forms from the Cretaceous of North America, European Russia, and from the earliest formations of Cretaceous of Eastern Asia (Transbaikalia) have been described. The localities of fossilized Cretaceous resins in Canada, in Lebanon, and in polar Siberia promise to provide a variety of information on the composition of the Cretaceous fauna.

Most extensive are the data on insects of Paleogene and Neogene, to which numerous investigations by scientists of various countries were devoted. The study of Cainozoic insects in this century acquired a different character than before. The necessity of precise comparison of fossils similar to modern forms demanded specialization on the part of investigators. The participation of entomologist-specialists on taxonomy of separate groups of insects in these works became almost a rule. Purely paleoentomological works, dealing with the fauna of a given location as a whole, became rare. The other feature of these new works on Cainozoic forms is revision of the earlier described forms in their second study in detail. Works on Cainozoic fossils of this kind are of special importance.

There is no possibility of considering fully all the data on paleogenous and neogenous insects obtained by the 1970s. The main works dealt with the faunas of North America and Northern Europe; Eastern Asia and South America were considerably less studied. The data on Central and South Asia and Africa was very rare.

Consideration of paleoentomological research on Cainozoic faunas shows clear predominance of special studies on separate orders and even families of insects. A greater part of the studies is devoted to three orders of holometabolous insects: Coleoptera, Diptera, and Hymenoptera. The overwhelming preponderence of the research on these major, most diverse, and usual orders of insects was devoted to the description of separate families.

Because of a lack of material, somewhat fewer investigations were carried out on the orders Heteroptera, Isoptera, Odonata, and Lepidoptera. Few works dealt with the orders of Ephemeroptera, Psocoptera, Thysanoptera, Orthoptera, Trichoptera, and Homoptera, and even fewer were devoted to Dermaptera, Blattodea, Mecoptera, Aphaniptera, Collembola, Diplura, and Thysanura. The orders of parasitic insects, Mallophaga, Anoplura, and Protura remain up to now fully unknown as fossils. All these features of paleoentomological data on separate orders of insects only partly reflect the real diversity and abundance of the given insects in the fauna. The main reason is, of course, the different conditions of fossilization of insects.

Mention should be made of the success in the study of insects of the Holocene, the systematic composition of which fully corresponds to the modern one: faunistic composition of Holocene complexes reflect geographic and climatic changes of the territories under study. It is obvious that these works were made exclusively by zoologists, specialists in the taxonomy of corresponding groups of insects. In the nineteenth century Quaternary insects were studied only in Europe by F. Sordelli (1882) and A. M. Lomnicki (1894). From the beginning of this century, the number of works gradually increased; mainly Coleoptera are being studied, remnants of which are most numerous as a result of their hardness; such are the works on northern and western Europe.

Among various groups of insects of the Holocene of Europe, Trichoptera, Diptera, and Lepidoptera were described in some works; others were devoted to Heteroptera or to Diptera of Europe. Works on the Holocene fauna of California and Alaska in North America should be noted. Cicadidae and Papilionidae were described from the Holocene of Japan. Finally one may note the survey of the Holocene fauna of Scandinavia by K. L. Henriksen (1933), and of other such faunas in general (D. G. Frey, 1965).

At present it is possible to draw some conclusions in the history of paleoentomology, namely regarding the work carried out for almost 200 years, its results, techniques, and its prospects. The data on the insects of the geological past cover the period from the late Devonian to the Modern epoch. However, the data on the faunas of the past are not consistent. The fossil materials of the fauna of early Carboniferous and of the insects of Devonian are very few or are almost lacking. The faunas of Trias, Cretaceous, early Paleogene are insufficiently studied. Vast territories of the earth still remain practically unexplored—such as the ancient Gondwanaland and the American, African, and South Asian regions. The Mesozoic fauna is known one-sidedly, only through its Eurasian and Australian regions; the whole of America and Africa remains almost completely unstudied with the exception of some data on the Cretaceous fauna. Cainozoic faunas are very little known from Asia and the territory of Gondwanaland.

Paleoentomological data obtained make it possible to judge with relative completeness the composition of the faunas of the greater part of geological periods and upon the system of major superorders of the class Insecta. Thus, great differences among the fauna of the Carboniferous epoch, containing peculiar ancient orders not represented in late epochs, are established. Peculiar is the fauna of the Permian period, characterized by the emergence of a number of new orders and having, on the whole, greater similarity to the faunas of the early Mesozoic period, Triassic and Jurassic, including representatives of a majority of modern orders. The Cretaceous fauna differs much from the Jurassic one by the disappearance of many ancient, early Mesozoic, and relict Paleozoic orders. The Cretaceous fauna is characterized by the emergence of a great number of Cainozoic families, and thus, by the acquisition of a general, purely Cainozoic appearance: the line between the Jurassic and Cretaceous periods turned out to be more distinct than between the Cretaceous and Paleogene periods. The whole of the Tertiary fauna does not, in its turn, differ by its composition from the modern one.

Thus, the Carboniferous-Permian and Jurassic-Cretaceous lines are very distinct. Classic lines of the Paleozoic, Mesozoic, and Cainozoic eras of the faunas of insects turned out to be displaced and not reflect generally accepted ideas.

The technique of paleoentomological study is determined by the originality of one or another fossil. If the general practical sides of the work are now quite clear, the very methods of investigation of fossils have changed very little and some new optical and chemical methods have not yet received wide application.

The need for specialized comparative study of separate groups of insects is the main condition determining the success of an investigation. Such will be true for the relatively monotonous orders (for example, Ephemeroptera), for some separate suborders or superfamilies (for example, Coleoptera, Archostemata, or Heteroptera Nepomorpha), and even for separate families or subfamilies (for example, Staphylinidae or Donaciinae).

The observation of the conditions of specialization is necessary for the study of Cainozoic and Cretaceous insects which are relatively close to recent ones and demand, by definition, the knowledge of the composition and organization of the forms now alive, and within the limits of the whole modern fauna, especially within rich and often very little studied tropical and subtropical regions.

A bit different are the conditions for the study of Mesozoic (Jurassic and Triassic) and especially of Paleozoic faunas, in which cases the knowledge of the general systematics of the orders and of the whole class is necessary. These ancient insects so differ from modern ones that they do not require full knowledge of specific and generic composition of the groups now alive and, as a rule, allow one to use more general data, approximately on the level of modern families of the class.

The foregoing stipulates participation of a great number of paleoentomological investigations. The speciality paleoentomology nowadays is becoming an anachronism. Descriptions of the whole faunistic complexes by one person, such as the works carried out by founders of paleoentomology in the past and at the beginning of the present century are no longer realistic. Attempts to carry out studies of such a scale lead inevitably to low quality descriptions of the fossil remains and thus to errors in the conclusions.

Special claims are laid to faunistico-paleoecological investigations of separate faunistic complexes. In such cases, the division of labor among several scientists is inevitable. Most of the effort must go into the taxonomic aspects of the work, that is, the precise definition of separate taxa. Thus is established the condition for the collective study of the whole faunistic complex, and the further analysis of ecological peculiarities of the whole biogeocoenoses can be carried out successfully and precisely.

Concluding the article on the history of paleoentomology, it is necessary to touch upon the immediate tasks of this science. First of all it is necessary to search for, obtain, and study fossils from the least studied deposits (from Devonian, early Carboniferous, Triassic, Cretaceous, and early Paleogene), to find new localities of fossils in the very little studied regions (as in the territory of ancient Gondwanaland), and to intensify the phylogenetic study of separate groups of insects, while carrying out revisions of imperfect descriptions of fossils.

I have not treated the tasks and peculiarities of carrying out phylogenetic-systematic studies which are only partly paleoentomological. The use of data on paleontology under conditions in which phylogenesis is treated turns out to be only one aspect of the matter, as the greater part of phylogenetic investigation constitutes and concerns peculiarities of modern animals.

LITERATURE CITED

Alexander, C. P. 1931. Crane-flies of the Baltic Amber. *Bernstein Forsch.* 2:1–135

Bagnall, R. S. 1924. Fossil Thysanoptera. II–III. *Entomol. Mon. Mag.* 60: 130–33, 251–52

Bauer y Landauer, J. 1935. Un manuscrito sobre Polonia en la biblioteca de Don Pedro de Aragon. *Bol. Soc. Geogr. Nat.* 75:634

Becker-Migdisova, E. E. 1940. Iskopaemaya permskaya tsikada semeistva Probolidae s reki Soiany. *Tr. Paleontol, Inst.* 9, 2, 5–79

Becker-Migdisova, E. E. 1949. Novoe permskoe semeistvo Boreoscytidae i vopros o filogenii predkov Homoptera. *Tr. Paleontol. Inst.* 20:171–82

Becker-Migdisova, E. E. 1951. Svoeobraznye novye iskopaemye nasekomye iz gruppy Paraneopter. *Dokl. Akad. Nauk SSSR* LXXVIII, 6:1207–10

Becker-Migdisova, E. E. 1960. Novye permskie ravnokrylye Evropeiskoi chasti SSSR. *Tr. Paleontol. Inst.* 76: 1–112

Becker-Migdisova, E. E. 1967. Tertiary Homoptera of Stavropol and a method of reconstructing continental palaeobiocoenoses. *Palaeontology, Cambridge,* 10, 4:542–53

Becker-Migdisova, E. E. 1968. Protopsillidiidy i ikh morfologiya (Protopsyllidiidae); Jurassic insects of Karatau. *Nauka, Moscow,* pp. 87–98

Bolton, H. 1925. Insects from coal measures of Commentry. *Brit. Mus. Fossil Insects* 2:1–56

Brauer, F., Redtenbacher, J., Ganglbauer, L. 1889. Fossile Insekten aus der Juraformation Ost-Sibiriens. *Mem. Acad. St. Petersburg* 7, 36, 15: 1–15

Brodie, P. B. 1845. *A History of the Fossil Insects in the Secondary Rocks of England, Accompanied by a Particular Account of the Strata in which They Occur, and of the Circumstances Connected with Their Preservation.* London: van Voorst. 8:1–130, I–IX

Brongniart, Ch. 1893. *Recherches pour servir à l'histoire des insectes fossiles des temps primaires précédées d'une étude sur les nervation des alles des insectes. Saint Etienne,* pp. 1–493

Brues, C. T. 1910. The parasitic Hymenoptera of the Tertiary of Florissant, Colorado. *Bull. Mus. Comp. Zool. Harvard Univ.* 85, 6:1–26

Brues, C. T. 1933. The parasitic Hymenoptera of the Baltic Amber. *Bernstein Forsch.* 3:4–175

Brues, C. T. 1939. Fossil Phoridae in Baltic Amber. *Bull. Mus. Comp. Zool. Harvard Univ.* 85, 6:1–26

Carpenter, F. M. 1930. The fossil ants of North America. *Bull. Mus. Comp. Zool. Harvard Univ.* 70, 1:1–66

Carpenter, F. M. 1939. The Lower Permian insects of Kansas. Pt. 8. Additional Megasecoptera, Protodonata, Odonata, Psocoptera, Protelytroptera, Plecoptera, and Protoperlaria. *Proc. Am. Acad. Arts Sci.* 73, 3:29–70

Carpenter, F. M. 1943. Studies on carboniferous insects from Commentry, France. Pt. 1. Introduction and families Protagriidae, Meganeuridae, and Campylopteridae. *Bull. Geol. Soc. Am.* 54:524–54

Carpenter, F. M. 1947. Lower Permian insects from Oklahoma. Pt. 1. Introduction and the orders Megasecoptera, Protodonata, and Odonata. *Proc. Am. Acad. Arts. Sci.* 76, 2:25–54

Carpenter, F. M. 1967. Studies on North American carboniferous insects. 5. Palaeodictyoptera and Megasecoptera from Illinois and Tennessee, with a discussion of the order Sypharopteroidea. *Psyche* 74, 1:58–84

Carpenter, F. M. et al 1937. Insects and arachnids from Canadian Amber. *Univ. Tronto, Geol. Ser.* 40:7–55

Carpenter, F. M., Richardson, E. S. 1968. Megasecopterous nymphs in Pennsylvanian concretions from Illinois. *Psyche* 75, 4:295–309

Cockerell, T. D. A. 1909. A catalogue of the generic names based on American insects and arachnids from the Tertiary rocks with indications of the type species. *Bull. Am. Nat. Hist.* 26:78

Cockerell, T. D. A. 1924. Fossils in the Ondai Sair Formation, Mongolia. *Bull. Am. Mus. Nat. Hist.* 51, 6: 129–44

Coope, G. R. 1970. Interpretations of Quaternary insect fossils. *Ann. Rev. Entomol.* 15:97–120

Davis, C. 1942. Hemiptera and Copeognatha from Upper Permian of New South Wales. *Proc. Linn. Soc. N.S.W.* 67:3–4

Deichmüller, J. V. 1886. Die Insekten aus dem lithographischen Schiefer im Dresdener Museum. *Mitt. Kgl. Mineral. Geol. Mus. Dresden* 7:1–84

Demoulin, G. 1954. Quelques remárques sur les Archodonates. *Bull. Ann. Soc. Entomol. Belg.* 90, XI–VII:327–37

Demoulin, G. 1955. Contribution à l'étude des Éphémeroptères jurassiques d'Europe centrale. III. Phylogénie et zoogéographie. *Mem. Soc. Entomol. Belg.* 27:176–83

Demoulin, G. 1968. Deuxième contribution à la connaissance des Ephéméroptères de l'ambre oligocène de la Baltique. *Deut. Entomol. Z., N. F.* 15,I–III:233–76

Demoulin, G. 1970. Remarques critiques sur des larves "Ephemeromorphes" du Permien. *Bull. Inst. Roy Sci. Nat. Belg.* 46, 3:1–10

Dlussky, G. M. 1967. Muravji roda Formica Baltijskogo jantarja. *Paleontol. Zh.* 2:80–89

Emerson, A. E. 1933. A revision of the genera of fossil and recent Termopsinae. *Univ. Calif. Publ. Entomol.* 6:165–96

Evans, J. W. 1956. Palaeozoic and Mesozoic Hemiptera. *Aust. J. Zool.* 4, 2:165–68

Fujiyama, I. 1969. A miocene cicada from Nasu, with an additional record of a pleistocene cicada from Shiobara, Japan. *Bull. Nat. Sci. Mus. Tokyo* 12(4):863–73

Geinitz, F. E. 1884. Über die Fauna des Dobbertiner Lias. *Z. Deut. Geol. Ges.* 13:566–83

Germar, E. F. 1839. Die versteinerten Insecten Solenhofens. *Acta Acad. Leopold Carol.* 19, 1:187–222

Giebel, Ch. G. A. 1857. Zur Fauna des lithographischen Schiefers con Solenhofen. *Z. Ges Naturwiss.* 9:373–88

Goldenberg, F. 1873, 1877. Fauna saraepontana fossilis. Die fossilen Thiere aus der Steinkohlformation von Saarbrücken. *Saarbrücken* 1:1–26; 2:1–54

Guthörl, P. 1934. Die Arthropoden aus dem Carbon und Perm des Saar-Nahe-Pfalz Gebietes. *Abt. Preuss Geol. Landesanstalt* 164:5–219

Guthörl, P. 1939. Zur Arthropodenfauna des Karbons und Perm. 9. Palaeodictyoptera, Mixotermitoidea, Miomoptera und Blattaria. *Senkenbergiana* 21, 5/6:314–29

Hagen, H. A. 1854. Ueber die Neuropteren der Bernsteinfauna. *Verhandl. Wien. Zool. Bot. Verein* 4:221–32

Hagen, H. A. 1862. Ueber die Neuropteren aus dem lithographischen Schiefer in Bayern. *Palaeontographica* 10:95–145

Handlirsch, A. 1906–1908. *Die fossilen Insekten und die Phylogenie der rezenten Formen.* Leipzig: Engelmann, pp. 1–1433

Handlirsch, A. 1911. New palaeozoic insects from the vicinity of Mazon Creek, Illinois. *Am. J. Sci.* 31, 28: 297–326, 353–77

Handlirsch, A. 1937–1939. Neue Untersuchungen über die fossilen Insekten mit Ergänzungen und Nachträgen sowie Ausblicken und phylogenetiche, paläogeographische und algemein biologische Probleme. *Ann. Nat. Mus. Wien* 48:1–140; 49:1–240

Heer, O. 1847, 1853. *Die Insectenfauna der Tertiärgebilde von Oeningen und Radoboj in Croatien.* Leipzig: Engelmann. 1:1–229; 2:1–269; 3:1–138

Heer, O. 1870. Die miozäne Flora und Fauna Spitzbergens. *Kgl. Svenska Vet. Händl.* 8, 5, 7:1–98

Heer, O. 1879. *Die Urwelt der Schweiz.* Zurich, 2nd ed., pp. 1–713

Heie, O. E. 1967. Studies on fossil aphids, especially in the Copenhagen collection of fossils in Baltic Amber. *Spolia Zool. Mus. Hauniensis* 26: 1–274

Heie, O. E. 1971. The rediscovered types of the fossil aphids described by Germar and Berendt in 1856. *Deut. Entomol. Z.* N.F. I–III:251–64

Hennig, W. 1969. *Die Stammesgeschichte der Insekten.* Frankfurt a. Main: W. Kramer, pp. 1–436

Heyden, L. v. 1870. Fossile Dipteren aus der Braunkohle von Rott im Siebengebrige. *Palaeontographica* 17: 237–66

Hull, F. M. 1945. A revisional study of the fossil Syrphidae. *Bull. Mus. Comp. Zool. Harvard Univ.* 45:249–355

Iablokoff-Khnzorian, S. M. 1961. Predstaviteli semejstva Helodidae (Coleoptera) iz baltijskogo jantarja. *Paleontol. Zh.* 1:108–16

Jaczewski, T. 1949. Uwagi o wioslakach (Corixidae) XXIV–XXVIII. Polskie Pismo-Entomologiszne *(Bull. Entomol. de la Palogne)* 19, 3–4:103–13

James, M. T. 1937, 1939. A preliminary review of certain families of Diptera from the Florissant Miocene Beds. *J. Paleontol.* 11, 3:241–47, 13:42–48

Jessen, K., Milthers, V. 1928. Stratigraphical and paleontological studies of interglacial fresh-water deposits in Jutland and Northwest Germany. *Danm. Geol. Unders., III Raek* 48:1–379

Kozlov, M. A. 1968. Yurskie Proctotrupidea (Hymenoptera). Jurassic Insects of Karatau. *Nauka Moscow,* pp. 237–40

Kukalová, J. 1964. Permian Insects of Moravia. Pt. II–Liomopteridea. Sborník geolog. věd. *Paleontologie P* 3: 39–118

Kukalová, J. 1968. Permian Mayfly Nymphs. *Psyche* 75, 4:310–27

Kukalová, J. 1970. Revisional study of the order Palaeodictyoptera in the upper carboniferous shales of Commentry, France. Pt. III. *Psyche* 77:1, 1–44

Kusnezov, N. J. 1941. A revision of the Amber Lepidoptera. *Akad. Sci. URSS* Moscow, Leningrad. 1:1–35

Laurentiaux, D. 1953. Classe des Insectes. In *Traité de Paléontologie,* ed. J. Piveteau, 3:397–527

Laurentiaux-Vieira, F., Ricour, J., Laurentiaux, D. 1953. Un Protodonate du Trias de la Dent de Villard (Savoie). *Bull. Soc. Géol. France* 6, 2:319–24

Loew, H. 1861. Ueber die Dipteren-Fauna des Bernsteins. Bericht über die 35. *Versammlung Deut. Naturforsch.* 88–98

Lomnicki, A. M. 1894. Pleistoceńske owady z Boryslawia. *Mus. im. Dzieduszyckich, Lemberg* 4:1–127

Lomonosov, M. W. 1794. Slovo chetvertoe, o rozhdenii metallov ot" tryaseniya zemli, sentiabria 6 dnia 1754 goda goverennoe. Polnoe Sobranie soch. M. V. Lomosova, izd. 2, III:1–261

Martynov, A. B. 1925. Résumé des recherches les insectes fossiles des couches jurassiques du Turkestan. *C. R. Acad. Sci., Russie* 105–8

Martynov, A. B. 1927. Über eine neue Ordnung der fossilen Insekten, Miomoptera nov. *Zool. Anz.* 72, 3/4: 99–109

Martynov, A. B. 1928. Permian fossil insects of Northeast Europe. *Trav. Mus. Géol. Acad. Sci. URSS* 4:1–118

Martynov, A. B. 1932. New Permian Palaeoptera with the discussion of some problems of their evolution. *Trav. Ins. Paléozool.* 1:1–44

Martynov, A. B. 1937. Wings of termites and phylogeny of Isoptera and of allied groups of insects. Sbornik in honour of N. V. Nasonov. *Acad. Sci. URSS* 83–150

Martynov, A. B. 1937. Liassic insects from Surab and Kisyl-Kiya. *Trav. Inst. Paléontol. Acad. Sci. URSS,* VII 1:1–232

Martynov, A. B. 1938. Review of localities of fossil insects in USSR. *Trav. Inst. Paléontol. Acad. Sci. URSS,* VII 3:7–28

Martynov, A. B. 1938. On a new Permian order of orthopteroid insects Glosselytrodea. *Bull. Acad. Sci. URSS, Cl. Math. Sci. Nat.* 1:187–206

Martynov, A. B. 1938. Etudes sur l'histoire géologique et de phylogénie des ordres des insects (Pterygota). l-e partie Palaeoptera et Neoptera-Polyneoptera. *Trav. Inst. Paléontol. Acad. Sci. URSS* VII 4:1–150

Martynov, A. B. 1940. Permian fossil insects from Tshekarda. *Trav. Inst. Paleontol. Acad. Sci. URSS* XI 1: 1–62

Martynova, O. M. 1948. Materialy po evoliutsii Mecoptera. *Tr. Palaeontol. Inst.* 14, 1:1–76

Martynova, O. M. 1949. Mezozoiskie setchatokrylye SSSR. *Tr. Palaeontol. Inst.* 20:150–70

Martynova, O. M. 1952. Permskie setchatokrylye SSSR. *Tr. Palaeontol. Inst.* 40:197–237

Martynova, O. M. 1953. Nastavlenie dlia sborov iskopaemykh nasekomykh. *Izd. Paleontol. Inst. Akad. Nauk. USSR, Moscow* 1–14

Martynova, O. M. 1961. Sovremennye i vymershie verbliudki (Raphidiioptera). *Palaeontol. J.* 3:73–83

McAlpine, J. F. 1970. First record of calypterate flies in the Mesozoic era. *Can. Entomol.* 102, 3:342–46

McAlpine, J. F., Martin, J. E. H. 1966. Systematics of Sciadoceridae and relatives with descriptions of two new genera and species from Canadian Amber and Erection of family Ironomyiidae. *Can. Entomol.* 98, 5:527–44

Medvedev, L. N. 1960. Zhuki-listoedy iury Karatau (Chrysomelidae). Jurassic Insects of Karatau. *Nauka Moscow,* pp. 155–65

Oppenheim, P. 1888. Die Insektenwelt des lithographischen Schiefer in Bayern. *Palaeontographica* 34:215–47

Oustalet, E. 1870. Recherches sur les insectes fossiles des terrains tertiaries de la France. 1-re Part. Insectes fossiles de l'Auvergne. *Ann. Soc. Géol.* 2,3:1–178

Panfilov, D. V. 1968. Ekologo-landshaftnaia kharakteristika iurskoi fauny nasekomykh Karatau. Jurassic Insects of Karatau. *Nauka, Moscow,* pp. 7–22

Pierce, W. D. 1919. The comparative morphology of the order Strepsiptera together with records and descriptions of insects. *Proc. US Nat. Mus.* 54, 2242:391–501

Pierce, W. D. 1963. Fossil arthropods of California. 25. Silicified leaf-hoppers from California mountain nodules. *Bull. S. Calif. Acad. Sci.* 62: 69–82

Ping, C. 1928. Cretaceous fossil insects of China. *Palaeontol. Sinica, B* 13, 1: 1–56

Piton, L., Théobald, V. 1935. La faune entomologique des gisements miopliocène du Massif Central. *Rev. Sci. Nat. Auvergne, Clermont-Ferrand* 1: 65–104

Pongracz, A. 1935. Die eozäne Insektenfauna des Geiseltales. *Nova Acta Leopold, F* 2:483–572

Ponomarenko, A. G. 1969. Istoricheskoe razvitie zhestkokrylykh-arkhostemat. *Tr. Palaeontol. Inst.* 125:1–238

Popov, Yu. A. 1968. Nastoiashchie poluzhestkokrylye yurskoi fauny Karatau (Heteroptera). Jurassic Insects of Karatau. *Nauka, Moscow,* pp. 99–113

Popov, Yu. A. 1971. Istoriuskoe razvitie poluzhestkokrylykh ihfraotriada Nepomorpha (Heteroptera). *Tr. Paleontol. Inst.* 129:1–230

Pritykina, L. N. 1968. Strekozy Karatau (Odonata). Jurassic Insects of Karatau. *Nauka, Moscow,* pp. 26–54

Rasnitsyn, A. F. 1965. Materialy do poznania biologii, systematyki i filogenezy rośliniarek z podrodziny Xyelinae (Xyelidae). *Polsk. Pismo Entomol.* 35:2, 483–519

Rasnitsyn, A. P. 1969. Proiskhozhdenie i evoliutsiia nizshikh pereponchatokrylykh. *Tr. Palaeontol. Inst.* 123:1–184

Richardson, E. S. 1956. Pennsylvanian invertebrates of the Mazon Creek Area, Illinois. *Fieldiana Geol.* 12:1–4, 3–76

Riek, E. F. 1953. Fossil mecopteroid insects from the Upper Permian of New South Wales. *Records Aust. Mus.* 23, 2:55–87

Riek, E. F. 1970. Fossil history. In *The Insects of Australia,* pp. 168–86. Melbourne: Melbourne Univ. Press

Riek, E. F. 1970. Lower Cretaceous fleas. *Nature* 227:746–47
Rohdendorf, B. B. 1946. Evoliutsiia kryla i filogenez dlinnousykh dvukrylykh. Oligoneura. *Tr. Paleontol. Inst.* 13:2, 5–108
Rohdendorf, B. B. 1957. Paleoentomologcheskie issledovannia v SSSR. *Tr. Palaeontol. Inst.* 66:1–102
Rohdendorf, B. B., Becker-Migdisova, E. E., Martynova, O. M., Sharov, A. G. 1961. The palaeozoic insects of Kuznetsk Basin. *Tr. Palaeontol. Inst.* 85:1–705
Rohdendorf, B. B. et al 1962. *Osnovy Paleontologii. Arthropoda, Tracheata and Chelicerata,* pp. 1–560. Moscow: Acad. Sci., USSR
Schlee, D., Deitrich, H. G. 1970. Insektenführender Bernstein aus der Unterkreide des Libanon. *Neues Jb. Geol. u. Paläöntol. Monatsch.* 1: 40–50
Scudder, S. H. 1879. Palaeozoic cockroaches: a comparative revision of the species of both worlds with an essay toward their classification. *Mem. Boston Soc.* 3:23–134
Scudder, S. H. 1886. Systematic review of our present knowledge of fossil insects. *Bull. US Geol. Survey* 21:1–128
Scudder, S. H. 1890. Tertiary insects of North America. *Rep. US Geol. Survey., Terr.* 12:1–734
Scudder, S. H. 1891. A classed and annotated bibliography of fossil insects. *Bull. US Geol. Survey* 69:1–101
Scudder, S. H. 1891. Index to the known fossil insects of the world including myriapods and arachnoids. *Bull. US Geol. Survey* 71:1–744
Scudder, S. H. 1900. Adephagous and clavicorn Coleoptera from the Tertiary deposits of Florissant, Colo., etc. *Monogr. US Geol. Survey* 40: 1–148
Sellards, E. H. 1909. Types of Permian insects, part 3. Megasecoptera, Oryctoblattinidae and Protorthoptera. *Am. J. Sci.* 4, 25:151–73
Sendelius, N. 1742. *Historia Succinorum corpora aliena involventium et naturae opere pictorum el caelatorum ex Regiis Augustorum cimeliis Dresdae conditis aeri insculptorum conscripta.* Lipsae, Gleditsch, 1–10: 1–328, Tab. 1–13
Serres, P. M. T. de. 1829. *Géognosie des terrains du midi de France, un tableau des principaux animaux invertébrés des terrain marins tertiaires de la France.* Montpellier: Durville. 8:1–92, 1–276
Sharov, A. G. 1957. Svoeobraznye paleozoiskie beskrylye nasekomye novogo otriada Monura (Apterygota). *C. R. Acad. Sci. URSS* 115, 4:795–98
Sharov, A. G. 1968. Filogeniia ortopteroidnykh nasekomykh. *Tr. Palaeontol. Inst.,* pp. 1–218
Snyder, T. E. 1949. Catalogue of the termites of the world. *Smithson. Misc. Collect.* 112:1–490
Statz, G. 1944. Neue Dipteren aus dem Oberoligocän von Rott. *Palaeontographica* 95, A, 1–6: 1–185
Sukatsheva, I. D. 1968. Mesozoic caddisflies from Transbaikalia. *Palaeontol. J.* 2:59–75
Teixeira, C. 1939. Insectes de Estefaniano do Douro Litoral. *Publ. Mus. Lab. Min. Geol. Ci. Porto.* 9:1–20
Théobald, N. 1937. *Les insectes fossiles des terrains oligocènes de France.* Thèse, Universitè de Nancy, pp. 1–473
Tichomirova, A. L. 1968. Zhuki-stafilinidy iruy Karatau. (Staphylinidae). Jurassic Insects of Karatau. *Nauka, Moscow,* pp. 139–54
Tillyard, R. J. 1918–1919. The panorpoid complex. *Proc. Linn. Soc. N.S.W.* 43:286–319, 626–57; 44:533–718
Tillyard, R. J. 1922. Mesozoic insects of Queensland. No. 9. Orthoptera and additions to the Protorthoptera, Odonata, Hemiptera and Planipennia. *Proc. Linn. Soc., N.S.W.* 47, 4:447–70
Tillyard, R. J. 1928. Some remarks on the Devonian fossil insects from the Rhynie Chert beds, Old Red Sandstone. *Trans. Entomol. Soc., London* 76:65–71
Tillyard, R. J. 1932. Kansas Permian insects. Parts XIV, XV. The orders Neuroptera and Plecoptera. *Am. J. Sci.* 23:1–30, 97–134, 237–72
Tillyard, R. J. 1937. Kansas Permian insects. Part XX. The cockroaches or order Blattaria. I, II. *Am. J. Sci.* 34: 169–202, 249–76
Timon-David, J. 1943. Insectes fossiles de l'Oligocène inferieur des Camoins (Bassin de Marseille). I. Diptères Brachycères. *Bull. Soc. Entomol. France* September: 128–34
Tshernova, O. A. 1961. O sistematicheskom polozhenii i geologicheskom vozraste podenok roda Ephemeropsis Eichw. *Entomol. Obozr.* 25, 4:358–69
Tshernova, O. A. 1965. Some fossil

May-flies (Misthodotidae) from Permian beds of the Urals. *Rev. d'Entomol. URSS* XLIV:353–61

Tshernova, O. A. 1970. On the classification of the fossil and recent Ephemeroptera. *Rev. d'Entomol. URSS,* XLIX:1, 124–45

Ulmer, G. 1912. Die Trichopteren des Balticher Bernsteins. *Beitr. Naturk. Pr. physik. ökonom. Ges. Königstein* 10:1–380

Usinger, R. L. 1940. Fossil Lygaeidae from Florissant. *J. Paleontol.* 14, 1: 79–80

Ussatshev, D. A. 1968. New Jurassic Asilomorpha (Diptera) in Karatau. *Entomol. Obozr.* 47:617–28

Vishniakova, V. N. 1968. Mezozoiskie tarakany s naruzhnym iaitsekladom i osobennosti ikh razmnozheiya (Blattodea). Jurassic Insects of Karatau. *Nauka, Moscow,* pp. 55–84

Weyenbergh, H. 1874. Enumération systématique des espèces qui forment la faune entomologique de la période mésozoique de Bavière. *Period. Zool. Argent.* 1:87–106

Wickham, H. F. 1914. New miocene Coleoptera from Florissant. *Bull. Mus. Comp. Zool. Harvard Univ.* 48: 423–94

Wickham, H. F. 1920. Catalogue of fossil North American Coleoptera. In Lang's *Catalogue of Coleoptera North America,* pp. 347–65

Wilson, F. O., Carpenter, F. M., Brown, W. L. 1967. The first mesozoic ants, with the description of a new subfamily. *Psyche* 74, 1:1–19

Wootton, R. J. 1963. Actinoscytinidae (Hemiptera, Heteroptera) from the Upper Triasic of Queensland. *Ann. Mag. N.H.* 13, VI:249–55

Zalessky, G. 1939. Quelques nouveaux réprésentants des Protohymenoptera, Homoptera, Hemipsocoptera, Psocoptera Protoperlaria, Isoptera et Protoblattoidea. *Problems Palaeontol.* 5: 33–91

Zalessky, G. 1955. Novye predstaviteli paleoentomofauny Urala i ikh geologicheskii vozrast. *Ezhegodnik Vses. obva.* 15:274–304

Zeuner, F. E. 1939. Fossil Orthoptera Ensifera. *Brit. Mus. London,* pp. 1–321

Zherichin, V. V. 1970. Zoogeograficheskie svyazi paleogenovykh nasekomykh. *Doklady na XXII ezhegodnom tshtenii pamjati N. A. Cholodkovskogo 14 Apr. 1969, Leningrad,* pp. 29–88

EVOLUTION AND PHYLOGENY

Herbert H. Ross
Department of Entomology, University of Georgia
Athens, Georgia

Evolution and phylogeny are inextricably intertwined. Both were born together in the earliest part of the nineteenth century when Jean Lamarck (18) published his *Philosophie Zoologique* in 1809. Before that time, it was generally accepted that organic life was the product of special creation, each species having been created by a divine hand and that these species had not changed since. The idea of the immutability of species was as firmly entrenched as the idea of divine creation. A few eighteenth century naturalists had argued that species were not immutable but represented products of change with time. Such ideas were outlined by George Buffon timidly and by Erasmus Darwin vigorously in the last two decades of the eighteenth century. Lamarck presented a theory of the course of evolution remarkably similar to that held today, embodying the ideas of species change with time and that life began with simpler forms and evolved to more specialized types. In his 1809 publication he included probably the first published family tree, simple in the extreme but nevertheless a picture delineating his idea of the course of evolution.

The inclusion of the family tree by Lamarck emphasizes the distinction between evolution and phylogeny. Evolution is the general theory of the change of life with time, including not only the changes that occur or have occurred but also the mechanisms that cause these changes. As part of this concept we assume that the life of any age evolved from previous ancestral forms, dating back to the origin of life. Expressing these thoughts in the other direction, we believe that there is a family tree that represents the evolution of all types of life. If some gnome or pixie had lived through the three billion years that life has been in existence and recorded all these past events so that we could review them, we would know what this family tree had been; but we have no such little friends to enable us to see this past panorama. Instead we try to decipher this past history from all the bits of evidence that we can bring to bear on various parts of it, including present life and the fossil record, and evidence from comparative anatomy, genetics, evolutionary theory, geology, and anything else we can find that will help explain the picture. The family tree that we deduce is our concept of the course of evolution, with the full realization that it may be incomplete or that part of it may be incorrect. This places phylogeny as a conceptual, hypothetical field within the study of evolution as a whole.

Interestingly enough, in his early years as a botanist Lamarck showed no evolutionary leanings. It was only after he became a zoologist that he began expressing his ideas about changes in life with time. It is highly likely that his studies in Crustacea, which have marvelous sequences of step-by-step change, may have triggered his evolutionary thoughts. It is just a fluke that he did not study insects, which would undoubtedly have elicited the same idea. But evolution and phylogeny did not affect entomological thought for several decades. The reason was the professional jealousy of Lamarck's fellow countryman Georges Cuvier, who became the leading comparative anatomist of the world. Lamarck had the basic idea of the course of evolution but for a mechanism he postulated the inheritance of acquired characters. Cuvier seized on many ways to refute this mechanism and virtually ridiculed Lamarck and the entire idea of evolution to the point that both were almost in oblivion for nearly half a century.

When in 1858 Charles Darwin and A. R. Wallace proposed the theory of natural selection as the mechanism of evolution, both were influenced by evidence from insects. Wallace was a tremendous insect collector and presumably had an excellent knowledge about many of the forms he encountered. Darwin's study of the intricate nature of orchid pollination by insects and the remarkable adaptations of orchid flowers as insect attractors gave great support to his other data on which he erected his theories. Unlike Lamarck, neither Darwin nor Wallace were much interested in more than general ideas of phylogeny. The person who welded the link between evolution and phylogeny was the German biologist Ernst Haeckel. In 1866 (9) he provided the scientific world with its first comprehensive family tree. In it the insects were part of the Tracheata Articulata.

Haeckel's work sparked many subsequent phylogenetic studies. In the early 1900s, two of unusual interest were those of the American paleontologist H. F. Osborn working with vertebrates and those of the German entomologist-paleontologist Anton Handlirsch (10) working with insects. Both reviewed the known fossil record, combined it with the living species, and constructed a phylogeny of the groups they studied. In spite of its defects, Handlirsch's family tree of insects is a classic, and set the stage for all later serious work in insect phylogeny. Handlirsch based much of his information on material gathered by several German expeditions that combed the geological outcrops known to contain fossil insects in many parts of the world. They assembled a remarkable collection of insect fossils ranging from older ones occurring in the Paleozoic to later ones approaching the present time.

The efforts of Osborn and Handlirsch met different fates. Osborn's ideas of vertebrate phylogeny were based chiefly on skeletal parts and often on series of fossils from successive periods of geologic time. As a result, Osborn's phylogenies became the backbone of vertebrate classification. Handlirsch had to rely almost entirely on evidence from insect wings, which we now know are prone to exhibit convergent evolution, as was pointed out by his detractors. Whereas Osborn's work attracted a battery of eager scientists to the field of vertebrate paleontology, Handlirsch stirred no such movement. In the first place, finds of good new insect fossil beds were few and far between. In the second place, systematic entomologists who might have been interested in fossils were deluged with thousands and thousands of new living species and genera to describe and name. Even so, notable contributions were made by the Australian entomologists R. J. Tillyard and Edgar Riek, the Russian entomologist A. B. Martynov and his successors, the American entomologists C. T. Brues and F. M. Carpenter, and a few others. In recent decades insect paleontology has blossomed again and made substantial contributions to phylogeny.

After Handlirsch's classical 1908 treatise, the next major step in the development of insect evolution and phylogeny accompanied the great advances being made in genetics. Until the rediscovery in about 1900 of Mendel's demonstration of unitary genetic inheritance, little was known about the fundamental aspects of inheritance and the application of these concepts to evolutionary processes. But after the turn of this century, genetic investigations opened up an amazing vista of theory concerning genetic change and inheritance, and from this matrix new ideas and concepts arose concerning evolution and phylogeny. To entomologists it is gratifying that insects contributed greatly to this advance in genetic knowledge. The discovery of the giant salivary chromosomes of the fly genus *Drosophila,* plus the ease of rearing certain members of the genus in the laboratory, plus the generation period of only a few weeks, made it an ideal experimental animal. *Drosophila* genetic centers arose in many parts of the world and these contributed to an expanding knowledge of genetic theory.

First to attempt a synthesis of the new genetic knowledge of evolution and phylogeny was the American entomologist A. C. Kinsey (16, 17). His studies were embodied in two large reports on the gall wasp genus *Cynips* published in 1930 and 1936. In these books Kinsey, later known as the "sex man," introduced several novel ideas, one of which was that of species as genetic systems. In this context he proposed that geographic populations of different genetic constitution, even though possessing overlapping characters, were distinct species. This concept was soon attacked by Theodosius Dobzhansky. Originally a student of the systematics of Coccinellidae, Dobzhansky had early turned to the study of genetics using *Drosophila* as a tool. His 1937 synthesis of genetics and evolution in *Genetics and the Origin of Species* (3) not only set a completely new concept of species in a biological and genetic sense, but demolished Kinsey's concept of a species.

Kinsey's second novel idea was that evolution proceeded step-wise, one character at a time. With this idea he fared better, but not during his lifetime. He observed that in *Cynips* it appeared that first one character changed, then another, and so on, with the result that the phylogeny could be depicted as a definite progression of steps, one character at a time. In symposia held in 1937, his ideas were hotly contested by the American vertebrate paleontologist G. G. Simpson, who contended that essentially all characters change gradually and most of them in unison. Simpson marshalled so much material from the vertebrates that for many years his ideas were accepted as of general application and little credit was given to Kinsey's ideas, so much so that Kinsey's titles were soon dropped from the "Literature Cited" section of many evolutionary books. Since Kinsey's death it has become increasingly apparent that in many groups of organisms evolution did indeed proceed step-wise, as he postulated for *Cynips*.

Phylogeny

Before and during Kinsey's time, many entomologists constructed family trees for genera and higher categories on the basis of the comparative morphology of living species, in groups for which no fossils were known. Family trees constructed on this basis were formerly branded by paleontologists as being illogical, irrelevant, and worthless, and these criticisms were sustained by many taxonomists of neontological species. Opposition to these criticisms came primarily from entomologists who pointed out that among living insect species we often have a sufficient representation of a group, that some existing species are so primitive that they are essentially living fossils, and that other species represent intergrading steps leading to highly derived groups. By precept, Kinsey cast an affirmative vote with these efforts.

The argument between paleontologists and certain neontologists eventually led the latter to propound logical criteria upon which phylogenies should be based (11, 12, 23). The German dipterist Hennig in particular explained in great detail the philosophical problems involved. He and Ross emphasized the importance of establishing the ancestral state as contrasted with the derived states of characters studied, and the necessity of erecting a family tree or phylogeny such that derived character states arise from ancestral states and not vice versa. This concept is gradually leading to a harmonizing of views by the paleontologists and neontologists, indicating that certain ideas once thought to be contradictory were simply semantiological. It is gradually being recognized that fossils do not come labeled as to what they are, but like a specimen of a living species they must be placed phylogenetically according to criteria of comparative morphology.

In the 1950s considerable interest developed in using quantitative approaches to classification. One of the first definitive results was a study by the Americans C. D. Michener and R. R. Sokal comparing quantitative and inductive methods as applied to a group of bees. Later these methods were extended on a broader basis by Sokal and the Briton P. H. A. Sneath (30). Many taxonomists felt that these numerical methods, called Adansonian or numerical taxonomy, would supplant the older ideas concerning the preparation of phylogenetic trees. Practical difficulties soon arose, however, in the application of numerical taxonomy to phylogeny. First, there was the problem of which kind of numerical taxonomy portrayed the correct result. Second, there arose problems in deciding what groups of characters to use. For example, the American F. J. Rohlf (22) discovered that in the mosquito genus *Aedes* a system based on only larval characters gave one result, one based on only adult characters gave a second, and a combination of both gave a third. The result of the claims of numerical taxonomists has had one extremely useful result. It literally forced the inductive phylogenists to re-examine their hypotheses and practices and express them more clearly, in other words, to tighten up the ship. This process is proceeding more vigorously as time goes on.

A novel contribution to phylogeny was made in 1965 by the American J. H. Camin and by Sokal (2). They prepared a statistical model for deriving phylogenies on the basis of ancestral and derived states of characters and devised some interesting tests for arriving at decisions involving convergent or parallel evolution. Although this method has not been used extensively to date, it should prove to be a valuable tool in problem cases involving large numbers of characters and unusual difficulties in establishing their ancestral states.

The Species Problem

The differing genetic viewpoints concerning "What is a species" expressed by Kinsey and Dobzhansky appear to have triggered a widespread reappraisal of this old question. In 1942, American ornithologist Ernst Mayr (19) summarized most of the literature dealing with animal species and in 1950 the American G. L. Stebbins Jr. (32) did the same for plants. Both were concerned with the evolutionary question: How do new species come into existence? This is a question with which few entomologists had been deeply concerned. Those who were did not have the abundance of collected material over much of the ranges of the species involved to permit incisive study of the question. Most entomological systematists were each struggling to keep abreast of new findings in genera or families embracing hundreds to thousands of species. As a result, most of the theorizing concerning the species problem was done by workers in smaller groups represented by an abundance of available material.

A notable exception was the team of geneticists working with *Drosophila*. In search of new material for genetic and evolutionary studies, they left the laboratory early and sought species in the wild. With the Americans A. H. Sturtevant, J. T. Patterson, and W. S. Stone as its guiding lights, by 1950 this worldwide group had collected and described over 600 species in the genus, reared them in the laboratory, and assembled a tremendous amount of information about their morphological, distributional, and genetic at-

FIGURE 1. Distribution patterns of the species of the American silkworm genus *Platysamia*. The species *kasloensis* is considered to be a hybrid species whose parents were *euryalis* and *gloveri* (33).

tributes. In the process, much came to light as to the evolution of species and their isolating mechanisms (21). At the present time the modern version of this team, with the American dipterists D. E. Hardy and M. R. Wheeler as the systematists, has unearthed an amazing endemic *Drosophila* fauna of over 600 species on the Hawaiian Islands. As these are cultured and studied in detail, we may expect new insights into the subject of the evolution of species.

Hybrid Species

In many respects botanists have been more advanced than zoologists concerning theoretical evolution. This is certainly true with regard to the origin of species by hybrid origin. The phenomenon has long been recognized by botanists as an important mechanism of evolution. On this score entomologists have made outstanding contributions. In his study of the American silk moths, Sweadner in 1937 (33) established *Platysamia kasloensis* as a species that arose by hybridization from two species occurring to the southeast and southwest of it, respectively (Figure 1). On genetic grounds, the American geneticist W. P. Spencer in 1938 postulated the origin of *Drosophila americana* as a hybrid entity between *Drosophila novamexicana* and *D. texana*. Since then, insect examples have added many instances to the probable occurrence of animal species involving hybrid origin (24).

Cryptic Species

A young bombshell was thrown into the evolutionary apple cart by the discovery of two populations of *Drosophila* that could not be differentiated by the systematists but which did not interbreed with each other, although fully interfertile within each population. Additional examples of the same phenomenon were discovered later in other groups of *Drosophila*. The geneticists recognized these reproductively isolated populations as species but the idea caused a great furor among the general rank and file of systematists who worked primarily with dead material. The view of the geneticists was established by the studies of the American H. T. Spieth who in 1952 pointed out that most sympatric species of *Drosophila* have a distinctive behavior pattern associated with mating and that the male of the wrong species does not have the proper behavioral cues and is repelled by a female of another species when mating is attempted (31). These findings established that the cryptic species differed physiologically as well as genetically.

In the same year, the American B. B. Fulton (7) established evidence for essentially an identical situation in four different field crickets sympatric in eastern North America. Gifted with a keen ear, Fulton had identified four distinctive songs produced by male field crickets that were then considered to be all one species. By the use of ingenius techniques he was able to establish experimentally that these crickets represent four populations completely isolated reproductively, to the point that it was virtually impossible to achieve even stress matings between males and females of different populations. Fulton did not name these species officially. This was done in 1957 by his compatriot R. D. Alexander when he made recordings of the various songs and transposed these to visual representations (1).

Together with examples from other groups of plants and animals, these and similar insect examples establish beyond a doubt the existence of many species that are perfectly distinct from a reproductive and evolutionary point of view but which cannot as yet be diagnosed morphologically by systematists. On further study, some of the early *Drosophila* instances proved not to be cryptic. In one group, excellent characters for identifying the supposedly cryptic species were found in the male genitalia. But in many instances the species still remain cryptic. This provides an interesting challenge for students of evolution, because for most species of insects our collections comprise only dead specimens. How many cryptic species might be present among these?

Evolutionary Concepts

Although few entomologists were involved in the early development of evolutionary concepts, insects as study organisms contributed greatly to the field. After genetics blossomed early in the twentieth century, organisms that could be readily reared were selected for study. Insects naturally were among the prime targets for this type of investigation. Especially notable were the studies of the German geneticist R. Goldschmidt on Eurasian moths. He developed the idea that the lower categories of species and subspecies evolved by a process of small genetic changes that he termed microevolution, but that higher categories of taxa differed by genetic changes of a larger magnitude, which he termed macroevolution. He worked also with other insects, including *Drosophila,* in which he was intrigued by the appearance of homeotic mutations, in which a certain character comes under the control of different genetic determinants and suddenly assumes an entirely different appearance. An example is the mutant *aristipedia,* in which the terminal portion of the antenna is a typical leg instead of its usual antennal form. In 1940 Goldschmidt expressed the thought that large changes of this type had played an important role in the evolution of higher categories (8). His contention met immediate resistance, but phylogenetic studies of insects are now providing excellent evidence that many apparent morphological puzzles are best explained on the basis of Goldschmidt's ideas.

The fly genus *Drosophila* has been the primary study organism in many genetic laboratories and has probably contributed more early basic information on evolutionary theory than any other group. During more recent years Culicidae and Chironomidae have been used extensively in genetic laboratories in various parts of the world, as have various species of beetles, black flies, scale insects, and others.

Signal contributions to an understanding of pseudoalleles and their action was made by the American P. W. Whiting (35), using the parasitic wasp genus *Mormoniella* as the experimental animal. In this genus he found that eye color inheritance was an extremely complex phenomenon controlled by two sets of these pseudoalleles, various combinations of which produced a dozen or more distinct shades of eye color.

In the area of cytogenetics and its implications of evolution, in 1956 the Australian M. J. D. White (34) assembled the world literature on animal cytology. It is amazing how much of this was based on insect examples. Since then many laboratories have made extensive contributions in this field, exemplified by the investigations of J. N. Smith of the Canadian Science Service.

The Interpretive Value of Evolution and Phylogeny

In the last few decades there has been an increasing interest among entomologists in deducing phylogenies. The reason for this interest lies in the fact that if the phylogeny of a group is adduced it may contribute numerous types of ideas as to the probable evolution or dispersel of the various taxa and their ancestral forms. Phylogenies based only on neontological evidence may, in this fashion, contribute evidence of past evolutionary happenings for which no fossil evidence is known.

The idea of using phylogenies in this interpretive fashion was first introduced by Kinsey in his 1930 work in *Cynips* (16). He hit on the idea that if the geographic distribution was plotted on the phylogeny, past dispersal patterns of ancestral forms could be adduced (Figure 2). Soon after, Hennig presented an extention of these ideas including the patterns of host transfers by ancestral forms of existing species (11). In 1952, the British entomologists M. Rothschild and T. Clay presented additional examples of host transfers, and pointed out that the phylogeny of certain parasitic insects contributed possible answers to the phylogeny of their bird hosts whose phylogenies were themselves not well established (26). Five years later the American

FIGURE 2. Phylogeny and distribution of the *etrusca* group of the gall wasp genus *Cynips,* indicating the probable paths of geographic dispersal (16).

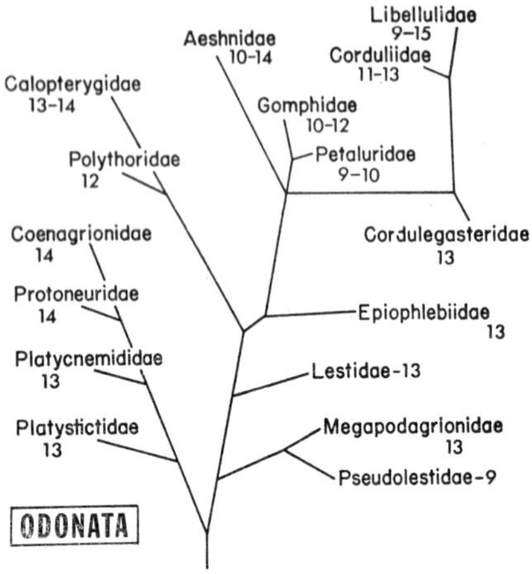

FIGURE 3. Chromosome numbers of certain groups of Odonata plotted against the phylogeny of the group (15).

entomologist C. H. Seevers worked out the phylogeny of beetles that live with termites, and when this was superimposed on the phylogeny of the termites previously worked out by his professor and compatriot A. E. Emerson, a remarkable picture of the ecological evolution of the termites and their beetle guests was apparent (27). Since that time entomologists have added numerous contributions to probable paths of dispersal and probable host transfers.

In 1967, the Dutch cytogeneticist Kiauta (15) used phylogenies of Odonata and Trichoptera to deduce their chromosomal evolution (Figure 3). In these two groups the chromosomes have diffuse centromeres and to date no methods have been devised to adduce the phylogeny from such chromosomes. In this interesting case, available phylogenies helped to solve a genetic problem.

Several parameters of ecological distribution have been inferred from phylogenies of species within various genera. These include the evolution of temperate species from tropical progenitors, of tropical species from temperate progenitors, of pond species from running water species, and switches in seasonal appearance from a situation in which overwintering is in the egg stage from progenitors in which overwintering was in the pupal stage.

Behavioral evolution has long intrigued entomologists, whose interest has been whetted by the early accounts of insect behavior given by such observers as the Frenchman J. H. Fabre, the Englishman John Lubbock, the Austrian M. Maeterlinck, and the American W. M. Wheeler (see chapter by G. Richard). Since then, phylogeny has been used as a means of deciphering the evolution of behavior in many groups of insects. Pioneer studies include the correlation of termite behavior with termite phylogeny by A. E. Emerson (4) and the correlation of spider wasp and digger wasp behavior by the American H. E. Evans (5, 6).

An interesting example concerns the caddisflies belonging to the family Phryganeidae. The Canadian entomologist G. B. Wiggins noticed that the primitive genera of the family had typical stout sickle-like pupal mandibles, similar to those in many other caddisfly families and used for cutting out the anterior sieve of the pupal case, thus allowing the emergence of the pupa. More specialized genera have shortened fleshy mandibles that would appear to be of little use in cutting a strong sieve. Curious about this relationship, Wiggins made a special study of pupal cases in the Phryganeidae and discovered that in the genera having long pupal mandibles, the larvae spin a typical strong sieve across the anterior end of the case but that in the derived genera having reduced pupal mandibles, the larvae spin no sieve but simply pull in a mass of plant fragments that form a loose plug (36). Later, in 1964, the present author employed the phylogeny of the entire order Trichoptera as a means of deducing the evolution of case-making by caddis larvae (25).

In another extremely interesting study, the American R. B. Selander discovered that many species of blister beetles exhibit distinctive mating patterns. When these were plotted against the phylogeny of the species, it was discovered that many differences in mating patterns had evolved between sister species that had become sympatric. In nearly 300 species of blister beetles studied by Selander and his colleague Juan Mathieu of Mexico, each has a distinctive pattern of sexual behavior. In numerous instances, observations of reared material have shown that in the adults the ontogenetic behavioral patterns are inherited, species-specific, and involve little if any components of learning. Numerical parameters of behavior were found to be subject to great variation with different temperatures or physiological states, although the quantitative aspects were remarkably constant throughout the life of the beetles (28, 29). Because it is difficult to be sure which observed units of insect behavior represent inherited differences and which units represent learning, we can expect to achieve more definitive conclusions concerning the evolution of behavior patterns if these can be plotted on a highly probable phylogeny based on other heritable characters.

Unquestionably the same techniques will lead ultimately to new concepts in biochemical evolution. Efforts to adduce phylogeny on the basis of biochemical compounds themselves have been devised for several chemical families, including cytochrome-c and some of the hemoglobins. The methods so far devised share certain inherent logical defects akin to those present in serological phylogeny, chiefly because they lack a satisfactory inductive basis. The wealth of biochemical evidence now being found in insects, if combined with highly probable phylogenies, could well open up avenues of testing various methods of deducing biochemical evolution. Exploratory studies along this line are being followed for proteins by the American team of J. L. Hubby and L. H. Throckmorton (13, 14), and for smaller molecules by the Australians B. P. Moore and B. E. Wallbank (20).

It appears that studies in insect phylogeny have the promise of unveiling innumerable questions on a most diverse front concerning geographic dispersal, biochemical evolution, ecological differentiation, and other evolutionary divergences that have occurred within specific groups. Amalgamating and extending this information will eventually give us a vast store of information for unraveling the evolution of ecological communities.

LITERATURE CITED

1. Alexander, R. D. 1957. The taxonomy of the field crickets of the eastern United States (Orthoptera: Gryllidae: Acheta). *Ann. Entomol. Soc. Am.* 50:584–602
2. Camin, J. H., Sokal, R. R. 1965. A method for deducing branching sequences in phylogeny. *Evolution* 19:311–26
3. Dobzhansky, T. 1937. *Genetics and the Origin of Species.* New York: Columbia Univ. Press
4. Emerson, A. E. 1938. Termite nests—a study of the phylogeny of behavior. *Ecol. Mongr.* 8(2): 247–84
5. Evans, H. E. 1953. Comparative ethology and the systematics of spider wasps. *Syst. Zool.* 2:155–72
6. Evans, H. E. 1957. *Studies on the Comparative Ethology of Digger Wasps of the Genus* Bembix. Ithaca: Comstock
7. Fulton, B. B. 1952. Speciation in the field cricket. *Evolution* 6: 283–95
8. Goldschmidt, R. 1940. *The Material Basis of Evolution.* New Haven: Yale Univ. Press
9. Haeckel, E. 1866. *Generelle Morphologie der Organismen,* II. Berlin: Georg Reiner. 462 pp.
10. Handlirsch, A. 1925. Kapitel: "Die systematischen Grundbegriffe und Stammeschichte." In *Handbuch der Entomologie,* ed. C. Schroder, Band II. Jena
11. Hennig, W. 1950. *Grundzüge einer Theorie der Phylogenetischen Systematik.* Berlin: Deutscher Zentralverlag
12. Hennig, W. 1966. *Phylogenetic Systematics.* Transl. D. Dwight Davis, Rainer Zangerl. Univ. Illinois Press
13. Hubby, J. L., Throckmorton, L. H. 1965. Protein differences in *Drosophila*: II. Comparative species genetics and evolutionary problems. *Genetics* 52(1):203–15
14. Hubby, J. L., Throckmorton, L. H. 1968. Protein differences in *Drosophila*: IV. A study of sibling species. *Am. Nat.* 102(no. 925): 193–205
15. Kiauta, B. 1967. Considerations of the evolution of the chromosome complement in Odonata. *Genetica* 38:430–46
16. Kinsey, A. C. 1930. The gall wasp, genus *Cynips,* a study in the origin of the species. *Indiana Univ. Waterman Inst. Sci. Res. Publ.* 42:1–577
17. Kinsey, A. C. 1936. The origin of higher categories in *Cynips. Indiana Univ. Publ. Sci. Ser.* 4:1–334
18. Lamarck, J. B. 1963. *Zoological Philosophy.* New York: Hafner. (Reprint of a translation of Lamarck's *Philosophie Zoologique,* 1809. Lamarck's first detailed statement of his theory.)
19. Mayr, E. 1942. *Systematics and the Origin of Species.* New York: Columbia Univ. Press
20. Moore, B. P., Wallbank, B. E. 1968. Chemical composition of the defensive secretion in carabid beetles and its importance as taxonomic character. *Proc. Roy. Entomol. Soc. London* (B)37: 62–72
21. Patterson, J. T., Stone, W. S. 1952. *Evolution in the Genus* Drosophila. New York: Macmillan
22. Rohlf, F. J. 1967. Correlated characters in numerical taxonomy. *Syst. Zool.* 16:109–26
23. Ross, H. H. 1937. A generic classification of the nearctic Sawflies (Hymenoptera, Symphyta). *Illinois Biol. Mongr.* 15(2), 173 pp.
24. Ross, H. H. 1962. *A Synthesis of Evolutionary Theory.* Englewood Cliffs, NJ: Prentice-Hall
25. Ross, H. H. 1964. Evolution of caddisworm cases and nets. *Am. Zool.* 4:209–20
26. Rothschild, M., Clay, T. 1952. *Fleas, Flukes and Cuckoos.* New York: Philosophical Library
27. Seevers, C. H. 1957. A monograph on the termitophilous Staphylinidae. *Fieldiana, Zoology* 40:1–334
28. Selander, R. B. 1964. Sexual behavior in blister beetles (Coleoptera). The genus *Pyrota. Can. Entomol.* 96:1037–82
29. Selander, R. B., Mathieu, J. M. 1969. Ecology, behavior, and adult anatomy of the *Albida* Group of the genus *Epicauta* (Coleoptera, Melidae). *Illinois Biol. Monogr. 41,* 168 pp.

30. Sokal, R. R., Sneath, P. H. A. 1963. *Principles of Numerical Taxonomy*. San Francisco: Freeman
31. Spieth, H. T. 1962. Mating behavior within the genus *Drosophila*. *Am. Mus. Nat. Hist. Bull.* 99:395–474
32. Stebbins, G. L. Jr. 1950. *Variation and Evolution in Plants*. New York: Columbia Univ. Press
33. Sweadner, W. R. 1937. Hybridization and the phylogeny of the genus *Platysamia*. *Carnegie Mus. Ann.* 25:163–242
34. White, M. J. D. 1954. *Animal Cytology and Evolution*. Cambridge: Cambridge Univ. Press, 2nd ed.
35. Whiting, P. W. 1955. Some interactions of mutant eye-color factors in the wasp *Mormoniella*. *Pennsylvania Acad. Sci. Proc.* 29:242–45
36. Wiggins, G. B. 1960. The unusual pupal mandibles in the caddisfly family Phryganeidae (Trichtoptera). *Can. Entomol.* 9:449–57

ANATOMY AND MORPHOLOGY

A. Glenn Richards
Department of Entomology, Fisheries, and Wildlife
University of Minnesota, St. Paul, Minnesota

A listing of the names of all of the persons who have published on one facet or another of insect anatomy since the eighteenth century would use up all the allotted space without allowing anything to be said about anyone! Obviously, only outstanding persons can be mentioned, and I feel sure no two would agree entirely on who should be included. To warrant a place one should make an innovation or produce a large volume of reputable work or best of all develop a new concept. Perhaps it is not fair to emphasize innovations but they are the landmarks of progress. As an openly egotistical biologist once said, "There are two kinds of biologists: architects and bricklayers. I am an architect." There is truth in this statement, and history is indeed made more by the innovators (architects) than by the mere gatherers of data even when the advance is due more to the introduction of a new instrument (e.g. the electron microscope) than to the user.

The several published volumes labeled histories of entomology by Howard, Essig, Weiss, etc have little on anatomy. And the general histories of biology by Locy and Nordenskiöld mention primarily only persons whose impact was on more than just entomology. Thus, Nordenskiöld's big book treats Lyonet, Leydig, Owen and Weismann, barely mentions Lubbock in passing, and omits entirely Savigny, Dufour, Straus-Durckheim and many others important to the history of entomology. Short historical chapters are given in the treatises by Henneguy (1904) and Berlese (1909). Berlese's chapter covers 26 pages and does quite well up to about the middle of the nineteenth century after which it, understandably, degenerates into lists of names. The bibliographies in Kolbe (1893), Packard (1898), and Henneguy (1904) are rich sources for references of the nineteenth century; the first two contain chapter bibliographies giving references by subjects (and roughly chronological), the third contains a single alphabetized bibliography. Analysis of these bibliographies, quod vide, assisted materially in the selection of names for the present treatment. In contrast, the more important treatises of the twentieth century, such as those of Snodgrass (1935) and Wigglesworth (1939 et seq), are concerned with the subject matter and not with history—as is indeed necessitated by the explosive increase in the number of names. Only reference sources such as the above are included in the bibliography to this volume.

Modern Beginnings

Insect anatomy had its beginnings in the work of Malpighi and Swammerdam in the seventeenth century. Réaumur was the most important name of the eighteenth century. These workers have been treated in a preceding chapter. In the middle of the eighteenth century Pierre Lyonet (a contemporary of Linnaeus) published his famous anatomy of a caterpillar (1760). Later, posthumously, there was published an incomplete and now largely forgotten treatment of the pupa and adult. Lyonet could be called a dilettante, but with no more than a simple magnifying glass held on a frame and crude instruments (illustrated in 1762 edition; available magnifications 7–104 X) he made beautiful dissections from which he prepared gross anatomical illustrations of such excellent quality that one could use some of them today. Lyonet tallied the muscles he illustrated, stated a total of 1647 for the somatic musculature of the body, and made quite a point of the fact that this is three times as many as the 529 muscles he says are to be found in the human body. Later in the volume he tallied 228 muscles for the head, 1647 somatic muscles, and 2186 splanchnic muscles associated with the esophagus, stomach, and intestines, a grand total of 4061 muscles delineated (a sum specifically stated as not including the tiny muscles associated with moving the antennae, palpi, etc, and seemingly not the muscles associated with the heart). Lyonet also illustrated the prothoracic gland, the function and significance of which was then incomprehensible (he called it the granulated vessel); also what we now call imaginal discs. However, it was still the work of an amateur rather than a trained scientist. The 616 pages of text are little more than an expanded explanation of the 18 plates; there was no attempt to analyze the data presented or to fit it within some framework of our interpretation of nature. But Cuvier, the Father of Comparative Anatomy, had not yet been born, training was self-education, and Lyonet's anatomical studies were the hobby of a lawyer-diplomat-translator.

Nordenskiöld considered that modern biology began with the nineteenth century. Such a statement fits better for anatomy than for many other subjects. In the Napoleonic era at the beginning of the nineteenth century, Cuvier rose to prominence, the field of comparative anatomy was founded, and the French educational system was reformed and expanded. In 1815 a Dr. Herold of the University of Marburg published inaccurate drawings of the changes in the gut and central nervous system as the larva of *Pieris brassicae* metamorphosed to the adult. He did not even recognize that the ventral nerve cord was double in the larva but fused to single in the adult. But he did perceive the important point that metamorphosis involves a progressive series of changes in internal organs. (Herold used only his surname on title pages. He is variously cited as E. or M. or M.J.D. or J.M.D. or just—; take your pick.) Herold also published on the embryology of spiders, and is recorded by Packard as having been the first to report seeing the dorsal vessel pump blood (1823). But the question of the reality of blood circulation in insects was unsettled for nearly two centuries. Malpighi presumed circulation when he discovered and described the heart and used the term systole (1669); Lyonet explicitly described pulsations in 1760 but queried the idea of circulation; others were pro or con. The question was whether or not the dorsal vessel of insects pumped in a circulating fashion something that could properly be called blood. This is a good example of the difficulty of assigning credit when certain authors are cautious and others presumptuous. The super cautious workers were not convinced that the dorsal vessel of insects functioned as a heart until Verloren's work in 1844 (175 years after Malpighi's opus).

While Herold is to be credited with having shown that metamorphosis involves a progressive series of internal changes, George Newport is to be given credit for appreciating the significance of such changes. In 1832 and 1834 Newport published a more accurate description of the changes in the central nervous system of a sphinx moth, and applied these facts to a consideration of the number of segments in the insect head. Thereby Newport introduced the use of embryological data for anatomical interpretations in entomology.

Of even more significance was the appearance in 1816 of the important treatise by Jules-César Savigny, *Memoires sur les Animaux sans Vertèbres.* This is a 117-page treatment that starts by stating a proposition: whatever their superficial appearance and mode of action, the mouthparts of diverse insects are all composed of the same elements. At the end of the memoir he tabulates what we now call the serial homology of the segmented appendages of insects, myriapods, crustacea, and arachnids, and compares the terminologies previously used by Fabricius, Cuvier, and Latreille. One may wonder about Savigny's method of presentation. Did he really deduce this important generalization and then collect the evidence, or did he arrive at this as a conclusion from comparative examinations—subsequently stating the conclusion as a proposition? Be that as it may, the memoir served the valuable function of unifying consideration of segmental appendages despite their superficial diversity. Henneguy, Berlese, and some others considered this contribution so important that they refer to it as the theory of Savigny.

Cuvier himself did only a little with insects but near the end of his life an associate (or perhaps student), Hercule Straus-Durckheim, produced the first comprehensive anatomy and physiology of an insect—prefaced by a dedicatory eulogy to Cuvier. This was Straus-Durckheim's exhaustive treatise on the adult of the beetle *Melolontha vulgaris* (1828). Chapters are given to the integumentary, muscular, digestive, reproductive, respiratory, circulatory, nervous, and sensory systems. While *M. vulgaris* is the main species described there are comparisons at various places to some two dozen other beetle genera of various families, and occasionally to species of other orders or even to crustacea, myriapods, and annelids. He not only considered such things as how individual muscles operate to move body parts in locomotion but also patterns of movement in space, a mathematical treatment of the forces involved in jumping, and the differences in wing movements during moving versus hovering flight. One difference from current point of view was his recognition of four primary body divisions: head, corselet (prothorax), thorax (meso- and metathorax), and abdomen.

The most prolific author of the first half of the nineteenth century was Leon Dufour. Beginning about 1820 and ending about 1860 he published several dozens of papers on external anatomy, the gut, fat body, circulation, respiration, and reproduction. In the introduction to one series he wrote: "In coordinating the material of my work I have sought above all to present a maximum of facts with a minimum of words while carefully avoiding the confusion which might result from obscure brevity." Scanning through his hundreds of pages as he describes, for instance, the anatomy of Orthoptera in general and then family by family the modifications of the various groups, one must admit that he achieved the recording of a laudable amount of fact per page. Dufour attempted to formulate a general insect anatomy and did indeed lay the foundation for such. His work is seldom cited today but it was commonly cited in a laudatory manner in the latter part of the nineteenth century (except for Miall & Denny who ". . . complain that his sagacity does not do justice to his diligence."). In the sense of seeking to homologize parts I am inclined to suggest that Dufour occupied in the nineteenth century a position comparable to that of Snodgrass in the twentieth century.

Up to about 1840 all microscopic work was limited by the chromatic aberrations inherent in simple lenses. In 1827 Amici invented an achromatic lens system; during the 1830s microscopes containing achromatic optics became available at the larger European universities; in 1831 Brown described and named the cell nucleus; in 1836 C. T. von Siebold used one of these "expensive" instruments and described hair-like sperm developing in packets in seven orders of insects; in 1838–1839 Schleiden and Schwann formulated the cell theory; and in 1840 Friedrich Stein of Berlin began to study insect microanatomy. In 1847 Stein published on the ovaries and egg development in beetles. Little time, then, elapsed between the development of a usable compound microscope and its exploitation by an entomologist—as a century later little time was to elapse between the development of the electron microscope and its exploitation by another entomologist. But Stein's contribution was not exciting. He worked with whole mounts (the microtome was not yet invented) which, to judge from his illustrations, were unstained. Mildly interesting are his description of ovarian sheaths as acellular (he did not use that term), and the origin of the egg shell from secretion of the follicular epithelium. Perhaps Stein should be forgotten in entomological history; he had a powerful new instrument but he missed the opportunity of establishing insect histology as a profitable field; this was to come a decade or more later with the work of Leydig and others.

Richard Owen was a professor of anatomy, also head of the British Museum of Natural History. His long life spanned most of the nineteenth century. In 1855 his 689-page *Lectures on the Comparative Anatomy and Physiology of the Invertebrate Animals* was published. Owen had a pompous oratorical style but his work is vastly important as the first consistent treatment covering the entire animal kingdom. Nine of the 24 lectures are given to the annelid-arthropod series (244 pages); 3 lectures (98 pages) to the insects alone. He deplored such inconsistencies as the usage of the vertebrate terms maxilla and mandible being reversed when applied to insects (the term chorion is another deplorable misapplication). Owen was a staunch supporter of the idea that "ontogeny recapitulates phylogeny" (Haeckel's biogenetic law), and, most importantly, developed and demonstrated the importance of the concepts of analogy and homology. For this, his position is secure in any anatomical history of any major part of the zoological realm. Incidentally, his lectures conclude with an important bibliography of the 404 references cited throughout the text.

The insect reproductive system naturally attracted interest early because of the controversy over spontaneous generation. The first sound treatment for the class however did not come until the work of J. A. Palmén in 1883–1884. Palmén pointed out that *Lepisma* and the mayflies have paired reproductive outlets as do most worms, some myriapods, crustaceans, etc, whereas higher insects do not. Some insect orders (e.g. stoneflies) are intermediate. This is more of a landmark for phylogeny than for anatomy but it is, of course, important to both. Of purely anatomical interest was his demonstration that the reproductive ducts are of dual origin: part associated with the gonads and part developed by invagination of the integument.

The last name I will mention in this section is that of Sir John Lubbock (Lord Avebury): banker, philanthropist, politician, and naturalist. His numerous entomological publications were from about 1860 to about 1890. He is remembered more as an experimentalist and analyst than as an anatomist but his contributions to anatomy were considerable. Perhaps of most importance is his small book, *On the Origin and Metamorphosis of Insects,* published in 1874. In this book he was predecessor to the Berlese theory. He pointed out that larvae differ from adults because they hatch in an immature condition, and that their structure is related to their immediate needs rather than to their final form. Also that metamorphosis only appears to be abrupt because it happens inside a hard skin. Snodgrass said much the same 80 odd years later. In his well-known book *On the Senses, Instincts and Intelligence of Animals* (1891) Lubbock has some memorable comments: "The clue [to function] afforded by anatomy is very imperfect, and sometimes almost misleading." And for all authors of reviews; "With this object I had to look up a great number of memoirs, in various languages, and scattered through many different periodicals; and it seemed to me that it might be interesting, and save others some of the labour I had to undergo myself, if I were to bring together the notes I had made, and give a list of the principal memoires consulted." Perhaps it was modesty that kept him from saying that while the above is a big chore, the importance is the emerging interpretation, and foremost of all Lubbock was a master of original and analytical thought (an architect of progress). However, one wonders what he might say if he had to review a field today.

General Treatises

Most advances have been announced in special research papers or monographs. But the large general treatises (variously called textbooks, manuals, introductions, principles, Lehrbuch, Handbuch, Grundriss, or just The Insects) fulfill two important functions: first, they allow the author to make reinterpretations and new interpretations of previously recorded data, which can involve the elaboration of concepts and even the propounding of new concepts, and second, they or the most recent of them are the sources from which beginning students get their initial background.

The first was *An Introduction to Entomology* by William Kirby and William Spence published in four volumes in 1816–1826. This was primarily a systematic survey of the insect orders with a reasonable treatment of external anatomy and some internal anatomy. This type of treatment was subsequently used by H. Burmeister, J. O. Westwood, J. T. Lacordaire, V. Graber, and D. Sharp in the nineteenth century and by J. H. Comstock and A. D. Imms in the first quarter of the twentieth century. More recently this style has been carried to an extreme where functional anatomy is virtually eliminated and the presentation is little more than a taxonomic and life history survey (e.g. the books by E. O. Essig and by D. J. Borror and D. M. DeLong).

Near the end of the nineteenth century another type of general treatise appeared from the pen of H. J. Kolbe, curator of Zoology of the Royal Natural History Museum of Berlin. This was the 700-page *Einführung in die Kenntnis der Insekten* published in 1893. Kolbe said in his preface that there was no pattern for him to follow, clearly then he did not approve of the taxonomic survey approach of his predecessors. He listed 14 aspects with which an introductory entomology should deal, and then went on to give a treatment of insect structure and function that set a pattern that was closely followed by Packard (1898) and less closely by Henneguy (1904), Berlese (vol. 1, 1909), Schröder (1928), and Weber (1933). Kolbe's book dealt with the form and characteristics of the body and its parts, the form of internal organs, and the functions of the body and its parts. This was an innovation: an introductory book treating only form and function.

The teaching of entomology in the US for the first half of the twentieth century was dominated by the influence of J. H. Comstock. This influence fostered taxonomy, phylogeny, and life histories. A similar statement could be made for the influence of A. D. Imms' textbook in England and the British Empire. At least for the subject of the present chapter this was unfortunate. Following the pattern developed by Kolbe, the American A. S. Packard wrote a *Text-Book of Entomology*, published in 1898, which the present author has found profitable to consult more than all other such books up to the time of appearance of Weber's *Lehrbuch* (1933) and Snodgrass' *Principles* (1935). Present and future students are urged to familiarize themselves with Packard's volume for presentation of the status of subjects at the end of the nineteenth century, as well as for the citation of more than 1000 titles. Packard's perspicacity is well illustrated by his apt reference to cerci as abdominal antennae (1871); his scholarship is shown by his large taxonomic monographs of moths, by his numerous and diverse anatomical papers, and by the compilation of his textbook. He deserves a high place in our esteem.

During the twentieth century general treatises have become less general. They recognize the virtual subdivision of entomology into systematic, functional, and applied. And now more and more presentations are to restricted subheadings within one of these three. For interest to anatomy and morphology there should be listed first and foremost R. E. Snodgrass' *Principles of Insect Morphology* (1935) which together with Snodgrass' numerous little monographs and his *A Textbook of Arthropod Anatomy* (1952) has dominated the field since 1935. (For those who wonder why Snodgrass never revised his *Principles,* he once told me that the publisher would continue to reprint the book but refused to let him prepare a revision that would require re-setting of type.) Snodgrass' *Principles* has been immensely influential but his *Textbook* seems to have made little impact, perhaps because in the US invertebrate zoologists tend to avoid insects and too many entomologists ignore all classes of arthropods except insects. But the smaller *Manual of Insect Morphology* by E. M. DuPorte (1959), containing many sets of directions for the dissection of various insects and insect parts, has been used extensively in the teaching of insect anatomy in North America. In regard to teaching one should also mention *The Embryology of Insects and Myriapods* by O. A. Johannsen & F. H. Butt (1941), an updating of the 1899 compilation by Korschelt and Heider, and for anatomy at the organ and tissue level *The Principles of Insect Physiology* by Sir Vincent B. Wigglesworth (1939 et seq).

Special Monographs

The books by Lyonet, Straus-Durckheim, and some others already mentioned are special monographs. There have been many others, a few of which warrant mention in an entomological history. L. C. Miall and A. Denny published *The Structure and Life-History of the Cockroach,* subtitled *An Introduction to the Study of Insects* in 1886. This exemplifies the selection of one member of a group and presenting it as a type species for teaching beginning students. A few years later (1900) Miall with A. R. Hammond followed the above with the *Life History of the Harlequin Fly (Chironomus).* This was followed by the three times more voluminous *The Anatomy, Physiology, Morphology and Development of the Blow-Fly* by B. T. Lowne (1890–1892). These three books were extensively used, especially by beginning workers, during the first quarter of the twentieth century. A somewhat similar monograph in French, including a comparative taxonomic treatment, was *Les Insectes Vésicants* by Henri Beauregard (1890). Such monographs continue to be produced occasionally for some important species or group but they no longer constitute an advance to the general field.

Eugen Korschelt was one of the important early workers on insect embryology beginning shortly after the introduction of the microtome (1870) and serial sectioning techniques. A decade later he collaborated with Karl Heider to produce the well-known two-volume work on the comparative embryology of invertebrate animals in 1891–1893 (English translation, 1895–1900; revised by Korschelt alone in 1936). Another decade or so and he began, about 1910, the ambitious project of directing research for a complete monograph on the diving beetle, *Dytiscus marginalis*. Students and younger faculty at the University of Marburg were induced to take as doctorate thesis topics or other research projects subjects which would eventually become chapters in this two-volume, 1827-page monograph. The student would finish the problem, publish it (usually in *Z. wiss. Zool.*), and then, when all had been finished, Korschelt put the manuscripts together and arranged for their publication under the sweeping title *Bearbeitung einheimischer Tiere* (1923–1924). He listed 42 publications by 20 authors (one being himself) as going into production of this monograph, and mentions as especially helpful Messrs. Alt, Bauer, Blunck, Euscher, Holste, and Oberle. (Hans Blunck published a number of papers in the period 1912–1924 but I have seen no reference to papers by him or by any of the other students since 1924.)

For the past 50 years the term special monograph has more often referred to a functional system rather than to a particular insect or insect group. The past half century has seen an explosive growth of the above under several formats. First, there are the several review journals or review series which contain a certain fraction of papers containing material relevant to anatomy and morphology. Second, there are the volumes representing papers presented at symposia. And, third, there are the many books, usually small, dealing with one system or another. Books which seem likely to have continued value for anatomy, especially as starting points for further compilations, include: Reinhard Demoll, *Die Sinnesorgane der Arthropoden, ihr Bau und ihre Funktion,* (1917); Bertil Hanström, *Vergleichenden Anatomie des Nervensystems der wirbellosen tiere,* (1928); Takadi Maaki, *Studies on the Thoracic Musculature of Insects,* (1938); August Krogh, *The Comparative Physiology of Respiratory Mechanisms,* (1941); A. G. Richards, *The Integument of Arthropods,* (1951); Paul Buchner, *Endosymbiose der Tiere mit pflanzlichen Mikroorganismen,* (1953); Otto Pflugfelder, *Entwicklungsphysiologie der Insekten,* (1953 & 1958); V. B. Wigglesworth, *The Physiology of Insect Metamorphosis,* (1954); J. W. S. Pringle, *Insect Flight,* (1957); V. D. Dethier, *The Physiology of Insect Senses,* (1963); K. D. Roeder, *Nerve Cells and Insect Behavior,* (1963); Ryuichi Matsuda, *Morphology and Evolution of the Insect Head,* (1965); Ryuichi Matsuda, *Morphology and Evolution of the Insect Thorax,* (1970); Franz Engelmann, *The Physiology of Insect Reproduction,* (1970). Obviously, 'anatomy' comes under many titles!

Embryology

Consideration of insect embryology at the descriptive level began in the first half of the nineteenth century. Ignoring the earlier arguments concerning spontaneous generation, perhaps the first important name was that of Rudolph Leuckart who recognized the micropyle of the insect egg and its relation to the entrance of sperm (1850–1860). But Albert Kölliker had described the blastoderm in 1842, and several beginnings had been made before this. At this time all embryology was done with whole mounts, at least usually unstained. Embryos might be dissected from eggs and their developing form recorded, or focal planes could be examined in transparent eggs mounted whole. Relatively little useful information was forthcoming.

The most important name preceding development of the microtome and reasonable staining procedures was that of August Weismann. With only transparent whole mounts (where he commonly watched development continuously in a living egg) he described the development of the egg in 1863, and then postembryonic development in 1864. He recognized the pole cells and traced them to the gonads. This gave rise to the concept of germ cell determinants and the idea that very early in embryonic development there is a complete segregation of the germ cell line from the soma. This continuity of the germ plasm looked good in the Diptera and certain Coleoptera (and let us not worry about the groups in which it is not clear). One is no longer concerned over what we might call the sanctity of the germ line or of germ layers but the vigorous promotion of these concepts gave purpose to the research of many contemporary and subsequent workers. In his later papers Weismann recognized that the tissues of the maggot become lost and that the adult organs come from imaginal discs. These are striking discoveries for the time. Weismann went on to become one of the important names in the early development of genetics but before that he had made an indelible name in entomological history.

Weismann's work was followed two decades later (1882) by a 348-page monograph on the metamorphosis of higher Diptera by Henri Viallanes. This monograph employed the then advanced techniques of fixation, sectioning, and differential staining, however, it provided little more than improved documentation for Weismann's conclusions. [Viallanes did later (1884–1888) make important contributions on the internal anatomy of the central nervous system.] Then 84 years later Poulson (1947) showed that while some of the pole cells do indeed reach the gonads and become germ cells, many become lost enroute and become differentiated as a recognizable cell type of special function in the epithelium of the midgut. Clearly, the fate of a particular cell depends on its environment.

The microtome was invented in 1870, by which time reasonable staining methods were beginning to be developed. These methods were immediately applied to the study of insect embryos by Kowalevsky (or Kovalevskii) (1871) who promptly homologized the embryonic layers of insect embryos with those of vertebrates. This led to the idea of an insect being an upside down vertebrate (or a vertebrate an upside down insect). Thereafter rapid progress was made as a number of persons described embryonic and postembryonic stages.

Near the end of the nineteenth century Lubbock preceded Berlese with the idea that larvae are precociously hatched embryos (1874); Paul Hallez pointed out that the orientation of the future embryo is determined during development of the egg within the ovary, the so-called Hallez's Law of Orientation (1885–1886); and numerous workers published descriptions of the embryo of one insect or another, the most often quoted names being those of Richard Heymons and W. M. Wheeler, the latter being the same who later became an authority on ants and social life among the insects. At the end of the nineteenth century Korschelt and Heider brought this literature together in one coherent treatment of invertebrate embryology as already mentioned. In 1941, Johannsen and Butt brought this up to date including detailed historic treatments of each topic (e.g. gastrulation, segmentation, etc).

More descriptions but little innovation appeared until the middle 1920s when Friedrich Seidel began to publish a series of papers in experimental embryology (1926–1935). Experimental embryology using insects was not new, but Seidel, following the ideas propounded by Roux and Spemann, developed new techniques which were powerful tools. As a result he formulated the concepts of embryonic developmental centers: the activation center and the differentiation center, and categorized the so-called indeterminate-determinate series of developmental types. It is presumably unfortunate that Seidel then shifted to the study of mammalian embryos; we will continue to wonder what further he might have done. In the same period another German, Alfred Kühn and his colleagues (1926 to about 1950), did pioneer work in developmental genetics, especially with experimental analyses of the wing patterns of moths. The same basic elements were found for wing patterns irrespective of whether the data were from genetic effects, surgical interferences, or comparative anatomy. This was an early approach in the direction of biochemical genetics. As an aside, one may well wonder how Kühn worked with his once close colleague, Karl Henke, when Kühn was a master of the simple statement whereas Henke was a master of beautiful but most difficult German. Kühn's *Vorlesungen über Entwicklungsphysiologie* (1955) is still fascinating reading for the interpretation of experimental data (a recent somewhat revised English translation is entitled *Lectures on Developmental Physiology*).

The preceding types of embryological work now await new ideas, new approaches. Recent developments deal mostly with hormones in postembryonic development, chromosome puffing and the DNA–RNA picture, and other things beyond the scope of this chapter. Descriptive work, however, does continue. Outstanding are the several large papers by O. W. Tiegs on the histology of metamorphosis of a chalcid wasp and a weevil (1922 and 1935), the analysis by H. E. Hinton leading to the concept of pharate (hidden) stages in development and their significance (1946), and papers by various workers, some with electron microscopy, on the degeneration and regeneration of muscles during growth and metamorphosis.

Histology and Cytology

Histology, and especially cytology, developed somewhat later than embryology. Even so histology was rather well established at the middle of the nineteenth century, the first compendium being in 1852. For perspective, one should remember that achromatic compound microscopes became available in the 1830s; staining was seldom used until the haphazard application of carmine and other natural dyes in the 1850s (although, to be sure, Leeuwenhoek mentioned staining with saffron in 1714); the systematic development of dyes and metallic precipitates occurred in the 1860s (hematoxylin in 1863 and 1865); the microtome was developed about 1870; the aniline dyes came in the 1870s; apochromatic lens systems came at the end of the century; and histochemistry and certified dyes not until well into the twentieth century.

Franz Leydig is credited with introducing the study of histology to entomology. His papers covered the period 1848–1891, i.e. from before the use of the microtome and differential staining to after them. He published approximately two dozen titles involving examination of insect material (especially gut, glands, sense organs, eggs, and sperm) in addition to papers on worms, mollusks, and vertebrates. His *Lehrbuch der Histologie des Menschen und der Tiere* (1857) was remarkable for paying as much attention to invertebrates as to vertebrates. One of his later titles was *Beiträge zur Anatomie und Histologie der Insekten* (1887). One could readily call Leydig not only the first significant insect histologist but also the founder of comparative animal histology. Incidentally, he introduced use of the term fine structure but obviously in a less fine sense than the current usage by electron miscroscopists.

In the latter half of the period when Leydig was active some 25–30 papers were published by Félix Plateau on a variety of subjects: digestion, excretion, respiration, muscle, flight, and vision (1865–1890). One may justify inclusion of his name here because of the successful combination of experimental and histological methods. Other commonly cited histologists of the last quarter of the nineteenth century were Arthur van Gehuchten (muscle, but especially holocrine secretion, 1890), Gustave Gilson (1888–1897), and Enrico Verson (1887–1894), the last two both for epidermal glands and for spermatogenesis. Also in this period came the several papers of Sigmund Exner (1875–1891). Few if any works have ever made more impression on the subsequent 75 years of a special topic than Exner's *Die Physiologie der facettierten Augen von Krebsen und Insekten* (1891). To be sure, vision is classified as physiology but in his suggestion of lens cylinders Exner postulated an anatomical basis for some of the facts of arthropod vision. This hypothesis has been repeated in almost all relevant books since then. Ironically, now that we have optical methods for adequate examination of insect lenses several workers have failed to find a basis for Exner's postulate. "A beautiful hypothesis spoiled by a simple fact."

In the early years of the twentieth century insect histology was dominated by three Frenchmen: Charles Janet, Charles Pérez, and A. C. Hollande. Janet was the first and foremost of these. He published about 30 papers on the higher Hymenoptera, especially ants, in the period 1893–1911 and then an epilogue in 1923 (ignoring papers on subjects such as the decimal classification, geology of the Beauvais area, and comments in 1882 on the possibility of a tunnel under the English channel). His innovations make a formidable list. Clearly, much of the anatomy he described was based on the then novel technique of examining serial sections. He described in detail the regression of flight muscles of queen ants after the nuptial flight, the nature of articular membranes, a description of what Hinton has since termed the pharate stage, detailed discussion of the problem of segmentation of the insect head, and even anatomy of wasp nests. In his final paper entitled *Revendications* (claims) he most graciously thanks Henneguy, Escherich, Berlese, Wheeler, and Schröder for using his illustrations with correct citation of credit, and then chides Prochnov for crediting one of his figures to another author, and sternly rebukes Forel for transgressing no less than nine times in his *Le Monde Social des Fourmis.*

In 1910 Pérez published the single monograph on which his reputation rests: a 274-page treatment of the metamorphosis of *Calliphora.* This is one of the best of the old histological treatments that attempted to cover everything. He gave a good treatment of the insertion of muscles into cuticle including the fact that the sarcolemma becomes continuous with the basement membrane of the epidermis, described the intracellular formation of tracheoles, etc. Hollande published a series of papers on insect blood and associated cells (oenocytes and paracardial cells) in the period 1909–1922.

About this time Hirowo Ito published, mostly in Japan but in the English language, a series of papers on the histology of the silk glands, gut, malpighian tubules, etc of *Bombyx mori* (1915–1920). Perhaps most important was his report that the corpora allata have a glandular structure (1918). Lyonet gave the first gross description of an endocrine gland; Ito pointed out that one, a different one to be sure, was indeed glandular; a few years later Stephan Kopec suggested the existence of hormones in insects, and this was documented by Wigglesworth and by Fraenkel two decades after Ito's report. More years elapsed before specific hormones were linked to specific glands; thus our stepwise advances.

In the period 1913–1924 another remarkable entomologist appeared and then as far as entomology is concerned mostly disappeared. This was David Keilin who left an indelible name for his work on the respiratory system, especially of Diptera. Reputedly, and confirmed by Keilin's own historic account in *The History of Cell Respiration and Cytochrome* (1966, posthumous), work on the respiration of larvae of *Gasterophilus* from a horse's stomach led to the spectroscopic examination of the red pigment of this larva, thence to the discovery of the cytochromes and to Keilin becoming a biochemist of international stature. But he remained editor of the journal *Parasitology,* and as late as 1944 published a large paper on respiratory adaptations of dipterous larvae.

For the nervous and sensory systems special methods are advantageous.

The pioneer work by S. R. Cajal (1890–1923) using methods developed by Golgi is still cited and its illustrations copied. Somewhat later this work was continued by Bertil Hanström (1925–1930) who, however, worked more with other groups of invertebrates. The light microscope classic on sensilla, still commonly cited, was by Fontzou Hsü of the University of Louvain in 1938.

The extremely thin outer layer of the insect cuticle which we now call epicuticle was recognized in 1823 by Odier who described chitin, but its importance was not recognized for over a hundred years. Its importance was suggested in 1928 by Wilhelm Kühnelt (who subsequently shifted to ecology) and documented two decades later by Wigglesworth and his associates.

For more current presentations of the histology of the various organs one cannot do better than consult *The Principles of Insect Physiology* by Sir Vincent B. Wigglesworth (1939 et seq). Wigglesworth's own personal contributions to the cellular anatomy of organs during development and during functioning have been both considerable and diverse (note how many of the illustrations in his *Principles* are copied from his own original papers). For those fortunate enough to have access to a copy, approximately 4000 titles can be found in *References for an Outline of Insect Histology* compiled by M. F. Day and privately distributed in 1948.

The development of cytology of insects has largely followed cytological studies on mammalian cells. An outstanding exception concerns cytogenetics where the correct interpretation of the giant polytene chromosomes of the salivary glands of dipterous larvae, by T. S. Painter in the early 1930s, provided a powerful tool for genetic studies. Some years earlier, C. E. McClung, using insect material, had developed the theory of sex determination involving sex chromosomes and autosomes. The development of polytene chromosomes and polyploidy has been correlated with histiogenesis in various insects but why this is so remains a mystery. Currently, puffing, a change in appearance of a particular chromosomal region, first reported by Wolfgang Beermann in 1952 in *Chironomous* larvae, is being examined in studies relating chromosomal activity to development.

Since the advent of a practical transmission type electron miscroscope in 1940, a new type of cytology (fine structure or ultrastructure) has developed. This extends direct visualization of structure down to macromolecular levels. The instrument was applied to insect cuticle (where problems of fixation and thin sectioning were not overly serious) by A. G. Richards in 1941 with publications beginning in 1942. The helical structure of pore canals, the presence of taenidia in tracheoles, and the structure of iridescent butterfly scales were among the objects then examined. Some of the discoveries were automatic results of the exploitation of this new instrument. With the subsequent development of adequate preparative techniques electron microscope work on tissues has become routine the world over. Nearly 20 years elapsed after development of the transmission electron microscope to the appearance of a practical scanning electron microscope (Stereoscan). When it did become available it was promptly applied to the study of plastrons, diffraction gratings, etc by H. E. Hinton (1960 et seq).

Anatomy and Morphology

Within the term anatomy one considers the structural details of insects from egg to adult and from molecular to macroscopic. Within the term comparative anatomy one compares structures as found in various species and higher taxa either to seek the evolutionary origin of modified structures (homologies) or to deduce the evolutionary development of a group (phylogeny). Within the term functional anatomy one considers how a particular structure acts for the insect to perform its functions. Within the term morphology one considers the significance or interpretation of anatomical details. Under morphology there are three different approaches: 1. the interpretation of anatomy relative to phylogeny as well exemplified by Crampton who was called a morphologist and by many who are primarily thought of as systematists; 2. the interpretation of developmental and comparative facts for deducing homologies and the probable evolutionary origin of modified structures (in practice commonly not distinct); and 3. the interpretation of anatomy relative to function as well exemplified by Snodgrass' various treatments of skeletomuscular mechanisms and by many workers who are usually thought of as physiologists. [Most anatomical studies are observational, descriptive, and deductively analyzed, but there is some experimental anatomy with analytical analysis (e.g. regeneration, some of functional anatomy), and even some attempts at theoretical anatomy (e.g. for homoeosis, allometric growth).]

In a once widely used manual for teaching insect anatomy to beginning students, Macgillivray (1923) wrote in his Preface, "That a thorough knowledge of the external anatomy of insects is fundamental to their taxonomy is self evident. . . . The goal sought [in writing this book] was . . . an introduction to the study of systematic entomology." This biased point of view is only one facet of anatomy. However, Macgillivray's statement is no more biased than the point of view of some physiologists who think of anatomy only as the parts of a functioning machine, or of some biochemists whose primary distinction is into soluble and insoluble fractions. In a sense all of these points of view are valid because anatomy is as fundamental to systematics and to function as arithmetic is to mathematics. Anatomy is chiefly used today in relation to something else, a descriptive prelude necessary for the formulation of working hypotheses. Attempts by G. F. Ferris and his students (1939–1958) to establish a morphology independent of function and adaptation of the parts foundered on facts marshalled by E. M. DuPorte, R. E. Snodgrass, and others. The only service performed by the dogmatic and contentious Ferris was forcing well-grounded morphologists to expressly document the view that insect anatomy cannot be satisfactorily interpreted without consideration of adaptations for function.

Comparative anatomy and phylogeny.—The topic dates from the time of the founding of comparative anatomy by Cuvier in the early years of the nineteenth century. The work of J. A. Palmén has already been treated. Since a considerable percentage of systematists treat phylogeny, the number of names involved is large. To mention a few, there was J. H. Comstock and J. G. Needham who, beginning near the end of the nineteenth century, elaborately developed systematic seriation of wing venations both for purposes of homology and of phylogeny. This culminated in Comstock's *The Wings of Insects.* (1918). The well-known R. J. Tillyard reviewed the theories of insect phylogeny in 1930 and then proposed his own. Snodgrass, while better known for functional anatomy, also treated phylogeny (1938 et seq). It is routine for authors of textbooks and large taxonomic monographs to include a treatment of phylogeny, their own or someone else's.

There is, however, one outstanding man who was little if anything other than a phylogenist. This was G. C. Crampton whose numerous papers (1914–1944) dealt not with either the phylum Arthropoda or the class Insecta but only with supposedly evolutionary lines within and between certain orders, especially orthopteroid and panorpoid orders.

For further treatment of this aspect of anatomy see the chapter on Evolution and Phylogeny, the systematic literature, and compendia such as that by Babcock, Durham & Myers (1955), where under the title *A Century of Progress in the Natural Sciences* nothing other than systematics and phylogeny is mentioned!

Functional anatomy.—Straus-Durckheim not only produced the first significant comparative anatomy, he also was the first to attempt to treat functioning (see earlier section). Kolbe and subsequent similar general textbooks treat functions.

In the latter part of the nineteenth century such advance as there was mostly concerned locomotion. Vitus Graber, after earlier studies on the mechanics of circulation, published an article entitled *Über die Mechanik des Insektenkörpers* in 1884. This was particularly good for his analysis of walking (a carabid beetle) and swimming (a diving beetle). He drew diagrams of models to show the components of all movements. This was a prelude to what is now called skeleto-muscular mechanisms. Flight is a more complicated phenomenon, and the understanding of it at this time was less complete. E. J. Marey summarized his work in *La Machine Animale* in 1874. He described the figure 8 trajectory of the wing tip and some other phenomena but, as Pringle says in his monograph cited above, his work "provided only incomplete information which is inadequate for an aerodynamic analysis. . . ." True, but do we laud for some achievement or condemn for incompleteness?

Félix Plateau, already cited, made early observations on breathing (ventilation of the tracheal system) but the analysis then possible (1884) was necessarily superficial. Similarly, the histologist Gutave Gilson made preliminary observations on the spinning of silk (1890).

Coming to the twentieth century, Hermann Weber, in addition to his *Lehrbuch der Entomologie* (1933), and several editions of *Grundriss der Insektenkunde* (1938 et seq) and *Biologie der Hemipteren* (1930), produced a number of large anatomical treatments, especially dealing with the thorax.

Incidentally, his accomplishments as an illustrator were superb (and painting was one of his hobbies; mountain climbing was another until his weak heart forced cessation of strenuous exertions).

This brings us to the dominant figure of insect anatomy and morphology, at least dominant for the twentieth century. Robert E. Snodgrass dominated studies of the homologies of parts and of skeleto-muscular mechanisms from 1924 to well past his death in 1962. He published a few unimportant papers dating back as far as 1896 but his important works began with *The Anatomy and Metamorphosis of the Apple Maggot* in 1924 and *The Anatomy and Physiology of the Honey Bee* in 1925 at which time he was almost 50 years old. I knew him well in the latter years of his life and he once told me that it took him that long to develop the background for what he wanted to do. In his earlier years his work on anatomy was limited to evenings and weekends; after getting an international reputation he was allowed to do much as he wished. Although officially retired for almost 20 years, he was working on his next paper at the time of his sudden death at the age of 87. He published about 90 papers, monographs, and books with a total of about 6000 pages. His series of 28 monographs totaling 2502 pages in the Smithsonian Miscellaneous Collections (1926–1962) are known almost as well as his books. I have reason to know that he verified much of what he cited from others but these monographs are written in such a way that it is commonly difficult to determine just what he did, what he verified, and what he merely cited. In toto, one could say that these works cover the insect's skeleton from head to tail and from birth to death. They emphasize skeleto-muscular mechanisms, the insect as a functioning organism, and some phylogeny and evolution. He was a real master of deductive logic and unexcelled at the analysis of descriptive anatomical data (he never did engineering type analyses involving consideration of stresses, forces, etc). Was he an innovator? Yes, to some extent, but he was more a keen observer and an analytical compiler whose important innovations were deductions including comprehensive concepts as to how an insect manages to be a functional animal.

An almost complete bibliography and a short biography are given in the *Snodgrass Anniversary Volume* published by the Smithsonian Institute in 1959. Ten more titles were published subsequent to this (two posthumously).

Having known Snodgrass so well, I could reminisce at length but this would not be fair to his predecessors whom I did not know. Like Lyonet, Weber, and Ferris, Snodgrass was an outstanding illustrator. I have been told, and he confirmed, that as a young man he had red hair and a fiery temper. By the time I knew him he was white-haired and mellowed, and while capable of being angered, no longer lost his temper. Occasionally he would become whimsical in his writing, as, "In truth we may say that the morphology of the insect abdomen is a field in which no angel foot has trod, and is, therefore, one in which the fool unhindered may achieve his destiny." And, on revisions, ". . . friends . . . expressed surprise that the bee should have changed so much in fourteen years . . ." Or, when in a cynical mood, "The arthropods are a group of related invertebrates; arthropodists, for the most part, are a group of unrelated vertebrates." Fittingly, on the stone slab sealing the crypt where his ashes repose there is engraved the figure of a honey bee.

Anatomy at ultramicroscopic and submicroscopic levels.—There is no clear break in continuity as one goes from macroscopic to microscopic to ultramicroscopic to submicroscopic. Even though I know some will object to extending the term anatomy to molecular levels, a few notes will be given on minute anatomy.

The electron microscope was mentioned under cytology. Hopefully, electron microscope studies of ultrastructure will become more interpretable with time. At present, for instance, the large amount of work on sense organs cannot be interpreted beyond the fact that chemoreceptors always have minute pores through their cuticle as revealed by the extensive work of Eleanor H. Slifer first with deductions from dye penetration seen with a light microscope (1954) and a few years later visualized with electron micrographs (1959).

Submicroscopic anatomy may be deduced from a variety of techniques. The use of polarized light analyses, dating back more than 50 years, has shown little of interest other than that linear molecules can become oriented by stress, that the sheaths of insect nerves resemble those of vertebrate nerves, and recently (1962), the demonstration by A. C. Neville that some insects deposit cuticle differently during the day than at night. Of even more interest are the analyses with X-ray diffraction methods by Rudall (1940 et seq) showing that chitin occurs in three different crystallographic forms, that silk is many different things, and that collagen occurs in insects but not in association with chitin (his earlier work was with wool).

As a byproduct of a biophysical analysis of flight, Torkel Weis-Fogh discovered (1960) an extraordinary elastic protein which he named resilin. This is protein biochemistry but also concerns the substance which supplies the elastic component for the flight machine, another example of anatomy intergrading to chemistry.

EPILOGUE

In conclusion, I do not feel like a mortician in writing on anatomy and morphology. These subjects are not dead; they are only no longer distinct. Anatomy will continue to be studied but what is studied will, even more than in the past, be determined mostly by persons who are interested in more and who use techniques originally developed in the also no longer distinct fields of physiology, biochemistry, biophysics, systematics, ecology, etc. Such persons will be working with the mechanism of flight, the mechanism of secretion, the nature of pheromone-controlled behavior, the evolution of halictine bees, or some such thing. In other words, the old terms anatomy and physiology, as labels for fields of endeavor, have outlasted their usefulness. In entomology today, as in biology in general, subdivisions such as cellular, developmental, environmental, functional (the living insect), systematic, etc would be more useful because they are more representative of what most individuals do.

Postscript

Anatomy and morphology are international and no attention was paid to nationality when selecting names for inclusion. Some persons may be interested in knowing what countries are represented. In a few cases I am uncertain of the nationality and citizenship of individuals. But tabulating as to the country where they worked or published we find for the preceding treatment 94 entomologists (not counting the six students listed under Korschelt or the several historians and other nonentomologists mentioned), or at least 94 persons who have worked with insects whether or not they would call themselves entomologists. In order of frequency these represent Germany (21), US (20), Great Britain (17), France (15), Belgium (4), Australia (3), two each for Austria, Denmark, Italy, and Japan, and one each for Canada, Finland, Holland, Russia, Spain, and Sweden.

About one quarter of the names above are of persons who compiled large textbooks, etc. Of the 25 persons cited for such compilations, one third or 8 warrant inclusion for other publications (Berlese, Comstock, DuPorte, Graber, Packard, Snodgrass, Weber, and Wigglesworth). Correcting for those mentioned only for a compilation, there would be a total of 77 instead of 94. All of the countries would still be listed, and in the same sequence, the 17 eliminations being distributed among the "big four".

LITERATURE CITED

Berlese, A. 1909. *Gli insetti, loro organizzazione, sviluppo, abitudini e rapporti coll'uomo. Embriologia e Morfologia.* Vol. 1. Milan: Soc. Ed. Lib. 1004 pp. ["Brief History of Entomology", pp. 5–30]

Henneguy, L. F. 1904. *Les Insectes. Morphologie — Reproduction — Embryogenie.* Paris: Masson et Cie. 804 pp.

Johannsen, O. A., Butt, F. H. 1941. *The Embryology of Insects and Myriapods.* New York: McGraw-Hill. 462 pp.

Kolbe, H. J. 1893. *Einführung in die Kenntnis der Insekten.* Berlin: Dümmler. 709 pp.

Locy, W. A. 1910. *Biology and Its Makers.* New York: Henry Holt. 469 pp.

Nordenskiöld, E. 1928. *The History of Biology. A Survey.* New York: Tudor. 629 pp.

Owen, R. 1855. *Lectures on the Comparative Anatomy and Physiology of the Invertebrate Animals.* London: Longman, Brown, Green & Longmans. 2nd ed. 689 pp.

Packard, A. S. 1898. *A Text-Book of Entomology.* New York: Macmillan, 729 pp.

Smithsonian Institution, 1959. *Studies in Invertebrate Morphology.* Published in honor of Dr. Robert Evans Snodgrass on the occasion of his eighty-fourth birthday, July 5, 1959. *Smithson. Misc. Coll.* Vol. 137, 416 pp.

Snodgrass, R. E. 1935. *Principles of Insect Morphology.* New York: McGraw-Hill. 667 pp.

Wigglesworth, V. B. 1939 et seq. *The Principles of Insect Physiology.* London: Methuen. 546 pp.

THE HISTORY OF INSECT PHYSIOLOGY

V. B. Wigglesworth
Department of Zoology, University of Cambridge
Cambridge, England

Our knowledge of the physiology of insects is a product mainly of the nineteenth and twentieth centuries, with an explosive expansion during the past forty years. The course of this recent expansion is clearly documented in current research papers and reviews. The purpose of the present survey will be to give some indication of the history of insect physiology up to 1930.

The arrangement of the sections is that adopted in the *Principles of Insect Physiology* (London: Chapman & Hall, 1972) by the present author, and many of the references quoted will be found in that book. All such references have been studied in the original; but there are others, chiefly older references, which have been taken from secondary sources, notably from: H. Burmeister, *Manual of Entomology* (Transl. W. E. Shuckard; London, 1836); George Newport, "Insecta" in R. B. Todd's *Cyclopaedia of Anatomy and Physiology* (London, Vol. 2, 1839, 853–994); H. J. Kolbe, *Einführung in die Kenntnis der Insekten* (Berlin, 1893); L. F. Henneguy, *Les Insectes* [Paris, 1904 (a particularly good source for the older references)]; A. Berlese, *Gli Insetti* Milan, Vol. 1, 1909); P. Marchal, "Physiologie des Insectes" in C. Richet's *Dictionaire de Physiologie* (Paris, Vol. 9, 1910, 273–386).

The history of insect physiology has, of course, been dependent on the development of physiology in general. In the early nineteenth century it was dominated by the search for organs and systems analogous to those of man. By the end of the century advances in organic chemistry and in histology had led to a much deeper understanding. In the present century, developments in cell biology, electrophysiology, and biochemistry have transformed the subject.

Physiology in the Egg

The nonchitinous nature of the egg-shell was recognized by Verson (1884) and by Tichomiroff (1885) who named the substance in *Bombyx* chorionin. Pantel (1919) had shown that in phasmids the shell is often impregnated with lime. That the main protective membrane of the egg may be a cuticle secreted by the serosa had been demonstrated by Blunck (1914) in *Dytiscus*. Leuckart (1855) had seen that in most insects the shell is pierced by ducts which convey oxygen to a respiratory meshwork that forms its lining and fills with air before the egg is laid (Pantel, 1913). It had been shown by Réaumur in the eighteenth century and by Rathke (1844) that many eggs in moist surroundings absorb water and swell before development begins.

Descriptive embryology was well advanced during the nineteenth century, but developmental physiology, inspired by the work of Boveri on other invertebrates, did not begin until Hegner (1908) made the classic observation that germ cell determinants reside in the cortical plasma at the posterior pole of the egg and induce the arriving cleavage nuclei to form the pole cells, the antecedents of the gonads. This led in the early 1920s to the demonstration by Seidel (1924 et seq) of an activation center, which initiates embryonic development at the posterior pole in *Pyrrhocoris* and *Platycnemis,* and of a differentiation center in the prospective thoracic region of many species. Cappe de Baillon (1920 et seq) had found that oocytes in phasmids could fuse completely to give normal embryos or partially to give double monsters; whereas Pauli (1927) had shown that the eggs of muscids are of the mosaic type with the regional determination of the embryo fixed in the cortical plasma before the cleavage nuclei arrive, so that regulation of this sort during development cannot occur.

As noted by Wheeler (1890) there is often a more or less prolonged pause during the blastokinetic movements of the embryo; he named this the diapause, and this term was given a wider use by Henneguy (1904) to describe any prolonged state of dormancy in the insect. In the eggs of Lepidoptera the winter arrest occurred at different developmental stages in different species (Tutt, 1888). The persistent dormancy of hibernating eggs had been known for centuries, notably in the silkworm *Bombyx.* As early as 1839 Moeglin and others (as quoted by Henneguy, 1904) had observed that diapause in the winter eggs of *Bombyx* could be broken by chilling. The details were worked out by Duclaux (1869) who showed that several months at around 1°C were necessary. The Japanese had selected strains of silkworms which had a varying number of broods before dormant overwintering eggs were laid. Umeya (1926) and other authors showed that clear-cut segregation of these strains did not occur because voltinism was determined by the mother; a circulating diapause factor determined whether the eggs in the ovaries were of the dormant sort or not. The extensive research of Bodine on the similar, but less resistant, diapause in the egg of *Melanoplus* had already begun in the 1920s. The dormancy of the fully developed young larva of *Aedes* within the egg had long been known and widely studied.

The use of hatching spines (van Emden, 1925) and other methods for rupturing the egg had been described; as had the practice of swallowing the amniotic fluid before hatching (Balfour-Brown, 1910) and of swallowing air after escape from the egg.

The Integument

That the cuticle of arthropods is the product of a single layer of epidermal cells, the hypodermis, was first recognized by Schmidt (1845) and by

Haeckel (1857) who speaks of the chitinogenic cells. Many descriptions of the cuticle were published and various descriptive terms proposed, notably by German authors. The nomenclature was reviewed by Campbell (1929) who recommended the terms epicuticle, exocuticle, and endocuticle. Lamination of the endocuticle was early described and the separation of exocuticle into wedge-like blocks at the intersegmental membranes was noted by Plotnikov (1904). As was first observed by Meyer (1842) in the elytra of *Lucanus* the endocuticle is often made up of colorless rods, the so-called Balken (beams), running parallel in a single lamina but at different angles in successive layers. As Biedermann (1903) pointed out, this arrangement is common in the skeletal structures of many animals.

In 1837 it was noted by Valentin that vertical processes could be seen running through the cuticle. This was confirmed by Leydig (1855) in the elytra of beetles; he believed these tracks to contain a nutritive fluid and called them pore canals. The contents of these canals are often continuous with vertical striae in the epidermal cells (Nordenskiöld, 1908). The pore canals may follow a spiral course through the cuticle (Plotnikov, 1904). It was believed by Holmgren (1902) that their contents became transformed into cuticle. Although Bütschli (1898) supposed the cuticle to have a foam-like structure, it was more generally believed to be fibrillar. The criss-crossing Balken in beetles are strongly birefringent (Biedermann, 1903). They show positive form-birefringence in the direction of the long axis and consist of crystallites (micellae) oriented along the strands (Gonell, 1926), with a tensile strength exceeding that of drawn copper (Schulze, 1926). A single isolated Balkenlage gives an excellent fiber diagram by X-ray diffraction (Künike, 1924).

By treatment with hot caustic alkali Odier (1823) succeeded in isolating and naming chitin from the elytra of *Melolontha*. Bütschli (1874) stressed the resemblance of chitin to cellulose; and in 1907 it was shown by Offer to consist of polymerized acetylglucosamine. By heating in strong alkali Hoppe-Seyler converted chitin to chitosan, which consists of partially de-acetylated chitin (Araki, 1895; Meyer & Mark, 1928). The color reaction of chitosan with iodine was developed by von Wisselingh as a test for chitin in fungi and was first used on the cuticle of arthropods by Wester (1910). Odier (1823) had noted that chitin makes up only a relatively small part of the cuticle which was mostly composed of protein. Chitin indeed was found always to be associated with protein (Krawkow, 1893). The iridescent colors of insects are almost always due to interference in the reflection of light from multiple thin plates (Onslow, 1921). The two types of lamination in the scales of *Urania* and *Morpho,* respectively, were demonstrated by Süffert in 1924.

The protein nature of the horny component of cuticle was recognized by Odier (1823) but subsequently forgotten. The so-called Inkrusten were considered to be carbohydrate in nature by Schulze (1922–1926) and Kühnelt (1928). Mirande (1904) had observed the reducing properties of the cuticle (probably due to diphenols) but he attributed these to reducing sugars. That the hardening of the cuticle was an oxidative process was recognized by Gortner (1911) and Dewitz (1916). The occasional use of lime in the hardening of insect cuticle was pointed out by Leydig (1860), Viallanes (1882), and others.

Interest in the impermeability of the cuticle to water loss was stimulated by Hazelhoff's (1926) theory of spiracular function (see Respiration). Kühnelt (1928) showed that the epicuticle was primarily responsible and suggested cholesteryl esters as the substance concerned. Keilin (1913) had suggested that dermal glands in tipulid larvae secrete a fatty covering that restricts evaporation, and Woods (1929) made the same proposal for larvae of the chrysomelid *Altica.*

There was prolonged controversy during the nineteenth century as to whether the cuticle was secreted and discharged by the epidermal cells, as claimed by Haeckel (1857), Kölliker (1857), Bütschli (1894), and Tower (1906); or whether it is formed by transformation of the cell substance as claimed by Vitzou (1882), Chatin (1892), Holmgren (1902), and Hass (1916). Vignon (1901) and Plotnikov (1904) showed that both processes could be involved; and Biedermann (1903) emphasized that there was no fundamental difference anyhow. A similar controversy centered around the mode of formation of the balken, as between an origin by spontaneous splitting of a homogeneous sheet (Kölliker, 1857) or by the coordinated activity of the epidermal cells (Kapzow, 1911; Korschelt, 1923). It was pointed out by Braun (1875) that two thirds or more of the thickness of the cuticle is laid down after molting.

The dermal glands, discovered by Verson (1890), were generally believed to secrete the molting fluid between the old cuticle and the new, which had already been noted by George Newport early in the century. But Verson himself, and von Buddenbock (1930) in Lepidoptera, Blunck (1923) in *Dytiscus,* and Poyarkoff (1910) in *Galerucella,* noted that the vacuoles in the glands disappeared only just before or immediately after the molt. The molting fluid itself was generally regarded as a lubricant to facilitate the shedding of the old skin. But Plotnikov (1904) and Tower (1904) considered its function to be the digestion and solution of the inner layers of the cuticle, thus making possible the rupture of the old skin at the ecdysial lines. They likewise regarded the fluid as being the product of the dermal glands; but Blunck (1923), Poisson (1924), and von Buddenbock (1930) considered the fluid to be mainly a product of the epidermal cells.

Growth and Metamorphosis, Polymorphism, Etc

The histological changes that lead to molting were fully described in the early years of this century; notably in *Leptinotarsa* (Tower, 1904), *Galerucella* (Poyarkoff, 1910), and aquatic Hemiptera (Poisson, 1924). The digestion of the old cuticle was reviewed in the preceding section. Until 1904 (Tower) it was supposed that the old cuticle split because it dried and became brittle.

The mechanics of ecdysis were clearly understood in the nineteenth century. Réaumur (1738) had already observed the action of the ptilinum of muscids in the emergence of the adult fly. Weismann (1864) showed that it was distended not with air but with haemolymph, under pressure created by contraction of the thoracic and abdominal muscles as fully described in *Schistocerca* by Künckel d'Herculais (1890). It was Jousset de Bellesme (1877) who first recognized, in *Libellula,* the role of swallowed air in distending all the external organs with blood to produce their definitive form; the pressure being sustained until the new cuticle was hard (Shafer, 1923).

The purpose of molting was seen as a method of growth in an animal with an unstretchable cuticle. But Pantel (1898) pointed out that molting was necessary also for the development of new form and new cuticular structures. The amount of growth achieved at a single molt was defined by certain empirical laws (Dyar, 1890; Przibram, 1912).

The physiological regulation of molting was wholly obscure until Kopeč (1917, 1922) observed that caterpillars of *Lymantria* failed to pupate if their heads were tied off with a ligature or if their brains were excised; whereas mere section of the nerve cord did not have this effect. Hence he inferred that a hormone produced by the brain was probably necessary for the initiation of molting. Buddenbock (1930) suggested that Verson's dermal glands were the source of the molting hormone in caterpillars.

Of the many theories of the nature of insect metamorphosis (reviewed by Henneguy, 1904) that which comes closest to our present ideas is that of Pérez (1902); that metamorphosis is a crisis in reproductive activity, in other words, the changes at metamorphosis are comparable with those at puberty. Many authors recognized metamorphosis in holometabolous insects only. But Henneguy (1904) and Handlirsch (1926) considered that there was no difference in principle between the relatively gradual development of Paurometabola and the more abrupt changes in Holometabola.

Lyonnet (1762) had observed in the caterpillar of *Cossus* the structures which Weismann (1864) named and interpreted as imaginal discs. He pointed out that their number and position gave rise to the suspicion that they might be the beginnings (principes) of the wings. Weismann pointed out that in *Calliphora,* as the discs grow at metamorphosis the larval tissues undergo histolysis to produce a nutritive brei for the growing organs of the adult.

Künckel (1875) proved the epidermal origin of the discs (histoblasts) in syrphids. Kowalevsky (1885, 1887) and Rees (1888) applying to insects the new theories of Metchnikoff on the phagocytic role of the blood cells, claimed that this process played an essential role in histolysis. But the subsequent research of Anglas, Berlese, Karawaiew, and others led to the conclusion that physiological changes in the larval tissues always precede phagocytosis. At the turn of the century there were extensive investigations of all the histological changes at metamorphosis. The papers of Pérez (1902, 1910, 1911) on *Formica, Calliphora,* and *Polistes* are outstanding examples.

Forms intermediate between larva, pupa, and imago (to use the terms proposed by Linnaeus) had been reported by Majoli (1813) in silkworms exposed to high temperatures; and the terms prothetely (Kolbe) and metathetely (Strickland, 1911) were proposed to describe these abnormalities. But no satisfactory theory of the control of metamorphosis emerged. Thought was concentrated on factors that might activate the imaginal discs or induce histolysis of larval tissues. For example, Bataillon (1893) proposed that asphyxia is the immediate cause of degeneration of tissues; and Dewitz (1902) that metamorphosis is due to oxidative enzymes (tyrosinase) set free into the blood. Nabert (1913) and Ito (1918) gave detailed descriptions of the corpora allata and the former noted their similarity to some endocrine organs in mammals, but no relation to metamorphosis emerged. Goldschmidt (1923) put forward some suggestive ideas on differential rates in different developmental processes as a possible factor in metamorphosis. Pérez (1902) had pointed to the contrast between the large cells of the ant larva and the small cells of corresponding organs in the adult; and Tiegs (1922) suggested in *Nasonia* that cell size might set a limit to larval growth.

It was found by Herold (1815) that sex in Lepidoptera is already established at hatching from the egg. Oudemans (1898) and Crampton (1899) showed that there was no effect on secondary sexual characters, color pattern, etc, when male and female pupae were cross-circulated or when the gonads were excised or transplanted to the opposite sex. Similar experiments were carried out with negative results by many authors early in the twentieth century (Kellogg, 1904; Meisenheimer, 1907–1910; Kopeč, 1911); and Klatt (1919) showed that there was no somatic induction when ovaries of one strain of *Lymantria* were transplanted to another. These observations concurred with the well-known examples of gynandromorphs in which half an insect would be male and the other half female. Sexual characters were not determined by circulating hormones. Sex was determined genetically (Doncaster, 1910).

On the other hand Goldschmidt (1916 and later) showed that sexual intermediates (intersexes) resulted if certain races of *Lymantria* were crossed; they resulted from a disturbance of the balance between male-producing and female-producing factors carried by the sex chromosomes. Comparable results were obtained in *Pediculus* (Keilin & Nottall, 1919), in *Drosophila* (Bridges, 1925) and in *Solenobia* (Seiler, 1929).

Other forms of polymorphism more directly under physiological control came to light. Leuckart (1858) first believed that the larval food of the honey bee was the regurgitated contents of the crop; later (1868) he concluded that it was the product of salivary glands. Planta (1888) showed that both views were correct depending on the stage of growth and the caste of the larva. The determining influence of the food was proved by Zander & Becker (1925). Lubbock (1889) had already proved that animal protein fed to ants gave greater numbers of queens. Poulton (1887, 1904) found that the adaptation of lepidopterous larvae and pupae to the color of their background was controlled by light receptors in the general integument. Uvarov (1921) discovered that the change of phase in successive generations of locusts depended on their degree of mutual crowding. Daylength would control the sexual forms in aphids (Marcovitch, 1924; Davidson, 1927, 1929). Seasonal dimorphism in *Araschnia* was controlled to some extent by temperature (Süffert, 1924). The form of certain endoparasitic Hymenoptera could be strikingly influenced by the species of host (Thompson, 1923). Intercastes in ants could be induced by the presence of *Mermis* parasites (Wheeler, 1910, 1928; Vandel, 1930).

It had long been known that hibernating insects are in a state of arrested development (diapause). In the European corn borer *Pyrausta (Ostrinia)* there are races which differ in voltinism, so that the incidence of diapause is genetically controlled (Babcock, 1924). Discussion on the physiological nature of diapause generally turned on a supposed depression of metabolism [e.g. Heller (1926) in *Celerio* pupa]; or as a kind of developmental fatigue (asthenobiosis) resulting from a surcharge of excretory products (Roubaud, 1919, 1927). But it was realized that an arrest of development could be due to various deficiencies; thus persistent underfeeding led to a prolonged arrest in the mosquito larva *Megarhinus* (Wigglesworth, 1929).

Muscular System and Locomotion

On account of their large size and well-developed striation the muscle fibers of insects have provided the material for most of the descriptions of muscle structure in textbooks of mammalian physiology, notably the classic work of Engelmann (1869) on the visible changes in muscle during contraction; the papers of Kölliker (1888); the fine details of striation by Hürthle (1909) on *Hydrophilus;* and the distribution of sarcosomes by Jordan (1919, 1920). That sarcosomes were probably mitochondria was indicated by Morison (1927, 1928).

Nerve endings in muscle were described by Orlov (1924) who observed Doyère's hillocks or endplates in the esophageal muscles of *Oryctes*. Mangold (1905) had seen the double innervation commonly present in insect muscle but had been unable to interpret its function.

The large masses which insects can lift in relation to their size had given rise to the belief that their muscles possessed extraordinary powers; but when the absolute power was related to the cross sectional area it was found that there was no great difference between muscles of insects and vertebrates (Camerano, 1893; Kahn, 1916). Likewise the properties of isolated muscle in respect to latent-period speed of response, summation of contractions, development of tetanus, etc (Rollet, 1884, 1887; Kahn, 1916) proved very similar to those of vertebrate muscle. In the flight muscles of most insects the speed of response was found to be adequate to explain the rate of wing beat by conventional nerve muscle stimulation. But the control of wingbeats in higher Diptera and Hymenoptera, where the rate may go up to several hundreds per second, remained a mystery. It was suggested by Fredericq (1928) who measured the chronaxie of Lapique in the wing muscles of

Calliphora, etc that perhaps the muscles served to set in vibration some elastic intermediary structure. Von Buddenbrock (1920) had observed that muscles going into a state of tonic contraction showed no measurable increase in metabolism.

The mechanism of walking in the six-legged insect, on an alternating triangular support, was described by Johannes Müller; and the mechanics of the muscles involved was worked out in detail by Graber (1884). The means by which the so-called adhesive organs on the feet of insects operated was a frequent object of speculation: did they act as suckers dependent upon atmospheric pressure (Simmermacher, 1884; West, 1884), by surface tension (Rombouts, 1884), by cohesion or stickiness of fluid (Dewitz, 1894), or by adhesion, that is, by molecular forces of attraction (Dahl, 1884; Arnhardt, 1923).

Much had been accomplished in the understanding of flight. Poujade (1884) and Voss (1913, 1914) noted that in such insects as *Agrion* the two pairs of wings operated independently. The nature of the wing movements was studied by Marey (1874) by observing the trajectory of the gilded wing tips in stationary flight and their rate by recording the wing tip movement by kymograph. Bull (1904–1910) and Kielich (1918) used cinematography; and Stellwaag (1910, 1916) observed the position assumed by the wings in the dead insect.

The need for a stiff front margin of the wing was indicated by Girard (1862) and the role of air pressure in inducing rotary movement of the wing was recognized by Bull (1904, 1910) who also showed that muscular control and the action of the basal articulation were important. The propeller-like function of wings was clearly described by Demoll (1918); the steering mechanism in butterflies by von Uexküll (1924); in *Agrion* by Bull; and in the bee by Stellwaag.

Samuel Pepys records that Robert Hooke in explaining to him in conversation the nature of musical sounds had pointed out that the rate of wing beat in a fly could be estimated by listening to the note which it emits. Marey (1874) and Voss (1913, 1914) obtained values ranging from 9 per sec in *Pieris* to 330 in *Musca.* The rate could be controlled by loading the wing (von Buddenbrock, 1919; Roch, 1922).

Jousset de Bellesme (1879) considered that the halteres were necessary to maintain equilibrium during flight: their removal led to the insect plunging to the ground; but if a small weight was attached to the abdomen normal flight was restored (cf. sense organs: mechanical)

Nervous System

The general anatomy of the central nervous system was described from the fine dissections of Newport in the 1830s, and later systematized by Brandt (1876) and others. Remak (1843) began the study of the internal structure; but the foundations were laid by Leydig (1862) who recognized the peripheral location of the ganglion cells and nerve fibers, with the central lacework of fine fibrils, the neuropile or Punktsubstanz. Binet (1894) confirmed by experiment the earlier conclusions of Newport (1838), Faivre (1864), and others that each ganglion contains both motor and sensory components, the motor nerves coming from the dorsal lobe and the ventral lobe receiving the sensory nerves. Benedicenti (1895) discovered the giant fibers running the entire length of the nerve cord in *Bombyx*.

After Dujardin (1850) and Leydig (1864) the internal structure of the brain was extensively studied. Dietl (1876) first used serial sections; but Viallanes (1883–1892) laid the foundation of present knowledge. He recognized the optic and olfactory lobes, the division of the cerebron into proto-, deuto- and tritocerebron, and also observed and named the two calices. This work is the source of the figures and descriptions that still appear in textbooks. Another classic source is the detailed study of the ganglia of *Aeschna* by Zawarzin (1924) using methylene blue staining.

The entire conception of the nervous system was, of course, profoundly influenced by the neurone theory of Ramon y Cajal. It was early recognized that the cell bodies of the motor neurones were confined to the central nervous system; only in comparatively recent years has it been accepted that the primary sense cells of insects are always peripheral and that the sensory axons grow inwards to reach the central nervous system. That integrity of the motor nerves was necessary for the maintenance of tonus as well as active contraction was shown by Kopeč (1912, 1918) and Sasse (1912); and the inhibitory effect of nerves appropriately stimulated was shown by Matula (1911) in the leg muscles of *Libellula*.

The idea of the simple reflex was adopted early as a basis for the analysis of insect movements, as in the sting in *Apis* (Bethe, 1897), the oviposition in *Bombyx* (McCracken, 1907), and grasping by the prolegs in caterpillars (Kopeč, 1912). Such movements were pictured as made up of an orderly succession of reflexes controlled by coordinating centers which would be either stimulatory or inhibitory, as in respiration (Alverdes, 1926) or cleaning movements (Szymanski, 1918). The plasticity of these reflex responses (with or without the brain) was stressed by Bethe (1930); likewise the tendency for impulses from strong stimuli to spread and produce responses in many different reflex arcs (von Uexküll, 1908; Szymanski, 1918; Weyrauch, 1929). The fluctuating courses that coordinating stimuli can follow was demonstrated by von Buddenbrock (1921) by cutting through the ganglia of *Carausius* in many different positions without impairing the ability to walk. Similar results were obtained after removal of legs in either walking or swimming insects; and the importance of peripheral stimulation was recognized (Ten Cate, 1928; Bethe, 1930; Matula, 1911; von Buddenbrock, 1921).

Besides being the coordinating center for most of the sensory stimuli received by insects the brain was recognized as having an important inhibitory function (Bethe, 1897; Sasse, 1912): in its absence reflex responses become exaggerated (von Uexküll, 1908). On the other hand it had long been known that piqûre of one half of the brain induced continuous circus movements towards the sound side (Faivre, 1857; Binet, 1894). Many authors ascribed this change to increased tonus and contractile power on the sound side (Bethe, 1897; Kopeč, 1912); but since the changed activity affects limbs on both sides of the body it was inferred by Baldus (1924, 1927) that the effect is central and, as Faivre had believed, each half of the brain directs movements chiefly to its own side.

The visceral nervous system, discovered in part by Swammerdam and studied by Cuvier, had been well described by the end of the nineteenth century; but apart from the fact that elimination of the frontal ganglion of Lyonnet abolished swallowing in *Dytiscus* (Faivre, 1957), nothing was known of its functions. The swellings on the side branches of the ventral sympathetic nerve were observed by Newport (1836) and figured as ganglion-like swellings by Leydig (1864).

Sense Organs: Vision

The structure of the compound eye was studied histologically by Leydig (1864); and Grenacher (1879) clearly recognized the segments of the rhabdom as the differentiated product of the innervated retinula cells and as being the light receptor. Grenacher classified compound eyes into acone, pseudocone, and eucone; and noted that in many eyes of acone type there was no fused visual rod but each retinula cell carried its own rhabdom. The mode of action of the compound eye as forming a mosaic image of the field of vision was first indicated by Hooke (1665) and later elaborated by Müller (1826).

The classic formulation of the mosaic theory was by Exner (1891) whose conclusions have dominated the subject for more than seventy years. Basing his experimental evidence mainly on the exocone eye of *Lampyris,* he conceived a new type of action which is found in night-flying insects in the dark adapted state, with the curtains of pigment in the pigment cells withdrawn. In such insects the apex of the rhabdom is widely separated from the crystalline cone, and in place of the classic apposition type of mosaic image there is a superposition type. By virtue of the peculiar refractive properties of the lens system a single rhabdom can receive light from a number of adjacent facets.

Exner demonstrated the occurrence of these pigment movements in response to light; he also described the iris tapetum, and the passage of light to and fro along the rhabdom acting as a wave-guide even when the rod is curved. He postulated that it is during this passage that the visual stimulus is given, the retinula cells being the intermediaries between rhabdom and nerve. The function of tracheae in the role of the tapetum was described by Bugnion & Popoff (1914). Demoll (1909–1917) in butterflies and Bedau (1911) in *Notonecta* observed pigment migrations in the retina which influenced the pseudopupil reflection (Leydig, 1855) from the compound eye. Pigment movement in the iris can affect the color banding of the eyes (Friza, 1928); the daily rhythm of such movements may be secondary to a rhythm of general activity (Demoll, 1917).

The electrical response of the retina to illumination was described by Hartline (1928), who showed that this response conforms to the Bunsen-Roscoe law of photochemistry, as indeed Patten (1914, 1916) had demonstrated for the photic responses of muscid larvae. In *Eristalis* (Dolley, 1929) and other insects the eye is capable of extensive adaptation to light intensity. The appreciation of luminosity is influenced by simultaneous contrast with the adjacent part of the visual field, as shown in *Macroglossa* by Knoll (1921, 1926).

The extent of the visible spectrum was first studied by Lubbock (1876) who showed that ants are highly sensitive to ultraviolet light; by covering the eyes, Forel (1886) confirmed that they are responsible. Lubbock inferred that by reason of this sensitivity the colors of flowers must be quite different from those perceived by the human eye and this was demonstrated photographically by Lutz (1924) who showed that many flowers have honey guides invisible to man but characterized by their failure to reflect ultraviolet light. It is the retina which is directly sensitive; it is not dependent on fluorescence (Merker, 1929). At the red end of the spectrum bees and wasps are insensitive (Lubbock, 1882; von Frisch, 1914) but some butterflies have an actual preference for red (Ilse, 1928).

It had been assumed since Sprengel in the eighteenth century that the colors of flowers were attractive to insects. This was first proved experimentally by Lubbock (1882) who trained bees to visit honey smeared on glass plates laid on differently colored papers. They could not be deflected by white papers. Von Frisch (1914) improved these experiments by using a range of grey shades as well as white, and Kühn (1927) used spectral colors so as to exclude the influence of ultraviolet reflection. By fatiguing the eye to particular wave-lengths Hamilton (1922) showed that *Drosophila* can detect intervals ranging from 25 nm in the ultraviolet to 50 nm in the blue-green. As a pure hypothesis, Hanström (1927) suggested that different retinal cells in a single ommatidium might have different color sensitivities. In 1929 Hertz had begun to analyze the limits of form perception in the honey bee, which shows a natural preference for highly divided figures; but in life, insects use elaborate visual patterns (Forel, 1910; von Frisch, 1914; Baldus, 1924). Visual acuity in the honey bee was explored by Baumgärtner (1928) and Hecht & Wolf (1929). Exner (1891) considered that the compound eye is concerned with the perception of movement rather than with form. He also appreciated the capacity of the compound eye for the exact determination of distance in the visual field (cf. Demoll, 1917; Baldus, 1926).

The function of lateral ocelli (stemmata) had received little experimental study, but de Lépiney (1928) had shown that caterpillars of *Lymantria* could recognize and seek out vertical silhouettes. The light sensitivity of apparently eyeless muscid larvae had long been known (Pouchet, 1872). Studies on the structure of the dorsal ocelli led to the conclusion that they were adapted to the perception of small changes in light intensity (Hess, 1920; Homann, 1924) and the experiments of Bozler (1925) had led to the suggestion that they serve in *Drosophila* as stimulatory organs which increase photokinesis. Finally the sensitivity of the general body surface, dermal light sense, had been described by Graber (1883) in *Blattella* and by Lammert (1925) in caterpillars.

Sense Organs: Mechanical and Chemical

Each of the little peripheral sense organs was named a sensillum by Haeckel; the associated sense cells being derived from the epidermis. In addition Holmgren (1892) described free nerve endings in the integument and joints. The tactile hair is the simplest type of sensillum (Leydig, 1860); the many modified forms were named and classified by O. von Rath (1888–1894). Their function was inferred from their structure and from general observations of behavior. Thus Hicks (1857) believed the campaniform organs at the base of the halteres to be olfactory organs and this view was adopted by Lowne (1870), Graber (1882), Bolles Lee (1885), etc; whereas Leydig (1860) held them to be auditory. Weinland (1890) considered that they responded to changes in the plane of vibration of the halteres and were thus able to regulate the steering of the fly. In other words that they are what Sherington named proprioceptive organs, responsive to the bending of the cuticle (cf. Demoll, 1917); just as tactile hairs (postural hairs or Stellungshaare) could indicate the flexion of joints (Lowne, 1892; Doflein, 1910) and curiously modified tactile hairs in *Nepa* could serve as static organs under water (Baunacke, 1912).

The abdominal tympanum of acridids which had been thought by Latreille and Burmeister to be a sound producing organ was recognized by Müller as an organ of hearing. It was fully investigated by Graber (1881) who had already discovered the chordotonal organs and regarded them as auditory sensilla. This was supported by Hurst (1890) and Child (1894) in describing Johnston's organ (1855) of culicids as an organ of hearing, and by von Adelung (1892) who gave a full description of the tympanal organs in the tibiae of Tettigoniidae, which had been discovered by Müller. Regen (1912–1926) analyzed the hearing and behavior of *Gryllus* and *Thamnotrizon,* with proofs of the auditory function of tympanal organs, showing that *Thamnotrizon* could perceive frequencies up to 25,000 cycles per sec (cf. Auger and Fessard, 1928 on *Acridium*). Eggers (1929) showed the widespread occurrence of tympanal organs.

Chordotonal organs are widespread in the legs and abdomen of insects which show no capacity for hearing (Lubbock, 1878) and they also came to be regarded as responsive to movements of the parts of the body (Demoll, 1917). A. M. Mayer (1874) had noted the harmonic virbration of the individual hairs of the antenna of the male mosquito and considered them to be the sound receptors. [At the time of Burmeister's book (1836) antennae in general were regarded as organs of hearing: "we may almost consider it as settled".] Many years later body hairs of caterpillars (Minnich, 1925; Abbott, 1927) *Blattella,* etc (Baier, 1930) were shown to react to sound.

Landois (1867) gave a full account of sound production, including the timbals of cicadas whose function had been recognized by Aristotle (see Prochnow, 1910).

Likewise the chemical senses of taste and smell were inferred largely from anecdotal accounts of insect behavior. Smell was located in the antennae for distant perception; in the palps for use at close quarters (Kolbe, 1892).

In his experiments demonstrating color vision, Lubbock (1882) introduced the procedure of training the insect to associate a reward or punishment with a particular response, and thus to elucidate the range of its sensory perceptions. Taken over from the methods of showmen this procedure, termed dressur, was used extensively by the German school led by von Frisch. This, together with other experimental procedures, had led to immense advances by 1930. It made possible the use of natural odors in place of the potent chemicals that had given misleading results in the past. Thus von Frisch (1920–1926) had analyzed in great detail the sense of smell in the honey bee, and had located the receptors as the poreplates on the antennae. In most insects the thin walled hair sensilla were associated with olfaction (Vogel 1921–1923, etc). Minnich (1921–1929) had demonstrated and analyzed contact chemoreceptors on the proboscis of *Calliphora,* the tarsi of butterflies and muscids and the antennae of Hymenoptera, and shown their very high sensitivity. The four taste qualities were shown to be distinguished in *Dytiscus* by sensilla on the palps and in the mouth (Schaller, 1926). The olfactory function of the palps was established in *Periplaneta* (Glaser, 1927), *Gryllus* (Sihler, 1924), and *Dytiscus* (Schaller, 1926). An important discovery by Verschaeffelt (1911) was the specific attraction of plant feeding larvae to characteristic components in their food plants: mustard oils for *Pieris,* etc.

A sense of temperature was found to be generally spread over the body, as in grasshoppers (Geist, 1926), *Liogryllus* (Herter, 1923, etc); but the antennae were highly sensitive, particularly in blood-sucking insects (Fricklinger, 1916; Weber, 1929, etc)

Behavior

The physiological study of behavior dates from the early years of this century. But the pioneer in this field was John Lubbock. In his *Ants, Bees and Wasps* (1884), which records experiments made over many years, he initiated the training (dressur) method for the study of the senses and in many of his experiments on behavior he attempted to define the mechanisms by which the sense organs are used in orientation (another major preoccupation of investigators, notably in Germany, from 1900–1930). Sherrington's analysis of reflex action likewise led to many attempts to describe insect behavior in terms of defined reflexes. The work of Loeb, which was brought together in 1918 in his book *Forced Movements, Tropisms, and Animal Conduct* attempted a synthesis of these methods of approach, which were dealt with also in another influential synthesis by Alfred Kühn in *Die Orientierung der Tiere im Raum* (1919).

All that can be done here is to note some of the principles that emerged, with specific examples. Loeb stressed the importance of kinesis, the increased activity in the nervous system produced by sense organs which serve as stimulatory organs. We have already noted the ocelli; the halteres were shown by von Buddenbrock (1919) and Ulrich (1930) to be important. Loeb supposed that forced movements were induced by differences in muscle tonus on the two sides of the body, as the result, for example, of different illumination in the two eyes: phototonus as seen in *Ranatra* (Holmes, 1905), *Proctacanthus* (Garrey, 1918), and *Eristalis* (Mast, 1923). This was the basis of the tropisms described by Loeb. Thigmotaxis or stereokinesis is the induction of sleep or akinesis by contact, as seen in the earwig (Weyrauch, 1929) or other insects creeping into crevices. An extreme was reflex immobilization or hypnosis. Such reactions could be induced by light, contact, chemical senses, humidity, etc.

The image-forming eyes permit the development of a quite new response which depends on a tendency of the insect to move so as to maintain a constant pattern on the retina. Rádl (1902) has shown that flies on a slowly rotating turn-table move so as to face a constant direction, the optomotor reaction. Lubbock had shown that ants orient themselves by maintaining constant the direction of incident light; the so-called sun-compass reaction further investigated by Santschi (1911) and Brün (1914). This response was termed menotaxis by Loeb. It was shown by von Buddenbrock (1917) that if the insect uses this type of orientation in respect to a lighted candle, instead of the distant sun or moon, it will follow a logarithmic spiral into the flame. A similar response is seen in the dorsal light reflex in some aquatic larvae which maintain their attitude by orientation to overhead illumination (Alverdes, 1927).

Respiration

Since Malpighi (1669) described the anastomosing tracheal system, an immense amount of study has been given to insect respiration; most of our present knowledge had been established by 1930. The spiracular valves were described by Burmeister (1832). The taenidial thread (so named by Packard) was claimed by Dujardin (1849) to arise as a spiral folding of the intima later filled by the horny substance. Chitin was found in the larger tracheae by Wester (1910). The tracheal end cells where the tubes break up into tracheoles were found by Meyer (1849) and Leydig (1851). Controversy centered around the tracheole endings. Anglas (1904) and Pérez (1910) observed tracheal cells entering developing muscle fibers. Holmgren (1907) supposed that they made connection with a pre-existant trophospongium. The prolonged controversy as to whether the endings contained air or liquid was resolved by the transillumination of living insects which showed that some contain air as far as they can be resolved; others contain fluid, the meniscus moving up and down in response to changes in osmotic pressure around the endings (Wigglesworth, 1930). The development of tracheae by inward growth was described by Weismann (1863) who also demonstrated absorption of the contained fluid by the tissues so that the tubes filled with air.

It was Thomas Graham (1833) who first considered that diffusion alone could probably account for the respiratory exchanges of insects. By measurement and calculation Krogh (1920) proved that diffusion would indeed suffice to convey to the tracheal endings the oxygen actually consumed. Krogh overestimated the partial pressure of oxygen in the tissues, for it was not until Hazelhoff (1927) put forward his theory of diffusion control that it was realized that the spiracles are normally kept closed, being opened only just enough to supply the insect with oxygen.

Superimposed on diffusion is mechanical ventilation. Respiratory movements had been observed by Severinus (1645); their rate was found by Sorg (1805) to be proportional to activity in *Lucanus*. They were described in detail by Rathke (1862) and Plateau (1884). Dujardin (1849) pictured expansion and compression of tracheae in their long axis, as was proved by Demoll (1927) and others. Flattened tracheae (Portier, 1911) and air sacs (Demoll, 1927) increase the volume of tidal air and can be compressed by remote action via the blood (Graber, 1877; Cortejean, 1890). Nitsch (1808) had observed the compression of the air sacs in grasshoppers during expiration. In some insects the vital capacity could equal one third of the total capacity of the system (Krogh, 1913; Demoll, 1927). It was suggested by Rathke (1862) that spiracles could open and close in a timed sequence so that a directed stream of air was driven through the system. Observations on the rhythm of movements of different spiracles (von Buddenbrock & Rohr, 1922; Du Buisson, 1924) supported this theory; but by 1930 it still lacked good experimental proof.

By 1842 Marshall Hall had already shown that isolated abdominal segments of *Libellula* could perform respiratory movements and had compared the ganglia with the medulla oblongata of vertebrates. Faivre (1860, 1875) in *Dytiscus* and Wallengren (1913) in the larva of *Aeschna* obtained good evidence of an overall respiratory center in the thorax, controlling the primary centers in the ganglia (Stahn, 1928). It was long claimed that these centers responded only to oxygen lack; but Krogh (1913) showed that at high levels carbon dioxide is strongly stimulating. Carbon dioxide was shown by Hazelhoff (1927) to be of prime importance in the regulation of diffusion control by the spiracles. Dewitz (1890) had shown that carbon dioxide will pass through the cuticle. This would provide an explanation for the deficiency of carbon dioxide in the gas of the tracheal system (Krogh, 1913), though Krogh favored an escape from the blood through the tracheal walls throughout the system.

An immense amount was published on the respiration of aquatic insects, dealing with the closed tracheal system and tracheal gills of many aquatic larvae, and with adaptations for the breathing of atmospheric air. Most authors came to accept the view of Dutrochet (1837) and Oustalet (1869) that the gill membrane is a passive diffusion barrier. Wallengren (1915) got good experimental support in *Aeschna* larva. Cuticular gills, which can function both in air and water, were seen by Pulikovsky (1927) as an adaptation to mountain streams liable to dry up.

Air-breathing aquatic insects are dependent on hydrofuge structures to make contact with the air. This property had long been known; it was most clearly analyzed by Brocher (1909). Comstock (1887) had suggested that air stores might act as physical gills; the mechanism of this process in the air bubbles carried by many insects was worked out by Ege (1915), a student of August Krogh. And Brocher (1909) explained the methods by which films of air are held so firmly and permanently by bent semihydrofuge hairs as to make possible what he called plastron repiration in *Haemonia,* etc.

Only in the case of hemoglobin-containing chironomid larvae was evidence obtained that the blood had a respiratory role. It had been shown that the red *Chironomus* larvae were less susceptible to oxygen want; and Leitch (1916) proved that at low oxygen tensions the hemoglobin is indeed acting as an oxygen carrier.

Circulatory System and Associated Tissues

The pulsating dorsal vessel of insects with its alary muscles had been seen by Harvey, Malpighi, and many others, and the arrest and occasional reversal of beat noted; but the existence of a complete circulation was first established by Carus (1827) and reinforced by Straus-Durckheim's (1828) description of the interventricular and ostial valves of the heart. Accessory pulsatile ampullae were discovered in the legs of Hemiptera by Behn (1835) and later at the base of the antennae of Lepidoptera (Burgess, 1881), Ephemeroptera (Vayssière, 1882), and various Orthoptera (Pawlowa, 1895). The way in which the heart is suspended was described by Graber (1873–1876) who gave the first clear description of the histology of the heart and the mechanism of the circulation, including the role of the dorsal and ventral diaphragms in regulating the backward flow in the abdomen, substantiating the conclusions of Newport (1832) in *Sphinx,* (cf. Kowalevsky, 1894). The mechanism of circulation was further described in Ephemeroptera (Popovici-Baznosanu, 1905–1910) and *Apis* (Freudenstein, 1928). But some of the most informative work is to be found in the series of papers by Brocher (1916–1929) dealing with a wide range of insects, showing the part played by differential pressures in the body sinuses in controlling circulation through wings and legs, and describing the accessory pulsatile organs in the thorax of many insects, which pump blood from the wings and return it to the aorta. Circulation in the wings had been observed by Henry Baker as early as 1744.

The pulse rate in *Sphinx ligusti* was studied by Newport (1837): it varies during the larval stages; in the pupa it falls, and almost ceases during hibernation; in the adult it varies enormously with activity. The accessory hearts behave similarly (Brocher, 1916, etc; Crozier et al, 1927). The property of rhythmical contraction resides in the heart muscle which continues to beat in isolation in the absence of nerve cells (Du Buisson, 1929, 1930); contractions are influenced by, but are not dependent on, muscular or elastic tension on the heart wall (Graber, 1873–1876). Studies on the reversal of heart beat by Gerould (1929–1930) emphasized that the pacemaker is normally at the hind end, and that the heart functions as a single tube, not as a series of chambers. Dual innervation from the ventral nerve cord and from the stomatogastric system was described by Zawarsin (1910) and Alexandrowicz (1926). The heart beat is modified by nervous action (Marchal, 1910; Lasch, 1913).

The chemistry of insect blood had been studied chiefly in Holometabola: *Apis* larva (Bishop, 1922–1923), *Dytiscus* and *Hydrophilus* (Barrett & Arnold, 1910), and Lepidoptera (Brecher, 1925, 1929). In general the very high content of nonprotein nitrogen, mainly amino acids, was noted, as was the low content of chloride; Ca, Mg, and phosphate high as compared with mammalian plasma; pH just on the acid side of neutrality; osmolarity in terrestrial insects usually exceeding that of 1% sodium chloride.

As recently as Graber (1877) there was doubt as to whether the corpuscles in the blood (Carus, 1827) were true cells. But Dewitz (1889) showed that they were capable of independent movement and Schäffer (1889) described concentrations of cells (Blutbildungsherde) differently situated in different groups. Their tendency to form phagocytic organs was stressed by Cuénot (1891, 1895) who also advanced the view that they produce the proteins of the hemolymph. Phagocytosis has been discussed (p. 207). Poyarkoff (1910) and Hollande (1911) classified the hemocytes into proleucocytes, phagocytic amoebocytes, nonphagocytic oenocytoids in many insects, and sometimes adipocytes. It had been shown that the blood of some insects, such as honey bee larvae, does not clot (Bishop et al, 1926). In others (cockroach, etc) it clots very readily particularly in the presence of blood cells (Muttkowski, 1924). Lazarenko (1925) observed that blood cells around foreign bodies break down to produce a capsule of connective tissue; he therefore suggested that hemocytes might normally be concerned in connective tissue formation.

The oenocytes, originally discovered by Fabre, were so named by Wielowiejski (1886) on account of their wine-yellow color. They arise from the epidermis (Koschevnikov, 1900).

The pericardial cells were first recognized by Leydig (1866) scattered along the heart and aorta. It was discovered by Kowalevsky (1889) that they specifically take up ammonia carmine injected into the blood; they were therefore termed nephrocytes and were long regarded as concerned in storage excretion.

The fat body had been compared to the liver by Treviranus (1816); it had long been recognized as a well-tracheated storage organ (Landois, 1865) in which glycogen, fat, and protein were laid up, notably before metamorphosis (Pérez, 1910). Leydig (1857) and Marchal (1889) stressed its importance in the deposition and storage of uric acid; Wielowiejski (1886) and Schäffer (1889) noted its close relationship with the blood cells, and considered it to occupy a central position in metabolism.

Finally, the fact that the luminous cells of fireflies, etc are modified fat body cells was well recognized (Leydig, 1863; Wielowiejski, 1882). Dahlgren (1917) and many others ascribed the light production to the activity of symbiotic bacteria; a view finally demolished by Harvey (1929). Verworn (1892) had studied the nervous control of luminosity and Dubois (1886) had described the oxidative nature of the process brought about by the enzyme he named luciferase acting on the substrate luciferin; a mechanism confirmed by Harvey (1916).

Digestion and Nutrition

At the time Burmeister (1836) and Newport (1839) were writing, digestion in insects was thought of as the simple extraction of nutritious juices from the food. The general structure of the digestive system was well described. The crop of the honey bee was named by Burmeister the sucking stomach and belief in this function was still sustained by Breitenbach (1881) for Lepidoptera, although Ramdohr (1811) had already clearly recognized that it is a reservoir for food. The cibarial pump was discovered by Graber and its structure described in Lepidoptera by Burgess (1880) and in Diptera by Meinert (1881) and Dimmock (1881), etc. The proventriculus had been regarded as a gizzard, but Plateau (1874) concluded that it served as a filter only. Later observations, however, by DuPorte (1918) and Eidmann (1924) established that at least in *Gryllus* and in *Periplaneta* the original belief was correct.

The esophageal invagination (rüssel) was described by Schneider (1887) as having a delicate chitinous extension (trichter) investing the contents of the mid-gut. But it came to be realized that this was none other than the peritrophic membrane which had been observed by Lyonnet (1762) in the caterpillar and which occurs in most insects. Balbiani (1890) regarded this membrane as being shed off the surface of the mid-gut epithelium. Van Gehuchten (1895) and Vignon (1901) showed that in Diptera it arises from a ring of cells at the cardiac end of the mid-gut and is molded by an annular press in the proventriculus; and Wigglesworth (1930) showed that both methods of formation occur and that in both types the membrane is chitinous.

The histology of the gut was described by many authors from Leydig (1857) onwards. Basch (1858) and others much later wrote of glandular follicles in the mid-gut of many insects; but Balbiani (1890), Miall & Denny (1886), and others recognized that the small cells were young replacement cells. Bizzozero (1892) and Rengel (1898) observed that in Hydrophilidae there is complete replacement of mid-gut cells every few days. Histological changes during secretion and absorption led to unresolved controversy as to whether these changes were artifacts or not. Mid-gut epithelium of distinctive type in different regions had been described in nematocerous larvae (van Gehuchten, 1890, and others) and in Hemiptera (Locy, 1884, and others). In the tsetse fly *Glossina* such changes were correlated with the rapid absorption of excess fluid in the enzyme-free anterior region, on the one hand, and the blackening of the ingested blood in the posterior region where proteolytic enzymes occur, on the other (Wigglesworth, 1929). Mark (1876) and Witlaczil (1885) described the short circuiting of the mid-gut by way of the filter chamber in coccids and other Homoptera (cf. Licent, 1912).

The biochemistry of digestion was worked out in the classic papers of Plateau (1875–1877) who reproduced on a small scale all the contemporary work on vertebrate digestion and established the presence of the main types of digestive ferments, including amylase in the salivary gland secretion. The gut contents were found usually to be just on the acid side of neutrality; more acid in the crop of Orthoptera as the result of bacterial fermentation (pH 4.8–5.9, Bodine, 1925; Wigglesworth, 1927); always strongly alkaline (pH 9.0–9.4) in the mid-gut of caterpillars (Jameson & Atkins, 1921; Swingle, 1928). By 1930 it had been shown by a wide range of authors that the relative activity of the major enzymes ran parallel with the composition of the diet; and the properties of the enzymes in relation to temperature, salts, pH, etc closely resembled those of vertebrates (Wigglesworth, 1927, 1929).

The role of the hind-gut, notably of the rectal epithelium and rectal glands, had been reviewed in detail by Chun (1876) and in later years many secretory functions were proposed. Bounoure (1919) and Krüger (1926) had pointed to the exceptionally thin cuticle over the rectal glands; the balance of evidence suggested an absorptive function, perhaps primarily of water, but this interpretation had not been clearly formulated.

The role of microorganisms in the breakdown of indigestible materials had often been considered in relation to cellulose digestion in the fermentation chamber of lamellicorn larvae (Mingazzini, 1889; Werner, 1926; Wiedemann, 1930). Cleveland (1925, 1926) had established the importance of flagellate protozoa in the hind-gut of wood-feeding termites: termites, able to survive long on pure cellulose, could no longer do so if deprived of their protozoal fauna; but they could still survive on fungus-digested cellulose.

It had been established that insects could survive for many months on a nitrogen-free diet: cockroaches (Zabinski, 1929); termites (Cleveland, 1925); but for growth they needed organic nitrogen, sulfur, phosphorus, and salts. Essential amino acids had been defined; and several factors extractable from yeast had been found necessary in *Drosophila* (Bacot & Harden, 1922). It was established that microorganisms were an important source of vitamins in the nutrition of *Drosophila* and other flies (Glaser, 1924; Baumberger, 1919).

Many insects had been found to contain microorganisms that were constantly transmitted from one generation to the next, and were regarded as symbionts (Buchner, 1930). Association of symbionts with the gut suggested that they might aid in digestion (Roubaud, 1919, in *Glossina;* Reichenow, 1922, in other blood-sucking insects). But in *Glossina* digestion does not begin until the blood has passed far beyond the site of the mycetome carrying the symbionts (Wigglesworth, 1929). The fact that among blood-sucking insects symbionts occur only in those groups which take no food but blood in all stages of their life history led to the suggestion that they provide an endogenous source of vitamins (Wigglesworth, 1929).

Excretion

Early in the nineteenth century insect physiology consisted largely of the search for correspondences between insects and mammals. Since their discovery by Malpighi, the tubules which bear his name (as proposed by Meckel) had been regarded as hepatic in function. Herold (1815) and Brugnatelli (1816) had found uric acid in them and this led Meckel (1826), Müller, and others to advocate a renal function. Newport (1839) adopted an equivocal position but still inclined to regard them as biliary organs and agree with Burmeister (1832) that the true urinary organs are the anal glands as seen in Carabidae. Kölliker (1858) demonstrated calcium oxalate in the Malpighian tubules, as well as sodium and ammonium urates; and the classic papers of Plateau (1874) and particularly Schindler (1878) firmly established their excretory function. Among much else Schindler demonstrated their specific uptake and discharge of indigocarmine.

It was agreed that uric acid was generally the main form of nitrogen excretion; but Weinland (1906) discovered that ammonia is the chief product in the meat-eating larvae of *Calliphora* and *Lucilia*. The virtue of solid uric acid as a vehicle for nitrogen excretion in insects living in dry environments was recognized (Babcock, 1912) and also during development in the closed systems of the egg or the pupa (Needham, 1929). Granules of lime had been observed in the tubules of *Eristalis* by Lyonnet and were reported in many other dipterous larvae.

The anatomical and histological characters of the Malpighian tubules were extensively studied and there was much inconclusive controversy about the nature of the striated border and the visible processes of secretion. It was noted that the urine as first formed is a clear fluid; concretions and crystals separate out later. This was attributed to the reabsorption of water by the cells (Bugajew, 1928). A dimorphism in the cells of the tubules had been observed in Coleoptera (Schindler, 1878), Lepidoptera (Metalnikov, 1908), and Diptera (Pantel, 1914). The cryptonephridial arrangement around the rectum had been described in Lepidoptera, Coleoptera, etc (Möbusz, 1897; Sedlaczek, 1902; Ishimori, 1924). Metalnikov (1908) had shown that dyes injected into the body cavity are not taken up by the enclosed segments of the tubules; he supposed that these regions remove toxic substances absorbed from the rectum; in scolytids, Sedlaczek (1902) regarded them as helping in conservation of water.

An entirely separate system, homologous with the cephalic nephridia of other arthropods, was discovered in the Apterygota (Bruntz, 1908; Philiptschenko, 1907). The evidence for their excretory function rested on the uptake of indigocarmine and ammonia carmine by different parts of these labial kidneys. More important was the recognition of storage excretion in insects. It was Fabre (1863) who discovered that in many caterpillars uric acid accumulates in the epidermis and in the fat body and is transferred to the Malpighian tubules and meconium in the pupa. Storage excretion of this kind was found in many insects. Hopkins (1895) considered the white material in the wing scales of Pieridae to be uric acid and the yellow pigments derived purines; he described this as an example of storage excretion. Estimated as uric acid, about one quarter of the total in the pupa of *Pieris* was found to be deposited in the wings; in *Vanessa* none (Wigglesworth, 1924). Wieland and Schöpf (1926) found that most of the material in the wings was in fact a white pterine pigment which thus played a substantial role in the storage excretion of nitrogen.

Metabolism

The quantitative changes in the major organic constituents of the insect body, fat, glycogen, and protein, had been mapped out during the last years of the nineteenth century and in the early years of the twentieth; though based upon a rather small sample of species. Kellner (1884) and Bataillon (1893) had followed these changes in *Bombyx* during growth and metamorphosis; Tichomiroff (1885) and Farkas (1903) in the egg and pupa. Similar results had been got by Sieber & Metalnikov (1904) on *Galleria,* Weinland (1906–1909) on *Calliphora,* Tangl (1909) on *Ophyra,* von Kemnitz (1916) on *Gasterophilus,* Heller (1926) on *Deilephila,* and Rudolfs (1926–1929) on *Malacosoma.* Slowtzoff (1904–1909) had followed the changes in a wide variety of insects during starvation.

The characterization of insect lipids by Timon-David (1927–1928) showed that the liquidity of fats at normal temperatures was assured by the high degree of unsaturation or, notably in aphids, by the presence of short-chain fatty acids (butyric, caprylic, etc). Slifer (1930) noted that the overwintering eggs of grasshoppers had fats of lower melting point. Glycogen had been first demonstrated in large amounts in the maggots of flies by its discoverer Claude Bernard; the nature of the soluble carbohydrate in the blood, which was clearly not glucose, remained a mystery.

Some of the familiar products of insect glands had been subject to chemical study. The amino acid composition of silk fibroin was worked out by Abderhalden (1909) and glycine, alanine, and tyrosine shown to be major components. The composition of insect waxes and lac was virtually unknown. Venoms had been studied; there was a strong suspicion that the venom of the bee and of myrmicine and dolichoderine ants contained toxic proteins (Stumper, 1922). Formic acid in camponotine ants had been known from Roman times; in *Formica rufa* Stumper found quantities up to 18% of the total body weight in some samples. The products in repellent scent glands were mostly unknown but salicylaldehyde as an end product of the breakdown of salicin in the foodplant was recognized in *Melasoma* (Garb, 1915) and *Aromia* (Hollande, 1909; Smirnoff, 1911) and the scents emitted by *Papilio* larvae were thought to have a similar origin (Schulz, 1912; Wegener, 1923). Butyric acid from *Carabus* (Marchal, 1910) and oxalic acid from mycetophilid larvae (Mansbridge, 1933) were other defensive or offensive products of known composition.

Of insect pigments melanin had been well studied and shown to arise by oxidation and ring closure from tyrosine (Schmalfuss, 1929; Thunberg, 1930). Its formation was arrested by lack of oxygen or by destruction of tyrosinase (Gortner, 1911; Henke, 1924). Pterine pigments had been identified in the wings of Pieridae by Wieland & Schöpf (1926). Carotinoids were found to reach a concentration in the blood of *Leptinotarsa* equal to that in many leaves (Knight, 1924) and to be deposited in the tissues of the elytra of *Melasoma* (Kremer, 1917), the fat body of *Pyrrhocoris* (Schulz, 1913), the blood of caterpillars (Meyer, 1930; Gerould, 1927), and in the cocoon of *Bombyx* (Vaney & Pelosse, 1920). Anthocyanins (Hollande, 1909) and flavones (Thomson, 1926) were also occasionally taken up from plants. The green pigments were still not identified, but were shown not to be chlorophyll (Toumanoff, 1927). Melanization in relation to temperature (Schlottke, 1926: on *Habrobracon*) and background (Poulton, 1890; Dürken, 1916; Brecher, 1919: on *Pieris* and *Vanessa*), and physiological and morphological color change in *Carausius* (Schliep, 1910; Giersberg, 1928) and *Mantis* (Friza, 1928) and their apparent control by hormones, had been extensively studied.

The cytochrome system, the universal pathway for electron transfer during oxidative metabolism, was first discovered by Keilin (1925) in the insect. This series of hemochromogens was found to be present in muscles in proportion to their activity. In the living thoracic mucles of *Gelleria* it could be seen to undergo continual oxidation and reduction. Lacking hemoglobin, insects were long known to be resistant to carbon monoxide, but at high concentration (80%) it poisons *Galleria* by combination with cytochrome oxidase (Haldane, 1927). Oxygen consumption varies with activity; as shown by Batelli & Stern (1913), it is greater in the larva of flies than in the pupa, and greater still in the adult and in each stage it rises with temperature. Indeed Regnault & Reiset (1849) and Bert (1885) had noted the fall in respiration of *Bombyx* at pupation. Kalmus (1929) had found a 37-fold increase in CO_2 output in *Deilephila* on going from rest to flight. In a given insect species, e.g. *Melanoplus* (Bodine, 1921–1929), respiration is proportional to the weight with an exponent of about ⅔ (0.67) as though it were determined by surface area.

Oxygen consumption remains constant down to quite a low level of oxygen tension, e.g. 5% 0_2 in *Tenebrio* pupa (Gaarder, 1918). Development in the pupa of *Drosophila* was prolonged at 3% oxygen and was arrested at 2% (Kalmus, 1929). At very low oxygen tension in the *Tenebrio* pupa unoxidized metabolites accumulate and lead to increased oxygen requirements on return to air (Gaarder, 1918). A similar oxygen debt builds up in *Blatta* which is just about equivalent to the lactic acid accumulated (Davis & Slater, 1926, 1928); also in grasshoppers (Bodine, 1921–1926) and *Chironomus* larvae (Harnisch, 1930). At these low oxygen tensions the respiratory quotient was found to rise as the result of tissue acidity driving off carbon dioxide (Thunberg, 1930; Harnisch, 1930). During development in the egg of *Bombyx* the high oxygen uptake falls after the blastoderm stage, remains low during the diapause period, and rises rapidly again in the spring (Luciani & Piutti, 1888). A similar U-shaped curve was found by Weinland (1906–1909) in *Calliphora* and in *Tenebrio* by Krogh (1914), *Ephestia* by Taylor (1927), *Deilephila* by Heller (1926), etc, whether a diapause occurred or not. But at the corresponding stage of development respiration was much more strongly depressed in the diapausing insect: egg of *Melanoplus* (Bodine, 1921–1929), *Deilephila* pupa (Heller, 1926). This led to the suggestion that metabolic rate was the causative factor in diapause.

Water and Temperature Relations

Serious study of the water relations of insects dates from the 1920s and arose largely from the ecological interests of applied entomologists. There is, therefore, little to report within our period. The fall in water content on entering hibernation (to 65%) was noted in *Chortophaga* and the rise (to 75%) on restoration of activity (Bodine, 1921, 1923); also the retardation of egg development in dry air (Andersen, 1930) and of pupae sometimes in saturated air (Headlee, 1917). It had been noted that under very dry conditions (in hibernating *Chortophaga*) the rate of water loss would begin to fall off (Bodine, 1921, 1923). The importance of metabolic water in such insects as *Tineola* (Babcock, 1912) and *Tenebrio* (Berger, 1907; Buxton, 1930) in the maintenance of water content during starvation was recognized.

As Girard (1861) showed, in the insect at rest the body temperature is the same as that of the surroundings: heat production balances heat loss. Newport (1839) already had a sound grasp of the nature of animal heat; he noted that in resting Hymenoptera the body temperature was only 3–4°F above the surrounding medium, when moderately active and breathing rapidly it was 15–20°F above. Less heat was generated in the fasting insect. But Newport supposed that the reaction with oxygen occurred chiefly in the circulating blood, and that increased ventilation was the cause of increased heat production. The gain of heat by fluttering was demonstrated by Girard (1861), Bachmetjev (1899), and Dotterweich (1928). The importance of evaporation in loss of heat was emphasized by Pirsch (1923) and Necheles (1924). The utilization of solar radiation with positioning of the body was stressed by Fraenkel (1929) in locusts; and the controlled use of water and fanning to cool the brood in social Hymenoptera by Steiner (1929, 1930) and notably, of course, in the honey bee (Krogh, 1916; Schulz, 1927).

The resistance of insects to low temperatures had been studied by Bachmetjev (1899) who demonstrated their variable capacity for supercooling. Cold hardiness is mainly prevention of freezing (Payne, 1927); it is increased during hibernation, often in association with loss of water (Bodine, 1921). In *Leptinotarsa,* Tower (1917) had found these capacities to vary in geographic races.

Attempts were made to define the optimum temperature for different species, but this depended largely upon the criteria employed (Krogh, 1916; Blunck, 1914). During the 1920s an immense amount of work was done on the relation of different physiological processes in insects with temperature, and the mathematical description of these relations; ranging from the linear relation of Krogh (1916) concentrating on the steeply rising section of the S-shaped curve of Andersen (1930) and many others, by way of thermal summation of Blunck (1916) and Peairs (1927), the Q_{10} rule of van't Hoff and Arrhenius, and the temperature characteristic of Crozier (1924, etc), to the caternary relationship of Janisch (1928).

Reproduction

The anatomy of the reproductive system in both sexes was well described by Burmeister and by Newport in the 1830s, and by the end of the century the varied types of male and female systems had been fully classified. Entomologists of the nineteenth century built up an extensive body of knowledge about the reproductive habits of insects: mating, copulation, numbers of eggs, and methods of egg-laying. The work of Künkel d'Herculais (1894) on the mechanics of egg-pod deposition in the soil by grasshoppers is classic; but almost all groups of insects were intensively studied and most of our present knowledge in this field had been established by 1900.

Oögenesis became a focus of interest in the 1920s. The chromidia or nebenkerne of Blochmann (1884) were described in many insects and their relation to yolk formation, along with the nurse cells and follicular cells, was a source of much controversy. The spermatozoa of insects were described by von Siebold (1836), who noted the feathered spermatodesms in grasshoppers; later in the century, spermatozoa were studied in great detail by the Ballowitz brothers (1890) and many others. The massive literature on germ cell formation in both sexes was reviewed by Depdolla in Schröder's *Handbuch* (1926). The micropyles were discovered by Leuckart (1855), and Meissner and he established the entry of spermatozoa by this route in Diptera. Blochmann (1889) was the first to observe the male nucleus in the egg. The formation of polar bodies and the details of fertilization in a wide range of insects was the subject of Henking's great monographs (1890–1892).

The role of scent organs in mating had long been known, notably in Lepidoptera; Dubois (1895) and others had tried unsuccessfully to get male *Bombyx* to hybridize with other species by applying to them the female scent of *Bombyx*. The experiments of Mayer (1900) and Kellogg (1907) on the mating of male silkmoths with the isolated abdomen of the female, etc are well known. The scent scales, or androconia, of many male Lepidoptera were generally regarded as having an aphrodisiac function (Freiling, 1909), though experimental evidence was wanting.

Spermatophores were described in the cricket by Yersin (1852) and Lespès (1855) and in grasshoppers by von Siebold. In *Dytiscus* (Blunck, 1912) and Lepidoptera (Eidmann, 1929) the spermatophore was thought to be emptied by external pressure; but in crickets the sperm are forced out by the swelling of the gelatinous body (Regen, 1924). Transfer to the receptaculum was variously ascribed to a mechanical suction pump action (Adam, 1912 in Hymenoptera; Breslau, 1906 in *Apis;* Eidmann in Lepidoptera) or chemotaxis toward the secretion of the receptacular glands (Ludwig, 1926 in *Lygaeus,* et al). Both mechanical action and chemotaxis were thought to be involved in migration into the egg. The theory of the voluntary control of sex in the honey bee by liberation of sperm from the spermatheca was put forward by Dzierzon (1848). The probable importance of excess sperm, or hypergamesis, as a source of nutriment for the female was stressed by Berlese. In *Cimex* it was estimated by Cragg (1920, 1923) that the sperm may provide nutrition for about one third of the eggs laid.

The most important effects of environmental factors on reproduction had been recognized; not only does low temperature delay egg production, but at temperatures only slightly below normal, egg production is arrested altogether in *Locusta* (Pospelov, 1926), *Pediculus* (Sikora, 1916), etc, without any notable effect on other organs. In the male, a slight rise in temperature (to 32°C in *Drosophila)* was found to cause temporary sterility (Young & Plough, 1926). Nutrition had been found to have little effect in the male; but in the female protein was usually essential for egg production, as in *Musca* (Glaser, 1923; Roubaud, 1922). Among Lepidoptera the necessary protein was carried forward from the larval stage (Eidmann, 1929); among social Hymenoptera the deflection of protein to brood feeding was regarded as the cause of sterility in workers (Marchal, 1910). Lack of impregnation caused reduced egg production in Lepidoptera (Eidmann, 1929); in *Cimex* no eggs developed (Cragg, 1920). Often it was oviposition that was arrested; in *Lymantria,* Klatt (1920) had suggested that the moving spermatozoa themselves provided a tactile stimulus acting through the nervous system. In the ovaries of worker ants and bees the oocytes had been found to develop up to the point of yolk formation and then to be re-absorbed (Weyer, 1928). The cause of this was unknown. The arrest of reproduction during diapause had been studied particularly in terms of gonotrophic dissociation in mosquitos (Swellengrebel, 1929).

The general outline and much detailed knowledge of all the special modes of reproduction that occur among insects had been established by 1930; polyembryony by Marchal (1906), was one of the more recent to be discovered. Viviparity and parthenogenesis in aphids were first described by Bonnet (1745) at the age of twenty. His work was inspired by Réaumur. The grades of viviparity and ovoviviparity in Diptera were well studied by

Keilin (1916). Placental nutrition of the embryo was found in *Hemimerus* by Heymons (1912); milk gland nutrition in *Glossina* by Roubaud (1909). Pedogenesis was discovered by Wagner in 1861 in the larvae of *Miaster,* but his paper was originally refused acceptance in the *Zeitschrift für wissenschaftliche Zoologie* by von Siebold. Extensive studies were carried out by Springer (1915) and Gabritschevsky (1928, 1930). The types of parthenogenesis had been recognized in the nineteenth century. The classic work of Hughes-Schrader on parthenogenesis and hermaphroditism in *Icerya* was published in 1930 and Seiler's work on sexual races in psychid moths (*Solenobia,* etc) was getting under way in the early 1920s.

THE HISTORY OF INSECT ECOLOGY

H. G. Andrewartha and L. C. Birch
Zoology Department, University of Adelaide and School of Biological Sciences University of Sydney, Australia

INTRODUCTION

Most textbooks in ecology define their scope broadly as the science of the relationships between living organisms and their environment. This is in the tradition of the classical foundations of the subject as they are found, for example, in the writings of the great naturalists of the eighteenth and nineteenth centuries. Réaumur, Buffon, and Haeckel. Consequently, the laws governing communities and ecosystems were thoroughly discussed in these texts and much was generally agreed upon. Plants and animals live in communities in which can be recognized food-chains and ecological webs. And energy, trapped as light by the autotrophs, flows upward and outward through the food-chains in the ecological webs until ultimately it is lost as heat.

Elton (1927) greatly influenced the scope of modern ecology by his argument that ecology was properly concerned with the distribution and abundance of animals in nature. As this idea was developed, it gave rise to the separate discipline of population ecology which has a theory of its own. The theory of population ecology is consonant with the classical theory about communities and ecosystems (because all of organic nature is organized in this way) but it is not directly derived from them. Population ecology has been stimulated by problems associated with the control of pests and the management of useful species. Recently, some entomologists have begun to think about the management of pests. The main stream of insect ecology has always been population ecology.

It is characteristic of natural populations of insects that they fluctuate, often widely. For example, a population of *Thrips imaginis* that was sampled daily for 14 years (by counting the number of thrips in a sample of 20 roses) maintained an average population of 11,492 thrips in 20 roses during November 1938 compared with an average density of 16 in 20 roses during August 1935. The numbers during November varied from 720 in 1933 to 11,492 in 1938 but there was no evidence of a secular trend in the mean density during the 14 years of this study (Davidson & Andrewartha, 1948).

The aim of population ecology is to explain the numbers that have been counted, their fluctuations, and trends, if any. The simplest approach is to consider changes in the rate of increase $r = b - d$ where b stands for birthrate and *d* for death rate; *r* may be positive or negative depending on whether *b* is greater or less than *d*. Both the fecundity of the individual and its expectation of life are age specific. For example, most insects have a long juvenile stage during which they do not reproduce, followed by an adult reproductive stage which is often short lived. Also, with many insects, the young larvae or nymphs and the older adults are much more likely to die than the intermediate ages. Consequently, to be realistic, it is necessary to take into account the age of the individuals in the population when considering the influence of the environment on birthrate and death rate. Strictly speaking, it is the individual that has the environment. And to understand how an individual's fecundity, speed of development, or expectation of life is influenced by its environment calls for knowledge of the animal's physiology and behavior.

It used to be fashionable to call these studies autecology. Today, we speak of ecological physiology to embrace broadly the physiological response of the whole animal to stimuli from its environment. Ecological physiology and behavior are recognized as important extensions of population ecology because they are the foundations of ecological explanations.

ECOLOGICAL PHYSIOLOGY

Ecological physiology is the study of particular aspects of physiology as affected by environment such as diapause, the influence of temperature and moisture on behavior, birthrates and death rates, the way temperature and moisture have their effect, and the role of light on numerous functions. There was a stage in the development of insect ecology when these aspects of ecology were studied almost to the exclusion of others. Shelford's (1929) *Laboratory and Field Ecology* was mostly ecological physiology and set the stage for the 1930s and 1940s. Chapman's (1931) *Animal Ecology with Special Reference to Insects* had a similar emphasis though it included some community ecology and anticipated the development of population ecology by translating, in an appendix, Volterra's theories and classic equations on competition and predation.

Speed of Development and Fecundity as Influenced by Environment

Réaumur's (1735) proposal that the time of ripening of fruit is related to the sum of daily average temperature initiated much work on temperature and rate of development, not only of plants, but of poikilothermic animals, especially insects. The practical objective of these studies was to estimate how weather affected the time these organisms took to complete their development. To this end a search began to find a mathematical expression relating speed of development to temperature.

The many equations that were eventually proposed were classified by Davidson (1944) as either theoretical or empirical. The theoretical equations were those of van t'Hoff and Arrhenius. Both these equations implied that the proportional increase in speed of development produced by a given difference of temperature was constant throughout the range at which the animal may develop. In practice it has been shown that this is not the case. Bělehrádek (1935) quoted many authors who had demonstrated this. The result was hardly surprising when one considers that the van t'Hoff and Arrhenius coefficients were originally conceived to describe the relationship of temperature to a single chemical reaction.

The purposes of ecology were better served by seeking an empirical equation. An early one was the straight line relationship between speed of development and temperature proposed by Oettingen (1879). A convenient and much used measure of the speed of development is the reciprocal of the duration of development. Reibisch (1902) plotted the reciprocal of the time required for the development of fish eggs against temperature, drew a straight line through the observed points, and defined the point where this line cut the abscissa as the threshold of development. Sanderson & Peairs (1913) used the same method to determine what they called the development zero for 12 species of insects. However, Krogh (1914) showed that the developmental zero, so determined, was not necessarily a true threshold since quite appreciable development might occur well below this temperature. He also showed that the speed of development at high temperatures near the upper limits of the favorable range was appreciably slower than might be expected from an extrapolation of the straight line. What Krogh indicated at this early stage has indeed been vindicated since and it is now well established that in properly conducted experiments the straight line relationship holds only for a very short range of temperature, departing widely from it at either end of the favorable range. The "threshold" determined by extrapolation has no biological meaning.

Simpson (1903) was probably one of the first to develop the concept of the thermal constant expressed in units of day-degrees. This follows from the straight line relationship thought to hold between speed of development and temperature. Speed of development was taken to be the reciprocal of duration of development, so when duration of development was plotted against temperature the points would fall on a hyperbola if the relationship between speed of development and temperature was a straight line. The equation of the hyperbola is $K = y(x - a)$ where 'y' is the duration of development, 'x' is temperature, 'a' is the developmental zero, and K is the thermal constant. As a crude approximation, a day was taken as the unit for 'y' and the mean daily maximum and minimal temperatures were substituted for 'x'. This was the basis for the practice of temperature summation, a method used by Glenn (1922, 1931) for the codling moth. Glenn made allowances for departure from linearity at high temperature. Shelford (1927), in a monumental study of the same insect, made allowances for departures from linearity at both ends of the range. He introduced the developmental unit to replace day-degree, the chief refinement being that he substituted the hour as the unit for which temperature was measured, rather than the day. For every additional degree a constant amount of development occurred. The sum of the developmental units that completed development was the developmental total.

Refinement in techniques of measuring speed of development in relation to temperature was to show that the relationship was not linear but curvilinear over the whole range. This meant, of course, that temperature summation based on the linear relationship would be in error. The search for a more accurate mathematical expression proceeded. Janisch (1925) proposed the mathematical expression of the asymmetrical catenary curve to describe the relationship. Bělehrádek (1935 and earlier papers) proposed an exponential expression which became linear on a logarithmetic scale. Yet none of these empirical relationships followed the observed data as closely as did the empirical Pearl-Verhulst logistic curve which Davidson (1944) used to describe the trend in a number of experiments he selected for their precision. Andrewartha & Birch (1954, p. 149) in commenting wrote "The logistic curve is realistic, giving an easily comprehended picture of the trend of speed of development at different constant temperatures. It is easily calculated directly from empirical data, and, in addition, it has the merit of being the most adequate empirical description of the relationship that has so far been suggested. From now on it should be used in preference to all the older expressions." The use of the logistic equation in place of the straight line allows for the curvature of the temperature—speed of development relationship throughout the whole range of temperature. Andrewartha (1944) applied it for estimating the date of hatching of nymphs of the grasshopper *Austroicetes cruciata* in the spring of each year for 50 years. He was interpreting events in the past and for this it proved effective. Indeed, this sort of use has turned out to be the main use for these procedures. The measurements and calculations needed for accurate estimates of events in the field are very complicated and laborious and depend upon an accurate knowledge of temperature in the field. Rough predictions about weather are of little use in this. The objectives Shelford (1927) so strenuously aimed to achieve are not to be realized this way. Entomologists have found simpler ways of anticipating the hatching of grasshoppers or the date when moths begin laying their eggs.

In the background of these developments the question was frequently raised, e.g. Ludwig (1928) and Shelford (1929), as to whether speed of development measured at constant temperatures could be considered equivalent to speeds of development at fluctuating temperatures. Reviewing the work that set out to clarify this question, Andrewartha & Birch (1954) concluded that with certain qualifications, depending upon diapause and the extent of the range of the fluctuations, laboratory studies at constant temperatures could be safely used for interpreting development in the field in fluctuating temperatures.

The influence of temperature on fecundity received much less study than its effect on speed of development. Harries (1939) reviewed what was known and his several examples showed the same trend; a sigmoid curve which fell off at high temperatures. Ten years after his review there was a renewed interest in measuring fecundity when it was realized that this data was essential for estimating rates of increase of populations. The curves obtained are in essence similar to the sigmoid curves described by Harries.

The influence of moisture on speed of development has been primarily of interest because, with some insects, development in the eggs ceases almost completely in dry air. This increases the chance of survival of desert species such as, for example, the locust *Chortoicetes terminifera* (Davidson, 1936). The development of some insects is retarded at high humidities and for others speed of development is relatively independent of humidity. By contrast, moisture was found to have a considerable effect on fecundity. Hamilton (1950) showed that there was an optimum humidity for egg laying of *Locusta migratoria*. And for insects that live in stored products the moisture content of their food and surroundings has a major influence on fecundity (e.g. Birch, 1945).

The Favorable Range of Components of Environment

Shelford's (1911) paper on animal geography contains the idea that species have a tolerable range for the components of environment beyond which they cannot survive and reproduce. This is a simple concept, yet one which ecologists had hardly appreciated and had certainly done very little about. Shelford was an outstanding exception. In this paper he wrote of "the law of toleration of physical factors," of how the centers of distribution would be those places providing optimal conditions and at these centers he anticipated that abundance would be the greatest. He saw the geographic range of the species being limited by the fluctuations beyond the tolerable limit of one or more components of environment. His model was very much influenced by Leibig's so-called "law of limiting factors" which agriculturalists had taken to heart in considering the conditions limiting plant growth. It is curious that Shelford, having set the stage for a quantitative approach to distribution, provided so little information on the tolerable limits of any of the species he wrote about in his *Animal Communities in Temperate America* (1913). What Shelford had set as a goal, very few followed.

Nearly 40 years later, Moore (1940) tabulated the favorable range of temperature for 41 aquatic animals and just two terrestrial insects, commenting that very little data was available for insects and that for the few on which there was data the tolerable range was known to exceed 20°C. Perhaps the limiting factor in this slow progress was that no one had clearly set out how best to determine the tolerable range for any single component of environment. Too many made halfhearted attempts. Yet there was a method lying at hand and extensively used by toxicologists and pharmacologists; the experimental procedure of the dosage-mortality curve and probit analysis. All that the ecologists had to do in taking over this method was to regard lethal temperature or lethal dryness as a dosage. The tools were waiting to be picked up and used for their own purposes (see next section). From the 1950s onward accurate data were obtained for numerous species of insects; furthermore, some observations revealed that the favorable limits varied within the species depending upon the latitude from which the population was collected and so complicating Shelford's (1911) proposition. For example, Bateman (1967) showed that populations of the tephritid fruit fly *Dacus tryoni* from tropical Australia had a much narrower range of tolerable temperature than populations from temperate Australia. The task of the insect ecologist becomes enormously more complex when he finds that he is not studying just a species with definable characters but must study local populations because they differ. He is then confronted with the question—but what is a local population? How big, how broad is its latitudinal range, how stable are its characteristics? So, in a very interesting way, studies on tolerable ranges led to investigations of a genetical and evolutionary nature. Physiological ecology led directly to the need to study population ecology. In the case of species whose distribution covers a wide range of latitudes, we may now expect to find that the tolerable range will vary according to latitude. The species, as such, is no longer the unit for identifying ecological and physiological tolerances. The local population is. This is exactly what we should have expected from the neo-Darwinian understanding of evolution, but entomologists were slow in realizing this.

Lethal Influence of Temperature and Moisture

To establish the tolerable range we need to have precise estimates of the tolerable limits. It was indicated in the preceding section that the appropriate methods for doing this were not taken up by entomologists until quite recently. However, quite a lot of data were obtained before this in answer to specific questions of ecological importance about limits of temperature and moisture inimical to life. One question that had received much attention in Europe and North America was the survival rate during winter, since this may be important in determining the numbers of a particular pest in a certain area, Carter (1925) investigated the survival of the bean weevil *Bruchus obtectus* at low temperatures and found none of the stages could survive more than a few hours exposure at —19°C. Payne (1926, 1927) tried to make a more comprehensive study on the effect of cold on representative insects from three groups. The oak borers represented those exposed during winter many degrees below 0°C. Certain aquatic forms represented those that might experience temperatures as low as 0°C but not below. Certain insects considered to be subtropical in origin and found in stored products, represented those that live in places usually well above 0°C. Payne referred to what she called the quality factor as the lethal influence of low temperature on the third group because exposure to cold usually above 0°C was lethal only after it had been maintained for some days or weeks. On the other hand, the other two groups survived unharmed for many months at these temperatures but tended to die instantly when their tissues froze. The temperature required to bring this about is the measure of what Payne called the intensity factor.

This arbitrary division of quantity and quality factors seems to have dominated the thinking of most students on the subject until Salt (1936, 1950) pointed out that the division of insects into those that responded to the quantity factor and those which responded to a quality factor is unreal. Representatives of the nonresistant group also die after short exposure to temperatures below 0°C. Furthermore, representatives of the resistant group may succumb to exposures that do not result in freezing of their tissues. The consequences to ecology of the long predominance of this too simple hypothesis was that the lethal influence of freezing temperature was studied only in relation to temperature required to produce instantaneous freezing, whereas the effect of duration of low temperature was virtually ignored.

Johnson (1940) provided the needed revitalization of these studies in his classic study of the effect of low temperature on the survival of eggs of the bed bug of *Cimex lectularius*. He was able to vary temperature, moisture, age, and time of exposure. Furthermore, he analyzed his data as dosage-mortality data and subjected them to probit analysis. The procedures which he adopted have become standard where precision is an objective. From this point the subject was set on a sound basis.

The causes of death at freezing temperatures were studied primarily in relation to the phenomenon of cold-hardiness of insects that live in cold places. Réaumur's (1736) suggestion that death occurred only when all fluids were frozen was for long the only hypothesis. Cold-hardy animals could tolerate freezing of a large part of their body fluids but it was too simple a hypothesis to suppose that death occurred only when all fluids were frozen. The hypotheses elaborated since Réaumur to account for cold hardiness, and the cause of ultimate death at low temperatures are many and are primarily the concern of physiology. They were reviewed by Salt (1961) who indicated many, as yet unresolved problems.

The existence of seasonal variability in cold resistance of insects was demonstrated by Guyelard & Portier (1916) on *Cossus* and *Carpocapsa* and by Knights (1922) on the bug *Perillus bioculatus*. Such an acclimatization effect was later demonstrated for many more species, e.g. by Mellanby for *Blatta orientalis* (1939) and *Aedes aegypti* (1959).

The lethal effect of high temperatures on insects has received less study than the lethal effect of low temperature. With insects that live in air it is difficult to measure the effect of temperature independently of moisture. Moreover, danger from evaporation is more likely to be pressing than danger from heat. Much of the earlier work reviewed by Uvarov (1931) did not consider adequately time of exposure and the influence of acclimatization. The appropriate model for experiments on determining both lethal high and lethal low temperatures was provided by Fry's (1947) work on fish in which LD50s were determined with probit analysis. This type of experimentation and analysis has since been extended to insects and is now standard procedure (Andrewartha & Birch, 1954; House et al, 1958) (see preceding section). The lethal temperature of an insect may vary with the temperature the insect was previously subject to. Such an acclimatization effect has been shown for a number of insects (Bursell, 1964).

The lethal effect of dry air was initially investigated because of the need to know how long insects can survive in desiccating environments. The early work of Bacot & Martin (1924) on the flea *Xenopsylla cheopsis* was guided by the so-called "saturation deficit law." This was the assumption that the rate at which water was lost from an insect was proportional to the saturation deficit and independent of temperature, as is the case for evaporation from a free water surface in still air. Bacot & Martin (1924) sought a relationship between saturation deficit and longevity. If water loss is directly proportional to saturation deficit and if an insect dies after losing a certain proportion of its water, then longevity would be expected to be inversely proportional to saturation deficit. The graph relating longevity and saturation deficit would be a hyperbola. Bacot & Martin analyzed their data to test this hypothesis. However, they used an inappropriate test and incorrectly concluded that their data on *Xenopsylla* did not conform to a hyperbola (Andrewartha & Birch, 1954). Johnson (1940) did experiments with *Cimex lectularius* to test the hypothesis that Bacot & Martin had failed to test properly. He obtained hyperbolas when longevity was plotted against saturation deficit except at high temperatures. This suggested that the theoretical relationship held, except at high temperatures where some other influence was important. However, in experiments with other insects since then a linear relationship describes the trend better in some cases than the hyperbola. (Johnson, 1942; Clarke & Sardesai, 1959.) Other workers had pursued the "saturation deficit law" in relation to mortality rate. Maercks' (1933) data on *Habrobracon* did not conform to any simple relationship though he thought it showed a linear relationship between mortality and saturation deficit. Mellanby (1935) argued from the "saturation deficit law" that mortality would be proportional to the product of saturation deficit and duration of exposure. His arguments were, however, incorrect. Birch (1944) and Andrewartha & Birch (1954) showed that if the saturation deficit law held then the relationship should be sigmoid over the medial range of temperature. This conclusion was verified for death rate in eggs of *Calandra oryzae* (Birch, 1944). At each of six temperatures within the medial range he obtained six different sigmoid curves relating mortality and saturation deficit $\times$ time of exposure. Larsen's (1943) experiments on eggs of *Musca domestica* also showed a sigmoid relationship. We would expect this to be the case in general for stages in the life cycle that are inactive in the sense of not taking in water and food from outside.

With active stages that take in water and food the relationship is more complicated since the moisture content of the food is determined by relative humidity of the air but the water loss from the insect depends upon saturation deficit and temperature. Insects that live in stored products get their moisture from the food they live in. The complete information relating moisture and temperature to mortality can be most usefully obtained for them by expressing moisture as the moisture content of their food. Mortality can then be plotted on a three dimensional graph in which the moisture content of the food and temperature form the horizontal axes. The result is a nest of sigmoid curves, one for each temperature studied (Andrewartha & Birch, 1954).

A case of extreme resistance to desiccation has been demonstrated for the larvae of the chironomid *Polypedium vanderplancki* which can be virtually totally dehydrated, subject to 102°C for one minute or immersed in liquid air and still metamorphose when later moistened (Hinton, 1960).

Water Balance

The capacity of insects to survive in dry places is largely dependent upon their capacity to prevent water loss and to make up what is lost from water in their food, by direct absorption through the cuticle, or possibly also from water of metabolism. The way in which a balance is established between gain and loss has long been a source of enquiry of insect ecologists and insect physiologists.

That insects can absorb water through the cuticle was demonstrated by Salt (1946) for the sawfly *Cephus cinctus,* and by Lees (1947) for ticks. In the special case of eggs the hydropyle was first demonstrated to be a water-absorbing organ by Slifer (1938) in the grasshopper *Melanoplus differentialis.* Since then, this function has been demonstrated in many other grasshoppers and locusts. Nutman (1941) identified the ventral tube in the collembolan *Onychiurus* as its main water-absorbing organ.

Another step was the demonstration that some insects could absorb water from unsaturated air. Breitenbrecher (1918) demonstrated this for the beetle *Leptinotarsa decemlineata* and since then it has been demonstrated for other insects, e.g. *Cimex* (Wigglesworth, 1931), the grasshopper *Chortophaga viridifasciata* (Ludwig, 1937), and the flea *Xenopsylla cheopsis* (Edney, 1947). Lees (1946) extended these studies when he showed that the tick *Ixodes ricinus,* in absorbing water from unsaturated air, absorbs it against a vapor pressure gradient. He demonstrated further that the cells involved in this active secretion are the hypodermal cells immediately beneath the cuticle.

The question as to whether insects that live on dry food use metabolic water to maintain their water balance has been unresolved ever since Buxton (1930) proposed this for the mealworm *Tenebrio molitor.* Buxton's hypothesis was quite credible since a number of insects were known to maintain their body water, despite the remarkable dryness of their food. Mellanby (1932a, 1932b, 1936) was the first to attempt to test the hypothesis. He was unable to detect a high rate of metabolism in insects fasting in dry air. In any case, he doubted that additional respiration could be effective, arguing that water would be lost in the additional flow of air through the spiracles. Fraenkel & Blewett (1944) claimed to have provided conclusive evidence for the conservation of metabolic water. However, the design of their experiment was not adapted to distinguish between conservation of metabolic water and dehydration of undigested food.

On the debit side of the water balance sheet are excretion, egestion, and loss through the cuticle and through the spiracles. The main unknown was the extent to which water was lost through the cuticle. As late as 1934, Mellanby was arguing that none was lost through the cuticle. However, Ramsay (1935) showed that transpiration from the cockroach *Periplaneta americana* took place through both spiracles and cuticle. In some insects the cuticle is quite impermeable, e.g. larvae of *Xenopsylla cheopsis*. The first person to measure the effect of temperature on the rate of transpiration was Gunn (1933). He showed that the rate at which *Blatta orientalis* lost water was slow below 30°C and above 30°C the rate suddenly increased. He attributed this to what he thought was a change in type of respiration from diffusion to active ventillation. However, this conclusion was unconvincing to Ramsay (1935) who repeated Gunn's experiments on *Periplaneta americana* by blocking the spiracles with wax and still he found a sudden increase in rate of transpiration at 30°C. Ramsay noticed that drops of water on the cuticle were covered with a thin film which showed interference patterns. Above 30°C it underwent a change of phase. He suggested this might be a lipoid which at lower temperatures might help to make the cuticle impermeable. This observation was the starting point for Wigglesworth's many experiments which have led to a full understanding of the structure and function of the insect cuticle. Chance was prominent in the early stages of this story. Had Gunn and Ramsay chosen to work with any other common insect, neither the abrupt change in the rate of water loss nor the occurrence of the oil film would have been noticed at the temperatures they used in their experiments.

Wigglesworth (1945) measured the rate of transpiration at various temperatures for a variety of insects and found that apart from aquatic and burrowing insects, transpiration increased abruptly above a certain temperature which he called the critical temperature. Beament (1945) extracted the wax from the cast cuticles of the species Wigglesworth had worked with and found that they varied from the soft grease in *Blatta* to hard waxes in *Tenebrio* and *Rhodnius*. Their melting points were a few degrees above the critical temperature of the insects from which they had come. These and other experiments demonstrated the existence of a thin wax layer forming the outermost layer of the epicuticle of insects. The water-proofing properties of the cuticle largely resided in this layer (Wigglesworth, 1948). Returning to the comparative differences between cuticles which Wigglesworth and Ramsay had initiated, Lees (1947) showed that the critical temperature for a number of species of ticks was closely related not only to systematic classification but also to the humidity of the places where the species usually lived. The lowest critical temperature for the desert argasid ticks was 17°C above the highest for the ixodid ticks which inhabit cooler and moister places. Thus, a nice relationship was shown to exist between the properties of the wax, the systematics of the animal, and the environment in which it lived.

The Inception and Duration of Diapause

The first experimental demonstration of the phenomenon of diapause was made by Duclaux (1869). He noticed that the eggs of the silkworm *Bombyx mori* failed to hatch when kept in a warm room in winter. When he kept them in an icebox for 40 days and then brought them into the warmth, most of them hatched. Further experiments he did showed that the eggs of *Bombyx* required exposure to cold before they could develop. The most suitable temperature for this appeared to be a little above 0°C. These experiments passed unnoticed. Wheeler (1893) coined the term diapause to denote a particular stage in the embryonic development of a grasshopper but this term did not enter into embryological usage. Henneguy (1940) lifted the word from its embryological setting and used it to refer to the physiological state of dormancy or arrested development which he observed in a number of insects. In this usage it gained wide currency. Diapause is an adaptation that increases the chance of survival in an unfavorable season and at the same time keeps the life cycle in phase with the rhythm of the seasons. In insects with a typically univoltine life cycle the inception of diapause is independent of any stimulus from outside the animal and is said to be obligate. In all others, diapause is facultative and its inception is dependent upon an appropriate stimulus from outside the insect. The identification of the appropriate stimulus for the inception of diapause has been an important aspect of ecological investigations on diapause.

Readio (1931) found that rearing the nymphs of the bug *Reduvius personatus* at low temperatures induced diapause, and Ludwig (1932) was able to induce diapause in the larvae of the beetle *Popillia japonica* by keeping them at low temperatures. Since then, in most insects studied temperature plays an influential part in controlling the onset of diapause. Usually it is low temperature that favors the inception of diapause but there are some exceptions in which high temperature has been found to induce diapause (Andrewartha, 1952; Lees, 1955). The changing composition of the food with advancing season was found to be the chief cause of diapause in the larvae of *Platyhedra* and *Diatraea* (Squire, 1937; Kevan, 1944). Neither temperature nor quality of food are particularly reliable indicators of season because of their variability; length of day is. The first person to recognize length of day as inducing diapause was Kogure (1933) in his detailed story of the silkworm *Bombyx mori*. The effect of photoperiod in inducing diapause was subsequently demonstrated for many other insects, e.g. *Grapholitha molesta* (Dickson, 1949) and *Diataraxia oleracea* (Way & Hopkins, 1950). In all of these, short day induces and long day prevents diapause. *Bombyx* which Kogure studied is one of several well authenticated examples of a short-day species.

The termination of diapause is dependent upon stimuli received from outside the insect. The appropriate stimuli have been the source of study since Duclaux (1869) showed that exposure to low temperatures caused a termination of diapause in the silkworm *Bombyx mori.* The complexity of this process was little realized until Steele (1941) and Andrewartha (1943) showed that the influence of temperature is intimately related to the stage of the embryo and the history of the egg in a complex way. Because of this, Andrewartha introduced the idea that instead of regarding diapause as a state of blocked development that is broken by cold stimulus, it is more realistic to consider the phenomenon of diapause as a process of gradual development that is influenced by temperature and hence his term diapause development (Andrewartha, 1952).

Length of day was found to be the factor controlling length of diapause in the larvae of certain tree hole mosquitoes by Baker (1935) and in the scelionid *Telenomus laeviusculus* by Gayspitz & Kyao (1953) and in some Lepidoptera (Lee, 1955).

Preferred Range of Temperature and Moisture

In the earliest experiments to test the idea that insects, when given a choice, will tend to congregate at a particular temperature or within a narrow zone of temperature, the experimental insects were given a choice between two temperatures only. Shelford & Deere (1913) devised an apparatus in which three currents of air at different temperature and humidity were passed across a rectangular box, thus creating three zones of temperature and moisture with gradients between each. But clearly it was better to design a gradient of temperature which covered a wide range of temperature so that the insects could be given a wide choice. When this was done insects usually congregated between narrow limits in the gradient called the preferred temperature, a term proposed by C. B. Williams in 1922 (Deal, 1941).

That the preferred temperature depended upon the temperature to which the insect had been previously subjected was shown by Herter (1924) with *Formica rufa* and for *Tribolium confusum* by Bodenheimer & Schenkin (1928).

The preferred temperature might also depend upon the geographic location of the population and so would vary within a wide ranging species; *Carabus nemoralis* populations from warm districts had a higher preferred temperature as compared with those from cooler districts (Krumbiegel, 1932). Then Wilkes (1942) showed that with the chalcid *Microplectron* the population even from one area was not homogeneous with respect to the preferred temperature. He found a trimodal distribution along a temperature gradient. Furthermore, he demonstrated that the differences were genetic. By selecting from a modal population and inbreeding he produced a race with a unimodal distribution in the gradient. Insects also show a response to humidity gradients. Pielou & Gunn (1940) showed that *Tenebrio molitor,* given a choice, chose the dry end of a gradient. Roth & Willis (1951) showed that the preferred humidity for *Tribolium castaneum* and *T. con-*

fusum depended upon the moisture of the environment to which they had been subjected prior to the experiment.

The question as to whether temperature or humidity exerts the greater influence on the insect seeking the preferred zone is one not easily resolved in either the laboratory or the field. Wellington (1949) attempted to give an answer to this question for the larvae of the spruce budworm *Choristoneura fumiferana.* He provided them with a gradient of temperature associated with a gradient of evaporation and in a series of such gradients found that the larvae tended to congregate within one zone of evaporation rate irrespective of temperature. Apparent responses to temperature were, in fact, responses to evaporation. In the field, observations have tended to be related to temperature because it is easily measured. It then has to "speak for" humidity. This has been particularly true of the study of locusts in the field. The aggregation of the locust *Schistocerca gregaria* in the Sudan outbreak areas was determined largely by temperature (Kennedy, 1939). The same conclusion was reached in Australia for the locust *Chortoicetes terminifera* by Clark (1947).

Temperature and Light as Token Stimuli

Fraenkel & Gunn (1940, p. 190) introduced the concept of the token stimulus as applied to light and other components of weather when they are a sign or token representing something else. Light may indicate circumstances that are, for other reasons, favorable or unfavorable. Fraenkel & Gunn referred to the work in 1890 of Loeb on the larvae of *Porthesia* which move toward light when unfed, a response which brings them nearer food. This type of response has been confirmed by others. Wellington (1948) showed that the spruce budworm larvae are positively phototactic to diffuse light which tends to hold them in the outer periphery of the tree, where the chance of encountering fresh food is greater than in the middle. In some insects length of day has been found to be a token stimulus which indicates seasonal changes in moisture, food, and temperature. Dickson (1949) showed that the short photoperiod exposure by the early larval stages of *Grapholitha molesta* during the summer induced diapause in the larvae long before the temperature ever got low enough to inhibit active development. In general, light is a token stimulus whenever the inception of diapause is determined by photoperiod.

Temperature is a token stimulus for some insects that live on mammals. It is the stimulus guiding them to the host as has been shown for the bug *Cimex lectularius* (Rivnay, 1932; Sioli, 1937). Furthermore, it is appropriate as Lees (1955) points out, to regard temperature as a token stimulus when temperature is the component that induces diapause.

POPULATION ECOLOGY

It was mentioned in the introduction that the central operation in population ecology is to discover how environment influences an animal's chance to survive and reproduce. The concept of environment is the very essence of the theory of population ecology. It has been widely discussed especially since the time of Darwin.

Concepts of Environment

In chapter 3 of the *Origin of Species,* Darwin's account of the struggle for existence is documented by numerous examples of how survival rate depends on environment. (In an odd sort of way he seemed to take fecundity for granted.) Darwin emphasized that the struggle for existence was used in a "broad and metaphorical sense." His examples included animals' struggling against bad weather, shortage of food, predators and diseases, other individuals of the same species, and individuals of different species with overlapping requirements. Darwin was the first to put forward the idea of an environment (he called it conditions of life) composed of a number of components which might act separately or jointly to influence an animal's chance to survive and multiply. The idea is explicit in a number of passages and implicit in the whole chapter.

For example,

> But we have better evidence on this subject than mere theoretical calculations, namely the numerous recorded cases of the astonishingly rapid increase of various animals in a state of nature, when circumstances have been favourable to them during two or three following seasons.

Or,

> In looking at Nature it is most necessary . . . [to remember] . . . that every single organic being around us may be said to be striving to the utmost to increase in numbers; that each lives by a struggle at some period of its life; that heavy destruction inevitably falls either on the young or the old, during each generation or at recurrent intervals. Lighten any check, mitigate the destruction ever so little, and the number of the species will almost instantaneously increase to any amount.

Darwin's basic idea that environment determines rate of increase has been universally accepted. Darwin based his argument on rate of increase ($r = b - d$) but most of his examples referred to the influence of environment on death rate. Most of the theories that have emerged during the first half of the twentieth century have reflected Darwin's emphasis on survival rate.

For example, according to Howard & Fiske (1911), the environment may be analyzed into two sets of factors according to whether the harmful influence of the environment increases with the density of the population or is independent of the density. The latter were called catastrophic factors, the former facultative factors. According to this theory facultative factors are necessary to produce a natural balance; catastrophic factors are not sufficient. Their prime example of facultative factor was parasitism which causes an increase in death rate as the population of hosts increases in density. The definitive passage from Howard & Fiske's paper was:

> . . . it is necessary that among factors which work together in restricting the multiplication of the species [caterpillars of *Porthetria dispar* and *Euproctis chrysorrhoea* which defoliate forest trees] there shall be at least one if not more, which are here termed facultative (for want of a better name and which, by exerting a restraining influence which is more effective when other conditions favour undue increase, serves to prevent it. . . . Very few other factors [influencing the survival of the caterpillars], however closely they may be scrutinized, will be found to fall into the class with parasitism, which in the majority of insects, though not in all, is truly 'facultative.' . . . A natural balance can only be maintained through the operation of facultative agencies which effect the destruction of a greater proportionate number of individuals as the insect in question increases in abundance.

From this classical paper can be traced the ideas of environment which characterized a number of theories that arose during the first half of the twentieth century, notably those of Chapman (1931), Gause (1934), Nicholson (1933), Smith (1935), Varley (1947), Thompson (1939), and Bodenheimer (1938). Chapman thought of the population as having a *biotic potential* for increase which was rarely realized because it was suppressed by the environmental resistance—an idea which might have come straight from Darwin but which was also reflected in the writings of Howard & Fiske and which can be traced back to the writings of Quetelet in 1835 (Cole, 1957). Chapman recognized biotic and physical factors in the environment but did not specifically emphasize the homology between biotic and facultative factors or physical and catastrophic factors.

In Nicholson's theory of environment, reactive factors (alternatively governing or controlling factors) were analogous to facultative factors but the reactiveness was emphasized and defined more precisely; reactiveness was defined as a reaction (or feedback) from the density of the population. That is, a controlling factor not only pressed more heavily on (e.g. killed proportionally more of) a dense population, but a dense population also caused a controlling factor to increase its activity. Nonreactive factors were analogous to Howard & Fiske's catastrophic factors but the meaning was extended because nonreactive factors might include legislative factors which determine the level at which governing factors work. The idea of biotic potential and environmental resistance was implicit in Nicholson's theory of environment but it was not emphasized. On the other hand, the distinction between biotic and physical factors was carried further than in Chapman: physical factors could never give rise to reactive factors but biotic factors might.

Varley (1947) coined the term delayed density-dependent factor thereby making Nicholson's idea of reactiveness quite explicit (before Nicholson redefined his own terms in 1954). On the other hand, Smith (1935) used the terms density-dependent and density-independent as synonyms of Howard & Fiske's original facultative and catastrophic. Similarly, in the writings of Thompson who used the terms individualized and general, and of Bodenheimer (1938) and Uvarov (1931), the concept of density dependence was more prominent than the concept of reactiveness. Clark et al (1967) defined a life system as everything (including the subject population itself) that influences the abundance and evolution of the population; they defined the environment of the population as everything external (*sic*) to the population that influences its abundance and evolution. It would seem that the theory of the life system merely re-states Nicholson's theory of environment in different words, with density of the subject population determining the activity of the reactive factors of Nicholson. But a critical appraisal of the idea of life system may have to await a further exposition of it.

Chapman emphasized environmental resistance but not the reactiveness of biotic factors. Nicholson and those who followed him emphasized the nonreactiveness of physical factors but the idea of environmental resistance was also implicit in their theories. So it is convenient to recognize the affinities of this group of theories which all stem, through Howard & Fiske, from the original ideas of Darwin.

But Darwin also spelled out the components of environment which, he said, an animal might struggle against; and he included other individuals of the same species, as well as individuals of other species, food, predators, and weather. According to Andrewartha & Birch (1954) the environment of an individual animal may be analyzed into four components, weather, other organisms (of the same and different species), food, and a place in which to live. The four components were said to comprise everything that might influence an animal's chance to survive and reproduce. No distinction was made between direct and indirect influences so that the environment might extend into remote ramifications of the ecological web. Maelzer (1965) pointed out that this led to ambiguity and he suggested that only things that had a direct influence on the animal's chance to survive and reproduce should be included in its environment. Browning (1962) criticized the narrowness of the component which had been called food and the excessive broadness and ambiguity of a place in which to live as a component of environment. He proposed that environment should comprise five components: weather, resources, members of the same species, members of other species, and hazards. Andrewartha (1970) revised the original theory incorporating the criticisms of Maelzer and Browning. He recognized five components: resources; mates; predators, pathogens, aggressors; weather; and malentities. According to Andrewartha (1970) the environment is defined in terms of what influences the animal directly but the activity of any component of environment may be determined by reactions in the ecological web, which is thus considered to influence the animal indirectly.

The concept of environmental resistance is lacking from the theory of Andrewartha and Birch because the components of environment are said to influence the animal's fecundity, longevity, or speed of development either positively or negatively depending on the nature of the component and its activity. The concept of reactiveness which is central in Nicholson's theory of environment is covered by allowing other individuals of the same species, as part of the ecological web, to influence the activity of components of environment such as resources or predators. The two groups of theories of environment cover the same set of concepts but the concepts are organized differently. It would seem that ecologists who favor deterministic models of insect populations use Nicholson's theory of environment and those who favor stochastic models tend to use the theory of Andrewartha and Birch.

The Concept of Exponential Growth

The concept that populations could potentially grow geometrically, that is by successive doublings in successive equal time intervals, was implied by Botero (1588) and other early scholars of human population growth. John Graunt (1662) went a step further when he estimated that a human population tends to double itself every 64 years. Perhaps the first person to extend this idea of the potential of a population to grow geometrically to species other than man was William Derham (1713). He argued that the whole surface of the globe could not possibly support the exponential growth of all species and this is prevented by starvation and predation of one species on another. "Voracious birds and beasts" had long lives but their reproductive rate was small. Insects, he said, exemplify animals with great reproductive ability but short lives. Graunt and Derham were the pioneers of demography who paved the way for Malthus' *Essay on Population* in 1798 with its predictions about the consequence of geometric growth for human populations.

The rate of increase of a population which is growing geometrically is given by the differential equation:

$$\frac{\delta N}{\delta t} = r_m N$$

When N denotes population size at any time t and r_m is the infinitesimal rate of increase (= intrinsic rate of increase, innate capacity for increase, Malthusian parameter)

The integrated form of the equation is

$$N_t = N_0 e^{r_m t}$$

N_t = numbers in time t
N_0 = numbers in time zero
e = base of Naperian logs.

The accurate estimation of the value of r_m for any population depends upon a knowledge of birthrates and death rates, and for most animals these vary with the age of the animal. Therein lay a difficulty in estimating r_m from birthrates and death rates. This was overcome by Lotka (1925) who showed how this data could be used to estimate values of r_m for human populations. Its relevance for ecology was at least implicit in A. J. Lotka's book

Elements of Physical Biology published in 1925 and in R. A. Fisher's *The Genetical Theory of Natural Selection* published in 1930. However, to Leslie & Ransom (1940), must be attributed the credit of having adapted the statistic r_m for a more general use in animal ecology in their experimental determination of r_m for the vole *Microtus agrestis*. Its use was extended to insect populations by Birch (1948) and by Leslie & Park (1949) and since then by many authors to many more species.

The importance of the equation of exponential growth for insect ecology has been twofold. It has been used as the basis for more elaborate models, notably the logistic model and the models of competition and predation of Lotka and Volterra. Secondly, it has been used empirically as an esimate of the capacity of insect populations to grow in environments where resources were not limiting. Such a use was made by Birch (1953) to show the effect of temperature and moisture on the potential growth rates of the populations of the grain insects *Calandra oryzae* and *Rhizopertha dominica.* It soon became obvious that r_m was an appropriate statistic to measure in comparing populations within species which were genetically different in different climatic zones, e.g. Bateman (1967). Another extension of its empirical application was developed by Messenger (1964) to measure the relative advantage of predator and prey in different temperatures, and by implication in different climatic zones. Cole (1954) showed how different life history patterns (of survival and fecundity) could be interpreted in terms of how these patterns affected the value of r_m. This approach was developed further by Lewontin (1965) in his study of the sort of survival and fecundity pattern that might be expected to be best for colonizing species. The statistic r_m has in recent years been adopted by geneticists and evolutionists, particularly those working with insects, as one of the measures of Darwinian fitness, thus providing a much needed link between population ecology and population genetics. (Birch et al, 1963; Ayala, 1968; Dobzhansky, 1968.)

The Concept of Logistic Growth

The model of exponential growth of a population applies in principle to populations growing in unrestricted space and within unlimited resources. Quetelet (1835) proposed that population growth of man is the consequence of the potential ability to grow exponentially and resistance to population growth which increases as the square root of the rate of growth. The idea is implicit that exponential growth is inhibited by increasing density as numbers increase in a limited space and with limited resources. The relationship between rate of increase and density (numbers per unit space) for human populations was formulated by Quetelet's student, Verhulst (1838) in his equation for the logistic curve of population growth. The same mathematical expression of the logistic curve, which is a sigmoid curve with an upper asymptote, was independently derived by Pearl & Reed (1920) as an empirical expression of the growth of the population of the United States. A rational derivation of the curve was given by Lotka (1925), who extended the theory to include the growth of two species living together. Volterra (1931) without knowing of Lotka's work, covered the same ground.

Whereas the equation for exponential growth is

$$\frac{\delta N}{\delta t} = r_m N$$

In the logistic equation the rate of increase r_m is reduced by a constant amount c for every individual added to unit space. Hence:

$$\frac{\delta N}{\delta t} = N(r_m - cN) = rN$$

The rate of increase thus continues to be reduced as a linear function of N as N becomes larger, until at some maximal value of N, cN approaches r_m and $r_m - cN$ approaches 0. Pearl (1926) applied the logistic curve to an experimental population of *Drosophila melanogaster* in milk bottles. However, as Sang (1950) showed there are aspects of Pearl's experiments which made them an inappropriate test of the logistic theory. For one thing he did not have a single species in his culture bottles but a predator (*Drosophila*) and prey (growing yeast cells). The objections to Pearl's experiments did not apply to Gause's 1934 experiments on the growth of two species of *Paramecium* in small vials in which food was supplied at a constant rate. The growth of Gause's population conformed to logistic curves.

Many workers since Pearl have recorded the growth of populations of insects kept in limited space with food supplied at a constant rate and have tried to fit their results to the logistic theory. Gause (1931) did this for the growth of populations to *Tribolium confusum* using data obtained by Chapman (1931). Chapman's experiments were not continued long enough to find out what happened after the population completed its first upward surge, but in this phase of its growth the experimental points conformed to a logistic curve. Park (1948) showed quite clearly that in the later stages the density failed to stay steady around an asymptote as required by the theory. And in the experiments done since then with a variety of stored product insects the results are much the same as those Park obtained. An initial phase of growth, which may or may not conform to the logistic trend, depending upon how the experiment was done, is followed by long-term fluctuations. (See Andrewartha & Birch, 1954, pp. 358–60.) The growth of populations such as *Paramecium* that have simple life histories conform to the logistic theory. With animals such as insects with complex life histories the conformation is less close and the logistic theory fails to account for some of the observed data. Insects have more complex properties than the assumptions of the logistic theory imply. It is therefore hardly an appropriate model on which to base theories of competition and predation.

Conceptual Models of Insect Populations

Because natural populations do not grow exponentially or logistically except briefly, but characteristically decrease as often as they increase, the central problem for the theorist is to explain the absence of exponential growth. The alternatives that have been explored are broadly divisible into deterministic and stochastic models.

Deterministic models—Two brief quotations from Nicholson make clear the philosophical basis for the conceptual models of insect populations that have been made by Nicholson and others who are in broad agreement with his theory of environment.

> For the production of balance it is essential that a controlling factor should act more severely against an average individual when the density of the animals is high, and less severely when the density is low. In other words, *the action of the controlling factor must be governed by the density of the population controlled.* (Nicholson, 1933)
>
> Logical argument based on certain irrefutable facts shows, not merely that populations may regulate themselves by density-induced resistance to multiplication, but that this mechanism is essential. . . . Thus it seems impossible to conceive of any situation in which populations do not adjust their densities in relation to the environmental conditions by themselves inducing a change in resistance to multiplication, when their densities are inappropriate to the prevailing conditions. This is the basic justification for the belief that all *persistent* populations exist in a state of balance in their environments; for 'balance' refers to such a condition of corrective reaction to change which holds a system *in being.* [(Nicholson, 1958), our emphasis].

Nicholson (1933) calculated a series of simple arithmetic models to explore the conditions of balance between populations of predator and prey which obeyed the general theory that balance implies a steady-state but which were also subject to two simplifying assumptions (*a*) the predator searches at random; (*b*) the predator has a constant area of discovery (i.e. in each generation each predator destroys a characteristic proportion of the prey). Nicholson & Bailey (1935) extended the analysis using a more sophisticated algebra. This lead was not actively pursued perhaps because the hypothetical deterministic animals in the models were too unrealistic.

Nevertheless, Nicholson's conceptual model of balance in insect populations in its most general form has been widely accepted. In the second quotation given above, we emphasized *persistent* and *in being* in order to make the point that this conceptual model rests on the logically impeccable proposition that if the rate of increase in a population, during an indefinitely long period is, on the average, above zero only a reactive (density-dependent) factor can prevent the population from becoming indefinitely large. This conceptual model cannot be tested empirically because no experiment can afford to wait an indefinite time to discover what ultimately happens. Moreover, the model is not, as is frequently implied, necessarily confirmed by the discovery of a density-dependent factor which may be present but not important (den Boer, 1968).

It was argued by Andrewartha & Birch (1954) that attention should be focused on the actual rate of increase in the natural population. One need only reflect that the rate of increase, under the influence of nonreactive factors, might, during an important interval of time, have a mean value close to zero in order to reach the conclusion that a reactive factor was not necessary in order to keep the population in being (den Boer, 1968). A population may be compared with water ebbing and flowing in a reservoir. Consider a dam built across a stream. The dam makes a large reservoir but the stream is fed by a small and erratic rainfall. The rainfall is sufficient to fill the reservoir once in 50 years (one need only imagine a larger reservoir or a smaller rainfall to extend the period to any length). It remains true that once in 50 (or 1000) years the water would be expected to overflow the dam and thus a ceiling is set to the size of the reservoir. This is undeniably true but is it important to the practicing hydrologist who has, in the meantime, to manage the town's supply of water?

Similarly, if the probability of a population reaching a ceiling which provokes a "density-induced resistance to multiplication" is low enough this risk may seem unimportant to the practicing population ecologist who may prefer a stochastic model.

Whereas the deterministic models that have been discussed in this section have been concerned with such ideas as upper limits, balance, or the theoretical reasons (excluding chance) for nonextinction, the stochastic models that are discussed in the next section are concerned with such ideas as the mean, the variance, and the regression on time of the numbers which might be estimated by sampling empirically in a natural population exposed to all the vagaries of chance. It is our opinion that much of the skeptical criticism of the stochastic approach has stemmed from the failure of its critics to appreciate that the two approaches serve different ends. The controversy awaits a philosophical resolution.

Stochastic models.—The models put forward by Andrewartha & Birch (1954) were intended to explain the numbers in natural populations of animals. A natural population was thought of as occupying a substantial area such as would usually be chosen for a study area by an ecologist. Because any substantial area is inevitably heterogeneous it is realistic to consider that a natural population comprises a large number of subpopulations or local populations existing more or less separately from one another and waxing and waning more or less independently of one another. This condition occurs partly because of the heterogeneity of the terrain and partly because of chance events that fall unevenly on the local populations in the local situations.

Andrewartha & Birch analyzed the conditions of commonness and rareness in local populations, using a series of simple diagrams to trace the temporal changes in local situations in which a particular component of environment was assumed to be limiting in a particular situation. Four diagrams served to represent the conditions of growth for local populations limited by some nonexpendable resource, by a diminishing resource, by weather, predators, or a relative shortage of a resource and by the distribution and abundance of refuges.

A fifth diagram illustrated (in relation to weather) the operation of time in stochastic models. The timing of events in the environments of animals in natural populations was seen as an important part of the stochastic process. For example, a population growing during a favorable season or during a spell of favorable weather will be dependent on the duration of the favorable period. A local population might grow freely during the period before it has been discovered by a predator though it may be doomed to extinction once the predator has discovered it. The chance that an immigrant, dispersing from a local population, will survive to found a new colony may depend on the duration of the period that elapses before it discovers a suitable place to live. All these processes must be considered to have a large stochastic element in them because they are likely to be caused by events that are not likely to be precisely determined by the density of the population. Andrewartha and Birch thought that shortage of time was likely to be more important (in that it happened more often) than shortage of resource.

In Andrewartha and Birch's model the conditions of commonness and rareness in natural populations were described by simple two-dimensional diagrams (representing substantial areas) in which the third dimension (time) was represented by the shape of symbols borrowed from the diagrams that represented the conditions of commonness and rareness in local populations. This theory, being restricted by its graphical presentation, oversimplified the stochastic processes that might be concerned in determining the densities of natural populations. The authors admitted this but claimed that the theory was intended to help the student to analyze the complex systems which he finds in nature into their simpler components so that he might understand them better.

A more sophisticated theory taking into account not only the heterogeneity in space and time of the places that the population inhabits but the heterogeneity in the animals themselves was developed by den Boer (1968) and discussed in relation to the illuminating idea of spreading the risk.

The risk is experienced by each individual. Individuals are variable by virtue of their genotype, age, or experience; so they respond differently to the same environment. The environment is different in different places at the same time or in the same place at different times. For these reasons and also because any natural population comprises a large number of local populations (subpopulations) scattered over a habitat that is usually highly heterogeneous, not only in space but also in time, it is likely that the rates of increase and decrease in the local populations will be out of step with each other. This lack of synchronism in the local populations will tend to dampen out fluctuations in the natural population and so engender a measure of stability.

According to den Boer, when the risk is spread over many factors a stochastic model may be sufficient to explain the observed stability. After considering a hypothetical example in which the risk was spread through a number of meteorological factors he continued:

> In general, the number of "factors", i.e. the amount of variation and heterogeneity, tends to have a levelling effect on extremes and trends. If we imagine a population composed of two sexes, a number of phenotypes and a number of developmental stages (age-classes), divided into a number of sub-populations with migration between them, living in different microenvironments with different micro-weather varying in time, consuming different kinds of food, influenced by a number of different predators, parasites and other animals we can form some idea of the immense number of interacting factors (amount of heterogeneity and variation) influencing animal numbers in nature and obtain an impression of the probable stabilization resulting from it.

The occurrence of density-dependent influences in nature was recognized by den Boer but he considered that they are most likely to be important in homogeneous conditions such as a plantation or in the laboratory. He continued:

> This does not mean, however, that a heterogeneous population in a heterogeneous environment would always be exclusively stabilized by spreading the risk. It only means that the greater the heterogeneity (the greater the spreading of the risk) the smaller the effect of density-dependent processes will be and the smaller the significance of such influences on the stabilization of animal numbers.

Advancing Frontiers in Population Ecology

Trying to look forward we have picked out a number of recent developments in insect ecology which we think might be the beginning of important advances. The progress of scientific knowledge is like a spiral passing successively through fact, explanation, hypothesis, experiment, fact, explanation, and so on. The truly great leaps forward come during explanation when imagination may be given full play; but progress also depends on new techniques. We mention first two new techniques that are closely related.

Key factors and functional analysis.—It is true of many insect pests that most of the damage is done by one stage of the life cycle. It is also true that mortality often falls most heavily on one or several stages of the life cycle. The key-factor analyses (proposed by Morris in 1963) provides a convenient way to compare the mortality in different stages and to identify the stage that is most influential in determining the numbers that will survive until the final stage of the generation. For the purpose of economic entomology it is usual to arbitrarily define the most damaging stage as the end of the generation.

Thus, if K is defined as the difference between the logs of the numbers in the first and final stages in the generation then:

$$K = k_0 + k_1 + k_2 \ldots k_n$$

where k_0 represents the difference between the logarithms of the numbers in the first and second stages of the generation; similarly, k_1, $k_2 \ldots k_n$ measure the mortalities in successive stages after the first. This analysis indicates directly the age group on which mortality falls most heavily, enabling effort to be concentrated on the causes of mortality at this stage. Morris suggested that the key factor, by directing attention to the stages in the life cycle that were important for prediction, might lead to considerable economies of labor.

Holling (1963, 1966) recognized the importance of identifying the key factor and went on to argue that the next step was to have a thorough understanding of the population processes that caused the mortality in this particular stage of the life cycle. Holling's argument was quite general but he illustrated it with reference to predation which he said was one of a number of important population processes. He analyzed predation into a number of components such as density of prey, density of predator, satiation, interval required for digestion, and so on. He measured the functioning of these components empirically with respect to particular species of predator and prey and he incorporated the results into a comprehensive model which he called a functional analysis of predation. This seems to be a powerful technique which, thoroughly applied, might be expected to enlarge our knowledge of the functioning of population processes. However to make a functional analysis of all the important population processes in the ecology of any species would be time consuming.

Mathematical animals using computers to simulate stochastic growth in populations.—The behavior and ecological physiology of the flour beetle *Tribolium castaneum* are well known (Park, 1954). Niven (1967), making use of this knowledge, defined a *Pseudotribolium castaneum* which was a computer program containing information about the fecundity, length of life, speed of development, cannibalism, etc, of *T. castaneum*. The information about each quality was expressed not as a predetermined quantity but as a probability distribution. The result, after some 40 generations, was a population curve remarkably like the curve got by Park when he counted the numbers of the real *T. castaneum* living in vials of flour. With the computer's speed, many generations may be completed in a few minutes. So the conditions of life, the ecological physiology of the pseudobeetle may be varied and the consequences observed without delay. This is a powerful technique for making subtle hypotheses which would be impracticable without the aid of a computer.

Reddingius & den Boer (1970) simulated population changes for hypothetical animals whose rate of increase was determined by a number of factors taken directly from published meteorological records. They chose these records because they were obviously independent of the density of the population and because they were more realistic than series drawn from tables of random numbers. The models were set up to explore the general theory of den Boer which states that spreading the risk (by increasing the heterogeneity of the animals and their environments) tends to stabilize the fluctuations in the population. The computer models of Reddingius and den Boer may be regarded as hypotheses arising from the general theory of den Boer (1968). They investigated the logical consequences of 1. varying the number of factors influencing the net reproductive rate; 2. varying the heterogeneity of the habitat; 3. varying the age structure of the population. It turned out that spreading the risk by increasing the heterogeneity in any or all of these ways increased the stability of the population. For example, one simulated natural population comprised nine subpopulations; after reproducing, the animals had a chance to migrate to other subpopulations; in each subpopulation several age classes were represented. It turned out that increasing the number of age classes increased stability as did the chance of migration between subpopulations. Letting the exchange between subpopulations be density-dependent had some stabilizing influence also but not a conspicuous one. Another population was simulated in which the net rate of increase was slightly greater than one; there was a chance of a crash when the population reached a high density. The model predicted considerable stability for such a population even when the size of the crash did not depend on density, and the times when crashes occurred were chosen at random.

The role of genetical heterogeneity.—The working assumption of many ecological studies has been that any sample of a population of a species could be regarded as representative of the species. Genetical studies have shown this to be an untenable concept. The genetical constitution of the species is different from place to place and from one time to another with the consequence that physiology, behavior, birthrates, and death rates may also be different. The first rigorous documentation of the role of genetical heterogeneity in time and space was Dobzhansky's (1937) book *Genetics and the Origin of Species.* What has been achieved since then has again been fully documented by Dobzhansky (1970) in *Genetics of the Evolutionary Process.* The thirty-odd years between these two milestones has seen a coming together of genetics and ecology. This has been largely due to population geneticists whose major interest was evolution and who were therefore interested in effects of genetical change on the chance to survive and reproduce.

The genetical component must now be regarded as a critical aspect of any theories having to do with extinction and survival, fluctuations in numbers, the role of environmental heterogeneity, and dispersal in determining numbers and changes in distribution. The full significance of population genetics for all these aspects of ecology has yet to be worked out. The emphasis to date has been on the role of genetical heterogeneity on secular trends. The field waiting to be explored is the role of genetical heterogeneity in space, which will require a linking of genetical studies with dispersal. Much of the work on this up to 1965 was reviewed in *The Genetics of Colonizing Species* edited by Baker & Stebbins (1965). Since then, a new tool has become widely used for studies on insects which adds greatly to the possibilities of identifying genetical heterogeneity. This is protein electrophoresis. The work of Hubby & Lewontin (1966) and numerous other papers by them and others has provided a new thrust for studies on the role of genetical heterogeneity in ecological studies.

The theory of resources.—The idea of a relative shortage of a resource first appeared in ecology when Jackson (1937) observed that a tsetse fly *Glossina morsitans* might have little chance of finding enough food where antelopes were few and sparsely distributed, not because of a shortage of food (blood of antelope) in any absolute sense, but merely because the chance of finding an antelope at the right time was small. The shortage of food experienced by a tsetse fly is called an extrinsic relative shortage because the shortage is not caused by the activity of the flies. Antelopes are scarce because of some other component in their environment. Conversely, the relative shortage experienced by a successful agent of biological control, e.g. *Cactoblastis cactorum* or *Rodolia cardinalis,* is called an intrinsic relative shortage because the shortage is caused by the activity of the insect feeding on its host (Andrewartha, 1970). The ladybirds *R. cardinalis,* by virtue of their feeding, reduced what had been a large dense population of scale insects, *Icerya purchasi,* to a small sparse patchily distributed population. Once this condition had been established, many *R. cardinalis* in every generation died of starvation even though some food (which, by chance, had not been discovered) remained uneaten. It is now recognized that nearly all the shortages of resources that occur in nature are relative shortages, either intrinsic or extrinsic. Absolute shortages occur only in special circumstances, and then only locally or temporarily (Andrewartha & Browning, 1961). The idea which originated with Jackson has grown to be an important frontier in the theory of ecology. Moreover, the sparse patchy distribution of resources that is characteristic of a relative shortage is merely a particular example of the general condition of heterogeneity of habitat which is central to the stochastic theory of population stability which was outlined by Andrewartha & Birch (1954) and developed by den Boer (1968) and Reddingius & den Boer (1970).

Adaptation for dispersal.—Caterpillars of *Bupalis pinarius* are pests of *Pinus* in northern Europe. According to Gruys (1970, 1971), when two caterpillars meet they usually perform a head-waving dance and one drops on a silken thread out of the way of the other. After a short while it climbs back and continues with its normal feeding behavior. It seems that the immediate cause of the avoiding action is a pheromone present in the vomit that one caterpillar spits at the other. The long-term consequence of the encounter is that a caterpillar that has received the vomit on a number of occasions grows slowly and ultimately emerges as a lightweight moth with relatively large wings well adapted for dispersal. Gruys suggested that the selective advantage of this build for dispersal more than offsets the disadvantage of the lower fecundity associated with the light weight. That this is a valuable adaptation that is strongly selected is indicated by the fact that little stimulus is needed to produce the result. In the laboratory, 5 days' living together with other larvae out of a total larval life of 100 days will suffice. In nature, the effect is observed at densities well below those that are likely to cause a noticeable shortage of food. Moreover, the larvae also have the behavior of moving freely on the trees which enhances the chance of meetings even at low densities.

Similar adaptations which indicate the high selective advantage of behavior that leads to active dispersal from quite sparse populations were reported by Way (1971) for the aphid *Brevicoryne brassicae* and by Hassel (1971) for the ichneumonid parasite *Nemeritus canescens*. Davidson & Andrewartha (1948) described similar behavior in *Thrips imaginis* and pointed out that this behavior was an advantage to an insect that found its food in flowers that were transient; any source of food was likely to be temporary and the distribution of the food was constantly changing. Birch (1971) documented the extreme patchiness in space and time of food for the fruit fly *Dacus tryoni* and the moth *Cactoblastis cactorum*.

Progress in the theory of population ecology has often been advanced through the mutual interaction of ideas from ethology, ecological physiology, and population dynamics. The discussions on conceptual models illustrate this principle which was also mentioned in the introduction.

A false trail.—Southwood (1966, p. 298 et seq) summarized the literature on key-factor analysis as a test for density dependence. More detailed expositions may be found in papers by Varley & Gradwell (1963, 1968), Solomon (1964), and Klomp (1966). In its simplest form the test is made by plotting the regression of log N_{t+1} on log N_t. If b is less than 1 the test is said to have indicated the operation of a density-dependent factor. Alternatively, the regression of log N_{t+1} — log N_t on log N_t is calculated, and the same criterion ($b < 1$) applies. In both forms the test N_{t+1} and N_t stand for the numbers in successive generations or stages in the life cycle (or generation). In both forms of the test, the variates fail to satisfy the criteria of independence which are necessary for unbiased estimates of b and its variance. Andrewartha (1963) criticized the regression of log N_{t+1} — log N_t on log N_t. St. Amant (1970) showed, with some elegant algebra, that the expected value for b in the regression of log N_{t+1} on log N_t is less than 1 for a series in which log N_{t+1} is independent of log N_t. The magnitude of b depends on the size of the sample: b approaches 1 as the sample becomes indefinitely large; for small samples of the sort usually available to ecologists, the bias is likely to be large and important. Maelzer (1970) reached the same conclusion by simulating populations on a computer. It is now clear that these so-called tests for density dependence are spurious; they should be abandoned by ecologists.

The search for a test for density dependence may have seemed more important than it is because of the widely held misconception that to discover a density-dependent factor is to discover what controls the population. This misconception embodies two philosophical errors; one stems from mistaking a logical deduction for an empirical verification (Andrewartha, 1961); the other stems from inconsistencies implicit in control used in this context (den Boer, 1968). It should be easier to abandon these philosophical errors now that den Boer's important paper (1968) has crystallized the ideas in the stochastic theory of population stability.

COMMUNITY ECOLOGY

Insect ecologists have been far more preoccupied with problems in physiological ecology and population ecology than in community ecology. In so far as they have become involved in community ecology it seems to have been through interest in rather special problems such as those listed below.

Succession

The concept of ecological succession in communities was first developed by the botanists Warming, Cowles, and Clements. They gave little consideration to the role of animals in succession. Shelford (1907) studied insects on the Indiana sand dunes on the southern shore of Lake Michigan where Cowles (1899) had already studied plant succession. In particular, Shelford showed that the tiger beetles (Cicindelidae), in successive vegetation zones from the shore to climax forest, were of different species. Since the sequence of vegetation types represented a historical sequence on a retreating shore line, it could be reasonably assumed that the sequence of insects also represented historical sequence. Chapman (1931) recognized that the role of animals in succession might be passive, meaning that they are not causes of change but follow floral and climatic changes. He also recognized some animals as active in succession when they caused the changes. He gave as an example the larch sawfly which, he said, was largely responsible for the disappearance of larch and its replacement by other trees in certain bogs. In a series of papers in 1911 and 1912, Shelford gave many examples of passive succession which were later assembled in his book on animal communities (Shelford, 1913). More recent studies on animal succession show a rather incomplete association of plant zones and animal associations as, for example, the work in 1949 of Janetschek (summarized by Macfadyen in 1963) on the high Alps and Van Heerdt et al (1960) on sand dunes.

The succession of insects living in a carcass was analyzed in detail by Fuller (1934), Waterhouse (1947), and Nicholson (1950) and their colleagues in Australia. These studies showed that different species of blowflies and other insects were attracted to the carcass at different stages of its decomposition. And, secondly, they showed that the presence of one species could be detrimental to another if it happened to be there at the same time. This was the case with a number of blowflies that Waterhouse studied in the sheep carcass. His objective was to find out if the carcass was a major source of the sheep blowfly *Lucilia cuprina* which is a pest of living sheep. Despite the fact that *Lucilia cuprina* lays its eggs on the carcass, very few survive to produce adult blowflies. This was almost entirely due to the presence of other blowflies in the carcass. In Chapman's terms, these other blowflies were active in influencing succession.

Competitive Exclusion

Gause (1934), having done a number of experiments in which two species of yeasts and, in another series, two species of *Paramecium,* were reared together in a restricted space with food kept as constant as possible, and having found that one species always died out leaving the other to thrive alone, looked around to see if he could find a similar phenomenon in nature. He thought he had found similar examples in nature in some fishes and crayfish. Crombie (1945) confirmed Gause's experimental results, using a number of species of insects that live in stored products. Both Gause and Crombie considered that their experiments were confirmations of Volterra's competition equations. Crombie extrapolated his experimental results to nature with the following general statement: "It is easy to demonstrate theoretically that two species with identical ecological niches cannot survive together in the same environment unless density-dependent factors keep the population low enough to eliminate interspecific competition." Although the main stream of interest in this subject seems to have stemmed from Gause and Crombie, there were others who had discussed it much earlier. Chapter 3 of Darwin's *The Origin of Species* contains the idea that the struggle for existence would be greater between species of the same genus because their habits and requirements would be similar. Grinnell (1917 a and b) wrote about competitive displacement of one species by another; and others, as well, had written about the same idea well before Gause (see De Bach, 1966).

The laboratory experiments of Gause and Crombie were confirmed with many species of insects by numerous workers but by none so assiduously as Park (1948) and numerous later papers on two species of *Tribolium* beetles. Likewise, Crombie's generalization to nature was pursued by Lack in papers on birds, and by numerous other authors on other animals, though not many on insects (see Andrewartha & Birch, 1954). These papers reveal that the competitive exclusion principle, demonstrated in laboratory experiments, has been applied quite uncritically to nature and very often assumed by biologists, including entomologists, to have been demonstrated in nature. In his review of the subject, De Bach (1966) wrote "Perhaps the only thoroughly documented case of competitive displacement between true ecologically homologous insects involves the closely related species of *Aphytis* studied in southern California," work done by De Bach and his co-workers. Of course, there are a number of species that differ widely in their ecology and yet which cannot live together when one species makes the environment uninhabitable for the other (e.g. Brian, 1956). It is unfortunate that the concept of competitive exclusion has become so closely linked with the rather special case of species having very similar requirements.

The Niche

Grinnell (1904, 1917a, 1917b) was probably the first person to use the term niche in ecology (Udvardy, 1959). His meaning was essentially the same as Elton's (1927) who gave the idea wide currency in ecology. For Elton, niche was a term to "describe the status of an animal in its community, to indicate what it is *doing* and not merely what it looks like." One merit Elton saw in the concept was that it would aid in the identification of resemblances, from an ecological point of view, of different animals in different communities. It was used in this way by Allee et al (1949) in their Tables 35 and 40 of what they called stratal categories in grassland and in forest communities. However, their tables contain only vertebrates. The idea seems to have caught on less in entomology although entomologists do recognize similar roles of different insects in different countries such as, for example, the tree hole mosquitoes. The concept of niche has been little developed since Elton introduced it into the general discussion of ecology, despite attempts that have been made since then to quantify it (Hutchinson, 1957; Levins, 1968).

Food Chains, the Pyramid of Numbers and Community Energetics

Elton (1927) put the matter simply when he wrote "the animals at the base of the food-chain are relatively abundant, while those at the end are relatively few in numbers, and there is progressive decrease between the two extremes." This was Elton's pyramid of numbers which diagrammatically represented the numbers of organisms in successive steps of a food chain.

The step from pyramid of numbers to an energy pyramid of the successive trophic levels in a community is an obvious one. It was first made by the limnologist, Lindeman (1942) for a like community. He attempted to account for all the transfers of energy through living organisms from that trapped by sunlight that fell on the lake to that fixed in the bodies of the predators at the top of the food chain. Numerous studies have followed since. Their impact on insect ecology has been small, perhaps for no other reason than this, that insect ecologists being primarily interested to explain the distribution and numbers of particular species, have asked other sorts of questions than the ones from which community energetics may provide answers. An exception to this general statement is Varley's attempt (1970) to convert Elton's (1966) food chains in Wytham Woods, Oxford, to energy chains.

Stability and Vulnerability to Invasion by Species from Other Countries of Simple as Contrasted with Complex Communities

Elton (1958) argued that simple communities (i.e. those with few species) are less able to withstand invasion by foreign animals than complex ones. The reasons he gave were based on the familiar mathematical and laboratory models, the vulnerability of oceanic island communities, and the frequency of pest outbreaks in crops, as compared, say, with a rain forest. Elton (1966) re-emphasized the aspect of vulnerability and made a distinction between this and stability of a community by which he meant the continued existence of most of the species. In this sense of stability he did not think it could be shown that simple communities were less stable than complex ones. Others have defined stability in other ways, particularly with respect to the tendency of numbers of a species to fluctuate widely or to fluctuate narrowly. The subject is one of interest to insect ecologists because of the claim of some that single species crops and their associated animals are highly unstable communities. The theoretical basis of this concept has been brought into question by May (1971) and empirical evidence is lacking on which to make any strong claims one way or the other. The evidence, in fact, is conflicting. Watt (1964, 1968) analyzed records of the Canadian Forest Insect Survey and showed that the more competitors a herbivore had, the more stable its population, but the more host trees the less stable its population, a finding which, as he said, conflicted with the "traditional wisdom of ecologists." On the other hand, Southwood & Way (1970) gave evidence from the literature that, whereas pest and disease outbreaks are common in agroecosystems, they occur less frequently in secondary tropical forests, unstable arid ecosystems, and simple coniferous forests; they claim that pest outbreaks are unrecorded from tropical rain forests.

Literature Cited

Allee, W. C., Emerson, A. E., Park, O., Park, T., Schmidt, K. P. 1949. *Principles of Animal Ecology*. Philadelphia: Saunders

Andrewartha, H. G. 1943. Diapause in the eggs of *Austroicetes cruciata* Sauss. (Acrididae) with particular reference to the influence of temperature on the elimination of diapause. *Bull. Entomol. Res.* 34:1–17

Andrewartha, H. G. 1944. Air temperature records as a guide to the date of hatching of the nymphs of *Austroicetes cruciata* Sauss. (Orthoptera). *Bull. Entomol. Res.* 35:31–41

Andrewartha, H. G. 1952. Diapause in relation to the ecology of insects. *Biol. Rev.* 27:50–107

Andrewartha, H. G. 1961 *Introduction to the Study of Animal Populations,* London: Methuen

Andrewartha, H. G. 1963. Density dependence in the Australian thrips. *Ecology* 44:218–20

Andrewartha, H. G. 1970. *Introduction to the Study of Animal Populations.* London: Methuen. 2nd ed.

Andrewartha, H. G., Birch, L. C. 1954. *The distribution and Abundance of Animals.* Chicago: Univ. Chicago Press

Andrewartha, H. G., Browning, T. O. 1961. An analysis of the idea of "resources" in animal ecology. *J. Theoret. Biol.* 1:83–97

Ayala, F. S. 1968. Genotype, environment, and population numbers. *Science* 162:1453–59

Bacot, A. W., Martin, C. J. 1924. The respective influences of temperature and moisture upon the survival of the rat flea (*Xenopsylla cheopis*) away from its host. *J. Hyg.* 23:98–105

Baker, F. C., 1935. The effect of photoperiodism on resting tree-hole mosquito larvae (Preliminary report). *Can. Entomol.* 67:149–53

Bateman, M. A. 1967. Adaptations to temperature in geographic races of the Queensland fruit fly, *Dacus* (*Strumeta*) *tryoni. Aust. J. Zool.,* 15: 1141–61

Beament, J. W. L. 1945. The cuticular lipoids of insects. *J. Exp. Biol.* 21: 115–31

Belehrádek, J. 1935. *Temperature and Living Matter.* Berlin: Borntraeger

Birch, L. C. 1944. The effect of temperature and dryness on the survival of the eggs of *Calandra oryzae* L. (small strain) and *Rhizopertha dominica* Fab. (Coleoptera). *Aust. J. Biol. Med. Sci.* 22:265–69

Birch, L. C. 1945. The mortality of the immature stages of *Calandra oryzae* L. (small strain) and *Rhizopertha dominica* Fab. in wheat of different moisture contents. *Aust. J. Exp. Biol. Med. Sci.* 23:141–45

Birch, L. C. 1948. The intrinsic rate of natural increase of an insect population. *J. Anim. Ecol.* 17:15–26

Birch, L. C. 1953. Experimental background to the study of the distribution and abundance of insects. I. The influence of temperature, moisture and food on the inate capacity for increase of three grain beetles. *Ecology* 34:698–711

Birch, L. C. 1971. The role of environmental heterogeneity and genetical heterogeneity in determining distribution and abundance. In *Dynamics of Populations,* ed P. J. den Boer, G. R. Gradwell. Purdue: Wageningen. In press

Birch, L. C., Dobzhansky, T., Elliott, P. O., Lewontin, R. C. 1963. Relative fitness of geographic races of *Drosophila serrata. Evolution* 17:72–83

Bodenheimer, F. S. 1938. *Problems of Animal Ecology*. Oxford: Clarendon

Bodenheimer, F. S., Schenkin, D. 1928. Über die Temperaturabhängigkeit von Insekten. 1. Über die Vorzugstemperatur einiger Insekten. *Z. Vergl. Physiol.* 8:1–15

Botero, G. 1588. *A Treatise Concerning the Causes of the Magnificences and Greatness of Cities.* Transl. R. Peterson, 1606. Reprinted 1956. London: Routledge & Kegan Paul

Breitenbrecher, J. K. 1918. The relation of water to the behaviour of the potato beetle in a desert. *Publ. Carnegie Inst. Washington* 263:343–84

Brian, M. V. 1956. Segregation of species of the ant genus *Myrmica. J. Anim. Ecol.* 25:319–37

Browning, T. O. 1962. The environments of animals and plants. *J. Theoret. Biol.* 2:63–68

Bursell, E. 1964. Environmental aspects: temperature: In *Physiology of Insecta,* ed. M. Rockstein, Vol. I. New York: Academic

Buxton, P. A. 1930. Evaporation from the mealworm (*Tenebrio,* Coleoptera) and atmospheric humidity. *Proc. Roy. Soc. London* 106:560–77
Carter, W. 1925. The effect of low temperature on *Bruchus obtectus,* Say, an insect affecting seed. *J. Agr. Res.* 31:165–82
Chapman, R. N. 1928. The quantitative analysis of environmental factors. *Ecology* 9:111–22
Chapman, R. N. 1931. *Animal Ecology with Special Reference to Insects.* New York: McGraw-Hill
Clark, L. R. 1947. An ecological study of the Australian plague locust *Chortoicetes terminifera* Walk. in the Bogan-Macquarie outbreak area in N.S.W. *Bull. Counc. Sci. Ind. Res. Aust.* No. 226
Clark, L. R., Geier, P. W., Hughes, R. D., Morris, R. F. 1967. *The Ecology of Insect Populations in Theory and Practice.* London: Methuen
Clark, K. U., Sardesai, J. B. 1959. An analysis of the effects of temperature upon the growth and reproduction of *Dysdercus fasciatus Sign.* (Hemiptera, Pyrrhocoridae). 1. The intrinsic rate of increase. *Bull. Entomol. Res.* 50: 387–406
Cole, L. C. 1954. The population consequences of life history phenomena. *Quart. Rev. Biol.* 29:203–37
Cole, L. C. 1957. Sketches of General and Comparative Demography. *Cold Spring Harbor Symp. Quant. Biol.* Vol. 22
Cowles, H. C. 1899. The ecological relations of the vegetation of the sand dunes of Lake Michigan. *Bot. Gaz.* 27:95–117
Crombie, A. C. 1945. On competition between different species of graminivorous insects. *Proc. Roy. Soc. London, Ser. B* 132:362–95
Davidson, J. 1936. On the ecology of the black-tipped locust (*Chortoicetes terminifera,* Walk.) in South Australia. *Trans. Roy. Soc. S. Aust.* 60: 137–52
Davidson, J. 1944. On the relationship between temperature and rate of development of insects at constant temperature. *J. Anim. Ecol.* 13:26–38
Davidson, J., Andrewartha, H. G. 1948. The influence of rainfall, evaporation and atmospheric temperature on fluctuations in size in a natural population of *Thrips imaginis* (Thysanoptera). *J. Anim. Ecol.* 17:200–22
Deal, J. 1941. The temperature preferendum of certain insects. *J. Anim. Ecol.* 10:323–55
De Bach, P. 1966. The competitive displacement and co-existence principles. *Ann. Rev. Entomol.* 11:183–212
den Boer, P. J. 1968. Spreading of risk and stabilization of animal numbers. *Acta Biotheoret.* 18:165–94
Derham, W. 1713. *Physico-theology or a Demonstration of the Being and Attributes of God from his Works of Creation.* London: printed for A. Strahan, T. Cadell, and W. Davies. The Strand. New ed. 1798
Dickson, R. C. 1949. Factors governing the induction of diapause in the oriental fruit moth. *Ann. Entomol. Soc. Am.* 42:511–37
Dobzhansky, T. 1968. In *Evolutionary Biology,* ed. T. Dobzhansky, M. K. Hecht, W. C. Steere, p. 1–32. Amsterdam: North-Holland
Duclaux, E. 1869. De l'influence du froid de l'hiver sur le développement de l'embryon du ver à soie et sur l'éclosion de la graine. *C. R. Acad. Sci, Paris* 69:1021–22
Edney, E. B. 1947. Laboratory studies on the bionomics of the rat fleas *Xenopsylla brasiliensis* Baker and *X. cheopis* Roths. II. Water relations during the cocoon period. *Bull. Entomol. Res.* 38:263–80
Elton, C. S. 1927. *Animal Ecology.* London: Sidgwick & Jackson
Elton, C. S. 1958. *The Ecology of Invasions by Animals and Plants.* London: Methuen
Elton, C. S. 1966. *The Pattern of Animal Communities.* London: Methuen
Fraenkel, G., Blewett, M. 1944. The utilization of metabolic water in insects. *Bull. Entomol. Res.* 35:127–37
Fraenkel, G., Gunn, D. L. 1940. *The Orientation of Animals.* Oxford: Clarendon
Fry, F. E. J. 1947. Effects of the environment on animal activity. *Univ. Toronto Stud. Biol.* 55:1–62
Fuller, M. E. 1934. The insect inhabitants of carrion: a study in animal ecology. *Bull. Counc. Sci. Ind. Res. Aust.* No. 82
Gause, G. F. 1931. The influence of ecological factors on the size of populations. *Am. Nat.* 65:70–76
Gause, G. F. 1934. *The Struggle for Existence.* Baltimore: Williams & Wilkins
Gayspitz, K. F., Kyao, I. I. 1953. The

influence of the length of illumination on the development of certain braconids (Hymenoptera). *Entomol Oboz.* 33:32. In Russian

Glenn, P. A. 1922. Relation of temperature to development of the codling moth. *J. Econ. Entomol.* 15:193–98

Glenn, P. A. 1931. Use of temperature accumulations as an index to the time of appearance of certain insect pests during the season. *Trans. Illinois Acad. Sci.* 24:167–80

Graunt, J. 1662. *Natural and Political Observations Mentioned in a Following Index and Made upon the Bills of Mortality.* London: T. Roycroft

Grinnell, J. 1904. The origin and distribution of the chestnut-backed chickadee. *Auk* 21:364–82

Grinnell, J., 1917a. The niche-relationships of the California thrasher. *Auk* 34:427–33

Grinnell, J., 1917b. Field tests of theories concerning distributional control. *Am. Nat.* 31:115–28

Gruys, P. 1970. Growth in *Bupalus pinarius* (Lepidoptera:Geometridae) in relation to larval population density. *Verhan. No. 1 Res. Inst. Nature Management.* pp. 1–127

Gruys, P. 1971. On the influence of mutual interference on growth in *Bupalus pinarius.* In *Dynamics of Populations,* ed. P. J. den Boer, G. R. Gradwell. Purdue: Wageningen. In press

Gunn, D. L. 1933. The temperature and humidity relations of the cockroach (*Blatta orientalis*). I. Desiccation. *J. Exp. Biol.* 10:274–85.

Guyelard, F., Portier, P. 1916. Recherches sur la résistance au froid de *Cossus carpocapsa. C. R. Soc. Biol. Paris* 79:774–77

Hamilton, A. G. 1950. Further studies on the relation of humidity and temperature to the development of two species of African locusts. *Locusta migratoria migratorioides* R. and F., and *Schistocerca gregaria,* Forsk. *Trans. Roy. Entomol. Soc. London* 101:2–56

Harries, F. H. 1939. Some temperature coefficients for insect oviposition. *Ann. Entomol. Soc. Am.* 32:758–76

Hassell, M. P. 1971. Parasite behaviour as a factor contributing to the stability of insect host-parasite interactions. In *Dynamics of Populations,* ed. P. J. den Boer, G. R. Gradwell. Purdue: Wageningen. In press

Henneguy, F. F. 1904. *Les Insectes: morphologie, reproduction, embryogénie.* Paris: Masson et Cie

Herter, K. 1924. Untersuchungen über den Temperatursinn einiger Insekten. *Z. Vergl. Physiol.* 1:221–88

Hinton, H. E. 1960. Cryptobiosis in the larva of *Polypedilum vanderplanki* Hint. (Chironomidae). *J. Insect Physiol.* 5:286–300

Holling, C. S. 1963. An experimental component analysis of population processes. *Mem. Entomol. Soc. Can.* 32:22–33

Holling, C. S. 1966. The functional response of invertebrate predators to prey density. *Mem. Entomol. Soc. Can.* 48:1–86

House, H. L., Riordan, D. F., Barlow, J. S. 1958. Effects of thermal conditioning and of degree of saturation of dietary lipids on resistance of an insect to a high temperature. *Can. J. Zool.* 36:629–32

Howard, L. O., Fiske, W. F. 1911. The importation into the United States of the parasites of the gipsy moth and the browntail moth. *Bull. US Bur. Entomol.* No. 91

Hubby, J. L., Lewontin, R. C. 1966. A molecular approach to the study of genic heterozygosity in natural populations. I. The number of alleles at different loci in *Drosophila pseudoobscura. Genetics* 54:577–94

Hutchinson, G. E. 1957. Concluding remarks. *Cold Spring Harbor Symposia on Quantitative Biology.* 22: 415–27

Jackson, C. H. N. 1937. Some new methods in the study of *Glossina morsitans. Proc. Zool. Soc. London* 1936:811–96

Janisch, E. 1925. Über die Temperaturabhängigkert biologischer Vorgäng und ihre kurvemässige analyse. *Pflüger's Arch für die gesammte Physiol.* 209:414–36

Johnson, C. G. 1940. The longevity of the fasting bed bug (*Cimex lectularius* L.) under experimental conditions and particularly in relation to the saturation deficiency law of water loss. *Parasitology* 32:239–70

Johnson, C. G. 1942. Insect survival in relation to the rate of water loss. *Biol. Rev.* 17:151–77

Kennedy, J. S. 1939. The behaviour of the desert locust (*Schistocerca gregaria,* Forsk.) (Orthoptera) in an outbreak centre. *Trans. Roy. Ento-*

mol. Soc., London 89:385–542

Kevan, D. K. McE. 1944. The bionomics of the neotropical cornstalk borer *Diatraea lineolata* Wlk. (Lep. Pyral.) in Trinidad, B. W. I. *Bull. Entomol. Res.* 35:23–30

Klomp, H. 1966. The interrelations of some approaches to the concept of density dependance in animal populations. *Med. Landbow. Wageningen* 66:1–10

Knight, H. H. 1922. Studies on the life history and biology of *Perillus bioculatus* including observations on the nature of the colour pattern. *State Entomol. Minnesota, 19th Rep.* p. 50–96

Kogure, M. 1933. The influence of light and temperature on certain characters of the silkworm, *Bombyx mori. J. Dep. Agr. Kyushu Imp. Univ.* 4:1–93

Krogh, A. 1914. On the influence of the temperature on the rate of embryonic development. *Z. Allgem. Physiol.* 16: 163–77

Krumbiegel, I. 1932. Untersuchungen über physiologische Rassenbildung. *Zool. Jahrb. Abt. Syst.* 63:183–280

Larsen, E. B. 1943. The influence of humidity on life and development of insects. Experiments on flies. *Vidensk. Medd. Dansk. Naturh. Fören, Kbh.* 107:127–84

Lees, A. D. 1946. The water balance in *Ixodes ricinus* L. and certain other species of ticks. *Parasitology* 37:1–20

Lees, A. D. 1947. Transpiration and the structure of the epicuticle in ticks. *J. Exp. Biol.* 23:379-410

Lees, A. D. 1955. *The Physiology of Diapause in Arthropods.* Cambridge: The University Press

Leslie, P. H., Park, T. 1949. The intrinsic rate of natural increase of *Tribolium castaneum* Herbst. *Ecology* 30:469–77

Leslie, P. H., Ranson, R. M. 1940. The mortality, fertility and rate of natural increase of the vole (*Microtus agrestis*) as observed in the laboratory. *J. Anim. Ecol.* 9:27–52

Levins, R., 1968. *Evolution in Changing Environments.* Princeton: University Press

Lewontin, R. C. 1965. Selection for colonizing ability. In *The Genetics of Colonizing Species,* ed. H. G. Baker, G. L. Stebbins. New York: Academic

Lindeman, R. L. 1942. The trophic-dynamic aspect of biology. *Ecology* 23:399–418

Lotka, A. J. 1925. *Elements of Physical Biology.* Baltimore: Williams & Wilkins

Ludwig, D. 1928. The effect of temperature on the growth curves of Japanese beetles (*Popillia japonica,* Newman). *Physiol. Zool.* 5:431–47

Ludwig, D. 1932. The effect of temperature on the development of an insect (*Popillia japonica*). *Physiol. Zool.* 1:358–98

Ludwig, D. 1937. The effect of different relative humidities on respiratory metabolism and survival of the grasshopper *Chortophaga viridifasciata* De Geer. *Physiol. Zool.* 10:342–51

Macfadyen, A. 1963. *Animal Ecology. Aims and Methods.* London: Pitman. 2nd ed.

Maelzer, D. A. 1965. A discussion of components of environment in ecology. *J. Theoret. Biol.* 8:141–62

Maelzer, D. A. 1970. The regression of log N_{n+1} on $\log_n$ as a test for density-dependence: an exercise with computer-constructed density-independent populations. *Ecology* 51:810–22

Maercks, H. 1933. Der Einfluss von Temperatur und Luftfeuchtigkeit auf die Embryonalentwicklung der Mehlmottenschlupfwespe *Habrobracon juglandis* Ashmead. *Arb. Biol. Abt. (Anst.-Reichsanst.), Berlin* 20:347–90

May, R. M. 1971. Stability in model ecosystems. *Proc. Ecol. Soc. Aust.* 6: 18–56

Mellanby, K. 1932a. The effect of atmospheric humidity on the metabolism of the fasting mealworm (*Tenebrio molitor* L., Coleoptera). *Proc. Roy. Soc. London, Ser. B* 111: 376–90

Mellanby, K. 1932b. Effects of temperature and humidity on the metabolism of the fasting bed bug *Cimex lectularius* Hemiptera. *Parasitology* 24:419–28

Mellanby, K. 1934. The site of loss of water from insects. *Proc. Roy. Soc. London, Ser. B* 116:139–49

Mellanby, K. 1935. The evaporation of water from insects. *Biol. Rev.* 10: 317–33

Mellanby, K. 1936. Humidity and insect metabolism. *Nature* 138:124–25

Mellanby, K. 1939. Low temperature and insect activity. *Proc. Roy. Soc. London, Ser. B* 127:473–87

Mellanby, K. 1959. Acclimatization affecting the position of the cold and heat death points of larvae of *Aëdes aegypti* (L). *Bull. Entomol. Res.* 50: 821–23

Messenger, P. S. 1964. Use of life tables in a bioclimatic study of an experimental aphid-braconid wasp host-parasite system. *Ecology* 45:119–31

Moore, J. A. 1940. Stenothermy and eurothermy of animals in relation to habitat. *Am. Nat.* 74:188–92

Morris, R. F. 1963. Predictive population equations based on key factors. *Mem. Entomol. Soc. Can.* 32:16–21

Nicholson, A. J. 1933. The balance of animal populations. *J. Anim. Ecol.* 2:132–78

Nicholson, A. J. 1950. Population oscillations caused by competition for food. *Nature.* 165:476–77

Nicholson, A. J. 1958. Dynamics of insect populations. *Ann. Rev. Entomol.* 3:107–36

Nicholson, A. J., Bailey, V. A. 1935. The balance of animal populations. Part I. *Proc. Zool. Soc. London* 1935: 551–98

Niven, S. 1967. The stochastic simulation of *Tribolium* populations. *Physiol. Zool.* 40:67–82

Nutman, S. R. 1941. Function of the ventral tube in *Onychiurus armatus* (Collembola). *Nature* 148:168–69

Oettingen, A. J. von. 1879. Phanologie der Dorpater Lignosen. *Arch. Naturk. Liv-, Est-u. Kurlands* (2)8:241–352

Park, T. 1948. Experimental studies of interspecies competition. I. Competition between populations of the flour beetles *Tribolium confusum* Duval and *Tribolium castaneum* Herbst. *Ecol. Monogr.* 18:265–308

Park, T. 1954. Competition: an experimental and statistical study. In *Statistics and Mathematics in Biology.* Ames: Iowa St. Coll. Press

Payne, N. M. 1926. The effect of environmental temperatures upon insect freezing points. *Ecology* 7:99–106

Payne, N. M. 1927. Two factors of heat energy involved in insect cold hardiness. *Ecology* 8:194–96

Pearl, R. 1926. *The Biology of Population Growth.* New York: Knopf

Pearl, R., Reed, L. J. 1920. On the rate of growth of the population of the United States since 1790 and its mathematical representation. *Proc. Nat. Acad. Sci. USA* 6:275–88

Pielou, D. R., Gunn, D. L. 1940. The humidity behaviour of the mealworm beetle, *Tenebrio molitor* L. I. The reaction to differences in humidity. *J. Exp. Biol.* 17:286–94

Quetelet, A. 1835. *Sur l'Homme et le Développment de ses Facultés ou Essai de Physique Sociale.* Paris: Bachelier. Imprimeur-Libraire

Ramsay, J. A. 1935. The evaporation of water from the cockroach. *J. Exp. Biol.* 12:373–83

Readio, P. A. 1931. Dormancy in *Reduvius personatus,* L. *Ann. Entomol. Soc. Am.* 24:19–39

Réaumur, R. A. F. de 1735. Observations du thermometre. *Acad. Roy. Sci. Paris.* 1735:737–54

Réaumur, R. A. F. de 1736. *Mémoires pour Servir à l'Histoire des Insectes.* Vol. 2. 514 pp.

Reddingius, J., den Boer, P. J. 1970. Simulation experiments illustrating stabilization of animal numbers by spreading of risk. *Oecologia* 5:240–84

Reibisch, J. 1902. *Über* den Einfluss der Temperatur auf die Entwicklung von Fischeiern. *Wiss. Meeresuntersuch.* 2:213–31

Rivnay, E. 1932. The tropisms of the bed bug *Cimex lectularius. Parasitology* 24:121–36

Roth, L. M., Willis, E. R. 1951. Hygroreceptors in adults of *Tribolium* (Coleoptera, Tenebrionidae). *J. Exp. Zool.* 116:527–70

Salt, R. W. 1936. Studies on the freezing process in insects. *Tech. Bull. Minnesota Agr. Exp. Sta.* No. 116

Salt, R. W. 1946. Moisture relationships of the wheat stem sawfly (*Cephus cinctus* Nort.). II. Some effects of contact moisture. *Sci. Agr.* 26:631–39

Salt, R. W. 1950. Time as a factor in the freezing of undercooled insects. *Can. J. Res. D* 28:285–91

Salt, R. W. 1961. Principles of insect cold-hardiness. *Ann. Rev. Entomol.* 6:55–73

Sanderson, E. D., Peairs, L. M. 1913. The relation of temperature to insect life. *New Hampshire Coll. Agr. Exp. Sta. Tech. Bull.* Vol. 7. 125 pp.

Sang, J. H. 1950. Population growth in *Drosophila* cultures. *Biol. Rev.* 25: 188–219

Shelford, V. E. 1907. Preliminary note on the distribution of the tiger beetles (*Cicindela*) and its relation to plant succession. *Biol. Bull.* 14:9–14

Shelford, V. E. 1911. Physiological Animal Geography. *J. Morphol.* 22:551–618

Shelford, V. E. 1913. *Animal Communities in Temperate America. A Study in Animal Ecology.* Chicago: Univ. Chicago Press

Shelford, V. E. 1927. An experimental

study of the relations of the codling moth to weather and climate. *Bull. Illinois Nat. Hist. Surv.* 16:311–440

Shelford, V. E. 1929. *Laboratory and Field Ecology.* Baltimore: Williams & Wilkins

Shelford, V. E., Deere, E. O. 1913. The reactions of certain animals to gradients of evaporation power of air. A study in experimental ecology. *Biol. Bull.* 25:79–120

Simpson, C. B. 1903. The codling moth. *Bull. US Div. Entomol.* No. 41

Sioli, H. 1937. Thermotaxis und Perzeption von Wärmestrahlen bei der Bettwanze, *Cimex lectularis.. Zool. Jarb. Abg. Algem. Zool. Physiol.* 58: 284–96

Slifer, E. H. 1938. The formation and structure of a special water-absorbing area in the membrane covering the grasshopper egg. *Quart. J. Microbiol. Sci.* 80:437–57

Smith, H. S. 1935. The role of biotic factors in determining population densities. *J. Econ. Entomol.* 28:873–98

Solomon, M. E. 1964. Analysis of processes involved in the natural control of insects. *Advan. Ecol. Res.* 2:1–58

Southwood, T. R. E. 1966. *Ecological Methods with Special Reference to the Study of Insect Populations.* London: Methuen

Southwood, T. R. E., Way, M. J. 1970. Ecological background to pest management. In *Concepts of Pest Management,* ed. R. L. Rabb, F. E. Guthrie, p. 6–29. Raleigh: N. Carolina St. Univ.

Squire, F. A. 1937. A theory of diapause in *Platyedra gossypiella,* Saund. *Trop. Agr.* 14:299–301

St. Amant, J. L. S. 1970. The detection of regulation in animal populations. *Ecology* 51:823–28

Steele, H. V. 1941. Some observations on the embryonic development of *Austroicetes cruciata* Suass. (Acrididae) in the field. *Trans. Roy. Soc. S. Aust.* 63:329–32

Thompson, W. R. 1939. Biological control and the theories of the interaction of populations. *Parasitology* 31:299–388

Udvardy, M. F. D. 1959. Notes on the ecological concepts of habitat, biotope and niche. *Ecology* 40:725–27

Uvarov, B. P. 1931. Insects and climate. *Trans. Entomol. Soc. London* 79: 1–247

Van Heerdt, P. F., Bruyns, M. F. Mörzer, 1960. A. Biocenological investigation in the yellow dune region of Terschelling. *Tidschr. Entomol.* 103:225–75

Varley, G. C. 1947. The natural control of population balance in the knapweed gall-fly (*Urophora jaceana*). *J. Anim. Ecol.* 16:139–87

Varley, G. C., Gradwell, G. R. 1963. The interpretation of insect population changes. *Proc. Ceylon Assoc. Advan. Sci.* 18(D):142–56

Varley, G. C., Gradwell, G. R. 1968. Population models for the winter moth. *Symp. Roy. Entomol. Soc. London* 4:132–42

Varley, G. C. 1970. The concept of energy flow applied to a woodland community. In *Animal Populations in Relation to their Food Resources,* ed. A. Watson, Brit. Ecol. Soc. Symp. No. 10. Oxford: Blackwell

Verhulst, P. F. 1838. Notice sur la loi que la population suit dáns son accroissement. Correspondence mathematique et physique. In *Readings in Ecology.,* ed. J. R. Cafmondy, 10: 113–21. Reprinted in English. Englewood Cliffs, NJ: Prentice Hall. p. 64–66

Volterra, C. 1931. *Leçons sur la Théorìe Mathématique de la Lutte pour la Vie.* Paris: Gauthiers-Villars

Waterhouse, D. F. 1947. The relative importance of live sheep and of carrion as breeding grounds for the Australian sheep blowfly *Lucilia cuprina. Bull. Counc. Sci. Ind. Res. Aust.* No. 217

Watt, K. E. F. 1964. Comments on fluctuations of animal populations and measures of community stability. *Can. Entomol.* 96:1434–42

Watt, K. E. F. 1968. *Ecology and Resource Management.* New York: McGraw-Hill

Way, M. J. 1971. In *Dynamics of Populations,* ed. P. J. den Boer, G. R. Gradwell. Purdue: Wageningen. In press

Way, M. J., Hopkins, B. A. 1950. The influence of photoperiod and temperature on the induction of diapause in *Diatarapia oleracea* L. (Lepidoptera). *J. Exp. Biol.* 27:365–76

Wellington, W. G. 1948. The light reactions of the spruce budworm *Choristoneura fumiferana* Clemens (Lepidoptera:Tortricidae). *Can. Entomol.* 80:56–82

Wellington, W. G. 1949. The effects of temperature and moisture upon the behaviour of the spruce budworm

Choristoneura fumiferana Clemens (Lepidoptera: Tortricidae). I. The relative importance of graded temperatures and rates of evaporation in producing aggregations of larvae. *Sci. Agr.* 29:201–15

Wheeler, W M. 1893. A contribution to insect embryology. *J. Morphol.* 8:1–160

Wigglesworth, V. B. 1931. A curious effect of dessication on the bed bug (*Cimex lectularius*). *Proc. Roy. Entomol. Soc. London* 6:25–26

Wigglesworth, V. B. 1945. Transpiration through the cuticle of insects. *J. Exp. Biol.* 21:97–114

Wigglesworth, V. B. 1948. The insect cuticle. *Biol. Rev.* 23:408–51

Wilkes, A. 1942. The influence of selection on the preferendum of a chalcid (*Microplectron fuscipennis* Zett.) and its significance in the biological control of an insect pest. *Proc. Roy Soc. London, Ser. B* 130:400–15

THE HISTORY OF SERICULTURAL SCIENCE IN RELATION TO INDUSTRY

TADAO YOKOYAMA
Silk Science Research Institute
Tokyo, Japan

Sericulture is usually considered to include all phases of raw silk production, i.e. the mulberry culture, the silkworm egg production, the silkworm rearing for cocoon production, and the filature in which raw silk is made from the cocoon. In this review the development of the science and techniques which have been the technical basis of the silk-producing industry is briefly summarized. The description is limited mostly to the studies on the biology of the silkworm carried out in Japan.

Silkworm eggs and rearing techniques are said to have been first introduced into Japan from Korea in 195 A.D. (21). In the primitive stage of the industry, all phases of the industry were carried out in the same farmhouse. As the industry developed, however, differentiation occurred. Silkworm egg production first became a specialized industry ca 1500 (21). The egg producers not only produced the silkworm eggs but studied the nature of the eggs, and gave technical advice to the silkworm-rearing farmers who used their eggs (21). The egg producers were located in every district where there was cocoon production on a large scale (such prefectures as Fukushima, Nagano, Saitama, etc). Most improved silkworm varieties and new sericultural techniques originated in the districts mentioned above.

By the time of the Meiji Restoration in 1868, there had been appreciable progress in developing silkworm varieties, and rearing methods and in silkworm egg production, but still there were many cases of unscientific charms and personification of the silkworm interwoven in the techniques (21).

Sericulture in Japan began to adopt scientific methods at about the end of the nineteenth century. There had been no scientific study on silkworm diseases before 1873, when Nagaatsu Sasaki (1830–1916) went to the International Exhibition at Vienna as a national delegate; there he learned the inspection technique for pebrine disease and began the foundation of pathological research of the silkworm in Japan (39). The knowledge obtained with the newly introduced scientific methods has been gradually added to the traditional knowledge over a long period; thus the system of modern sericultural techniques was established.

Raw silk production in Japan was 7,300 metric tons in 1905, and 4,600 tons was exported (24% of the total exported in the world). In 1930, raw silk production was 42,500 tons, and 33,595 tons or 70.5% of the world export market was from Japan (30). The largest factor that contributed to such a remarkable development was the improvement and expansion of the summer-autumn rearing of the silkworm. If comparison is made between the cocoon production in 1905 and that in 1930, the annual production increased 3.9 times, the spring production 3.5 times, and the summer-autumn production 5.3 times. The factors which contributed to the development of summer-autumn sericulture are as follows: 1. The establishment of techniques of supplying the necessary amount of silkworm eggs and hatching them at any time through the summer-autumn rearing period from July to September. 2. The breeding of silkworm varieties which provided a stabilized yield of cocoons of excellent quality under the unfavorable environmental conditions of the summer-autumn season. 3. The advance of techniques of controlling the rearing environment, especially for young silkworms, and the development of techniques of mulberry cultivation for production of the required amount of mulberry leaves of good quality at the right time. 4. The practice of effective disinfection for all kinds of major diseases.

These factors were less important when only spring rearing was practiced, for in the spring silkworm eggs hatch in natural conditions and the mulberry leaves of good quality grow naturally. The silkworms grow in good health in the good environmental conditions of the spring. The silkworm varieties for spring rearing reached a high level, thanks to the efforts over a long period of time of specialized egg producers to answer the large demand for spring rearing.

Silkworm Eggs

Most silkworm eggs used in Japan at the beginning of the nineteenth century were of monovoltine variety for spring rearing. A small portion of the eggs was bivoltine and a still smaller portion was polyvoltine. The latter two kinds of eggs were produced in spring and in summer. There was nothing peculiar in the technique of the production of monovoltine eggs in spring. It was necessary only to raise healthy silkworms to get healthy moths which could lay good eggs in large number. The techniques were fixed as a routine system through the experience of a long time. The bivoltine silkworms were reared in spring mostly for obtaining mother moths which laid non-hibernating eggs for summer-autumn rearing. Although there had been various methods to supply the eggs for summer-autumn rearing, none of them were completely reliable (39). In order to discover a reliable technique to obtain the eggs for summer-autumn rearing, intense efforts were made in the studies on voltinism, cold storage, and artificial hatching. As a result, good practical techniques were established to get hatchable eggs in the necessary amount at any time of the year. Some of these techniques are briefly mentioned below.

Relationship between the incubation temperature and voltinism. A method of obtaining the eggs for autumn rearing in ca 1870 was as follows:

The hatching of silkworms which were designated to hatch in early summer instead of in spring was retarded by keeping the eggs in a cold chamber. The nonhibernating eggs laid by the silkworms reared in early summer were used for autumn rearing. The eggs laid by the silkworms of summer rearing often became hibernating eggs which did not hatch for autumn rearing. Jinzaburo Fujioka (1839–1892) in Nagano Prefecture discovered the method of obtaining nonhibernating eggs. He found that eggs taken out of cold storage in early summer produced silkworms which laid a high percentage of nonhibernating eggs. The silkworms grown from the eggs taken out of the cold storage later in summer produced a smaller percentage of nonhibernating (a larger number of hibernating) eggs. As the temperature in early summer was cooler than in the later summer, he came to the idea that a high temperature during incubation of eggs caused the production of hibernating eggs and a low temperature during incubation produced nonhibernating eggs. He repeated his experiments to prove his idea and found in 1875 that eggs which were kept at the temperature of 70°F in the period between their removal from cold storage and hatching, produced silkworms which laid nonhibernating eggs when they became adults (39). Kanji Watanabe (1887–1956) carried out detailed research on the relationship between the temperature and the voltinism in 1924 (51). He showed that the bivoltine strain of silkworms laid nonhibernating eggs if they were incubated at 15°C and hibernating eggs if incubated at 25°C. He also showed that the period which was quite sensitive to temperature during incubation was that just after the appearance of thoracic legs (Watanabe in 1922; 40). Thanks to his research, it was possible to adjust the external conditions that influenced the results of genetical experiments and the genetics of voltinism were elucidated. It was found that the hibernating eggs were resistant to cold storage for long periods, and the silkworms which hatched from high-temperature incubation produced a larger amount of silk than the worms which came from low-temperature incubation (Yokota et al in 1912; 40).

The illumination during incubation and voltinism.—Kohatsu Nagase (1892–1953) incubated silkworm eggs at 15°C in the summer of 1923 under electric light expecting to get nonhibernating eggs. The hatched larvae, however, laid hibernating eggs when they became adults. As it had been proved by Watanabe that the temperature of 15°C during incubation destined the hatched silkworms to lay nonhibernating eggs, the voltinism of the hibernating eggs which Nagase obtained might have been changed by illumination. After several experiments Nagase discovered that the effect on voltinism of incubating silkworm eggs at comparatively low temperatures under illumination was similar to the effect of incubation at high temperatures (1927) (51). In 1930 Makita Kogure (1899–1964) accomplished a detailed study on the relation between illumination and voltinism, showing the threshold of intensity and duration of illumination, as well as the sensitive stage in the development of the embryo (51). Thus Kogure was the first researcher in sericulture who elucidated photoperiodicity in silkworms.

On the other hand, the photoperiodism in *Antheraea pernyi* was studied by Tanaka (46). In *Antheraea pernyi* the influence of photoperiodicity is stronger in the feeding period than in the incubation period and a long-day feeding period produced nonhibernating pupae, while a short-day period produced hibernating ones. During the egg period the effect of illumination was smaller than in the larval period and a long-day period had the tendency to produce hibernating pupae, contrary to that in the long-day feeding period.

Embryonal development.—The embryonal development of the silkworm was studied first by Tichomiroff in 1879 (51) and then by Kamataro Toyama (1867–1918) in 1902 (51). Toyama elucidated the origin of germ layers and the formation of various organs and began the basis for later research in this field. In his research he first described the development of the prothoracic gland which later became the organ of great concern in the study of insect hormones. He named the organ the hypostigmatic cell. The important studies on embryology were: Takami's work in 1942 (25) proving the mosaic nature of the silkworm egg by cauterization methods; Miya's work in 1953 (51) showed the domain of the origin of reproductive cells in the embryo; Goldschmidt's and Katsuki's work in 1928 (51) was on the mechanism of formation of mosaic silkworms; and such work as that by Itikawa in 1952 (51) and Suzuki and Ichimaru in 1955 (51) concerned the embryology of various kinds of mutants.

As the resistance of the embryo to cold storage and its sensitivity to artificial induction of hatching varies according to the stage of development, it is necessary to be able to determine the stage of development of the embryo by external morphology. Standard scales of development shown by morphological characteristics were proposed by Mizuno in 1920 (51), Nakada in 1932 (32), Takami et al in 1960 (51), etc. Artificial hatching, cold storage, and the incubation of eggs have been adjusted, checking the stage of development of the embryo in reference to the standard scale.

Cold storage.—The cold storage of silkworm eggs was first tried between 1848 and 1867. It is not certain who initiated the cold storage of silkworm eggs, but the first large-scale practice was by Kisaburo Maeda (1840–1892) upon the request by Kamenosuke Tsutsui in 1865 (37). Maeda used his own "wind cave" which had been used for the storage of foodstuff. The wind cave is a natural cave found in mountainous regions, in which there is always a current of cold air and low temperatures throughout the year. The cold storage of eggs was improved by the use of the refrigerator in 1902, which made it possible to control the temperature (31). If the eggs were kept at a fixed low temperature, the latest date of hatching, without causing injury to the physiology of eggs, was the beginning of July. Chotaro Yokota (1871–1920) invented the complex cold storage method in 1917 (54) by which the storage could be prolonged to August or even later. This method consisted of two or three phases. In the first phase the eggs were stored at a low temperature of —2.5°C, until late May. The embryo was made to develop to another stage and the eggs were stored for a certain duration of time. Tatsugoro Mizuno (1880–1939) showed in 1920 that the safe duration of storage was dependent upon the stage of the embryo, the temperature of storage, and the characteristics of the strain of silkworm (51). Considering these factors he proposed several ways of storing for a long time without lowering the hatching rate. In the early days the cold storage of eggs was applied only to hibernating eggs to retard the hatching from the spring to summer or even to early autumn. After the improvement of artificial hatching, however, cold storage was widely practiced in combination with artificial hatching; today cold storage of hibernating eggs for a long period is no longer as important as before.

Artificial hatching.—1. Artificial hibernation: if the hibernating eggs are kept at a low temperature of about 5°C for about 2 months and then incubated, they hatch. This process is artificial hibernation and was studied by Yasushi Tsuchiya (1865–1938) in 1901 (40). 2. Artificial hatching with chemical reagents: the use of chemical reagents for hatching of hibernating eggs was discovered in France (26), but the practical technique widely used now was developed in Japan by Kozo Koike (1886–1919) in 1914 (40) and by Takeo Araki (1871–1955) and Eitaro Miura (1887–1964) in 1917 (40). There are two kinds of hatching in this category and both are practiced in combination with cold storage. The first one is the common acid treatment which is used to hatch the eggs within 30 days after egg laying. Treatment with warm hydrochloric acid is applied to eggs kept at 25°C for 20 hr after laying. The hatching occurs about 11 days after the treatment. As the hatching can be postponed as long as 20 days by cold storage of eggs 30 hr after treatment, rearing can be scheduled to begin at any time from 12 to 30 days after egg laying. The second is the acid treatment after chilling, studied by Araki and Miura in 1916 (40), which is used to hatch the eggs 40–50 days after laying. When the eggs are 40 hr old they are stored at a temperature of 5°C for 30–40 days; then, they are treated with warm hydrochloric acid, washed, and incubated.

Several methods of artificial hatching were invented in Europe by the end of the nineteenth century, but the selection among such methods from a practical point of view was not made in Europe, probably because in Europe most sericulture was practiced only in the spring season and artificial hatching was less important than in Japan. Another reason may be that, since the European monovoltine varieties were not so easy to hatch artificially, the reliability of the method was not established.

If the appropriate one of the four methods of hatching (natural hatching, common acid treatment, acid treatment after chilling, and artificial hibernation) is applied, the eggs can be hatched at any time of the year. The development of these hatching methods made it possible to positively obtain hatchable eggs for summer-autumn rearing, especially for autumn rearing, thus removing the largest obstacle in summer-autumn sericulture. It also contributed much to the industry of egg production, to silkworm breeding, and to scientific research.

As for the mechanism of artificial hatching, there is no conclusion so far. Adachi, however, made an observation that the serosa of the acid-treated egg showed heterogeneous structure when the treatment was effective (1).

Parthenogenesis.—Tichomiroff in 1902 and Toyama in 1909 reported that the unfertilized silkworm eggs were developed by various kinds of stimulation (47). Later, Harutaro Sato (1886–1971) in 1924 and 1934 and Eisaku Kawaguchi (1895–1953) in 1937 promoted its study (47). All the silkworms obtained by artificial parthenogenesis were diploid. As for the mechanism of the formation of diploid cells from haploid egg nuclei, the fusion of egg nucleus and polar body (Kawaguchi in 1934) or the union of two egg nuclei (Sato in 1924) was proposed (47). According to Astaurov (5) and H. Hashimoto (unpublished observations) silkworms developed by artificial parthenogenesis with heat treatment held the same genetical characteristics as the mother. When the mother was heterozygous in relation to a certain gene, the silkworms developed by artificial parthenogenesis kept the heterozygosity through many generations. If this holds in all cases, it may be used in the application of heterosis.

Besides the parthenogenesis resulting from the union of two haploid egg nuclei, the merogony in which two male nuclei conjugated and developed into silkworms independent of the egg nucleus was proved by Hashimoto in 1934 (11) and Astaurov in 1937 (47).

In vitro culture of the silkworm embryo.—Takami in 1957 (51) succeeded in culturing silkworm embryos in vitro. He found that the embryo itself was more important than the cytoplasm surrounding the embryo in the initiation of embryonic development.

Hormones and pheromones.—Bounhiol in 1936 (7) and Kin in 1939 (51) independently discovered that the silkworms deprived of the corpora allata in the third or fourth stage entered the pupal stage omitting further larval stages. Muroga in 1940 and Fukuda also in 1940 succeeded in proving the important role played by the prothoracic gland in pupation and later in emergence (51). Hasegawa and Fukuda independently found in 1951 that the hormone which controlled the diapause of eggs was secreted by the subesophageal ganglion (51). In 1957 Kobayashi (51) stimulated the prothoracic gland to cause the development of the brainless "dauerpupa" by the transplantation of the brain, suggesting the existence of a brain hormone, which he and his collaborators obtained later from the silkworm brain. Ito et al (20) showed in their experiments that the feeding of a molting hormone of plant origin accelerated the maturation of the silkworm. This experiment is an example of the application of hormone research to practical rearing of silkworms.

As for the chemical nature of the insect hormone, Butenandt and Karlson (51) showed the empirical formula of ecdysone $C_{18}H_{30}O_4$ in 1954, and the chemical structure was shown by Huber & Hoppe in 1965 (17). Williams et al (49) reported that the effective component of juvenile hormone was methyl-9,10-epoxy-hexadecanoate, and Röller et al (38) determined that it was methyl-10-epoxy-7-ethyl-3,11-dimethyl-2,6-tridecadienoate. There are two active components in the brain hormone, one is steroid (Kobayashi et al; 24) and the other is proteinous (Ichikawa and Ishizaki in 1961, 18; and Kobayashi in 1963, 23). Hasegawa extracted the effective substance which changed nonhibernating eggs into hibernating ones by injection into the pupa. The substance was a lipid in nature, 15 μg of which was able to change the voltinism of the egg. Bowers et al (8) extracted from balsam fir a substance effective as juvenile hormone for *Pyrrhocoris,* a methyl ester of todomatuic acid, and named it juvabione. Nakanishi et al (34) obtained ponasterone from *Podocarpus nakaii,* and Takemoto et al (44) iso-inokosterone and inokosterone from *Archyranthes fauriei.* These substances of plant origin were proved to have the same function as ecdysone.

Besides the hormones mentioned above, there were studies on so-called gene hormones. The enzymes determined by genes exude from various tissues and have physiological effects. Kikkawa in 1937 (47) could change the white egg of w_1 strain to a black egg by transplanting the ovary of a w^+ strain female into the w_1 female. In 1945 this study was advanced further to elucidate the relation between the egg color and eye color which was determined by various substances at certain phases in the metabolic cycle of tryptophane (51).

Pheromone is the word first used by Karlson and Butenandt in 1959 (6) to designate a substance secreted by an animal to influence the behavior of other animals. In 1941 Butenandt (51) obtained a sex attractant from a female silkworm moth and later, in 1961, the constitutional formula was determined and named bombycole (6). This study has been expanded to many kinds of insects and is introducing a new field of study in insect pest control (6).

Breeding of Silkworms

Silkworm varieties in Japan before the Meiji Restoration (1868).—As stated before, silkworm egg production became a specialized industry as early as 1500 (21). The egg producers made great efforts to get superior varieties of their own which would be a useful tool to expand their business. The quality of the varieties at about 1880 was not much inferior to that in modern stock races. For instance, the cocoon of Shiroaya variety produced by Kunitaro Sato (1882–1900) and exhibited in the Kanagawa and Three Prefecture Competitive Exhibition of Raw Silk, Cocoon, and Textiles held at Hachioji in 1881 had the cocoon fiber length of 950 m, silk amount 29.3 cg, and the silk ratio 19% (28).

Superior varieties of this age were Matamukashi bred by Hikojiro Ito in Fukushima Prefecture ca 1740, Akajuku bred by Tomonobu Sato in

Fukushima Prefecture ca 1780, Sekaiichi bred by Saheiji Nakamura in Fukushima Prefecture ca 1790, Koishimaru bred by Genemon Odanaka in Nagano Prefecture ca 1790, Hakuryu bred by the Kubota family in Nagano Prefecture ca 1790, Aojuku bred by Isaburo Tamura in Fukushima Prefecture ca 1852, and Shiroaya bred by Kunitaro Sato in Gumma Prefecture ca 1870, etc (13). As cocoon production was concentrated in the spring season in those days, the aforementioned improved varieties all belonged to monovoltine ones fit for spring rearing. Bivoltine varieties were also reared during this period in several districts; well-known bivoltine varieties were Omiko bred in Shiga Prefecture ca 1755, Okusa bred by Uhachi Sasaki in Nagano Prefecture ca 1790, Tanegashima bred in Hyogo Prefecture ca 1810, and Hakuryu, the origin of which was uncertain (13). A variety of this period which must be mentioned was Kakeawase, a kind of hybrid variety, which was first created by Tsunane (Jokatsu) Fujimoto (1815–1890) and Jukichi Nakayama in Nagano Prefecture ca 1845 (39). The original egg producers prepared Kakeawase eggs by crossing bivoltine females with monovoltine males. Then they were distributed to commercial egg producers. The commercial egg producers reared F_1 hybrids and produced F_2 eggs for farmers' rearing. Kakeawase was said to be resistant and easy to rear. As the principles of genetics and heterosis were not understood by egg producers, the eggs showed too wide a variation to be called one fixed variety. Though it was not used very widely at this time, it was a forerunner of the hybrid silkworms which completely displaced the old pure races (see below).

Practical use of F_1 hybrids.—Toyama in 1905 (54) reported that an F_1 hybrid between Japanese and Siamese varieties showed a 30% increase in cocoon yield and 3 times the production in silk ratio (compared with aboriginal Siamese variety). He insisted on the advantage of F_1 hybrid based on his study in genetics in 1906 (3). His opinion was accepted by the government and the rearing of hybrid silkworms was officially encouraged.

Stock races for preparing hybrid eggs, 6 Japanese, 6 Chinese, and 6 European races, showing 12 model types of crossing among them, were distributed from the national sericultural experiment station to prefectural stations. Since the F_1 hybrid had a short feeding period, high survival rate, long cocoon fiber, and a large amount of silk production, the rearing of hybrid silkworms increased rapidly and in 1930, 16 years from the first distribution, the hybrid eggs amounted to 99% of the total commercial eggs (3).

Shigetane Ishiwata (1868–1941) discovered the external characteristics of the female larva in 1904 (26), which contributed much to the production of hybrid eggs.

Continued improvement has produced strong and highly efficient silkworm varieties. Now the rearing of a highly productive variety in the summer-autumn season, which had been impossible before the hybrid silkworms were used (28), is safely carried out. As an example, to show the improvement of the silkworm variety, the change in the amount of silk production by a single worm from the beginning of the nineteenth century to the present time is shown in Table 1.

From Table 1 it can be seen that the improvement of the silkworm varieties, as revealed in the increase of silk weight of a cocoon, was slow until the adoption of the hybrid silkworm in 1914, but it was very remarkable after that, especially in the varieties for summer-autumn rearing. The same will be seen in other economically important characteristics.

One of the recent remarkable advances in silkworm breeding is the technique of distinguishing the sex of the silkworm by either egg color, larval pattern, or cocoon color. It is necessary to separate males and females to produce hybrid eggs. Sex distinguishing by Ishiwata's method frequently causes error. An easier and safer method was discovered by Tazima in 1941 (51). The principle is that the femaleness of the silkworm is determined by the presence of the W chromosome (Hashimoto in 1933, 47). Tazima succeeded in attaching a visible marker gene on the W chromosome. Later, in 1951, Tazima and his collaborators bred a strain of which the female eggs were dark-colored and the male ones white (47). Recently, Kimura, Harada & Aoki (22) got a strain of which the female cocoon was yellow and the male white.

Silkworm genetics.—The first report on silkworm genetics was made by Toyama in 1906 (47). Summarizing the research which has been carried out since then, it can be seen that most research has involved the characteristics which usually attracted the attention of silkworm-rearing people, consequently the characteristics in the egg, larva, and cocoon have dominated the research as seen from Table 2.

TABLE 1. Improvement in silk production by a single silkworm 1802–1929(39); 1939–1949(41); 1959–1969(29)

Period	Weight of silk (cg) of one cocoon		Index	
	Spring	Summer–autumn	Spring	Summer–autumn
1802–1803	16.1			
1830–1840	22.0			
1864–1867	22.1			
1877	20.6			
1887	22.6	12.5	100	100
1906	19.4	11.1	86	89
1913–1916	24.1[a]	—	107	—
1916–1919	—	19.5[a]	—	156
1929	32.8	28.0	144	223
1939	42.8	32.2	190	257
1949	46.2	40.2	204	321
1959	52.8	40.2	233	321
1969	55.1	42.4	269	337

[a] Denotes the beginning of hybrid rearing.

TABLE 2. Hereditary traits in *Bombyx mori* (50)

Characters in—	Number of traits
Egg	48
Larva	127
Pupa	6
Cocoon	16
Moth	14
Total	211

In silkworm genetics the following fields are regarded as advanced: sex determination, development, voltinism, moltinism, cocoon color, and egg color. From a practical point of view, the resistance to various diseases is the most important item. It was known that there was a genetical factor in the resistance to some kinds of diseases, as reported by Funada (9) on the genetics of the resistance of various silkworm strains to infectious flacherie.

In the early stage of silkworm genetics most studies were concerned with morphological characteristics. Since Kikkawa's research on the gene hormone came out in 1940 (47), biochemical items have been pursued by many researchers. These include Harizuka's study in 1960 (53) on the genetics of carotenoid substances in the cocoon, Tsujita's study in 1963 (53) on the pteridine metabolism, and Yoshitake's study in 1968 (58) on the classification of silkworm strains by genetical behavior of the isozyme.

A new field of silkworm genetics is artificial mutation. In 1927 Tanaka and Hashimoto (47), applying high and low temperature, centrifugal force, X-rays, etc to pupae and eggs obtained exceptional phenotypes for sex-linked inheritance; they also found 52 mosaics and gynandromorphs and 20 monsters among 12,185 silkworms derived from treated eggs or pupae. Later, in 1934, Hashimoto induced dispermic androgenesis with high temperature and in 1936 Kawaguchi got polyploids by centrifugation (47). Besides these, there were studies by Kogure, Aruga, Takasaki, Chikushi, Tsujita, Hirobe, etc (47). Tazima's study by which he obtained a strain of sex-limited pattern mentioned before attracted the attention of both the scientific and practical circles. In another interesting work, Tazima obtained, by X-ray irradiation, a silkworm strain which showed abnormality in feeding behavior in 1952 (50).

Silkworm Rearing

Ecological research.—There were various opinions about the optimum temperature of silkworm rearing. Yahei Tazima (1822–1898) expressed an opinion in favor of cool rearing (42), Tomehei Taguchi (1771–1818) recommended warm rearing in 1796 (39), and Chogoro Takayama (1830–1885) published a book on his cool-warm mixed rearing in 1868 (39). The sericulture thermometer was first made in Japan in 1842 (21) by Zenemon Nakamura (1806–1880) and was conveniently used to find the relation between the duration of the rearing period and temperature. Nakamura called the rearing at 80°F rapid rearing, rearing at 75°F medium rearing, and rearing at 70°F slow rearing. Air conditioning apparatus, from the principle of W. H. Carrier, was introduced from the US to the Sericultural Experiment Station, Tokyo, and was modified to be adjusted for silkworm rearing (Yanagimachi in 1925; 54). This apparatus has been used in several sericultural experiment stations for the physiological and pathological studies of silkworm. According to Kogure in 1940 (55), the optimum temperature for silkworm is 30°C from the first to the third instars, 25°C in the fourth, and 20–25°C in the fifth. Thus rearing at these regulated temperatures shortens the duration of the larval stage, decreases the mortality, and increases the weight of the cocoon and silk production. Generally speaking, the rearing at high temperatures in the younger stage and at lower temperatures in the older stage brings about high productivity of silkworm. The border between younger and older stage was conventionally considered to be between the third and fourth instars. Ueda et al, however, came to the conclusion in 1962 that the physiology of the fourth instar silkworm was rather similar to that of the third instar (51).

In 1928 Suemi Matsumura (1889–1958) tried to put a physiological basis to the optimum temperature of rearing by determining the optimum temperature for the function of physiologically important enzymes (51). As for amylase, the function was more active as the temperature rose between 20 and 60°C. As for tyrosinase, the function became active as the temperature rose until 24°C, above that it declined. This explains the phenomenon that high temperature incubation brings about the light-colored ant. Carotenoid pigments in the cocoon layer was greatest in amount when the silkworm was reared at 23°C and decreased at either 17.5 or 30°C (Harizuka et al in 1960; 53).

The younger silkworms are resistant to high humidity, and in high humidity the evaporation from the food leaf is small, thus it enables the silkworm to eat more leaves. The older silkworms, however, are not so resistant, and show greater mortality in high humidity, especially when combined with high temperatures (Matsumura et al in 1928; 51). High humidity in the spinning period has a remarkably bad effect on cocoon quality (Matsumura in 1949; 55). The bad effect of an environment of 30°C and 90% R.H. was partly improved if there was an air current of 0.2–0.3 m/sec (Okawa in 1925; 40).

Wakana in 1934 (55) showed that CO_2 gas above 1% in the rearing room had a bad effect on the physiology of silkworm. Suzuki et al showed in 1962 that the relation between the concentration of CO_2 gas in the rearing room and the mortality of the silkworm was expressed in a straight line (51). Ammonia gas produced in the rearing bed was shown to be injurious for silkworm physiology by Arima in 1952 (55).

The effect of illumination on the silkworm in the feeding period is slight, but the silkworm which is ready to spin a cocoon is positively phototropic as long as the intensity remains under 15 lux (Miyagawa et al in 1954; 51). Mature silkworms gather in the illumination of 550 mμ (Yagi in 1923; 55). Mature silkworms show distinct negative geotropism (Yokoyma et al in 1942; 51).

Nutrition.—As sericulture is an industry to change the mulberry leaf into silk by rearing silkworms, nutrition is the most important field of silkworm physiology. The scientific study of the nutrition of the silkworm was first carried out by Pelligot in France in 1852 (12). He made a study on organic and inorganic elements in various products of sericulture and explained the general process of the nutrition of the silkworm. As a visiting professor in the College of Agriculture of Komaba, Tokyo in 1884, Kellner (12) studied the composition of the food leaf, mode of digestion, and growth of body, as well as the chemical change during metamorphosis. Following Kawashima in 1902 and 1905 and Kawase in 1914, Hiratsuka's report came out in 1917 and has been looked upon as a standard work until now (12). In the first part of this report the detailed processes of ingestion and digestion of food, building of tissues, and production of silk and exuvia, as well as the consumption of matter for the maintenance of life, were revealed. In the second part there were the calorimetric studies on the income, storage, and consumption of energy. In Table 3 a summary of the first part is shown.

Hiratsuka found the difference in nutrition in the male and female. The quantitative difference is seen in Table 3. As for the qualitative difference, the female utilized comparatively more fat and less carbohydrate than the male in larval, pupal, and adult stages. More recently, the data concerning nutrition were re-examined using improved modern silkworm varieties as experimental material by Matsumura & Takeuchi (27) who showed that the increase in the amount of digested food was 55% and that of digestibility 4.7%. It was found that the quality of the mulberry leaf had a great influence

TABLE 3. Nutrition of the silkworm (12)

Stage	Dry matter ingested (g/1000 worms)	Dry matter digested (g/1000 worms)	Digestibility (%)
1st stage	2.66	1.245	46.89
2nd stage	14.46	6.43	44.47
3rd stage	81.05	30.47	37.59
4th stage	373.07	144.34	38.69
5th male	2401.73	866.44	36.08
5th female	2878.67	1086.06	37.73
5th male & female aver.	2640.20	976.25	36.98
All stages	3111.44	1158.74	40.92

on the healthiness and consequently on the survival rate of the silkworm by Watanabe in 1944 (51) and on the amount of silk production by Kanezaki in 1928 (40), and therefore many trials were made to improve the quality of mulberry leaves by the supplementation with, or taking away of various substances such as calcium salts (Hatano et al in 1915; 45), carbohydrates (Hiratsuka in 1925; 45), glycine with cane suger (Nakajima, 33), vitamin C (Gamo, 10), vitamin B_2 (Koyanagi in 1941; 51), choline (Yoshida et al, 57), soybean casein (Ito et al, 20), etc. As it was not expected to attain complete analysis of nutrition by the supplementation method using the natural leaf as a basis, experiments to find an artificial diet for silkworm were begun in 1954 by the research team lead by Sasaki (52). In 1960 Fukuda (51) succeeded in rearing the silkworm through all the stages with an artificial diet. Watanabe in 1958 (51) found that α, β-hexenal and β,γ-hexenol in mulberry leaves were attractants for silkworm. Hamamura and his collaborators added to them citral, linalyl acetate, linalol, terpenyl acetate, and β-sitosterol as stimulating substances in experiments in 1959 and 1961 (51). Ito elucidated the nutritional requirement of silkworms in a series of works beginning in 1960 (19). The improvement of the artificial diet reached such a level that the growth of the silkworm reared with it is the same as, or, in some instances, better than that with mulberry leaves, but the rate of silk production with an artificial diet is inferior to that with the natural food leaf (16). Recent development in this field is directed towards the elucidation of the mechanism of each substance in nutrition (14) and the balance among the component substances (15). Another approach to this problem is biological study to find silkworm strains which are fit for rearing with the artificial diet (56), or to determine the specialized diet in accordance with the change in the feeding behavior of the silkworm as it develops (35).

Silk secretion.—In former times, the sericultural technicians tried and, to some extent, succeeded in improving the quality or increasing the quantity of silk by breeding superior varieties. As seen from Table 1, in modern varieties the silk production by a single worm is 260% of that in 1887 in the spring variety and 337% in the summer-autumn variety. According to Hiratsuka in 1917 (12) one gram of silk in dry weight was produced for each 12.5 g in dry weight of mulberry leaves ingested by the male and 13.4 g by the female while in a modern variety 11.7 g were consumed by the male and 11.8 g by the female (16). Therefore, there was improvement in utilization of the mulberry leaves for silk production by about 6% in males and 12% in females, besides the absolute increase of silk production by a single worm. The silk producing efficiency is higher in the male. Hiratsuka (12) showed that 69.69% of the ingested nitrogen was changed into silk in the male and 61.03% in the female. Naturally an idea arose to rear male silkworms while discarding the females. Tazima et al in 1951 (50) bred a strain of silkworm in which female eggs were black and male eggs white, thus enabling the separation of both sexes in the egg stage.

Scientific studies on the various phases of silk secretion began at the end of the nineteenth century by Blanc in 1887–1888 and by Toyama in 1896 (45, 48), and the general morphology and embryonic development of silk glands were studied. The cocoon fiber is constructed with two components, the fibroin in the core and the sericin surrounding it. There were several opinions on the origin of the two substances; some considered that only fibroin was secreted at the posterior part of the silk gland and the peripheral portion of fibroin changed into sericin (Blanc in 1887; 45); others held that the two substances were secreted from different portions of the silk gland (Yamanouchi in 1922; 45). Jiro Machida (1885–1964) carried out an experiment which supported the different origin theory in 1926 (51). He transplanted separated sections of silk gland into the silkworm body and showed that fibroin was produced in the section of the posterior part and sericin in the section of the middle part of the silk gland. Besides the silk substance, coloring pigments of the cocoon are secreted in the middle part of the silk gland. The physiological-genetical analysis of the carotenoids in the cocoon was performed by Harizuka et al in 1960 (53), and on the flavonoid cocoon color by Fujimoto in 1958 (53). In 1955 Shimizu (51) found that cellulose was secreted in the middle part of the silk gland, causing a kind of silk defect. In the study of silk production by silkworms, radioactive tracers were used to follow the course of the metabolism (Fukuda in 1956; 53). The electron microscope was introduced for the study of silk secretion by Akai in 1963 (2).

Silkworm Diseases

Though silkworm diseases such as muscardine or silkworm parasites such as the kyoso fly were recognized in the early nineteenth century in Japan, there were no special measures to prevent them except trying to rear the worms in as good an environment as possible. To produce healthy silkworms there were theories about the rearing temperature as mentioned before.

The first experimental study on silkworm diseases was begun in 1869 (39) by N. Sasaki to clarify the life history and the course of parasitism of a parasitic fly, *Sturmia sericariae*. The results were reported in the fifth International Congress of Sericulture at Milan in 1876 (28). His son Chujiro Sasaki (1857–1933) furthered the study in 1883 (39). This parasitic fly lays eggs on the mulberry leaf. The egg is swallowed by the silkworm and the hatched larva of the fly lives as a parasite and kills the host silkworm. The fly seldom lays eggs on mulberry leaves growing on the banks of rivers where there is always a breeze. The mulberry leaf grown in an area where fly eggs were scarce was called buguwa. The silkworm egg producers in the past knew how to use buguwa to decrease the damage of the parasitic fly, though the true relationship between the fly and buguwa was not known (42).

Pebrine.—N. Sasaki introduced the method of pebrine inspection into Japan from Austria in 1873 (39). He made comparative studies on the rate of infection in the Japanese silkworm eggs and European ones in the newly established laboratory at Naitoshinjuku, Tokyo, in 1874 (39). The Japanese government established the Silkworm Disease Experiment Station at Uchiyamashitacho, Tokyo, in 1884 (3) on the recommendation of Sasaki, and in the next year issued the Silkworm Egg Inspection Regulation. As a result of these administrative and technical activities the pebrine infection rate in reproductive eggs in Japan decreased from 23.4% in 1898 to 0.02% in 1958 (36). In 1949 Oshima (51) improved the method of inspection by mathematical study.

Muscardine.—Muscardine of the silkworm was known as early as the sixteenth century. The silkworm bodies attacked by muscardine fungus were sold as medicine for human paralysis (39). There were many scientific works on muscardine from the late nineteenth century until now. Most works, however, had been concerned with the classification of pathogenic fungi and the symptoms of the diseases until Aoki's series of papers came out between 1939 and 1955. According to Aoki (4) there are more than 20 species of fungi which attack the silkworm. He elucidated the characteristic symptoms, pathogenic activity, and the incubation period of each important pathogen, as well as the relationship of the muscardine disease of wild insects and that of the silkworm. He showed that the conidia of muscardine first stuck to the skin of the silkworm and germinated. Then, the germinated tubes penetrated through the skin into the body. Therefore, disinfection must be done prior to the penetration of the tubes. In general, young worms were less resistant to muscardine. The important point of disinfection was to apply it to healthy silkworms, especially to young worms within a definite interval after hatching, if the environment of the rearing room was suspected to be contaminated with pathogens. This method was proved to be effective in several cases (4).

Flacherie.—In older books descriptions of silkworm diseases which show symptoms common to present day flacherie are found. Flacherie has been the most notorious of all diseases in Japan, causing about 70% of all disease damage in cocoon production (4). As the summer-autumn rearing of the silkworm increased, the damage by flacherie became more and more conspicuous. To stabilize the cocoon yield meant practically to control flacherie. The first step taken to control flacherie was to carry out bacteriological research (40, 54). Omori (1897), Ishiwata (1901), Honda (1920), Iwafuchi (1922), Chigasaki (1930), Katsumata (1931), and many other researchers reported the results of their experiments. However, no bacteria were found which definitively caused flacherie, except *Bacillus thuringiensis* var. *sotto*. This bacterium produces a potent toxin which has been utilized in biological control of insect pests in recent years (4).

The next step in solving the flacherie problem began about 1925. As mentioned before, air conditioned rearing rooms were constructed at national and several prefectural experiment stations mainly for the study of the relation between the environmental conditions and the occurrence of flacherie. As the result of experiments, it was found that silkworms were frequently attacked by flacherie if they were reared in high temperatures (above 28°C) and in high humidity (above 90%) (Matsumura in 1931; 40). Okawa in 1925 and 1926 (40) showed that the bad influence of high temperatures and high humidity which usually brought about flacherie was reduced if there was an air current of about 0.2–0.3 m/sec. Many experiments, including those mentioned above, showed that there was an intimate relationship between the occurrence of flacherie and the environmental conditions, but it was also found that the environmental conditions were not the original cause of flacherie.

The research on flacherie proceeded to the third step which was to study the relation of the nutrition of the silkworm and the occurrence of flacherie. In the spring rearing the damage of flacherie was far smaller than in the summer-autumn rearing. This fact was thought at first to be attributable to the bad environmental conditions in summer-autumn rearing. After finding, however, that the environmental conditions in themselves could not always cause flacherie, the deterioration of the quality of the food leaf was considered to induce flacherie. Several experiments by Watanabe 1944 (51) supported this view. The efforts to elucidate the physiology of nutrition and to improve the quality of the food leaf developed into the study of an artificial diet.

In the fourth step of the study of flacherie emphasis was put on the study of viral pathogens. Ishimori (1890–1961) found cytoplasmic polyhedrosis in 1934 (51). At first it was not thought to be an important disease in Japan. However, in 1955 Oba et al (51) and Kurisu (51) found that the disease occurred widely in Japan and was an important cause of flacherie. A few years later, Yamazaki et al in 1961 (51) proved the prevalence of infectious flacherie, another viral disease. Disinfection became the most effective measure to prevent flacherie from this period. Many experiments were accumulated to prove the effectiveness of disinfection (43).

Through all stages of studies on flacherie control, efforts to breed resistant varieties of silkworms were continued.

In practicing all measures combined, the damage of flacherie decreased rapidly in recent years in Japan. The average loss in cocoon yield owing to flacherie was 7430 tons, which was 6.6% of the total cocoon production, between 1955–1957, and 3302 tons, or 2.8% of the total yield, between 1967–1969 (30).

LITERATURE CITED

1. Adachi, M. 1947. On a mechanical effect of hydrochloric acid on the silkworm eggs in the artificial hatching. *Proc. Tohoku Branch Congr. Jap. Assoc. Sericult. Sci.* 1:34–36
2. Akai, H. 1963. Electronmicroscopical observation on the secretion of fibroin in the silkgland of the silkworm, *Bombyx mori. J. Sericult. Sci. Jap.* 33:22–24
3. Akashi, H. 1939. *History of the*

Advancement of the Modern Raw Silk Industry. Tokyo: Meibundo

4. Aoki, K. 1969. *Sericultural Pathology and Pathological Entomology*. Tokyo: Nippon Sanshi Kohokyokai
5. Astaurov, B. L. 1959. The present state of problems of artificial parthenogenesis in the silkworm. Diploid and polyploid parthenogenesis in *Bombyx mori* L., *B. mandarina* Moor and in their hybrids. *Atti Simp. Int. Biol. Sperimentale, Celebrazione Spallanzaniana*. 11:1–20
6. Beroza, M. 1970. *Chemicals Controlling Insect Behavior*. New York: Academic
7. Bounhiol, J. J. 1936. Metamorphose après ablation des corpora allata chez le ver à soie (*Bombyx mori*). *C. R. Acad. Sci.* 203:388
8. Bowers, W. S., Fales, H. M., Thompson, M. J., Uebel, E. C. 1966. Juvenile hormone: identification of an active compound from balsam fir. *Science* 154: 1020–22
9. Funada, T. 1968. Genetic resistance of the silkworm, *Bombyx mori* L., to an infection of a flacherie virus. *J. Sericult. Sci. Jap.* 37:281–87
10. Gamo, T. 1941. The relation between the development of the silkworm, *Bombyx mori* L. and vitamin C. I. On the variation of vitamin C content in the mulberry leaf. *J. Sericult. Sci. Jap.* 12:65–78
11. Hashimoto, H. 1934. Formation of an individual by the union of two sperm nuclei in the silk worm. *Bull. Imp. Sericult. Exp. Sta., Tokyo* 8:455–64
12. Hiratsuka, E. 1917. Researches on the nutrition of the silkworm. *Bull. Sericult. Exp. Sta.* 2:353–412
13. Hiratsuka, E. 1969. *Historical review of the silkworm varieties in Japan*. Silk Sci. Inst., Tokyo
14. Horie, Y. 1967. Dehydrogenase in carbohydrate metabolism in larvae of the silkworm, *Bombyx mori* L. *J. Insect Physiol.* 13: 1163–75
15. Horie, Y., Inokuchi, T., Watanabe, K. 1970. Effects of dietary amino acids balances on growth and cocoon quality in the silkworm, *Bombyx mori* L. *Bull. Sericult. Exp. Sta.* 24:345–65
16. Horie, Y., Watanabe, K. 1969. Energy requirement by larvae of the silkworm, *Bombyx mori* L., reared on mulberry leaves and artificial diet. *J. Sericult. Sci. Jap.* 38:377–85
17. Huber, R., Hoppe, W. 1965. Zur Chemie des Ecdysons. VII. Die Kristall- und Molekülstrukturanalyse des Insektenverpuppungshormons Ecdyson mit der automatisierten Faltmolekülmethode. *Chem. Ber.* 98:2403–24
18. Ichikawa, M., Ishizaki, H. 1961. Brain hormone of the silkworm, *Bombyx mori*. *Nature* 191:933–34
19. Ito, T. 1967. Nutritional requirements of the silkworm, *Bombyx mori* L. *Proc. Jap. Acad.* 43:57–61
20. Ito, T., Arai, N. 1961. Nutrition of the silkworm, *Bombyx mori*. V. Effects of supplementation with single nutrient to dried mulberry leaf powder as a basal component of diet. *Bull. Sericult. Exp. Sta.* 17:269–93
21. Japan Academy. 1960. *History of Sericultural Techniques before Meiji Era*. Tokyo: Nippon Gakujutsu Shinkokai
22. Kimura, K., Harada, C., Aoki, H. 1971. Studies on the W-translocation of yellow blood gene in the silkworm. *Jap. J. Breeding* 21:199–203
23. Kobayashi, M. 1963. The chemistry and physiology of the brain hormone. *Proc. Int. Congr. Zool., 16th, Washington, DC.* 4:226–33
24. Kobayashi, M., Kirimura, J., Saito, M. 1962. Crystalization of the brain hormone of an insect. *Nature* 195:515–16
25. Kume, M., Dan, K. 1957. *Invertebrate Embryology*. Tokyo: Baifukan
26. Maillot, E., Lambert, F. 1906. *Traité sur le Ver à Soie du Mûrier et sur Mûrier*. Montpellier: Coulet et Fils
27. Matsumura, S., Takeuchi, T. 1950. Experiments on the nutrition of silkworm. I. The nutrition of the modern hybrid races. *J. Sericult. Sci. Jap.* 19:192–200
28. Ministry of Agriculture and Forestry. 1955. *Nomutenmatsu*. (Tokyo) 3:1017
29. Ministry of Agriculture and For-

estry (Raw Silk Bureau, Japan). 1970. *Gijutsu Shiryo,* No. 46 (1959), 70 (1969), 71 (1970)
30. Ministry of Agriculture and Forestry (Raw Silk Bureau, Japan). 1939, 1966, 1970. *Sanshigyo Yoran* (Sericulture Statistics)
31. Mita, I. 1909. On the refrigerator. *Sangyo Shimpo* 197:23–26
32. Nakada, T. 1932. The resistance of silkworm embryo to abnormal high temperature and dryness and the external morphology of the embryo. *Bull. Fukuoka Sericult. Exp. Sta.* 1:1–26
33. Nakajima, S. 1939. On the symptom of flacherie in the silkworm caused by feeding with tender leaves of mulberry and a countermeasure by supplementation with some chemical substances. *Bull. Nagano Sericult. Exp. Sta.* 45:1–102
34. Nakanishi, K., Koreeda, M., Sasaki, S., Chang, M. L., Hsu, H. Y. 1966. The structure of ponasterone A, an insect moulting hormone from the leaves of *Podocarpus nakaii* Hay. *Chem. Comm.* 24:915–17
35. Niimura, M., Kirimura, J., Katsuya, N. 1969. On the change of the feeding habit in the silkworm. *Proc. Congr. Jap. Assoc. Sericult. Sci. Kanto Branch, 20th, Chiba-Togane,* p. 84
36. Nippon Nogyo Kenkyusho. 1969. *History of Post-war Development in Sericultural Techniques,* Vol. 7. Tokyo: Nippon Nogyo Kenkyusho
37. Nippon Norin Gyogyo Shinkokai. 1968. *Centenary of Meiji Publication on Meritorious People in Agriculture, Forestry and Fishery.* Tokyo: Nippon Norin Gyogyo Shinkokai
38. Röller, H., Bjerke, J. S., McShan, W. H. 1965. The juvenile hormone 1. Methods of purification and isolation. *J. Insect Physiol.* 11:1185–97
39. Sericultural Association of Japan (Dainihon Sanshikai). 1936. *History of Silk Industry in Japan,* Vol. 3
40. Sericultural Association of Japan (Dainihon Sanshikai). 1936. *History of Silk Industry in Japan,* Vol. 5
41. Sericultural Experiment Station. 1939, 1949. *Sanshishikenjo Shiryo,* 2 (1939), 11–12 (1949)
42. Tajima, Y. 1872. *Yosan Shinron.* Tokyo: Tokyo Shoin
43. Takasu, T., Takahashi, S. 1961. Survey of the sericultural farmers whose crop frequently fails. *Acta Sericol.* 37:73–95
44. Takemoto, T., Ogawa, T., Nishimoto, K. 1967. Isolation of an insect metamorphosing hormone from *Archyrantes fauriei* L. & V. *Jap. J. Pharmacol.* 8:325–27
45. Tanaka, Y. 1943. *Sangaku* (Sericology). Tokyo: Kobunsha
46. Tanaka, Y. 1951. Studies on hibernation with special reference to photoperiodicity and breeding of Chinese tussar silkworm I–V. *J. Sericult. Sci. Jap.* 19:358–71; 429–46; 580–90 (1950); 20:1–9; 132–38 (1951)
47. Tanaka, Y. 1952. *Silkworm Genetics Illustrated* Tokyo: Shokabo
48. Toyama, K., Ishiwata, S. 1896. *Experimental Silkworm Anatomy.* Tokyo: Fuzanbo
49. Williams, C. M., Law, J. H. 1965. The juvenile hormone–IV. Its extraction, assay and purification. *J. Insect Physiol.* 11:569–80
50. Yokoyama, T. 1959. *Silkworm Genetics Illustrated.* Tokyo: Nippon Gakujutsu Shinkokai
51. Yokoyama, T. 1963. Sericulture. *Ann. Rev. Entomol.* 8:287–306
52. Yokoyama, T. 1964. Artificial diet of the silkworm and future research on it. *Sanshi Kagaku To Gijutsu* 1964(5):10–19
53. Yokoyama, T. 1965. Physiology and pathology of the silkworm. Science in Japan. *AAAS Publication 79, Washington, DC*
54. Yokoyama, T. 1964, 1965, 1966. History of the development of sericultural techniques. *Sanshi Kagaku To Gijutsu* 3:11–12; 4: 1–12; 5:1–12
55. Yokoyama, T. 1968. Lectures on silkworm physiology. *Sanshi Kagaku To Gijutsu* 7(2):24–27; 7(4):18–21
56. Yokoyama, T. 1970. Occurrence of silkworms of abnormal feeding habit among silkworm varieties and their reaction to artificial diet. *Bull. Silk Sci. Res. Inst.* 18:16–23
57. Yoshida, T., Miyauchi, K. 1960. On the influence of choline and its derivatives in the mulberry leaves on the physiology of the silkworm. *Bull. Sericult. Exp. Sta.* 16:58–81
58. Yoshitake, N. 1968. Phylogenetic aspects on the origin of Japanese race of the silkworm, *Bombyx mori* L. *J. Sericult. Sci. Jap.* 37: 83–87

INSECT PATHOLOGY

J. W. MacBain Cameron
Insect Pathology Research Institute[1]
Canadian Forestry Service
Sault Ste. Marie, Ontario, Canada

According to the dictionary, a history of a subject is a systematic written account of events connected with the development of that subject. Unfortunately insect pathology did not develop systematically, and writing its history is perforce an effort to select those particular events, often seemingly unrelated, that taken together have led it to its current status as a specific section or discipline in the science of entomology.

Diseases of insects, like those of other animals and of man himself, were recognized and the symptoms described long before the cause was known. Pliny, in his *Historia Naturalis* written in the first century A. D., gave a very good description of a disease of honey bees that we call foulbrood and which we now know is caused by the bacterium *Bacillus alvei*. Diseases of the silkworm also were recognized early, and Vida mentioned them in 1527. It is not surprising that diseases of bees and silkworms should have been the first to be observed and recorded. The products of both insects were very important in the economy, and any agent having a deleterious effect on them was bound to have attracted attention.

As long as agriculture was diversified, with each individual farmer producing a little of each of a number of crops, insect pests were probably looked on just as pests rather than as economic problems. Within the lifetime of many people now alive, such things as wormy apples were considered a nuisance but not much more. Occasional outbreaks of such insects as grasshoppers occurred, but usually, except for the migrating locusts, they seem to have been relatively local, and disappeared after one or two years. In the light of our present knowledge it appears quite likely that many of these disappearances may have been largely, if not entirely, due to disease epizootics developing in the crowded conditions of the outbreak. But in the absence of means for finding and recognizing the effective pathogen the disappearance as well as the original appearance was probably considered to be "an act of God." It is very interesting that in 1664 John Evelyn (23) published in London a book titled *Sylva* in which he described a method for ridding trees of caterpillars by making a decoction of them and sprinkling it over the plants being attacked. The caterpillars are not identified, but it

[1] Contribution No.: 223.

appears entirely likely that they were a species that suffers from a virus disease, and that he was anticipating by 300 years essentially the method now recommended for making use of these pathogens.

Despite these individual references to diseases in insects in early entomological literature, the first attempt to present a comprehensive analysis and summary of them appears to be that of Kirby (35). He refers to diseases arising from some internal cause, and describes some of these as vertigo in ants and bees and convulsions in flies. He also described a disease of Diptera as follows:

> A white crust appears to be formed upon the abdomen both above and below, of a granular appearance, much resembling fine moist sugar. On the back of that part this crust does not cover the margins of the segments, which gives it the appearance of white bands; so that deceived by it, I have often at first flattered myself that I had met with some new species. The under-side of the abdomen is wholly covered by it, divided in the middle into two longitudinal masses, the anal segment being bare. De Geer has noticed this or a similar disease, which, when flies are attacked by it, causes the abdomen to swell so as even to burst, and the segments become dislocated. Upon opening the abdomen it is found filled with a white unctuous substance, which often accumulates (as above described) on its external surface. Dr. Host says that in this disease when the animal is dead, the wings, which were before incumbent, become extended, and its almost invisible pubescence grows into long hairs. De Geer seems to think that these flies are thus affected in consequence of having eaten some poisonous food; but I rather suspect, as I have observed it become prevalent chiefly in wet seasons, that it arises from a superabundance of the nutritive fluid, or of the fat, so that it seems to be a kind of *plethora*.

This description suggests that the condition is due to a fungus infection, and D. M. MacLeod of this Institute thinks it is probably a *Beauveria*. Although Kirby refers to the similarity to a condition described by De Geer [presumably the one listed by Steinhaus (56) as "probably the first description of what was undoubtedly an *Empusa* infection"], he apparently did not recognize it as a fungus because several pages later in the same chapter he has a section dealing specifically with fungi. He recognized that a larva killed by muscardine did not putrefy, a point also made by Bassi. Unlike Bassi, Kirby did not think muscardine was contagious, but that it was a result of moist heat.

It is not possible in a short paper to attempt to name all those who have contributed to the development of insect pathology, nor is it necessary. There are several works in which parts of the story is told, and reference to them for further details is strongly recommended. Unfortunately not all the publications referred to were available to the present author for original study, and there has of necessity been some dependence on the accounts and interpretations of others. For this reason many references are not cited, but in such cases most of them can be located by referring to one or other of the following references: 7, 15, 46, 48, 54–56.

The Beginning

Bassi (5) is generally credited with the first experimental demonstration that a mircroorganism, the fungus *Beauveria bassiana,* causes the infectious muscardine disease in the silkworm. According to his own account, Bassi devoted the years from 1808 to 1813 "wasting money and journeys and labour in vain" attempting to prove that the disease arose spontaneously. He came to the conclusion that the "disease was produced in the silkworm by an excessive proportion of acid substance introduced into the insect or generated in it by some cause." He believed this acid substance to be phosphate, and although by treatment with phosphoric acid he could cause the death of the insect, "their corpses remained uncorrupted like the worm killed by *calcinaccio* and almost as white, but not equally hard, so that the similarity between them and the true calcified worms was not perfect." Other treatments, including hanging worms in little paper bags at various heights in the chimney of a fireplace, also yielded specimens that closely resembled those produced by nature, but "I could not help noticing, to my inexpressible grief, that the former lacked the essential characteristic that marks the true muscardined worm, indeed the only property that can distinguish it from all the others that may exist with exactly similar appearances, i.e. it lacked the power of contagion—that of communicating the disease to other insects."

After thinking about these things apparently for two years, ("humiliated in the extreme, silent and idle, I wept over my lost laurels and bitterly lamented the adverse fate which had put me to so much study, expense, and labour in vain"), in 1816 Bassi decided that the disease "did not originate spontaneously in the insect, and that it needed an extraneous germ which entered the insect from the outside and caused the disease." The next eight years were spent in proving this new theory and in devising methods for handling and rearing silkworms that would prevent development of the disease. He attempted to capitalize on his findings, by keeping the essential information to himself, and "by means of public news-sheets, invited princes, agricultural and scientific societies and wealthy gentlemen to purchase his valuable discovery," but could not generate any interest. Finally, "tired of secrecy in a subject of such importance, and putting the interests of the public before his own, he determined to publish his discovery." He submitted it to the University of Pavia in 1834, and it was published in 1835 thus establishing the foundation of insect pathology as it is known today. In this treatise Bassi pointed out that "the contagion is communicated by food, by inoculation, and by the mere contacts of insects that have died of the terrible disease, of any infected object, and even of air contaminated by the disease-bearing germs." He stated that it is not peculiar to the silkworm, but occurs also in other caterpillars and can be transmitted to them in the same way. If it is then passed back to the silkworm, "it is always identical, and never in the least alters its nature or its behaviour." This seems almost to be a statement of one of what are now known as "Koch's postulates," although Koch was only born in the year after Bassi's work was published! Apparently Balsamo-Crivelli recognized that the causative organism of muscardine is a fungus, and in 1835 he described it and, recognizing the fundamental nature of Bassi's research, named it *Botrytis bassiana* (4). In 1912 Vuillemin erected the genus *Beauveria* and placed in it *B. bassiana* as well as other species that had been described as either *Sporotrichium* or *Isaria* (39). *B. bassiana* is still recognized as one of the most widely occurring entomogenous fungi, and is frequently the cause of more or less severe epizootics among numerous insect species.

Several other diseases of silkworm were of more or less economic importance, and came in for their share of attention. A number of names for these diseases are to be found in the literature, but because the etiology was not understood there is confusion about identity in different reports. Nevertheless it appears that the one now generally known as grasserie or jaundice was pretty uniformly recognized, as was pebrine. Cornalia (14) observed polyhedral bodies in the blood corpuscles of silkworms infected with jaundice, and described some of the pathological conditions associated with the disease. Maestri (43) observed these bodies in other tissues as well as in the blood, and noted that they occur in the nuclei of the cells. As various investigators studied the disease, the polyhedral bodies were interpreted in a number of ways—as inert crystals (9, 30), as protozoa (10, 44, 62); as degeneration products of the cell nuclei (51); and as an unknown virus for which the polyhedra acted as a carrier (22, 33).

The Early Years

Protozoan diseases.—The disease called pebrine probably occurred in silkworms from very ancient times. It was undoubtedly confused with various other diseases under any of a considerable number of names, but in 1859 the presently accepted name was given by de Quatrefages (16) because of the dark pepper-like spots that appear on the integument of infected insects. He gave a complete description of symptoms of the disease, both external and internal, referred extensively to the confusion that existed regarding different diseases, but did not reach a definitive decision regarding the actual cause, although he emphasized the appearance of peculiar globules in the blood of diseased worms.

Steinhaus (55) has given an excellent account of the work done by Pasteur in establishing that pebrine is caused by a protozoan, and also of Pasteur's studies of flacherie, and nothing can be added to that account. It amply supports Steinhaus' contention that Pasteur initiated the development of insect pathology as a science.

Fungal diseases.—Considering the tremendous economic losses in the silk industry due to the occurrence of disease, it is not surprising, once the nature of the diseases became at least partially understood, that there occurred the idea of turning them to advantage. It seems fairly certain that Bassi (6) was the first to make the suggestion, even though his actual recommendation would have been useless because he did not appreciate that the muscardine fungus was quite distinct from organisms of putrefaction. Immediately following Bassi several instances were reported of the transfer of muscardine to insects other than the silkworm, but the concept was not developed until after Pasteur's work had been published. LeConte in 1873 proposed at a meeting of the American Association for the Advancement of Science that there should be a "careful study of epidemic diseases of insects, especially those of a fungoid nature: and experiments on the most effective means of introducing and communicating such diseases at pleasure" (37). The same idea occurred at about the same time to Pasteur, who in 1874 published a suggestion that pebrine should be used to control the grape phylloxera, and that a search be made for a fungus capable of destroying the insect so that it could be introduced into vineyards infected by phylloxera (49). Apparently the idea was never actually tested, and in 1884 Balbiani (2) pointed out some of the problems, such as distributing the fungus and obtaining infection in a sucking insect—he seems to have assumed that infection must take place by ingestion.

It has not been possible to determine with certainty whether it was as a result of the suggestions mentioned or just a happy coincidence, but there is no doubt that it was at this time, i.e. in the late 1870s, that serious study of the potential use of fungi was undertaken by professional entomologists. Hagen (31), apparently influenced by reports published by Bail of Prussia recommended the use of yeast sprinkled on food plants to control currant worms and potato bugs, or more importantly the Colorado grasshopper. He suggested that it be used "to infect the newly-hatched brood, which live together in great numbers." He pointed out that there might not be great success, as the quantity to be applied could be determined only by experiment. But "the remedy proposed is very cheap, is everywhere to be had or easily to be prepared, has the great advantage of not being obnoxious to man or domestic animals." Thus, even though he was incorrect in suggesting that yeast might be used he was aware of the same basic principles that are a part of the philosophy of microbiological control even today.

The credit for actually initiating experimental work to demonstrate that disease may be intentionally caused in insect pests of economic importance is generally given to Metchnikoff. Chesnova (13) has given an account of Metchnikoff's work and references to his publications. He apparently recommended the techniques verbally in the fall of 1878 as a means of controlling the grain beetle *Anisoplia austriaca,* suggesting the use of a fungus he called green muscardine and which taxonomically is now known as *Metarrhizium anisopliae.* He reported to an entomological conference in Kharkov in 1882 on the results of further work. In collaboration with L. S. Tsenkovskii a method was devised for production of spores in large quantities by repeatedly adding healthy larvae to a box of earth containing fungus-killed ones. Later he described how, with the help of A. O. Kovalevskii, he grew the fungus on sterile beer wort, removed it and mixed it with sand so it could be spread on fields. He claimed that the fungus in this mixture was still reproductive after 11 months. He also claimed that the fungus was even more effective against the sugar beet curculio *Cleonus (Bothynoderes) punctiventris* than against the grain beetle. Others had also been investigating the subject, and the conference closed with a resolution that this work was very important and should be continued.

Apparently this new approach to insect control was not universally approved. Chesnova describes how Lindemann in particular refused to accept it, particularly as regards the mode of action. This opposition was expressed in publications in which it was claimed that infection could spread only through contact of healthy insects with diseased or dead ones, so that those living in protected habitats would be protected. Lindemann believed that the spores could penetrate only through injured cuticle, and in any case that the whole thing was too complicated and expensive. Apparently the controversy continued over a period of ten years or so. Metchnikoff continued his work, however, and eventually joined the Pasteur Institute in Paris where he directed the development of biological control of injurious insects.

Krasil'shick (or Krassilstschick) was one of the entomologists interested in using fungi, and in 1883 he went to the estate of a sugar-refiner, Bobrinskii, in Kiev province to supervise production and to check its usefuless in the field against *Cleonus*. The following year he set up a small production center. However, in 1886 he reported that the culture system developed in the laboratory was unsuitable for large-scale production. He designed special flat metal trays equipped with a system of ventilation that would control contamination, as well as an apparatus for harvesting the fungus. He was able to reduce the development time by about half. All these methods combined to produce about 200 gm of dry, ready-to-use material per square meter of culture surface, and altogether he produced about 55 kilograms in a 4-month period. The spores were mixed with fine sand and scattered about in field plots. Mortality of *Cleonus* amounted to as high as 80% after 10 to 15 days. This caused much interest, and a group of ten sugar refiners wanted to set up a large factory to produce green muscardine. Apparently the factory was built, but it developed that there was an overproduction of sugar beets, and further financial support for the project was not forthcoming.

A number of reports appeared during the next 20 years or so describing attempts to use fungi to control pests. Many of them are simply descriptions of individual trials, interesting because they demonstrate the wide appeal of this new approach, but because of the ad hoc nature not really contributing very much. Two groups of papers stand out, however, one by S. A. Forbes in Illinois and the other by F. H. Snow in Kansas (see bibliography in Steinhaus, 56). They were concerned with the use of white muscardine (*Beauveria globulifera,* not *B. bassiana* of the silkworm) to control the chinch bug, *Blissus leucopterus,* which Forbes (26) describes as a pest recurring at intervals of about five to eight years, and at the height of the population causing very serious crop destruction. The idea seems to have had considerable publicity, because Forbes reports that he had appeals for help from 610 towns, and most of these were requests for material with which to introduce the disease. Somewhat dubiously he decided to try to meet the requests, and obtained from the State Agricultural Experiment Station at Urbana an allotment "not to exceed" $450 to carry on the work He used the method of propagation devised by Metchnikoff and further developed at the Kansas University Station, namely the exposure of healthy bugs to those dead and covered with the characteristic fungous growth. His supply of healthy insects was obtained by offering to accept them from the public and to return infected ones in their place. The offer was widely taken up and he reports "we were very nearly overwhelmed—as were also the local express offices and the post office—by packages of chinch bugs arriving from all parts of the state, and in all imaginable conditions." Nevertheless he found that the contagion did not spread rapidly enough to meet more than a small percentage of the demand, this being due in part, as he later discovered, to an infestation of a mite that fed on the fungus as fast as it developed. He therefore supplemented the supply of material by collecting thirteen-year locusts, *Cicada tridecem,* which had died of the disease, as well as by fungus grown on a mixture of corn meal and beef broth. Both materials were equally as effective as the original, and the total effort resulted in distribution of over two thousand lots of material, each accompanied by a circular containing instructions for its use with particular emphasis on the fact that moisture is important and the best time for distribution is in the evening or following rain. It also suggested that the recipient should initiate a culturing system of his own (which was described) so that repeated field inoculations throughout the season would be possible.

Analysis of the results reported back by cooperators led to the general conclusion that the fungus is widely distributed wherever the insect occurs, that if the environmental conditions are favorable an epizootic will develop naturally, and that little is to be gained by artificial distribution of infectious material. Lugger (38) on the basis of limited tests had already reached essentially the same conclusion, and by 1902 the method appears to have been generally abandoned. Billings & Glenn (8) made a thorough study of the whole effort, and they confirmed that this conclusion was quite justifiable. Although there have been numerous attempts since that time to use fungi in pest management, including control of San Jose scale by *Sphaerostilbe coccophila* (50), of citrus whiteflies by *Aschersonia* spp. and others (45), of brown-tail moth by *Entomophthora aulicae* (53), and of European apple sucker by *Entomophthora sphaerospermum* and green apple bug by *Empusa erupta* (19–21), none have been particularly successful. Fawcett (24) explains that wind-borne spores are produced naturally in great profusion, and artificial distribution has little added effect. Steinhaus (55) however points out that it is dangerous to make sweeping generalizations regarding all insect mycoses on the basis of investigations of one fungus applied specifically against one insect in relatively restricted areas. In more recent research this point has been pursued further.

Bacterial diseases.—While these studies of entomogenous fungi were being pursued, attention was also directed to insect diseases caused by bacteria. Pasteur had already reported that flacherie of the silkworm is caused by two species of bacteria. Others disagreed, and various other causes were suggested until finally Paillot found what appears to be the true explanation—that the disease is initiated by a virus and that the bacteria are secondary invaders (55). Although the bacterium associated with true flacherie is *Bacillus bombycis* (Pasteur's vibrion à noyau), others, including Ishiwata's *Bacillus sotto,* have from time to time been suggested as the cause. This latter, like the others, seems to be only an accidental association, but nevertheless *B. sotto* has been of great importance in recent years in the development of insect pathology both as a field of research and as a means for regulating insect populations.

As in the silkworm, disease in the honey bee has been recognized for centuries. It was particularly noted that the brood suffered, and the general term foulbrood appears to have been used. Steinhaus (55) credits early use of the term to Schirach in 1771, but he also gives a quotation from Gandernackee published in 1682 in which the name is used. In 1885 Cheshire & Cheyne (12) isolated a spore-forming bacterium which is now generally recognized as the cause of one particular form of the disease called European foulbrood. In 1904 White (66) reported another spore-forming bacillus as the cause of American foulbrood. (It should be noted that the designations European and American are matters of convenience only, and bear no relation to the geographical occurrence of the diseases.) Bacteriologists still question the correctness of the taxonomic designations of these organisms, but there is no doubt that there are two specific diseases, as well as some other occasional conditions also loosely called foulbrood but with which neither of these bacteria is associated.

Although important bacteriologically as well as economically, the bee diseases do not appear to have had much impact on the early development of insect pathology as a science. Most of the studies appear to have been either to elucidate the taxonomy, or to find means to reduce the damage from the diseases, and success has been achieved on both fronts.

Metchnikoff was one of the earliest workers to suggest the possibility of utilizing bacteria to control insects. He had noted that *Anisoplia* larvae were subject to infection by a "putrefactive disease" caused by a bacterium which he named *Bacillus salutarius.* He described the symptoms in 1879, and on the basis of experiments he concluded that it was most contagious and could possibly be used to control the pest (13).

Forbes (25) described in great detail tests he made to determine infectivity of bacteria for several insects, including the silkworm. Although his experiments demonstrated infectivity, and he noted that the disease destroyed large numbers of green cabbageworms by natural epizootics and in at least one test he believed he had induced the disease among larvae in an area previously uninfected, this work does not appear to have been followed up. He also described the destruction of an outbreak of the forest tent caterpillar *(Clisiocampa sylvatica)* by the muscardine fungus, but apparently made no effort to test or utilize this pathogen.

Intensive study of the potential of bacterial diseases for insect control began with the investigations of the French bacteriologist d'Herelle. In 1909, he noticed that a migratory locust, originally thought to be *Schistocerca pallens* (17) but which he later stated was really *Schistocerca americana* (18), suffered a certain amount of natural mortality among the swarms arriving in the state of Yucatan, Mexico, from regions of Guatemala where they passed the winter. By 1911 the swarms were generally infected and in 1912 there was no invasion. He found the causative organism to be a bacterium which he described and characterized, and called *Coccobacillus acridiorum.*

D'Herelle (18) described in detail the experiments he performed to demonstrate the value of this bacterium. He applied the bacterium to locust-infested areas in Colombia, Argentina, Cyprus, and Algeria, different species of locusts being involved, and reported that he controlled the insects. However, he found it imperative to increase the virulence of the bacteria by passing them several times in rapid succession through the insect, using an injection technique. From 4 to 12 passages were usually enough, and they always had to be made using the species against which the bacterium was to be applied. When a satisfactory degree of virulence was attained—death 4 or 5 hr following injection—the body fluids could then be used to infect a culture broth to provide material for spraying on the food of the insects. It was necessary also to be certain that it was a pure culture that was used, as indicated by the lack of any disagreeable odor; any that were putrid were discarded.

Despite the fact that eventually this method was rather generally discredited, d'Herelle (18) made several general observations concerning the utilization of pathogens that are still frequently applicable in the light of our more advanced knowledge. For instance, he stated that since the epizootic does not kill the insects immediately, it is still necessary, at least for agricultural crops, that other means of protection be used. In uncultivated areas this need not be done. Defensive measures as commonly used must be repeated each year, whereas an offensive attack, such as the biological method, is not aimed at destruction of the species but at reducing the numbers to a level where the injury they do is negligible; this is the essence of the current concept called integrated control. He pointed out that weather and other environmental conditions have a great effect on results. For example, fungi will often work under humid conditions but are much less effective when the weather is dry, a generalization that is still widely accepted. On the other hand, the growth of bacteria within their host makes them much less susceptible to conditions once they have survived transferral. He also found that the bacteria were much more effective among those species that aggregate in swarms, move frequently and rapidly and are highly cannibalistic; especially in noncannibalistic species and in those that are sedentary, the epizootic developed much more slowly.

Sergent & l'Heritier (52) transferred d'Herelle's bacillus to the Moroccan locust *Stauronotus maroccanus,* and with successive passages (up to 100) reduced the average time to death from 23–36 hr to 4 hr. When this material was used to infect the nocturnal resting places of the locusts, high mortality occurred after some days of incubation. Transmission to other swarms was very poor. Generally, the tests raised more questions than they answered.

Other tests were made against grasshoppers and other insects in various parts of the world over the next few years, but none were very successful, and the idea seems to have been dropped by about 1917. Bucher (11) has made a critical analysis of these attempts, supported by his own investigations with bacteria isolated from grasshoppers from various areas as well as with cultures obtained from repository collections and reported to be the original d'Herelle species. He concluded that *C. acridiorum* was never a valid species, but was a Cloaca type A bacterium, a group commonly found in the guts of grasshoppers and not true pathogens. The claims of success were based on faulty techniques and uncritical assessment of results, with no adequate reference controls. This work of Bucher's deserves very careful study by anyone intending to attempt to establish bacterial epizootics for the regulation of insect populations.

Viral diseases.—In early papers on insect diseases (e.g. d'Herelle, 18) the word virus appears to have been used in the general sense of an infectious agent, even though the agent may have been a bacterium or a fungus. But early in this century the word began to be applied to infectious agents that there so small they could not be seen with a microscope, and did not lose their infectivity even after passing through bacteria-retaining filters. Steinhaus (55) has traced the development of this usage of the word, and it is now generally accepted in this more restricted sense.

Although the actual infectious agent (the virus particle or virion) cannot be seen with an ordinary microscope, two Italian scientists, Cornalia (14) and Maestri (43) recognized and described characteristic crystal-like bodies regularly to be found in tissues of diseased insects. These inclusion bodies, now known as polyhedra, are highly refractive. A succession of workers proposed various theories as to their identity and relationship to the disease. Bolle (9) at first thought they were simply crystals, but later (10) decided they were spores of a protozoan parasite. He showed that they can be dissolved in weak acid or alkaline solutions including the gut juices of the insect. Von Prowazek (62) also thought that the disease was caused by a protozoan which he named *Chlamydozoa bombycis,* and that the polyhedra were merely a byproduct of the disease, since he could demonstrate that diseased material was still infective after removal of the polyhedral bodies. Later, when filtration through a Berkefeld candle failed to remove the infectivity von Prowazek (63) questioned the chlamydozoan theory. Escherich & Miyajima (22) believed that the disease was caused by an unknown virus and that the polyhedral bodies were the carrier. Hayashi & Sako (33) concluded that the polyhedra carry the infectious agent. Glaser & Chapman (29) examined gypsy moth larvae infected by wilt disease and found that the polyhedra are nucleoprotein degeneration products and not organisms; Glaser (27) further showed that the effects are not due to an enzyme or a toxin and concluded "one cannot but feel inclined to adopt the view that one is dealing with minute parasitic forms. . . . We are justified at present however in not classifying such viruses either with the plants or animals."

Other than the silkworm, the earliest reports of insects suffering from viral diseases appear to have been two European forest pests, the nun moth *Porthetria* (=*Lymantria*) *monacha* and the gypsy moth *P.* (*L.*) *dispar.* In some of the early reports, as might be expected, there is some confusion regarding the exact identity of the disease or of the causative organism. There seems to be little doubt that in some cases in which the cause was stated to be a bacterium, in fact a virus was involved. At other times the term virus was used in an indefinite sense to describe the agent causing any one of a number of diseases.

The disease in the nun moth apparently became a widespread epizootic in 1889, and killed enormous numbers of the insect. It was called the wipfelkrankheit, because the sick larvae migrated to the tops (wipfeln) of the trees and died there. Several bacteria were isolated from the sick or dead larvae and named as the causative organism, but when the filterable nature of the pathogen was demonstrated the disease was recognized as being due to a true virus and is now known to be a nuclear polyhedrosis.

Meantime the gypsy moth had been introduced into the United States and had become a major pest in the forests of Massachusetts. About 1907 disease began to be noticed and caused much mortality among the half-grown caterpillars. The origin of the disease is uncertain, but it was probably introduced in importations of parasites from Europe where it was known to occur. Because of the symptoms, it was referred to as the wilt disease, and at first it was thought to be a result of overcrowding and poor food. It was soon shown to be infectious, and Glaser & Chapman (28) ascribed it to a virus. This disease too is a nuclear polyhedrosis. Despite the fact that it spread naturally, and eventually came to be recognized as the same disease that infects gypsy moth in Europe, it has not been effective in bringing about consistent regulation of populations in the United States.

Up to this time, that is about 1920, some 30 or so species of insects had been reported as susceptible to polyhedrosis virus. Paillot (48) published a monograph on insect pathology in which he reviewed developments to that time. He drew attention to the fact that not all virus infections are accompanied by production of polyhedral bodies. He had observed a disease in *Pieris brassicae* in which the bodies produced were granular, and later similar bodies were found in other insect species. Paillot also found in *P. brassicae* a disease in which the bodies are very irregular in form and, unlike the polyhedral bodies, do not appear to be directly associated with the infection because they can be removed from the blood by ordinary centrifugation but the blood is nevertheless still infectious. He made observations on the effects of such factors as insolation, air currents, and high and low temperature on the development of virus diseases, and stated that transmission from generation to generation occurs both through the egg and by contamination of the habitat.

Assessment.—In addition to the major efforts described, there are numerous reports in the early literature of individual attempts to use pathogens in a practical way to control insects; mostly these were of an ad hoc nature. But in 1927 the International Live Stock Exposition organized the International Corn Borer Investigations in an effort to discover and develop methods to eliminate the corn borer. Cooperation and active participation came from most countries of Europe as well as from the United States and Canada. The coordinated studies were a model for this type of project; including as they did the biology of the host insect in different areas, the occurrence of parasites and predators, the reaction and potential resistance of different strains and varieties (and even species) of host plant to the insect attack, various methods for using chemical insecticides, and the potential value of pathogens, both those found associated with the corn borer and those known to kill other insects. The study, under the direction of Tage Ellinger, extended over four years, and four reports were published, the last one in 1931. In the final report Vouk said of the work with bacterial pathogens ". . . hardly any doubt remains that bacteria can be applied successfully in the fight against the Corn Borer. The scientific problem has been completely solved. The practical development of the method is now merely a question of technical and economic character, the significance of which should certainly not be underestimated." And again, referring to tests with the fungus *Metarrhizium anisopliae* he said:

> The result was most satisfactory since a mortality of 98 to 99 per cent, as produced by this fungus, practically means a complete check of the Corn Borer infestation. . . . So far, the microbiological methods have given the best and most decisive results. In fact, their practical significance cannot yet be fully measured. It is true that these methods are not entirely new to science, but their application on the Corn Borer will undoubtedly mean a revival of this line of attack in the fight against noxious insects.

It has not been possible to find a document setting forth the terms of reference on which these investigations were founded, so whether the abrupt termination after four years was by original design, or because the results were accepted as being as satisfactory as the quoted reports indicate, or for some other reason, is not known. It is a fact that little or no work with pathogens of the corn borer has been reported since that time, nor is there any recommendation, based on the results, that the pathogens should be generally used. Possibly the parasite complex, coupled with cultural methods, reduced the seriousness of the problem.

Despite the number of individuals who had at one time or another investigated pathogens, there seems to have been very little formal or official encouragement up to this time. It is interesting that Howard, who was the Chief of the US Bureau of Entomology and Plant Quarantine and who in his earlier years published several articles on insect diseases, in his *History of Applied Entomology* (34) does not even mention the subject.

Whether or not it was as a result of the Corn Borer Investigations, the study of insect pathology began to be oriented and organized. No doubt the development of modern scientific instruments such as the electron microscope and the ultracentrifuge contributed largely to the resurgence of interest and made scientific study possible. Also there was realization of the fact that superficial testing is not enough, but the approach must be through an understanding of fundamental principles. Perhaps a great deal of the credit for the new approach can be given to Paillot (48) who says in the introduction to his monograph:

> In 1912, when my mentor, Prof. P. Marchal, asked me to study the diseases of insects, I had as a principal goal the utilization of microbial parasites in the battle against destructive insects. . . . However, following the revelations of bacteriology, one could believe that it should be possible to create at will artificial epidemics among insects, given the prodigious powers of multiplication of the lower organisms; lack of success was the rule, and the method has been discredited and generally abandoned. But what exactly do we know of the infectious pathology of Invertebrates? . . . One has thus attempted to use prematurely a weapon that one does not know. . . . Truly, the subject has been only skimmed and by a small number of authors who for the most part were not trained for this kind of research.

He goes on to quote Roger, speaking at the first International Congress of Comparative Pathology: "To build a true general pathology, it is necessary to consider the morbidities in the whole series of living beings beginning with the lowest, the Protozoa and Protophytes, and to follow progressively to Mammalia and Man," and says "It is in that spirit that I have pursued my research on the infectious pathology of Insects. The primitive utilitarian objective has been subordinated to the solution of that important question." For the next thirty years this was the attitude adopted by most insect pathologists and supported by those who provided the research funds, and very great progress has been made in all aspects of the subject, including practical application. Unfortunately there has developed of late a trend back to the utilitarian goal, and a tendency to de-emphasize the fundamental research. It is to be hoped that this situation will not be allowed to develop too far, or both scientific and practical progress will almost certainly be seriously curtailed.

Paillot reviewed the entire field of insect pathology as it was known at that time. He summarized the work on the four main classes of pathogens: protozoa (including the minor groups: gregarines and flagellates), fungi, viruses, and bacteria; as well as the work on immunity, and symbiosis in aphids, and concluded with chapters on the economic consequences of infectious pathology in insects, the utilization of microbial parasites in agriculture, and the role of insects in transmitting diseases of man, domestic animals, and plants. He presented this as "an introduction to a work of the greatest scope; a point of departure rather than a definitive goal", with lacunae requiring long research to supply. His successors have done much to fill the gaps.

The More Recent Developments

Because of the disruption of communications caused by World War II it is difficult to identify relationships between various developments that took place in the fifteen years following the International Corn Borer Investigations. They occurred in various parts of the world and as normal conditions returned and information began once again to flow and research results to be correlated, insect pathology made great strides forward.

F. T. Bird was working as an undergraduate student assistant in forest entomology in New Brunswick, Canada, when he became interested in studying the cause of a spectacular decline in populations of the sawfly *Gilpinia hercyniae* which had for a number of years been destroying large areas of forest in eastern Canada. He was encouraged and supported in this work by R. E. Balch, and they published (3) an account of their research including a description of the disease responsible and an assessment of its value and significance. During the same period Bergold and his associates had been studying the structure of polyhedral inclusion bodies found in jaundiced silkworms and had come to the conclusion that the polyhedron protein was the infective principle, a conclusion that Bergold later corrected when he was able to show that the virion is a definitive rod enclosed in the polyhedron and accounts for only 3 to 5% of the total mass. American workers (Glaser, Stanley, Wyckoff) also had been studying virus structure, and their findings too were important in elucidating the problems.

Development of research centers.—Steinhaus had studied bacteriology, but was also very interested in insects and eventually took a minor in entomology during his postgraduate training. From studies of the microbial flora of insects he moved to the US Public Health Service where he was specifically engaged in research on tick-borne diseases, but also continued studies of associations of microorganisms with other insects. In 1944 he went to the Department of Bacteriology in the University of California at Berkeley. He was soon transferred to the Division of Beneficial Insect Investigations, where in 1946 he founded the Laboratory of Insect Pathology and became its first Director. Here he developed the first formal course in insect pathology, for which, since no textbook was available, he found it necessary to write *Principles of Insect Pathology,* published in 1949 (55).

Although it was not his intention that it should necessarily be so, the fact that Steinhaus was in the Department of Biological Control inevitably led to an association of ideas, that insect pathology was primarily a means for controlling pests. This was undoubtedly encouraged by the already established use of milky disease to control Japanese beetle (67, 68), and by the discovery of the nuclear polyhedrosis of the alfalfa caterpillar and the experiments in using it to control the insect (57). Steinhaus also promoted the use of bacteria, particularly *Bacillus thuringiensis,* as insecticides, and stressed that fungi, protozoa, and nematodes should not be overlooked even though they could not be as easily manipulated as the viruses and bacteria. From this school there eventually emerged such alumni as K. M. Hughes, C. G. Thompson, Y. Tanada, M. E. Martignoni, and others.

An outbreak of spruce budworm, *Choristoneura fumiferana,* began to develop in northwestern Ontario, Canada, in the late 1930s and had become very serious by about 1943. J. J. deGryse was at that time the Chief of the Forest Biology Division, Entomological Branch, Canada Department of Agriculture. He conceived the idea of a forest entomology research center to be established in Sault Ste. Marie in which a group of specialists would be brought together to study all aspects of insect populations. The studies would embrace ecology (including effects of temperature, light, etc), pathology, genetics, toxicology, physiology, general biology, and surveys. Although the original concept was never completely realized, it did result in K. Graham being appointed in 1946 as the first insect pathologist. He resigned after a year to accept an appointment at the University of British Columbia, and Bird was transferred from Fredericton, New Brunswick, where he had done his research on the polyhedrosis of *G. hercyniae,* while Bergold was recruited from the Kaiser Wilhelm Institut in Tübingen. They both arrived in Sault Ste. Marie in 1947 and the program that eventually led to the establishment of the Insect Pathology Research Institute was begun. This is not the appropriate place to give a detailed account of the development of the Institute, save to say that it produced T. A. Angus, D. M. MacLeod, G. R. Stairs, A. M. Heimpel, G. R. Wyatt, P. Faulkner, and others well known for their contributions to the fundamentals of insect pathology.

The Sault Ste. Marie Institute is probably unique in that it arose de novo, whereas others seem to have developed out of organizations designed in the first instance for biological control of insects, i.e. by the use of parasites and predators. As noted above that was the way Steinhaus' program was initiated. It was true also of the Institute für biologische Schädlingsbekämpfung in Darmstadt headed by J. Franz, which was originally the Institut für biologische Bekämpfung und Kartoffelkafer-Forschung to study biological control—i.e. natural enemies, utilization of parasites and predators both native and imported and alone or combined with insecticides—as well as insect pathology. In addition to Franz, the staff included Krieg,

Müller-Kögler, and Langenbuch, later succeeded by Huger. Besides the research actually carried out, Franz made a great contribution to the field by compiling a bibliography of insect pathology, published annually in the journal *Entomophaga*. In Prague, J. Weiser was appointed head of the parasitology group in the Biology Institute of the Academy of Sciences when it was founded in 1951. The concern of the group was insect control by chemicals, use of insecticides in public hygiene, medical entomology, helminthology, and protozoology, but they also concentrated on evaluation of insect pathogens in insects of medical importance. A separate Laboratory of Insect Pathology was organized in 1954 with Weiser, Veber, and Samsinakova; Lysenko joined it in 1955, and Vankova shortly afterwards.

Much fundamental work on viruses was done at the Sericultural Institute in Japan by Yamafuji. He introduced the concept of occult or inapparent infection, capable of being stimulated by stress factors, specifically certain chemicals. These ideas aroused considerable controversy, and even yet there is not complete unanimity on the question. But they undoubtedly contributed significant impetus to the trend toward studies of chemical composition and molecular structure, and the part played by nucleic acids and subcellular components in cell metabolism. Much of this information has been of scientific value in the general field of molecular biology, and the insect viruses have been found useful as model systems for such studies.

During the late 1920s and up to 1940 Pospelov, in Russia, promoted the use of pathogens to control various insects. Some of his interpretations of the mode of action of the pathogens were not valid, as when he suggested that yeasts could disseminate virus causing jaundice in *Porthetria monacha*. Nevertheless his work resulted in the establishment of a laboratory of insect diseases, and his students have been active in study and application of insect pathology in the Soviet Union. These include Yevlakhova, Shvetsova, Vasilevskiy, and Pilat, among others. As in other establishments a number of pathogens (fungi, bacteria, and viruses) have been and are being studied. In 1956 Telenga, in Kiev, began testing combinations of fungi and very low dosages of chemical insecticides and found a synergistic effect.

Protozoan diseases.—It is an interesting fact that, although nosematosis of the honeybee and pebrine of the silkworm were among the earliest insect diseases recognized and studied, and Kudo & DeCoursey (36) described how they were able to infect fall webworm with *Nosema bombycis* from the silkworm, relatively little effort has been devoted to investigating the potential of protozoa as insect pathogens. Paillot (47) stated that protozoa play a much more important part than bacteria in the destruction of noxious insects, but he presents no data to substantiate the statement. Thomson (59, 60) studied the effects of a microsporidian on the spruce budworm, and concluded that it resulted in general debility and reduced fecundity. He recorded the development in an isolated population of the host, and showed (61) that the percentage infection increased naturally from less than 35%

to over 80% in a period of five years, and this was followed by a very great reduction in the population. No cause other than the microsporidian infection could be found to account for this. Weiser (64) describes a similar complete breakdown of a population of gypsy moth, and apparently successful attempts to introduce and spread microsporidia in several other insects (65).

The lack of more extensive research on entomogenous microsporidia may have been due in part to the difficulty in maintaining stocks for experimentation and study. They are practically all obligate parasites, developing and reproducing only within living host cells. Thus, in order even to study their biology it has been necessary to maintain cultures of the host insect. Moreover, while the statement by Paillot may indeed be correct, it is also true that the effects of protozoa on natural populations are almost never catastrophic as is frequently the case with fungi and viruses. Rather they tend to have a continuing depressant effect that is much more difficult to evaluate. And finally, the protozoa have a relatively short shelf life, that is to say quantities cannot be accumulated against the time that a control operation may be required. Recent developments in tissue culture now offer a possible solution to this problem.

Fungal diseases.—Comparatively little recent work in utilization of fungi has been attempted. Considerable effort has been devoted to an assessment of earlier work, and a major revision of the taxonomy of entomogenous fungi has been found necessary (39, 42). Early workers suggested that the infection and development processes in fungi are dependent on occurrence of proper ecological conditions (40). This is no doubt true, although there is still a great deal to be learned about the physiological requirements (41). When all the taxonomic points have been clarified, and more knowledge of fungal physiology has been accumulated, it is likely that management of fungi will become more practical.

Bacterial diseases.— There have been no recent attempts to utilize bacterial diseases to control insects by causing infection. But one of the most significant events in the whole development of insect pathology was the observation by Hannay (32) that sporulation of *Bacillus thuringiensis* results in the formation of a crystalline parasporal body within the sporangium, and the subsequent demonstration by Angus (1) and others that this body, now known to be proteinaceous in nature, is highly toxic for a considerable number of insect larvae of the order Lepidoptera. As a result, this bacterium is now in commercial production, and is being used in many countries for controlling various phytophagous Lepidoptera, especially on salad and other vegetable crops. It has been demonstrated that even though the crystal is highly toxic for many insects, it is completely harmless to other forms of life. Regulatory authorities have approved it for use without any restrictions regarding residue.

Most investigations and tests involving *B. thuringiensis* in its various strains, varieties, etc, indicate that it is the parasporal crystal, or in a few cases a water-soluble material, acting as a toxin that is the primary insecticidal agent. The bacterium itself seems to be only mildly infectious, if at all. For this reason preparations are used in practice in much the same way as are chemical insecticides; that is, they are applied when the insect begins to cause damage and applications are repeated when required as long as the crop needs protection. The one exception to this is described by Talalaev (58) who found that when he infected a population of *Dendrolimus sibiricus* by *Bacillus dendrolimus* (now recognized as a variety of *B. thuringiensis*) he created an epizootic that would recur in succeeding years. This appears to be directly attributable to the habits of the insect rather than being a particular property of this variety of the pathogen.

Viral diseases.—While many studies of viral structure and development were being undertaken, efforts to use them effectively were not forgotten. Several of these are described in the literature, some successful and others less so. Generally, the incubation period required before the insect dies has made viruses less attractive, especially in forage and vegetable crops. Again as our knowledge of the total biocoenosis improves, some of these shortcomings can be overcome. But generally most effort has been devoted to using virus against forest insects, where a moderate degree of damage can be tolerated. Several successful applications are well documented, although the gypsy moth in eastern North America is not yet included.

The Current Situation

It would probably not be profitable at this time to attempt to evaluate in any comparative way research in insect pathology during the last ten years; we are still too near the events. As the potential has become apparent the number of people involved has increased and their lines of investigation have diversified.

Recent studies of the viruses have tended to move from physical structure and electron microscopy to chemical and molecular structure and the involvement of nucleic acids. This may, indeed almost certainly will, lead eventually to development of methods for direct synthesis of viruses or at least viral analogs, and consequent lessening of dependence on producing them by rearing and infecting susceptible insects Even if direct synthesis is not attained for some time, use of tissue cultures as a substitute for living insects will certainly be developed further.

It has now been demonstrated that the endotoxin of *Bacillus thuringiensis* that affects Lepidoptera is only a fraction of the protein that comprises the parasporal inclusion body (the crystal). Doubtless its chemical structure will soon be elucidated, and again synthesis will follow. This can conceivably lead to development of a whole new family of synthetic chemicals of specific toxicity that can be used with impunity against individual insect pests in the full knowledge that nontarget insects, e.g. parasites and predators, will not be harmed, that warm-blooded animals and fish are not susceptible, and that biodegradation will eliminate any pollution problem.

Studies of fungi, which on occasion are so spectacularly effective, are gradually yielding information that will eventually make it possible to manage them at will. Although they appear even yet to be more dependent than other pathogens on such physical factors as temperature and humidity, their physiological peculiarities are becoming better understood and they too will make a manageable contribution to insect regulation.

Protozoa, where it all started a hundred or more years ago, and which have been largely ignored ever since, are making their presence in epizootics discernible and understandable as methods for population analysis are developed. Parallel studies of their biology and physiology will lead to methods by which they too can be produced and distributed at will. Since they, like other pathogens, have the power of self-reproduction they have potential for management as population regulators by encouraging their development early rather than late in the population cycle.

The first 70 years in the development of insect pathology were productive because it was a new field and trial-and-error methods had to win at times. But there were many disappointments because of misinterpretations due to lack of knowledge of fundamental principles. The next 30 years made up the golden age, when new concepts were developed, made possible by modern scientific instrumentation and by acceptance of the proposition that understanding of fundamental principles is basic to sound scientific progress. Unfortunately there appears to be a developing tendency in the last four or five years to insist on mission-oriented research. While recognizing that unapplied research may appear to be frivolous and of little value, one cannot refrain from concluding this account by referring again to Bassi, to Metchnikov and Krassilstschik, to Forbes and Snow, and to D'Herelle, all of whom tried to apply pathogens without clear understanding of what they were or should be trying to do, and none of whom succeeded in their specific aims. It is to be hoped that the lessons of history will be learned by those who make the management decisions so that they will not sell short the scientists who should and must make the research decisions. Otherwise we may lose the momentum we have gained at a time when we can least afford to do so.

APOLOGIA

This account is incomplete in several ways. Restrictions of space dictated that only those things should be included that influenced the course of development of insect pathology, and indeed not all of them have been covered. Numerous individual research programs have not been mentioned or described. Lack of access to literature in some foreign languages may have resulted in overlooking some studies that were of significance. For such omissions the author offers his sincere apologies, and expresses the hope that the next account that is written will make up the deficiencies.

ACKNOWLEDGMENTS

Thanks are expressed to the considerable number of people in different countries who were kind enough to provide literature references or translations, or to supply information from their personal recollections. It is hoped that the accuracy of this information has not suffered unduly in the condensation that was found necessary, and it is regretted that they cannot be individually acknowledged.

LITERATURE CITED

1. Angus, T. A. 1954. A bacterial toxin paralyzing silkworm larvae. *Nature* 173:545–46
2. Balbiani, E. G. 1884. *Leçons sur les Sporozaires*. Paris:O. Doin
3. Balch, R. E., Bird, F. T. 1944. A disease of the European spruce sawfly, *Gilpinia hercyniae* (Htg.) and its place in natural control. *Sci. Afr.* 25:65–80
4. Balsamo-Crivelli, G. 1835. Osservazione sopra una nuova specie di mucedinea del genere Botrytis, et cetera. *Biblioteca Italiana* 79:125
5. Bassi, A. 1835. Del mal del Segno, Calcinaccio o Moscardino, malattia che afflugge I bachi da Seta. Parte Prima. Dalla Tipografia Orcesi, Lodi. (Transl. P. J. Yarrow, Univ. of Exeter, as Phytopathological Classics No. 10, *American Phytopathological Society*. 1958)
6. Ibid 1836. Parte seconda
7. Bergold, G. H. 1958. Viruses of Insects. In *Handbuch der Virusforschung*, ed. Hallauer und Meyer, IV. Vienna: Springer
8. Billings, F. H., Glenn, P. A. 1911. Results of the artificial use of the whitefungus disease in Kansas. *US Dep. Afr. Bur. Entomol. Bul.* 107
9. Bolle, J. 1894. Il Giallume del Baco da Seta. Notizia preliminaire. *Atti e Mem. dell'I. R. Soc. Afr. Gorizia* 33:193
10. Bolle, J. 1898. *Der Seidenbau in Japan, nebst einem Anhang: die Gelbroder Fettsucht der Seidenraupe, eine Parasitare Krankheit.* Budapest, Wien und Leipzig: A. Hartleben
11. Bucher, Gordon E. 1959. The bacterium *Coccobacillus acridiorum* d'Herelle: The taxonomic position and status as a pathogen of locusts and grasshoppers. *J. Insect Pathol.* 1:331–46
12. Cheshire, F. R., Cheyne, W. W. 1885. The pathogenic history and the history under cultivation of a new bacillus (*B. alvei*), the cause of a disease of the hive bee hitherto known as foul brood. *Roy. Microsc. Soc. J.*, Ser. 2, 5 Pt II:581–601
13. Chesnova, L. V. 1962. *Essays from the history of applied entomology in Rusia. Academy of Sciences of the U.S.S.R. Institute of History, Natural Sciences and Technology, Moscow.* (Transl. Israel Program for Scientific Translations, Jerusalem. 1968.)
14. Cornalia, E. 1856. Monografia del bombice del gelso (*Bombyx mori* Linneo). *Memorie dell' I.R. Instituto Lombardo di Scienze, Lettere ed. Arti* 6:1–387. (Parte quarta: Patologia del Baco. pp. 332–66)
15. deBach, P. Ed. 1964. *Biological Control of Insect Pests and Weeds.* London: Chapman & Hall
16. De Quatrefages, A. 1859. *Études sur les Maladies Actuelles du Ver a Soie.* Paris: Librairie de Victor Masson
17. d'Herelle, F. 1911. Sur une épizootie dc nature bactérienne sévissant sur les sauterelles au Mexique. *C. R. Acad. Sci.* 152: 1413–15
18. d'Herelle, F. 1914. Le coccobacille des sauterelles. *Ann. Inst. Pasteur* 28:387–419
19. Dustan, A. G. 1923. The control of the European apple sucker by means of a parasitic fungus. *Ann. Rep. Fruit Growers' Assoc. Nova Scotia, 59th pp.* 100–4
20. Dustan, A. G. 1922–1923. The

natural control of the green apple bug (*Lygus communis* var. *novascotiensis* Knight) by a new species of Empusa. *Ann. Rep. Quebec Soc. Prot. Plants, 15th* pp. 38
21. Dustan, A. G. 1927. The artificial culture and dissemination of *Entomophthora sphaerosperma* Fres., a fungous parasite for control of the European apple sucker (*Psyllia mali* Schmidb.). *J. Econ. Entomol.* 20:68–75
22. Escherich, K., Miyajima, M. 1911. Studien über die Wipfelkrankheit der Nonne. *Naturwiss. Z. Forst.-u. Landw.* 9:381–402
23. Evelyn, J. 1664. *Sylva, or a Discourse on Forest Trees.* London
24. Fawcett, H. S. 1944. Fungus and bacterial diseases of insects as factors in biological control. *Botan. Rev.* 10:327–48
25. Forbes, S. A. 1886. Studies on contagious diseases of insects. *Illinois State Lab. Nat. Hist. Bull. 2,* Art. IV, pp. 257–321
26. Forbes, S. A. 1895. Experiments with the muscardine disease of the chinch-bug, and with the trap and barrier method for the destruction of that insect. *Univ. Illinois, Agr. Exp. Sta., Urbana, Bull. 38*
27. Glaser, R. W. 1918. The polyhedral virus of insects with a theoretical consideration of filterable viruses generally. *Science* n.s. 48:301–2
28. Glaser, R. W., Chapman, J. W. 1913. The wilt disease of gypsy moth caterpillars. *J. Econ. Entomol.* 7:479–88
29. Glaser, R. W., Chapman, J. W. 1916. The nature of the polyhedral bodies found in insects. *Biol. Bull.* 30:367–91
30. Haberlandt, F. 1871. *Der Seidenspinner des Maulbeerbaumes, seine Aufzucht und seine Krankheiten.* Wien: Carl Gerold's Sohn. 247 pp.
31. Hagen, H. 1879. Obnoxious pests—Suggestions relative to their destruction. *Can. Entomol.* 11: 110–14
32. Hannay, C. L. 1953. Crystalline inclusions in aerobic spore-forming bacteria. *Nature* 172:1004–6
33. Hayashi, D., Sako, W. 1913. *Recherches sur la Grasserie des Vers á Soie.* Lyons: Moniteur des Soies
34. Howard, L. O. 1930. *A History of Applied Entomology,* Vol. 84. Washington, D.C.: Smithson. Misc. Coll.
35. Kirby, W., Spence, W. 1826. *An Introduction to Entomology.* Vol. IV, pp. 197–232. London: Longman, Rees, Orme, Brown & Green
36. Kudo, R. R., DeCoursey, J. D. 1940. Experimental infection of *Hyphantria cunea* with *Nosema bombycis. J. Parasitol.* 26:123–25
37. LeConte, J. L. 1874. Hints for the promotion of economic entomology. *Am. Assoc. Advan. Sci. Proc.* 22:10–22
38. Lugger, O. 1888. Fungi which kill insects. *Univ. Minnesota Coll. Agr. Bull. 4*
39. MacLeod, D. M. 1954. Investigations on the genera *Beauveria* Vuill. and *Tritirachium* Limber. *Can. J. Bot.* 32:818–90
40. MacLeod, D. M., Cameron, J. W. MacB., Soper, R. S. 1966. The influence of environmental conditions on epizootics caused by entomogenous fungi. *Rev. Roum. Biol.* (serie de Botanique) 11: 125–34
41. MacLeod, D. M., Lougheed, T. C. 1963. Entomogenous fungi. In *Recent Progress in Microbiology, VIII,* ed. N. E. Gibbons, 141–50. Toronto: Univ. Toronto Press
42. MacLeod, D. M., Müller-Kögler, E. 1970. Insect pathogens: Species originally described from their resting spores mostly as *Tarichium* species (Entomophthorales: Entomophthoraceae). *Mycologia* 62:33–66
43. Maestri, A. 1856. Frammenti anatomici, fisiologici e patologici sul baco da seta (*Bombyx mori* Linn.). Fusi, Pavia: Fratelli. 172 pp.
44. Marzocchi, V. 1908. Sul parassita del giallume del *Bombyx mori* etc. *Arch. Parasitol.* 12:456–66
45. Morrill, A. W., Back, E. A. 1912. Natural control of white flies in Florida. *US Dep. Agr. Bur. Entomol. Bull. 102*
46. Ordish, G. 1967. *Biological Methods in Crop Pest Control.* London: Constable

47. Paillot, A. 1928. On the natural equilibrium of *Pyrausta nubilalis* Hb. *Int. Corn Borer Investig. Sci. Rep. 1927–1928,* ed. Tage Ellinger, pp. 77–106
48. Paillot, A. 1933. *L'Infection chez les Insectes.* Trevoux: G. Patissier. 535 pp.
49. Pasteur, L. 1874. On the use of fungi against Phylloxera. *C. R. Acad. Sci. Paris* 79:1233–34
50. Rolfs, P. H. 1897. A fungus disease of the San Jose scale. *Florida Agr. Exp. Sta. Bull. 41*
51. Sasaki, C. 1910. On the pathology of the jaundice of the silkworm. *J. Tokyo Coll. Agr.* 2:105–59
52. Sergent, E., l'Héritier, A. 1914. Essai de destruction des sauterelles en Algérie par le *"Coccobacillus acridiorum"* de d'Herelle. *Ann. Inst. Pasteur* 28:408–19
53. Speare, A. T., Colley, R. H. 1912. *The artificial use of the browntail fungus in Massachusetts.* Boston: Wright & Potter
54. Steinhaus, E. A. 1947. *Insect Microbiology*. Ithaca: Comstock
55. Steinhaus, E. A. 1949. *Principles of Insect Pathology*. New York: McGraw-Hill
56. Steinhaus, E. A. 1956. Microbial control—the emergence of an idea. *Hilgardia* 26(2):107–60
57. Steinhaus, E. A., Thompson, C. G. 1949. Preliminary field tests using a polyhedrosis virus to control the alfalfa caterpillar. *J. Econ. Entomol.* 42:301–5
58. Talalaev, E. V. 1960. The bacteriological method of control of *Dendrolimus sibiricus. Trans. Int. Conf. Insect Pathol. Biol. Control, Prague 1958.* pp. 51–57
59. Thomson, H. M. 1958. Some aspects of the epidemiology of a microsporidian parasite of the spruce budworm, *Choristoneura fumiferana* (Clem.) *Can. J. Zool.* 36:309–16
60. Thomson, H. M. 1958. The effect of a microsporidian parasite on the development, reproduction, and mortality of the spruce budworm, *Choristoneura fumiferana* (Clem.). *Can. J. Zool.* 36:499–511
61. Thomson, H. M. 1960. The possible control of a budworm infestation by a microsporidian disease. *Can. Dep. Agr. Bi-Mo. Prog. Rep.* 16(4):1
62. von Prowazek, S. 1907. Chlamydozoa. II. Gelbsucht der Seidenraupen. *Arch. Protistenk.* 10: 358–64
63. von Prowazek, S. 1912. Untersuchungen über die Gelbsucht der Seidenraupen. *Zentr. Bakt. Parasitenk. Infekt., I Abt. Orig.,* 87:268–84
64. Weiser, J. 1957. The diseases of *Lymantria (Porthetria) dispar* and *Nygmia phaeorrhea* caused by microsporidia. Vestnik Ceskoslovenske Zoologicke Spolecnosti (*Bull. Czech. Zool. Soc.*) 21:65–84
65. Weiser, J. 1957. The possibilities of biological struggle with *Hyphantria cunea* Drury. III. *Czech. Parasitol.* 4:359–67
66. White, G. F. 1904. The further investigation of the diseases affecting the apiaries in the State of New York. *NY Dep. Agr. Ann Rep. Comm. Agr., 11th, 1903,* pp. 103–14
67. White, R. T., Dutky, S. R. 1942. Cooperative distribution of organisms causing milky disease of Japanese beetle grubs. *J. Econ. Entomol.* 35:679–82
68. White, R. T., McCabe, P. J. 1951. Colonization of the organism causing milky disease of Japanese beetle larvae, 1939–1949. *US Dep. Agr., Bur. Entomol. Pl. Quarantine E-816*

AGRICULTURAL ENTOMOLOGY

D. Price Jones
Plant Protection Ltd., Jealott's Hill Research Station
Bracknell, Berks,[1] *Great Britain*

The following account of the history of agricultural entomology can deal with little more than the main trends and can mention only briefly some of the outstanding contributions. It is submitted that, however localized the origins of the subject, the significant advances have been global rather than national in character; what may have begun in one country has been continued or reflected in another. This fact, coupled with the vast literature, makes it impossible to apportion credit fairly for major contributions. The account is, therefore, frankly prejudiced to this extent: it reflects the author's main interests and experience in agricultural entomology and his limited linguistic facilities; it is heavily biased towards those developments reported primarily in English and to a much lesser extent in German and French. Nevertheless, it provides a framework by means of which developments in other countries can be placed in context.

Agricultural entomology may be defined in the first instance as the study of those insects harmful to or associated with agricultural and horticultural crops. This definition is normally extended to the study of the control of such insects and, almost imperceptibly, to the actual control process itself. It can be shown that this definition is too restrictive but for the present it may be accepted.

Other sections of this volume, for example those devoted to insect taxonomy, behavior, physiology, and toxicology, all have an important bearing on the development of agricultural entomology, but for the purposes of this chapter such aspects are referred to only to emphasize particular contributions to insect control. It follows, therefore, that in the special context of this history of entomology, the history of agricultural entomology must reduce largely to the tracing of man's progressive control over those insects harmful to his agricultural crops, and to those problems and structures associated with development of control measures.

Historical Phases

Sciences invariably pass through a descriptive phase before becoming quantitative and progressively theoretical. In this respect, agricultural entomology is no exception. However, as a biological science, it inevitably has

[1] Present address: 82 Shinfield Road, Reading RG2 7DA, England.

its roots in folklore, of which description is initially merely the written record—the occasional detectable peak on a vast mountain of imperfectly structured knowledge. Thus the depicting of locusts on tombs of the sixth dynasty at Saqqara in Upper Egypt dating from about 2470–2220 B.C. is of interest mainly because it signifies local knowledge of locusts. The book of Moses (ca 1500 B.C.), the Papyrus of Anastasi (ca 1200 B.C.), the decree of Emperor Shentsun of China (ca 1075) represent recordings of information and advice or mandatory instructions in all probability previously transmitted as folklore and mediated by both experience and religious or superstitious belief. The Greeks, Homer (950 B.C.), Democritus (460–377 B.C.), Aristotle (384–322 B.C.), and the Romans, Cato (234–149 B.C.), Virgil (70 B.C.), Marcus Pollio (13 B.C.), also appear to be drawing heavily on popular knowledge with some injection of personal experience and judgement (28). Certainly the daily business of insect control continued right up to the end of the nineteenth century to be largely a matter of traditional practices encoded in folklore and only slowly modified by experience. The existence of such a body of information was revealed by Li Schitsen (22) in China in 1596 in his pharmacology of herbs and, more recently, by McIndoo & Sievers (25) in their survey of records of the presumed insecticidal or repellent properties of a wide range of plants.

The beginnings of a truly descriptive (as opposed to a purely recording) approach are difficult to identify. Certainly the German, Albertus Magnus, Count von Bollstadt (1193–1230), managed to contribute what would appear to be significant original observations on pests and their control (28). In all probability, the printing press and the microscope invented in the fifteenth and seventeenth centuries, respectively, and Linnaeus' binary system of nomenclature published in the eighteenth century, were decisive factors in the progressive development of original description. Among the important contributors were Linnaeus in Sweden; Fabricius and Rostrup in Denmark; Börner and Appel in Germany; Kirby, Curtis, and Ormerod in England; Fabre, Trouvelot, and Réaumur in France; Riley, Fitch, and Glover in the USA.

The third phase, the emergence of agricultural entomology as an applied science with an injection of technology, was, in its multiple origins, almost contemporaneous with the descriptive phase, but its take-off was much later. The first (anonymous) textbook on pest science appeared in Nuremberg in 1709–1712; a patent on a process for impregnating wood with oil and spirit of tar was granted in South Carolina in 1716; Réaumur's observations on temperature and phenology in 1734 were followed by the rediscovery of derris by Rumpf in 1747 (28). And so it continued throughout the eighteenth and nineteenth centuries—isolated discoveries in biological control, of plant resistance to pests, of the significance of environmental factors, man-induced pest outbreaks, and side effects of insecticides.

The last decades of the nineteenth century saw the beginnings of substantial collation of this information in textbooks, in the transactions of learned societies, in university teaching, and in instructional bulletins. Agricultural entomology as a science was being rough-hewn from some extremely raw material. It was well into the twentieth century before more rigorous and more sophisticated analytical and experimental methods could accelerate advances in ecology, physiology, and behavior with concommitant contributions to agricultural entomology. Mathematics and chemistry, in a variety of ways, contributed greatly to these advances, and are discussed in greater detail below.

A Chronicle of Progress

A detailed chronology of crop protection, embracing pests, diseases, and weeds, has been provided by Meyer (28). This excellent compendium contains a great deal of historical information on insect pests, and sets this in an agricultural and social context. It is not proposed to repeat this exercise here. The intention, rather, is to outline the course of development of agricultural entomology throughout the ages and to comment, when appropriate, on the implications and some of the influences involved. The reader should refer to Mayer's compendium for further information on events up to the nineteenth century and also for the relevant literature references, unless other sources are indicated. Table I provides an evolutionary frame of reference for the brief chronicle and commentary that follow.

The advent of agriculture.—Archaeological research places the origin of settled agriculture at about 10,000 years ago, with at least some of the earliest settlements in a great arc in the Near and Middle East and leading ultimately to the development of such civilizations as those of Mesopotamia and Egypt. It is assumed that cereals provided the staple diet and that storage

TABLE 1. Agricultural entomology in an evolutionary context

Significant event	Years ago (1972)	Date
First land plants	400×10^{6}	
First insects	350×10^{0}	
First angiosperms	100×10^{6}	
First hominids	15×10^{6}	
First *Homo sapiens*	250×10^{3}	
First records of insects in human society	14×10^{3}	12,000 B.C.
Beginnings of agriculture	10×10^{3}	8000 B.C.
First record of insecticides	450×10	2500 B.C.
First descriptions insect pests	350×10	1500 B.C.
Burgeoning of descriptions	20×10	18th, 19th centuries
DDT and beginning of insecticide era	3×10	1939
Rachel Carson's *Silent Spring*	1×10	1962

from one harvest to the next was an essential factor in establishing a village and later an urbanized society. Archaeologists have identified the remains of granaries; for example, under Rameses II standard cereal silos were built throughout Egypt ca 1700 B.C. Many stored grain pests are known to attack standing crops in the Near and Middle East today. It is not surprising, therefore, that such pests attacked stored grain in the early years of civilization, and that they appeared to be sufficiently important to warrant special measures, including the use of insecticides, supported by legislation. Judging by the following species identified in the tomb of Tutenkhamun (ca 1350 B.C.), the pest fauna of grain could have been extensive: *Lasioderma serricorne, Giddium psylloides, Sitodrepa panicea, Oryzaephilus surinamensis, Rhizopertha dominica,* and *Tribolium castaneum.*

Locusts are also known to have been regarded as highly important pests. It is even suggested that the desert locust, *Schistocerca gregaria,* was probably the first insect recognized as a pest. At Seqqara in Upper Egypt there are carvings of locusts on tombs dating from 2470–2220 B.C. The Book of Exodus (ca 1300 B.C.) records the catastrophic damage caused by the desert locust swarms. Locust plagues were also recorded in China in 707 B.C. (42). There are some grounds for believing that the development of agriculture was partly responsible for the appearance of locusts as plagues.

According to Klemm (22), the Chinese must have acquired at least a modicum of entomology from their culturing of the silkworm about 3300 B.C.; 300 years later, silkworm culture was an important part of their agriculture. Around 1200 they used chalk and wood ash for the control of pests in enclosed spaces and insecticides of plant origin for treatment of seed. An outbreak of corn Pyralidae was recorded in 718. Parasitic insects were recognized as such about 400. In the third century, phenology was established as a science and it was known that the timely planting of crops ensured a measure of protection against pests. Arsenic was used as an insecticide about 200. Finally, by 100 B.C., entomology had advanced to the point where the development of dragonflies, midges, and cicadas could be described.

In these early civilizations there were therefore (*a*) a recognition of the importance of insect pests, (*b*) a recording of information, (*c*) advice and instructions on control measures, including the use of insecticides and the adoption of cultural practices, and (*d*) an incipient social infra-structure related to the requirements of control: truly, the beginnings of agricultural entomology.

The ascendancy of Greece and Rome.—To an ecologist, the Greek and Roman Empires had two features in common: they were based on the Mediterranean, with all that that implied in ecological terms; and they were founded and administered as city states, which meant that their food economy was based on importation, either from surrounding rural areas or from distant lands. It would be reasonable, therefore, to expect some common features in the insect pest fauna and major pest problems of the metropolitan areas of both empires and some similarity to those of the earlier civilizations.

Certainly locusts were an outstanding problem, as recorded in the writings of Homer (950 B.C.) and others. Caius Plinius Secundus (A.D. 23–79) reported the enactment of control legislation in Greece and of enforced collection in Lemnos.

According to Marcus Portius Cato (234–149 B.C.), granaries were sprayed with oil for protection against beetles. The Roman architect, Marcus Pollio (13 B.C.) recommended a specially constructed granary to prevent or retard heating of the grain, specifically as a protection against pests.

Daily contact with pest problems must certainly have generated ideas, accurate or inaccurate, about pests, the causes of pest attack, and the factors that influenced the activities of pests. Unfortunately the records are sparse and not very illuminating. However, there is evidence of an awareness of the importance of weather and the value of phenological observations.

It is possible to deduce that a great deal of practical application coexisted with religious or superstitious attitudes. Thus Homer records that fire was recommended for locust control and Straban that the intervention of Appollion Pomopion was invoked for the same purpose.

Insecticides and repellents were in use in this period and there is an unequivocal reference in the Talmud (A.D. 200–600) to legally imposed dosage rate limits for the control of stored product pests—a clear precedent for the residue tolerances of modern times.

In A.D. 304 the Chinese in Kwantung Province were deploying ants in the control of leaf beetles (*Clitea metallica*) on citrus. This is apparently the first record of the use of biological control (22).

The Dark Ages.—From the end of the Roman Empire until A.D. 1000, culture and agricultural entomology were both in eclipse. Doubtless, there was some connection. Between 571 and 630, followers of Islam, seeking protection against locusts, wrote the prayers of Mohammed on papers which they then displayed on poles in the fields. In 666, St. Magnus, Abbot of Flussen, repulsed locusts and other vermin with the staff of St. Columba. The success of these integrated techniques is not recorded. The scientific sceptic had not arrived.

About 900 an outbreak of *Agrotis occulta* occurred in Greenland, destroying both the pastures and the economy of the Norsemen colonizing the land.

The Renaissance.—As though to wash away the gloom of the European Dark Ages, soap was used in China in 1101 for the control of pests. Then, around 1200, Albertus Magnus, Count von Bollstädt, in one of the first manuscripts on natural science, provided a thorough description of pests known to him, and also instructions on pest control. The significance of this contribution lies in the inclusion of personal observations—from just recording to observing and recording, perhaps even to directed observation. Then, in 1320, a backward step: a lawsuit against the cockchafer was conducted before an ecclesiastical court in Avignon.

Forward again, in principle, in 1335: protective measures for insectivorous birds were promulgated in Zurich. In 1493, an artistic reminder appeared in Albrecht Dürer's 'Holy Family' that locusts continued to invade Central Europe; if not a reminder, perhaps a foreboding: in 1538 a locust invasion of Rumania resulted in a severe famine.

The period of the Renaissance in Europe had seen improvements in agricultural practice but not, in general, sufficient to meet the needs of a growing population. Towards the end of the period, the inevitable overseas expansion carried European agriculture to many parts of the world; with it went crop plants and, probably, some of the attendant insects, even if the pest infestations went unnoticed initially. A reciprocal effect of this expansion is seen in La Quintye's 1690 recommendation for the use of tobacco infusions for the control of bugs on pears.

In China, remote from the Renaissance in Europe, the control of locusts was legally enforced in 1182. In 1596 a detailed report was published on investigations of certain scale insects; also, Li Schitsen's pharmacology of herbs, in which insecticidal properties are recorded. Evidence of the continued interest in arsenic emerges from its use in seed treatment in 1630 (22).

However, the major contributions of the Renaissance to agricultural entomology had nothing to do with agriculture or insects. The printing press provided the means of recording and disseminating knowledge; the microscope extended the vision of man to the minute, and mathematics give him an opportunity to bring pattern into the variable and the mobile—all clearly of immense value in biology, including agricultural entomology, but their effects were to come much later.

The second agricultural revolution.—This historically evolutionary process (it was revolutionary only on an evolutionary time-scale) is commonly assigned to the eighteenth century but it had roots in the seventeenth and extended into the nineteenth. It was marked by the spread of new crops and cropping systems and by the introduction of new ploughing, drilling, and cultivating equipment. That this must have affected the insect pest fauna of agricultural crops is beyond question but there is very little information on the subject, beyond that relating to introduced crops, e.g. turnips into Britain. More significant, perhaps, was the spirit of enquiry that engendered the revolution—an enquiry that spread from the immediate crops and husbandry systems to the agricultural science behind them. Agricultural entomology was a small, but by no means insignificant, part of this movement.

To this period belong two eminent biologists who made major contributions to entomology. Réaumur, in France in the 1730s, began to publish the results of his memorable studies on insects, which included some original thinking on the subject of biocoenoses and the significance of host/parasite relationships in pest outbreaks (39). Most remarkable of all were his observations on the importance of temperature summation in determining a phenological period, an observation to be expressed two centuries later as Hopkins' bioclimatic law, and exploited in forecasting insect pest occurrence. In Sweden, Linnaeus advanced his concept of binary nomenclature and laid the foundations of a classification and information system that ultimately revolutionized biology; he also expressed views on biocoenoses that contributed to the beginnings of ecology.

Vallisneri's demonstration (1700) of the nature of insect parasitism was followed by much discussion of biological control and even some positive action, albeit of doubtful efficacy—German ordinances protected sparrows, woodpeckers, and ants. The possible resistance of winter wheat to Hessian fly, *Mayetiola destructor,* was discussed in the USA in 1785, and aphid resistance in the apple variety Winter Majetin was reported in England in 1831. Roberjet in 1787 used light traps for the control of the cochylis moth, *Eupoecila ambiguella.*

Progress with insecticides was signalled by the rediscovery or introduction of the botanicals, pyrethrum, derris, quassia, and tobacco leaf infusion. The hazards of pesticide use were underlined by the Frenchman Aucante in 1754, after observing arsenic poisoning in agricultural workers, but it was not until 1786 that France prohibited the use of arsenical and mercury steeps for seed treatment.

Industrial and imperial expansion.—By increasing agricultural productivity, the agricultural revolution released workers for other activities and, initially at least, provided the food to feed them. These were the contributions required of the agricultural economy before industrial development could begin its historic acceleration. Soon increasing industrial output demanded fresh markets, and the growth of the population, outstripping agricultural production, necessitated food importation. Imperial expansion was not a new phenomenon, but this time it had the drive of technology behind it, both in the weapons of conquest and the substance of commerce.

On the metropolitan front, advances in the more pure aspects of agricultural entomology from about 1850 onwards and well into the twentieth century, were mainly a continuation of trends detectable earlier—description, recording, and simple experimentation, all accompanied by increasing criticism. Perhaps criticism is the key to this development. The influence of superstition and religion had waned to the point where, within the field of applied biology, they were of very little significance. Already there was sufficient understanding of the climatic and other factors influencing pest outbreaks to make it unnecessary to seek explanation in supernatural intervention. The last animal lawsuit was a mere memory of 1830. In an allied field, Pasteur and de Bary had revealed the nature of infection, thus dispelling another source of evil spirits.

What biology, including agricultural entomology, lacked at the turn of the century was a well established tradition of critical appraisal. The pioneers were there and exemplary models were available, but it needed time to raise the general standard of biologists and to increase the output of reliable information. In particular, mathematics had scarcely begun to penetrate into biology.

Imperial expansion, aided and abetted by technology, had a profound effect on agricultural entomology. The movements of men and goods around the world ensured the spread of pests to new territories—San Jose scale from China to the USA, Colorado beetle from the USA to Europe, grass gall midges from the UK to New Zealand, and many others. These introductions, or the threat of them, stimulated the passing of quarantine laws, created opportunities for entomologists in official organizations, prompted the convening of international congresses, spawned learned and professional societies, and, inevitably, produced a spate of papers and bulletins.

It also had the effect of stimulating entomological work in countries hitherto lacking indigenous entomologists. French biologists were active in North Africa, West Africa, and Indo-China; the British in many parts of Africa, in the West Indies, India, the Far East, and, for a period, in Australasia; the Germans in various parts of Africa; the Dutch mainly in southeast Asia; the Italians in North Africa; and the Americans in Central and South America. It was not until the 1930s that nationals of many of the developing countries concerned began to play an important part in coping with their local problems and only with the demise of colonialism after World War II that a significant take-over occurred. Even in the past three decades, Europeans and North Americans have played a prominent, though decreasingly important, role in these countries through the entomological activities of the UN Food and Agriculture Organization and other international bodies.

Technology has contributed progressively to the advance of agricultural entomology by providing general and specialized equipment. The microscope was a major contribution: it led to significant advances in insect systematics, anatomy, histology, and physiology; the simple low-powered, wide-field binocular microscope became an indispensable adjunct to a great deal of ecological work, especially in relation to the soil fauna. Specialized equipment for insect extraction from plant material, soil, air, and water greatly facilitated sampling procedures. Progressive improvement in insectary equipment, particularly controlled-climate facilities, boosted investigations in behavior, physiology, and toxicology. The development of electronic equipment added further impetus. Analytical chemical methods, especially the physical methods developed in the past two decades, have transformed the approach to insecticides and to insect physiology and biochemistry. The introduction of desk calculating machines made it possible to bring mathematics into biology in an effective manner and the subsequent development of computers accelerated this process and also made a systems approach possible.

Technology has not only subscribed to agricultural entomology as an applied science, it has also contributed directly to the control of agricultural pests, partly through the provision of insecticides and the means to apply them, and partly through other economic inputs. The main developmental trends in insecticides are outlined below (pp. 325–27). The origins of application equipment must be sought in the vineyards of France, where the value of the crop and the severity of diseases and pests compelled resorting to control measures. In 1880, the firm of Vermorel in Villefranche produced what is claimed to be the first commercial spraying machine. The Swiss in 1886 and the Germans in 1888 followed with more advanced designs. The Germans are credited with the first powered sprayer, put into operation in 1904. In Europe generally, commercial sprayers were used almost exclusively on fruit crops, and vines. In the USA, cotton was also extensively treated but mainly with dusts. The advent of selective herbicides during World War II led to the development of so-called low-volume sprayers and the large-scale extension of spraying to field crops. This coincided with the appearance of a new generation of insecticides and the two factors together introduced what can fairly be called an insecticide era in the history of insect control. The use of aircraft was pioneered by the Americans in 1921 and adopted for an emergency operation against boll weevil on cotton in 1922. However, the major expansion of aerial spraying did not come until after 1945, when redundant military aircraft came into use. The introduction of ultra-low volume spraying from the air, followed some pioneering work in East Africa, in which special atomizing equipment was developed (43). Its extensive use came first in the USA, where malathion proved to be unexpectedly amenable to this form of application.

Industry was responsible for other socio-economic contributions to entomology. As the insecticide sector of the chemical industry developed, it began to employ an increasing number of entomologists. To supply those needs, academic departments developed entomological interests and were able to increase their research contributions to entomology. Industry supported much applied work at these centers. In due course, difficulties with insecticides stimulated academic research on side-effects and the search for alternative control measures.

Finally, agriculture, which was a prime contributor to the industrial revolution, was in turn affected by that revolution. Not only did industrialization and urbanization provide a convenient market for a wide range of agricultural goods, it also provided a range of inputs into agriculture, of which fertilizers, pesticides, and machinery were initially the most important. Later came capital and an injection of financial control into farm management. The mixed farms of Europe retained their general character until World War II and then began to change more profoundly. These changes are still taking place, farms and individual fields are getting larger, enterprises more specialized and larger areas are devoted to one crop, often one variety of that crop; also the market has come increasingly to demand high quality produce. All these changes are creating ecological and economic conditions that radically affect the practice of insect control. In the USA the same conditions arrived earlier and are apparently more severe.

The Advance of Numeracy

If description was an inevitable preliminary stage in the development of agricultural entomology, then numeracy was an essential prerequisite for the emergence of the subject as a mature applied science or technology. No doubt the need for quantification was felt in the early stages of civilization—a need certainly expressed in the early adoption of measures for grain, liquid, and land. Many references have already been made to the early recognition of the pest status of a number of insects. Such recognition implies, however faintly, some knowledge of, or belief in, the importance of the pests concerned; it also implies an awareness of the social implications, as testified by the frequent resort to legislation. For the historical record, the main task is, therefore, to trace the evolution of precision and confidence in quantification and the development of a philosophy of evaluation.

That this process of quantification must have progressed steadily, along with the spread and growth of arithmetical facility in commerce and everyday life, is a reasonable supposition. Even so, it had not reached a very high level by the middle of the ninteenth century. For example, John Curtis in his admirable record of pests of agricultural crops (11) is generally obliged to use vague phrases such as "a great many hundred acres (of turnips) were destroyed in Norfolk [by the turnip sawfly]". It comes as a surprise to see a financial assessment of loss as in "For so long back as 1786, Mr. Young stated that the turnip crop destroyed in Devonshire alone was valued at £100,000." One significant observation by the Swedish entomologist Bierkander, also reported by Curtis, reads: "I have often observed that a single worm [again the sawfly] has bitten from eight, twelve to twenty stalks in one place; if one destroys so much, what may not thousands do?" Entomological literature of the eighteenth and nineteenth centuries abounds in such attempts at quantification.

Howard (19) described the Rocky Mountain locust swarming flights:

> over the cultivated areas of the western States, including Kansas, Nebraska, Iowa, Texas, Oklahoma, a part of Missouri and other portions of the trans-Mississippi in the years 1874 to 1876. Migratory grasshoppers had been occasionally reported in that general region for many years, and in 1864 there were swarming flights . . . which did much damage and caused considerable alarm. Other flights were noticed in the intervening years between 1864 and 1874, but in the latter year the previous experiences dwindled into insignificance; growing crops in many states were devoured, farms were abandoned, the trend of settlers towards Kansas and neighboring States was stopped, and the direst consequences were predicted. Even the eastern newspapers contained startling accounts of the devastation. Thousands of people were said to be starving.

Such was the information available at the time. It hinted at quantities and intensities and it implied value judgements, but there was no real quantification. According to Riley, in a report published in 1877 (40), the damage done in 1874 amounted to many millions of dollars. It prompted him to suggest that a national commission should be formed to study the locusts and certain other pests. Such a commission was formed and later became the forerunner of the entomological service of the US Department of Agriculture.

In the first two decades of the twentieth century, biologists in general, and entomologists in particular, were struggling to come to terms with the variability of their material. This is illustrated by the plaintive report of Fryer (16): "As regards the different varieties of apple, it is not possible yet to say that any kind is either immune [to capsid injury] or specially susceptible, since facts obtained from one affected orchard were negatived by observations in the next." Even more illuminating is a description of *Kakothrips robustus* by C. B. Williams (55) in which measurements on antennae and antennal segments were recorded without a hint of mathematical control over the variation involved: this by one who twenty years later was pioneering the application of mathematical techniques to entomological material. Petherbridge (36) reported detailed investigations of the effect of numerous insecticidal formulations on apple sucker, *Psylla mali;* he employed replication in his experiments (quite a novelty at that time) but, in comparing the insecticidal efficiency of the formulations, he could use only the means of the replicates.

A further development was revealed in the methods employed by Dry (12) who attempted to measure the local and seasonal abundance of the swede midge in parts of Yorkshire, England, over the years 1912 to 1914. He observed percentage of plants attacked in two rows and accepted the figures only if they agreed reasonably closely; otherwise further counts were taken. This, of course, was merely an application of a traditional method used in the physical sciences. Dry was clearly analytical in his approach and was concerned with plant populations, including the effect of singling, the incidence of different broods, and the distribution of the midge in from the hedge, which he plotted graphically. He developed a swede midge index which he related to agronomic, ecological, and edaphic factors. Such then was the grappling with variability by British workers. Similar problems were engaging entomologists in many other parts of the world.

This variability of biological material and biological phenomena ensured that, until the development of concepts of frequency distribution (especially the normal curve and the binomial distribution) and of probability theory, little progress could be made in quantification. These ideas were conceived in another context in the eighteenth century and revived subsequently, but they had to wait until the 1920s for their first significant expression in agriculture. Contributions came from many quarters, but those of R. A. Fisher and his colleagues at Rothamsted Experimental Station in Britain were outstanding. Tests of significance were established for a variety of purposes—for sampling procedures, field observations, and laboratory results. Experimental designs were provided for field trials and these were elaborated and refined to make precision instruments for biological research. Similar work was undertaken in the USA but, rather surprisingly, was only slowly adopted in other parts of the world.

Another line of mathematical evolution in the field of entomology began to diverge in the 1920s. The contributions of Lotka (24) in the USA and Volterra (52) in Italy paved the way for subsequent theoretical exercises in population dynamics with important implications for the study of predator/prey relations and insect control: indeed, these initial exercises led ulimately to the more sophisticated computer models of the 1950s and 1960s.

The 1930s were also rich in new developments or elaborations of earlier innovations. Perhaps the outstanding advance was the development of probit analysis in toxicology, the contributions of C. I. Bliss (5) being paramount in this field. The technique was rapidly applied to the testing of candidate insecticides and formulations, and undoubtedly contributed directly to the intensification of insecticidal screening in the 1930s, and the subsequent flood of new products. In due course it also enabled official organizations to monitor the claims of commercial firms and to reject the less effective materials. It is tempting to speculate that the valuable tool created by Bliss and refined by others was misused to the extent that it elevated the concept of an LD_{50} or LD_{90} to a level that overemphasized the significance of kill in insect control.

In the 1930s also, C. B. Williams started a period of intensive research into the use of light traps in the study of insect migrations. His catches were so variable that the raw data were unsuitable for statistical analysis. He then began a study of various types of mathematical manipulations to render his data more amenable to analysis. He established a number of transformations that have been widely used since. When challenged about the mathematical validity of these transformations, he would reply that, if they brought order into chaos, that in itself was sufficient justification. It took courage to preach heresy when orthodoxy was the sanctuary of the majority—the mathematically inadequate. The application of transformations to the catches of light traps culminated in the use of the logarithmic series for representing the frequency distribution of species in communities, and provided the mathematical basis for the expression of diversity (56). The concept of an index of diversity contributed greatly to the subsequent emergence of a philosophy of ecosystem diversity and its application to pest management.

One intriguing investigation by Le Pelley (23) in East Africa in the 1930s may or may not have affected the attitudes of agricultural entomologists (citations have been few) but it certainly represents a modest landmark in the development of pest control and in the evolution of a mathematical approach. He studied the effectiveness of hand-collection on the control of *Antestia orbitalis* on coffee, showed how this tailed off asymptotically with continued picking and established a mathematical relationship. This experiment, supported by an analysis of estate records, demonstrated conclusively that hand-collection was ineffective and, even more important, it showed how a quantitative assessment could be used to determine policy, and how a demonstrable effect could be given mathematical support.

The 1940s saw the increasing quantification of many aspects of agricultural entomology, particularly those related to the use of insecticides. With the advent of relatively efficient pesticides, it became possible to demonstrate some of the losses attributable to pests and to express them in terms of lost yield, lowered quality, or financial loss. Much of this was done as a routine operation in the development of a new insecticide: it served to demonstrate what an insecticide could achieve on a given crop but could not be used to provide a reliable estimate of national losses. These latter remained highly speculative for many years—in fact are still largely speculative in the 1970s.

One interesting development in Britain emerged out of the need for an increased arable acreage during World War II, the ploughing out of old grassland and the consequent high incidence of wireworm (mainly *Agriotes*) damage. A wireworm survey was organized on a national scale, sampling methods were examined and eventually standardized, and the results pooled (13). Analysis showed what degree of reliability could be attached to the results—and how frequently a grower could be wrongly advised. It also provided information on the relationship between wireworm population and yield. Examination of survey costs, particularly in relation to density of sampling, costs of traveling and extraction, revealed in due course that to survey a field could cost more than to protect the crop with an insecticide. This is possibly one of the earliest examples of a systems approach to the control of a pest.

In Britain in the late 1940s, the extension service concerned with the wireworm survey, at that time recently reorganized as the National Agricultural Advisory Service, broadened its activities into damage assessment studies on a number of important pests, including the cabbage aphid, *Brevicoryne brassicae,* on Brussels sprouts and the cabbage root maggot, *Erioischia brassicae,* on cauliflowers. Such studies, continued in the 1950s and 1960s, largely under the direction of A. H. Strickland (50), provided much valuable information on the significance and nature of pest damage on a local and national scale.

Ordish (35) in 1952 published his *Untaken Harvest,* which broke new ground in the analysis of pest damage. He began to measure effects other than those on the immediate crop, effects that have since become known as external economies or diseconomies. This pioneering contribution had an important influence on pest assessment in official and industrial circles in Britain. In particular his concept of a cost/potential benefit ratio was rapidly adopted and developed, albeit with some qualifications.

The 1950s saw important developments in population ecology, derived in part from the progressive improvement in sampling techniques, attributable both to advances in sampling theory and to improved equipment. A growing interest in computer models subsequently strengthened this trend. Among the early contributors were Andrewartha & Birch (1) and Nicholson (34) in Australia; Huffaker (20) and Pimentel (37) in USA; R. F. Morris (30) in Canada; Milne (29) and Solomon (48) in UK; Franz (14) in Germany, and Bodenheimer and Schiffert (6) in Israel. Much of the output was controversial but there was no doubting that progress was being made in quantifying population dynamics. The greatest part of this effort was directed into agricultural and forest entomology.

The 1950s, too, saw the first gropings towards the application of systems analysis to the control of crop pests. K. E. F. Watt (53) and fellow Canadians were prominent pioneers in a movement that matured in the 1960s with the adoption of a broad philosophy of pest management. This philosophy stimulated the systems approach to pest problems in many parts of the world. The use of life tables, an idea borrowed from demography, was advanced by R. F. Morris (32); also, Morris (31), in 1959, introduced the concept of key factor analysis, later to be taken up and exploited by Varley & Gradwell (51). These developments were further pursued and elaborated in the next decade.

In the 1960s, environmental considerations became increasingly important, leading to tentative applications of cost/benefit analysis to the American scene by the Environmental Pollution Panel of the President's Scientific Advisory Committee (2) and, later, a notable attempt by Headley & Lewis (18) to develop an economic approach to public policy on the pesticide problem. Also, Simmonds (45) attempted a cost/benefit analysis on the work of the Commonwealth Institute of Biological Control. By the end of the decade, such considerations were becoming increasingly important, both in decisions on the funding of research work and in determining national policy in relation to particular pesticides and pesticides in general.

Finally, again in the 1960s, industrial research was coming increasingly under the influence of technological economics. Technological forecasting was being used to assess the prospects for particular lines of research or development in pesticides. Such techniques were also spreading into official agricultural research and administrative circles.

Numeracy in agricultural entomology was, as has already been mentioned, an inevitable concommitant of, and contributor to, the development of the subject. No significant advance, after the early descriptive stages, would have been possible without it. It does in fact reflect progress to maturity in agricultural entomology, both as an applied science and as a technology. Few advances, however, are achieved without a list of casualties. Entomologists are not equally endowed with mathematical facility, with the result that divergent views and even personal conflicts arose over the role of numeracy in the development of the science and its application. Regretfully, the development of numeracy is associated with abstraction and a certain remoteness from reality. The entomologist has, in the past 50 years, been translated from a natural historian to an applied scientist, and in the process may to some extent have lost his 'feeling' for insects in the field. Agricultural entomology rightly demands that its practitioners should be numerate and scientifically competent; perhaps it should also require that they remain true to their natural history origins. The agricultural entomologist with an ecological bent is well placed to exploit the best of both worlds.

The Contribution of Chemistry

According to Mayer (28) the acaricidal and insecticidal action of sulfur was known to the Sumerians about 2500 B.C.; chalk and wood ash were used for control of pests in enclosed spaces and botanicals for treatment of stored grain in China about 1200 B.C.; Democritus (460–377 B.C.) recorded the steeping of seed with infusions prepared from *Sedum acre;* Aristotle (384–322 B.C.) reported the long-established use of fumigants and Cato (234–149 B.C.) the use of oil sprays, oil and bitumen sticky bands and oil and ash, and sulfur and bitumen ointments; and in China, arsenic was already known as an insecticide in the second century B.C.

The frequency of references to stored grain pests in the first two millennia or so of recorded history suggest that such pests were critically important. It is tempting to speculate that, as a settled agriculture appears to have been essential for the development of civilization, the preservation of the crop between harvests was an inescapable requirement. The frequency of such references may, however, be explained in another way: the insecticides available were relatively ineffective and, requiring heavy dosage rates, were practicable only in special circumstances, as in granaries.

The historical significance of the so-called botanical insecticides is somewhat equivocal. Certainly the "discovery" of nicotine, derris, and pyrethrum from the sixteenth century onwards was merely an extension of a use which may well have originated some thousands of years earlier. As far as can be ascertained no special selection of insecticidally rich plants was practiced until comparatively recent times. This contrasts with, for instance, hemp, *Cannabis sativum,* in which high yielding narcotic varieties are vastly different morphologically from other varieties, indicating a long period of selection by man.

An inventory of insecticides in use in the nineteenth century would include sulfur, arsenicals, fluorides, soaps, kerosene, and various botanicals, of which nicotine and preparations from *Derris,* sabadilla, pyrethrum, and *Quassia* appear to have been most widely used. This was a transition period in which some deliberate, even if sporadic, attempts were made to standardize materials and to seek replacements with more desirable qualities.

The first four decades of the twentieth century can be regarded as a prelude to the subsequent explosive development of insecticides. These were years when crude materials began to undergo refinement, when standardization and specification became increasingly important, stimulating developments in bioassay techniques. This, too, was the time when a start was being made in the use of chemical synthesis to explore the relationships between structure and biological activity, a process almost certainly influenced by developments in the pharmaceutical field. In this way, insecticides such as DNOC and the organic isothiocyanates were discovered. Standardization of, for example, petroleum oils and botanicals was accomplished during this period (26).

The insecticide DDT was discovered by Dr. Paul Müller of J. R. Geigy A. G., Basle, Switzerland, in 1939 (27). Its existence was revealed to the Western Allies during World War II and its first important uses were for malaria and typhus control. It was, effectively, the first of the organochlorines and continued to be the most important at least for the next three decades. The discovery of DDT, and subsequently that of BHC (27), led to a spate of new insecticides derived from the chlorination of hydrocarbons. Many of these materials had relatively low mammalian toxicities and partly for this reason, but partly because of their persistence and wide spectrum, they came into widespread use quickly; some perhaps, too quickly, as later events were to suggest. Nevertheless, for better or for worse, these were the materials that opened up the commercial field for insecticides and it was with these that a whole host of formulations and methods of application were devised. It was with the aid of such materials, too, that many crop husbandry systems were established.

While the Western Allies were experimenting with DDT and BHC and even using them on a large scale, the Germans were making rapid progress with the development of organophosphates as insecticides. The discovery of insecticidal activity in the organophosphates, originally investigated as potential war gasses, and then as insecticidal substitutes for nicotine, is attributed to Dr. Gerhard Schrader and his colleagues in Farbenfabriken Bayer AG (27). Three of these materials, namely HETP, parathion, and schradan, despite their high toxicities, were subsequently used on a worldwide scale and pioneered a number of novel developments. In particular, the systemic activity of schradan pointed the way to an effective control of aphids and aphid-borne viruses, and, in fact, to the exploitation of systemically active materials for the control of a range of pests. The full development of the organophosphates was delayed initially by the high toxicities of the early materials but subsequently was accelerated by the side effects encountered with the organochlorines. The discovery that mammalian toxicity could be dissociated from insecticidal activity, despite the significance of cholinesterase inhibition in both groups of animals, gave added impetus to fresh exploration of organophosphates.

The existence of another class of insecticides, the carbamates, was foreshadowed by Swiss workers in the 1940s (17), but the first major success was the American carbaryl, which first appeared in the late 1950s (27).

Subsequently, many more products in this class have been marketed. Carbamates resemble the organophosphates both in being cholinesterase inhibitors and in possessing, as a group, a wide range of insecticidal characteristics. While not complementary to the organophosphates, they do at least facilitate the selection of insecticides appropriate to specific situations.

The early 1950s saw the first strong reactions to pesticide residues in food crops, mostly associated with DDT and the cyclodienes. The subsequent tightening of regulatory measures stimulated the refinement of analytical techniques and the development of biochemical methods for the detection of applied materials and their metabolites in plant and animal tissues and the soil. In the 1950s and 1960s, pesticide chemistry in all its ramifications was an important part of the burgeoning of organic chemistry and helped to stimulate, and subsequently benefited from, the concommitant developments in physical techniques. Paper chromatography and infrared and UV spectroscopy initiated the trend and the momentum was kept up by the introduction of thin layer chromatography, electrophoresis, gas liquid chromatography, mass spectroscopy, and nuclear magnetic resonance. These techniques supported synthetic effort and raised the efficiency of screening and also greatly facilitated residue, metabolite, and mode of action studies. Ironically, they also provided evidence subsequently used to restrict or terminate the use of certain insecticides.

Insects as Vectors of Plant Diseases

The detection and, in due course, control of insect-borne diseases of crop plants represent an important contribution from agricultural entomology to crop agronomy. One sector, that embracing the viruses, has been particularly attractive to entomologists and it is not surprising that some, in the discharge of their duties, have become translated from entomologists to virologists; K. M. Smith and Marion A. Watson are two eminent virologists who started their careers as entomologists.

Although diseases of plants, for example cereal rusts and mildews, were known from ancient times, no significant advances could be made in the study of these diseases until the microscope had been invented in the seventeenth century and used for the study of living material. Even then, it required the courage of a de Bary in the mid-nineteenth century to dispel the shiboleths of superstition and establish the significance of pathogenicity.

Insect-borne diseases were known long before the significance of insects as vectors was appreciated. Thus peach yellows was discovered in the USA in 1791 and potato leaf-roll in Germany in 1845 (28). Color break in tulips was known in the Netherlands even earlier but was not recognized as a disease. The role of insects as vectors of plant diseases was first reported by Takami in Japan in 1901 (9); he showed that the leaf-hopper *Nephotettix apicalis* could transmit stunt disease of rice. Subsequently, but without knowledge of Takami's work, Ball in the USA demonstrated in 1906 that *Eutettix tenellus* was responsible for transmission of curly top, a virus disease of sugar beet (3) but the etiology of this disease was not established until 1915 (46). Although subsequent work was to highlight the extraordinary importance of aphids as vectors of plant viruses, the first clear demonstration of virus transmission by an aphid was not achieved until 1920, when Brandes transmitted sugar cane mosaic with *Aphis maidis* (7).

From 1920 onwards, the work of Murphy (33) and Schultz & Folsom (44) incriminated aphids as vectors of potato viruses, and widespread investigations led to a better understanding of the significance of climate and ecological isolation in the production of healthy seed potatoes. Such work culminated in the establishment of seed production and certification schemes that contributed greatly to the advancing technology of ware potato production in the 1930s, 1940s, and 1950s. During the 1950s, insecticides, particularly systemically-acting organophosphates, eased the task of seed production and also made it possible to prolong the life of seed stock in ware production areas (8).

Sugar beet production has been severely affected by insect-borne viruses in most of the areas in which it has been practiced. Following the identification of curly top virus, intensive field work established the ecology of the vectors and in particular the significance of weeds in the initiation of outbreaks. In Eastern Europe *kräuselkrankheit* was found to be transmitted by the bug *Piesma quadrata* (54). In the Netherlands, Quanjer (38) identified sugar beet yellows as a virus disease transmitted by aphids. Subsequent work demonstrated the predominant role of *Myzus persicae* and led to some fascinating and highly productive studies on the inter-relationships between the viruses, vectors, and their host plants. The unique structure of the beet sugar industry was exploited in the development of effective control measures.

Some stimulating ideas on the ecological and evolutionary significance of insect-borne viruses were advanced by Kennedy in 1951 (21). In the late 1960s, some of the yellows viruses were recognized as mycoplasmas. However, the full significance of this distinction in relation to insect vectors has still to be determined.

Viruses and mycoplasmas are by no means the only plant pathogens transmitted by insects. Bacteria and fungi are also implicated. The whole field is well reviewed by Carter (9), who from 1930 onwards made major contributions to our knowledge of insects in relation to plant diseases (his investigations into mealybug wilt of pineapple, a systemic toxemia associated with mealybugs, saved the Hawaiian pineapple industry from total collapse).

The discovery of an elm disease (subsequently called Dutch elm disease) in the Netherlands in 1919 was followed by further reports from many European countries. The first report from England was in 1927 and from the USA in 1930. The introduction of the fungus *Ceratosomella ulmi* into the USA probably occurred in elm logs also infested with *Scolytus* beetles (9). The movement westwards in the USA attracted a great deal of DDT spraying in the 1950s, particularly in university campuses and civic amenity areas, leading ultimately to a high mortality of American robins and a reaction against such spraying.

The introduction of cocoa into West Africa in the late nineteenth century was followed in due course by its colonization by indigenous capsids, especially *Sahlbergella* and *Distantiella.* These capsids caused considerable injury, although the importance of this did not become clear until the late 1930s. In the 1940s, the association of the injury with the transmission of *Calonectria rigidiuscula* by capsids was established (10). The advent of gamma-BHC enabled a start to be made in controlling a pest/disease problem that threatened the commercial viability of cocoa production in the area. Ultimately control became a practical possibility when powered knapsack sprayers and gang spraying became available as state-aided services in the 1950s.

Developmental Patterns in Agricultural Entomology

Other chapters in this volume deal with the history of developments in insect systematics, physiology, ecology, behavior, and biological control, all of which have contributed immensely to the development of agricultural entomology. No significant progress in the accumulation of knowledge about specific pests could be made until reasonably accurate identification was possible and this in turn demanded a framework of systematics. Similarly, advances in physiology and ecology led to a much better appreciation of the relationship between particular insects and their environment. Studies in insect behavior provided more accurate information on which to base control measures, whatever their nature. The development of toxicology provided the platform for the blast-off of pesticides in the 1940s. Biological control, by its very nature an applied science, provided solutions to the control of certain severe pests not amenable to restraint by pesticides or other measures; it also assumed the dual role of foil and complement to pesticides, a role that ultimately gave the impetus to the integrated control movement.

In acknowledging these contributions, it is as well to record that they were rendered possible by the earlier developments in the physical sciences and scientific philosophy. These in turn benefited from, and contributed to, the industrial revolution, which was itself generated by an agricultural revolution. Thus, these vital contributions to agricultural entomology had their roots in increased agricultural productivity which inevitably created pest problems of increasing economic significance. These feed-back loops are detectable at all stages of development of agricultural entomology and, indeed, are manifestly operative at the present time.

I am tempted, as others must have been, to draw comparisons between the eighteenth, nineteenth, and twentieth centuries in the rate of development of agricultural entomology. If the comparison is based on the mere accumulation of knowledge, there can be little doubt that progress in the eighteenth and nineteenth centuries was lamentably slow compared with that in the twentieth century. However, if some weight be attached to the development of a scientific attitude and establishment of a technology, then the earlier centuries made some very considerable contributions. Nevertheless, one cannot but be impressed by the increasing tempo of development in the opening decades of the twentieth century and by the tremendous acceleration in the period 1930–1970. This latter period, which has been by far the most productive in the accumulation and structuring of knowledge and the developing of control measures, deserves more detailed consideration.

The *Journal of Economic Entomology,* published without a break from 1908 to the present time, provides an invaluable record of the development of applied entomology, including agricultural entomology, in the USA during that period. It is not proposed to analyze this record in detail, but an analysis of the main trends is both feasible and illuminating. Table 2 indicates the percentage of papers published in four main categories. Papers not specifically concerned with control measures were classified under general biology; those concerned with control measures were assigned to insecticides, biological control or other measures or, occasionally, apportioned to two categories. Some decisions were difficult and so could fairly be described as subjective. Nevertheless, with a little interpretation, again perhaps somewhat subjective, the trends reveal what most entomologists would probably regard as the developmental pattern over the past half century.

Papers on general biology (including records of pest incidence and damage, bionomics, ecology, and physiology) were frequent during the 1920s and 1930s but by the late 1930s were being outnumbered by reports on laboratory and field testing of insecticides, these latter heralding developments in the 1940s and 1950s, when pesticide papers clearly dominated the contents of the journal. In the 1960s the proportion of pesticide papers fell to pre-war levels but without any apparent increase in papers devoted to biology. During this period a significant proportion of papers on insecticides was concerned with the consequences of insecticide use (resistance, residues, wildlife effects) rather than the control measures themselves. The continued

TABLE 2. Patterns in applied entomological research over a period of 44 years, as reflected in *J. Econ. Entomol.*

	Papers (percent) 1927–1970									
	1927	32	37	42	47	52	57	62	67	70
General biology	45	40	27	28	13	13	23	16	20	22
Insecticides	44	46	58	60	76	79	64	62	42	43
Biological control	3	7	6	8	4	3	7	8	9	6
Other measures	8	7	8	3	7	4	6	14	29	29

low proportion of biological papers in the 1950s and 1960s is partly an artifact in that much of the biological work was specifically directed at one or other method of control and hence appears in one or other of those categories, possibly reflecting changes in methods of funding research projects. On the whole, the biological content of the journal increased greatly in the 1960s. The relatively low level of incidence of papers on biological control can be partly explained by the seduction of contributions by the *Journal of Insect Pathology* (later, *Journal of Invertebrate Pathology*) launched in 1959. It may also reflect the limited allocation of funds to such work, a state of affairs of which workers in biological control have long complained.

The outstanding trend in the 1960s was the increasing emphasis on novel methods of insect control. This trend was probably prompted by a number of factors. The rapid development of pesticide usage in the late 1940s and early 1950s outstripped the essential complementary biological work and resulted in many instances of pest resurgence, resistance, residues in food for human conumption, and effects on wildlife. The reaction was a diversion of effort to the study of other methods of control. The success of the screw-worm campaigns in Curacao and Florida probably reinforced the trend. It was certainly heavily financed by the US Department of Agriculture in the early 1960s. Then came the inevitable boost from the public controversy following the publication of Rachel Carson's *Silent Spring.*

The above analysis embraces medical and veterinary entomology, as well as agricultural entomology, but the main trends apply specifically to agricultural entomology, which is in any case the predominant interest in the journal.

That the pattern has been reflected in other countries, but with local variations, is incontrovertible. Outside the USA, however, the commercial development of pesticides was somewhat slower and less intensive. Side-effects were less severe and the reaction less vigorous.

The Social Infra-Structure of Insect Control

In the early days of civilization, the whole population was closely associated with the production of food and, therefore, potentially concerned with the control of crop pests. By way of comparison, the position in (for example) Britain in 1970 was that less than 3% of the population was engaged in food production and probably a much smaller proportion was directly involved with pest control in the field. This change in the involvement of the population was associated with the development of a progressively more complex structuring of that part of the population directly or indirectly engaged in pest control.

For many thousands of years, the individual cultivator was relatively independent, although subject to some limited measure of guidance or control from priest, administrator, or scribe. Occasionally, certain control measures were legally enforced. This condition persisted through into the Renaissance in Europe when the growth of commerce must have included trafficking in insecticides, among other remedies. Still later came an injection of science, with the subsequent involvement of entomologists in special commissions and in quarantine arrangements. In the nineteenth century, commercial participation increased markedly and so, too, did official intervention and academic involvement. In the twentieth century the commercial sector grew in size and in complexity and came to encompass production, distribution, and application and to include supporting research and advisory services. University courses developed to cope with increasing entomological knowledge and the greater demand for such knowledge. Official circles came increasingly to be concerned with the approval and registration of pesticides, control of their use, and the monitoring of their direct and indirect effects. One should not forget the popular educators nor the population at large that acquired instant knowledge and involvement with insect control; the wheel has almost come full circle.

The evolution of attitudes in the development of agricultural entomology is a subject well worthy of study. Here it can attract only a brief comment. It is suggested that the actions of primitive cultivators were determined by accumulated experience. Superstition may well have been a distortion of experience, subject to some cultural sophistication. While superstition and, later, organized religion mediated the attitudes of the people, progress towards an objective appraisal of pest problems was slow. The ultimate adoption of a scientific attitude was perhaps inevitable but it had to await the nineteenth century for substantial progress. The sharpening of this attitude is one of the characteristic features of agricultural entomology in the twentieth century but, lest it be thought that the development of the subject is now along purely objective lines, the progress of the integrated control bandwagon should be considered.

That sciences and technologies have fashionable areas of research from time to time is well known. Economic entomology has had nothing quite so fashionable as integrated control. With vague origins in the work of Ripper (41) and Bartlett (4) in the 1940s and 1950s, it was given definition and purpose by Stern et al (49) in 1959. In essence, it was the integration of chemical control with the natural biotic factors but the concept was later enlarged to include the integration of all elements in the environment determining pest abundance (15). This extension changed the status of integrated control from a practical technique to a philosophy but in no way impeded its development. Indeed it gathered strength from its very breadth: it was all things to all entomologists. To the pesticide entomologist, it was predominantly a reaffirmation of faith in biology; to the practitioner of biological control it offered an enlarged horizon and a place in the sun; to the ecologist it provided a positive approach as an alternative to negative, if justified, criticism; to the politician it presented an arena with popular appeal and room for maneuver. But, above all, it acquired the characteristics of a religious movement, with its own priesthood, faithful following, and body of doctrine. Such indeed was its strength.

This movement received support from the public reaction to *Silent Spring*; it won both acceptance and funds. It led ultimately to the formation of a special panel within FAO to coordinate such effort, particularly in developing countries (47). It also led to a broadening of the interests of the Organisation Internationale pour la Lutte Biologique and ultimately to the formation of a truly global organization, the International Organization for Biological Control.

Agricultural entomologists may be scientists. They are certainly human.

Acknowledgments

Valuable information on historical developments in Germany was supplied by Professor Dr. W. Madel, Cela Landwirtschaftliche Chemikalien GmbH, Ingelheim/Rhein, and on developments in France by Professor L. Bonnemaison, Director of Research, Central Station for Agricultural Zoology, National Center of Agricultural Research, Versailles.

LITERATURE CITED

1. Andrewartha, H. G., Birch, L. C. 1954. *The Distribution and Abundance of Animals.* Chicago: Univ. Chicago Press. 782 pp.
2. 1965. *Restoring the Quality of Our Environment.* Rep. Environ. Pollution Panel, President's Science Advisory Committee. Washington. 317 pp.
3. Ball, E. D. 1906. The beet leafhopper (*Eutettix tenella*). *Utah St. Agr. Coll. Exp. Sta., 16 Ann. Rep., 1904–1905*
4. Bartlett, B. R. 1956. Natural predators: can selective insecticides help to preserve biotic control? *Agr. Chem.* 11:42–44, 107–9
5. Bliss, C. I. 1934. The method of probits. *Science* 79:38–39
6. Bodenheimer, F. S., Schiffer, M. 1952. Mathematical studies in animal populations: I. A mathematical study of insect parasitism. *Acta Biotheor. Ser. A.* 10:23–56
7. Brandes, E. W. 1919. The mosaic disease of sugar cane and other grasses. *US Dept. Agr. Bull.* 829
8. Broadbent, L. 1969. Disease control through vector control. In *Viruses, Vectors and Vegetation,* 593–630. New York: Interscience. 666 pp.
9. Carter, W. 1962. *Insects in Relation to Plant Disease.* New York: Interscience. 705 pp.
10. Crowdy, S. H. 1947. Observations on the pathogenicity of *Calonectria rigidiuscula* (Berk. and Br.) Sacc. on *Theobroma cacao* L. *Ann. Appl. Biol.* 34:45–59
11. Curtis, J. 1860. *Farm Insects: Being the Natural History and Economy of the Insects Injurious to the Field Crops of Great Britain and Ireland and also Those Which Infest Barns and Granaries, with Suggestions for Their Destruction.* Glasgow: Blackie & Son. 528 pp.
12. Dry, F. W. 1915. An attempt to measure the local and seasonal abundance of the swede midge in parts of Yorkshire over the years 1912 to 1914. *Ann. Appl. Biol.* 2:81–108
13. Finney, D. J. 1941. Wireworm populations and their effect on crops. *Ann. Appl. Biol.* 28:282–95
14. Franz, J. M. 1950. Uber die genetischen Grundlagen des Zusammenbruchs einer Massenvermehrung aus inneren Ursachen. *Z. Angew. Entomol.* 31:228–60
15. Franz, J. M. 1962. Definitions in biological control. *Int. Congr. Entomol., 11th, 1960,* 2:670–74
16. Fryer, J. C. F. 1914. Preliminary notes on damage to apples by capsid bugs. *Ann. Appl. Biol.* 1:107–12
17. Gysin, H. 1954. Some new insecticides. *Chimia* 8:205–10
18. Headley, J. C., Lewis, J. N. 1967. *The Pesticide Problem: An Economic Approach to Public Policy.* (Resources of the Future Inc., Washington, DC) Baltimore: Johns Hopkins Press. 141 pp.
19. Howard, L. O. 1930. A History of Applied Entomology, (Smithsonian Misc. Coll.), Vol. 84. Washington, DC: Smithsonian Institute. 564 pp.
20. Huffaker, C. B., Kennet, C. E. 1953. Developments toward biological control of cyclamen mite on strawberries in California. *J. Econ. Entomol.* 46:802–12
21. Kennedy, J. S. 1951. A biological approach to plant viruses. *Nature* 168:890–94
22. Klemm, M. 1959. Entomologie und Pflanzenschutz in China. *Nachr. Bl. Deutsch Pfl. Schutz. d. Braunschweig* 11:121–24
23. Le Pelley, R. H. 1935. Observations on the control of insects by hand collection. *Bull. Entomol. Res.* 26:533–41
24. Lotka, A. J. 1923. Contribution to quantitative parasitology. *J. Wash. Acad. Sci.* 13:152–58
25. McIndoo, N. E., Sieverts, A. F. 1924. Plants tested for, or reported to possess, insecticidal properties. *US Dep. Agr. Dep. Bull. No. 1201,* 61 pp.
26. Martin, H. 1940. *The Scientific Principles of Crop Protection.*

London: Edward Arnold, 3rd ed. 385 pp.
27. Martin, H. Ed. 1971. *Pesticide Manual.* London: Brit. Crop Protection Council, 2nd ed. 495 pp.
28. Mayer, K. 1959. *4500 Jahre Pflanzenschutz. Zeittafel zur Geschichte des Pflanzenschutzes und der Schädlingsbekämpfung unter besonderer Berücksichtigung der Verhältnisse in Deutschland.* Stuttgart: Eugen Ulmer
29. Milne, A. 1957. The natural control of insect populations. *Can. Entomol.* 89:193–213
30. Morris, R. F. 1960. Sampling insect populations. *Ann. Rev. Entomol.* 5:243–64
31. Morris, R. F. 1959. Single factor analysis in population dynamics. *Ecology* 40:580–88
32. Morris, R. F., Miller, C. A. 1954. The development of life tables for the spruce budworm. *Can. J. Zool.* 32:283–301
33. Murphy, P. A. 1923. On the cause of rolling in potato foliage; and on some further insect carriers of leafroll disease. *Sci. Proc. Roy. Dublin Soc.* 17:163–84
34. Nicholson, A. J. 1933. The balance of animal populations. *J. Anim. Ecol.* 2:132–78
35. Ordish, G. 1952. *Untaken Harvest.* London: Constable. 171 pp.
36. Petherbridge, F. R. 1916. Spraying for apple sucker (*Psylla mali*). *Ann. Appl. Biol.* 2:230–34
37. Pimentel, D. 1961. Species diversity and insect population outbreaks. *Ann. Entomol. Soc. Am.* 54:76–86
38. Quanjer, H. N. 1934. Virus yellows of sugar beet. *Tijdschr. Pl. Ziekt.* 40:201–14
39. de Réaumur, R. A. 1734–1742. *Memoires pour Servir a l'Histoire des Insectes,* 6 volumes. Paris: Imprimerie Royale
40. Riley, C. V. 1877. *The Locust Plague in the United States.* Chicago: Rand, McNally. 236 pp.
41. Ripper, W. 1958. Experiments on the integration of biological and chemical control of insect pests. *Proc. Int. Congr. Entomol, 10th, Montreal, 1956* 3:479–93
42. Roffey, J. 1970. *The Anti-Locust Research Centre, London: A Concise History to 1970.* London: Anti-Locust Res. Centre. 36 pp.
43. Sayer, H. J. 1959. An ultra low volume spraying technique for the control of the desert locust, *Schistocerca gregaria* (Forsk.). *Bull. Entomol. Res.* 50:371–86
44. Schultz, E. S., Folsom, D. 1923. Transmission, variation, and control of certain degeneration diseases of Irish potatoes. *J. Agr. Res.* 25:43–117
45. Simmonds, F. J. 1967. The economics of biological control. *J. Roy. Soc. Arts* 115:880–98
46. Smith, R. E., Bonquet, P.A. 1915. Connection of a bacterial organism with a curly leaf on the sugar beet. *Phytopathology* 5:335–42
47. Smith, R. F., Reynolds, H. T. 1966. Principles, definitions, and scope of integrated pest control. In *Proc. FAO Symp. Integrated Pest Contr. October 11–15, 1965.* 1:11–17
48. Solomon, M. E. 1949. The natural control of animal populations. *J. Anim. Ecol.* 18:1–35
49. Stern, J. M., Smith, R. F., van den Bosch, R., Hagen, K. S. 1959. The integrated control concept. *Hilgardia* 29(2):81–101
50. Strickland, A. H. 1954. Assessment of cabbage aphid damage in commercial Brussels sprouts crops. *Plant. Pathol.* 3:107–17
51. Varley, G. C., Gradwell, G. R. 1970. Recent advances in insect population dynamics. *Ann. Rev. Entomol.* 15:1–24
52. Volterra, V. 1926. Variazioni del numero d'individui in specie animali conviventi. *Mem. Accad. Lincei, Ser 6* 2(3):31 112 (In *Animal Ecology with Special Reference to Insects,* transl. R. N. Chapman, 1931. New York: McGraw Hill. 464 pp.
53. Watt, K. E. F. 1961. Mathematical models for use in insect pest control. *Can. Entomol.* Suppl. 19, 62 pp.
54. Wille, J. 1928. Die durch die Rübenblattwanze erzeugte Gräuselkrankheit der Rüben. *Arb. biol. Reichsanstalt f. Land- u.*

Fortwirtschaft 16:115–67
55. Williams, C. B. 1915. The pea thrips (*Kakothrips robustus*). *Ann. Appl. Biol.* 1:222–46
56. Williams, C. B. 1964. ***Patterns in** the Balance of Nature and Related Problems in Quantitive Ecology*. London: Academic. 324 pp.

MEDICO-VETERINARY ENTOMOLOGY: A GENERATION OF PROGRESS

CORNELIUS B. PHILIP[1] AND LLOYD E. ROZEBOOM
California Academy of Sciences, San Francisco and
Department of Pathobiology, Johns Hopkins School of Hygiene and Public Health
Baltimore, Maryland

INTRODUCTION

Until it was discovered, roughly within the generation just passed, that the blood-sucking pests of man and his animals could carry disease-causing organisms, the chronicles of the previous centuries have recorded only the annoyance to the (nevertheless often ailing) hosts. About 550 B.C., Homer described an apparently sick dog: "there lies Argos, the hound, full of kynoraestes" (a name later Latinized in modern taxonomy for a genus of acarids; we suspect the poet alluded to the familiar castor bean or dog ticks of the Mediterranean area, now known to be vectors of disease agents).

A sometimes fatal "stranger's disease" in ancient Persia was reported in diplomatic dispatches to European courts to affect travelers (but not local residents) staying at caravansaries; such travelers also suffered severe attacks by the ill-reputed Miana bug or teigne (presumably an argasid tick). Since a virulent form of relapsing fever has been known in certain ornithodorine-infested caves in the Middle East in modern times (Adler, *in litt.*), it is possible that these early implications antedate by many decades the presumption that a similar disease is caused by tick bites in native huts in tropical Africa as described during Livingstone's missionary travels as discussed later.

Inevitably, many of the discoveries which quickly matured medico-veterinary entomology as a discipline have also been basic to ancillary entomological specialization considered in previous chapters, notably systematics, pathology, genetics, and behavior. This permits us to minimize significant, overlapping historical events, and to restrict this article to a requisite reasonable length. At best, space will permit our tracing progress of special importance to the advancement of knowledge of only a few of the known arthropod-carried pathogens.

A sad prefatory comment to our chapter relates to the timeworn observation that wars, like death and taxes, have inevitably accompanied the progression of civilization, and many of the monumental advances in our subject have been accelerated by military or quasi-martial needs. In part, this has

[1] Formerly Director, Rocky Mountain Laboratory, US Public Health Service, Hamilton, Montana

resulted from the movement of susceptible troops into health hazards in new environs. Examples are the military outbreaks in endemic foci of dengue, malaria, sandfly fever, and certain rickettsial infections. Ironically, troops in the Pacific and Oriental theaters during World War II simulated sentinel animals in discovering new foci of scrub typhus on remote islands and in jungles of many countries. Countermeasures required the rapid deployment of team efforts—the call upon many skills which widened the experience of professional entomologists.

The Golden Age: Discoveries of the Role of Arthropods as Vectors of Disease Agents

The beginnings of medical entomology are also those of medicine and parasitology. Primitive man attributed illness, pestilence, and other disasters to vengeful gods or malevolent spirits, beliefs which continued through the Middle Ages and persist to some extent even today. The theories of Hippocrates and Galen that a balance of the sanguine, phlegmatic, choleric, and melancholic humors of the body determined a person's state of health or disease represented a step forward in that it focused attention on the body rather than on supernatural forces. Also, the association of the larger parasites and clinical symptoms often were too obvious to be ignored. Hoepli (1959) writes: ". . . it is unthinkable that *Ascaris,* and also probably *Enterobius,* as well as ectoparasites . . . could have been overlooked. . . . The clinical symptoms connected with the infection by ascaris, enterobius, taenia, and hydatid cyst were well known from early times and are already mentioned in the Hippocratic collection." Hoepli states further that as the common estoparasites of man and animals were well known to primitive races of man, unquestionably this was true also of prehistoric man.

There were, however, other diseases, the causative agents of which could not be seen with the naked eye. With the increasing concentration of the human population in agricultural communities and cities, the way was open for epidemics. With no knowledge of the etiology of these diseases, it was inevitable that such catastrophes would be attributed to the machinations of malicious devils or a vindictive god. Geddes Smith (1941), who cautions us against smug self-satisfaction in comparing our state of knowledge against the ignorance of our ancestors, points out that even in early times there was one hard and obvious fact: whatever it might be that caused a disease to break out, it could spread from one person to another. Contagion could not be denied and the obvious question was: what was being passed from one individual to another? Evidence that the ancients had some understanding of contagion is found in Mosaic law. The Jewish priests had the power to inspect and isolate lepers. This practice was followed through the Middle Ages, and according to Newsholme (1927), this was "the first great feat of preventive medicine" in that it led to almost the complete disappearance of leprosy in Western Europe, except for Norway, where these measures were not followed. Fracastorius, a doctor in Verona, in his treatise *De Contagione* published in 1546, spoke of seminaria morbi which could reproduce themselves among human populations.

One of the most significant advances was Leeuwenhoek's (1632–1723) development of the microscope; with improvement, these lenses were able to magnify up to 300 diameters. He devoted the latter part of his life to the study of the microorganisms he discovered with his microscope, and in a letter written to the Royal Society in October 1676, he gave what appears to have been the first description of bacteria. One of his classics is his description of a flea as a creature "which keeps a dog from brooding on being a dog." Chandler (1944) calls Leeuwenhoek one of the greatest explorers of all time. It is of interest to note that Kircher was using simple lenses when Leeuwenhoek was still a very young man. In fact, it is incorrect to say that Leeuwenhoek invented the microscope; he improved it. The origin of the use of magnifying lenses is obscure.

In 1687, Bonomo and Cestoni showed that scabies is caused by mites which burrow into the skin and can be transferred from person to person. Chandler considered this observation of special significance because it was the first designation of a specific organism as the causative agent of disease. However, years before this in 1656, an Italian Jesuit, Anathasius Kircher, examined the blood of plague patients and found "worms". These probably were roulets of red blood cells. Kircher thought that innumerable animaculae were formed in the air and organic matter as a result of morbific exhalations which took place when Saturn and Mars were in conjunction. His important contribution was to focus attention on an hypothesis of contagium animatum as the cause of infectious disease.

Another fallacy which had to be laid to rest was that of miasmas as a cause of disease. This belief also did not die easily. For example, in 1884, Sternberg who had studied carefully Laveran's and Richards' descriptions earlier of the malarial parasites and the evidence that these parasites were responsible for the disease, insisted that "the cause of the periodic fevers—malaria—*is of telluric origin*. The evidence in favor of this assertion would fill volumes, and is beyond question."

In this brief review, one cannot do justice to the many great minds which were involved in the relays of thought which gradually converged to bring into focus the specificity of infectious disease agents, the basic concept for elucidation of vectorship mechanisms. For example, Livingstone wrote about the tsetse fly in 1850: ". . . it is well known that the bite of this poisonous insect is certain death to the ox, horse, and dog . . . We lost forty-three fine oxen by its bite . . . the poison germ contained in a bulb at the root of the proboscis, seems capable, although very minute in quantity, of reproducing itself . . ." This was 7 years before Pasteur (1822–1895) presented his papers on lactic acid and alcoholic fermentation, while Koch was only 7 years old, and a half century before Bruce's experimental transmission of nagana or the discovery of *Trypanosoma gambiense* by Dutton.

As recently as the 1930s, general microbiological agents were often referred to as viruses. Viruses have since arbitrarily been restricted to filter-passing agents. In Pasteur's laboratory, Chamberland devised a porcelain filter to exclude bacteria during passage of liquid media. William H. Welch (1850–1934) called Loeffler and Frosch's filtration of foot and mouth virus in 1898 to the attention of Walter Reed (1851–1902), who then, with Carroll and Agramonte (US Senate Doc., 1911), showed that yellow fever could be induced in a volunteer by the filtrate of a serum which had passed through a bacterial filter. The growth of the science of virology since that time has been phenomenal; the Arbovirus Committee's catalog listed 274 entries through 1971, more are being found each year and they are more often isolated first from arthropod parasites than from their vertebrate hosts. Historically, this development was promoted by adaptation of laboratory animals and of tissue culture techniques as recounted below.

As was true of the association of specific microorganisms with diseases, the ultimate proof that arthropods are carriers of pathogens was often predicated on direct observations by primitive or later lay people for deductions by sophisticated doctors and scientists. As early as 1810, residents in Akita Prefecture, northern Honshu Island, associated "tsutsugamushi" (disease bugs) with the disease later called tsutsugamushi fever near the end of the century. The Lumba people of Kenya in East Africa avoided certain areas because they were in danger of being blinded from the bites of black flies (causing onchocerciasis). African natives had long associated the house-dwelling *Ornithodoros moubata* with severe illness. Sleeping sickness and nagana were associated with tsetse flies by people living in endemic areas of both West and East Africa. According to Curson (1932), the word tsetse is of ancient Bantu origin and referred not only to the fly, but to the disease caused by it. The Mandingo people in Gambia were able to distinguish between the dangerous "fly of the river" (*Glossina palpalis*) and the "fly of the baboon" (*G. morsitans*), which was considered to be harmless. Some people reportedly fed tsetses to their animals hoping to protect them against the nagana. Other protective measures were to avoid fly-infested areas, or to traverse them only at night.

Certain observant, early stock ranchers who accused ticks of being responsible for Texas cattle fever and for human spotted fever in Montana and Idaho, were amply substantiated by later investigators. Ricketts inserted the term "(Tick Fever?)" in the title of his first paper in 1906 before he had even confirmed the tick transmission tradition! Of course, there are always competing theories among laity locally that have to be proven fanciful as well. Kilborne, himself, had wondered why the western stockmen's beliefs about tick fever had not sooner been put to the test than the initial classic observations with Smith in 1891 (not fully reported by them until 1893).

W. A. Riley (1910) called attention to a report, attributed by Kircher in 1658 to the Italian physician Mercurialis (1530–1607), of the death of a nobleman who was "stung by a wasp" which had sucked up the lethal poison from a corpse. Kircher also credits Mercurialis with the theory that ". . . flies feed on the internal secretions of the diseased and dying, then flying away, they deposit their excretions on the food in neighboring dwellings, and persons who eat it are thus infected."

Russel wrote in 1952, "Who first formulated the mosquito-malaria theory we do not know." Nuttall previously noted in 1889 that "the common people" in malarious areas believed mosquitoes were responsible and he quoted Koch's observation that natives of the Isambra mountains attributed their febrile illnesses, which they acquired when they visited the lowlands, to the mosquitoes which bit them there.

There were several medically trained nineteenth century observers who assembled what was known of the epidemiology of malaria and yellow fever, and from this drew the conclusion that mosquitoes were responsible for the transmission of these diseases. These were the people of whom Boyce (1910) said, "It is the rule that all great movements and discoveries are heralded by premonitory signs. In other words, there are always 'John the Baptists' who go before. . . ." Boyce designated Beauperthy as the father of the doctrine of insect-borne disease. In 1853 Beauperthy asserted that yellow fever was not a contagious disease, and that mosquitoes, especially "the *zancudo bobo* with legs striped with white", introduced into the body of man by their bites, a poison similar to that of snake venom. He thought that the mosquitoes obtained this poison from extraneous decomposing matter.

Other well-known epidemiological contributions are those of Josiah Nott in 1848 and A. F. A. King in 1883. The latter, a gynecologist and obstetrician, discussed his pertinent theory with C. V. Riley and L. O. Howard, but received no encouragement because, according to Howard (1930), "the idea appeared to be altogether too far fetched." Erasmus Darwin (1731–1802), the sagacious physician-grandfather of Charles, was remarkably close to guessing an association of lice with typhus in his under-appreciated *Zoonomia* (1794–1796).

A third body of evidence leading to the realization that arthropods were carriers of disease organisms was supplied by the early helminthologists and insect morphologists who, in the course of their dissections, found many parasitic insects and helminths in the specimens they were studying. Included in this group of workers were Roesel (1705–1759), De Geer (1720–1778), Réaumur (1683–1757). Rudolphi (1771–1830) and von Listow in 1878 and 1889 listed a great many. Rudolph Leuckart (called the father of modern helminthology) from 1858 to 1867 proved that *Cucullanus elegans,* a nematode of fish, underwent development in *Cyclops;* as well as observing in 1867 development of the mouse spirurid, *Protospirura muris,* in the mealworm, *Tenebrio molitor.* This led Fedeschenko to the discovery in 1870 that *Cyclops* was also the intermediate host of a parasite of man, *Dracunculus medinensis.* Leuckart and Melnikoff in 1869 showed that the dog tapeworm,

Dipylidium caninum, passed through a part of its life cycle in the dog louse, *Trichodectes canis.* This was definite proof of an insect intermediate host of a vertebrate parasite prior to the discovery of pathogenic microorganisms.

And so we arrive at the epoch-making discovery of Patrick Manson in 1877 of the development of *Wuchereria bancrofti* in *Culex fatigans.* T. R. Lewis, a medical officer who was sent to Calcutta in 1869, had found *"Filaria sanguinis hominis",* and associated these microfilariae with chyluria and elephantiasis. When Patrick Manson returned to London in 1874 from Formosa, where he had become deeply interested in diseases of the Chinese, he learned of Lewis' discovery. Manson tells us: "On my return to China in 1876, I endeavored to ascertain if these parasites occurred also in China." He found them, and reasoned that since the microflariae did not develop in the blood cells, they must be the young "of some other animal." This other animal, the adult worm, was found in the lymph vessels, but the mystery remained as to why there were no intermediate forms between the active microfilariae in the blood and the adult in the lymphatics. To quote Manson again: "How does the parasite continue to pass from one human being to another? . . . some other agent must intervene . . . one which is capable of piercing the skin of the human body. . . . Now, the agent which occurred to me as being the most likely to effect the necessary step in the translation of the filaria was the mosquito." His observation of nocturnal periodicity strengthened this view. He allowed *Culex fatigans* to bite a person with microfilaremia; and by a series of dissections observed the growth of the worm from about 1/100 to 1/16 of an inch, during which it developed a mouth and an alimentary canal. "Manifestly, it was on the road to a new human host." Manson's supposition that the worm escaped from the mosquito into drinking water and so gained entrance into the new human host in no way lessens his factual contribution. The importance of Manson's direct involvement in Ronald Ross' (1897–1898) discovery of mosquito transmission of the malarial parasite is an oft-told story. Meanwhile, Smith and Kilborne had carried out their classic studies on Texas fever. It is interesting to note that, according to Howard (1930), Ross knew nothing of the Texas fever tick-transmission work prior to or during his own investigations. H. R. Rosen (1929) considers that medically-oriented investigators have not done justice to M. B. Waite (a colleague of Smith and Kilborne in the US Department of Agriculture), who, in 1891, convincingly detailed the mechanical transmission of proven bacterial fireblight of fruit trees by bees. Nor was Mitzmain, isolated in the Philippines, aware of this pioneer plant-disease work, when he also demonstrated, in 1913, mechanical dissemination by *Tabanus striatus* of the important Oriental livestock trypanosome disease, surra.

The point we wish to make is that there must have been cross-insemination of thought which ushered in the exciting pioneer era which we would designate as the Golden Age of Medico-Veterinary Entomology, distinguished by the following especially important discoveries. The events of this active period can only be outlined with some pertinent comments; it is interesting that, to the turn of the century, the microorganisms comprised protozoa or bacteria visible under the light microscope and concerned ticks and mites as well as insects.

Chronology of Pioneer Events in Medico-Veterinary Entomology

1891—Smith (1859–1934) & Kilborne (1858–1936): Piroplasmosis of cattle transmitted by *Boophilus annulatus,* the first proven arthropod-borne, as well as transovarially-passed, microorganism (protocols, 1893). First proposal by Curtice (1868–1939) of disease control by vector eradication (ultimately successful throughout southern U.S.).

1896—Bruce (1855–1931): Nagana trypanosomes by *Glossina pallidipes.*

1897–1898—Ogata (1854–1919): Simond: Infection of rat fleas with plague bacilli, independently.

1897—Ross (1857–1932): Developing human malaria parasites observed in *Anopheles.*

1898—Ross: Avian malaria transmitted by *Culex.*

1898—Grassi (1854–1925): Bignami and Bastianelli: Development cycle of human malaria parasites in *Anopheles.*

1898—Veeder (1848–ca 1915); 1901—Reed, Vaughan, and Shakespeare: Mechanical transmission of typhoid bacilli by flies. The role of flies as mechanical carriers of enteric and other pathogenic organisms had been suspected for a long time, and Grassi and others had attempted transmission as early as 1883. L. O. Howard (1857–1950) in 1909 proposed that the common name of *Musca domestica* be stigmatized as typhoid fly rather than house fly.

1900—Manson (1844–1922): Infection in London of two human volunteers with malaria by bites of *Anopheles* brought from Italy where they had been fed upon a malaria patient.

1900—Reed (1851–1902), Carroll, Lazear, Agramonte: Yellow fever transmitted by *Aedes aegypti,* the first mosquito-borne (later labeled) virus. The hypothesis had been presented in 1881 by Carlos Finlay, an observant physician in Havana who supplied the Yellow Fever Commission with material to establish its own mosquito colony. Creditably, Finlay actually attempted some earlier transmission experiments, which, however, were uncontrolled and inconclusive.

1902—Graham: Dengue transmitted by mosquitoes. (It was later pointed out that he must have used *A. aegypti* in his Beirut tests.)

1903—Bruce and Nabarro: African sleeping sickness by *Glossina palpalis.*

1905—Liston (1873–1950): Plague transmitted by *Xenopsylla cheopis.* The classic experiments by second Indian Plague Commission (1906–1907) should be familiar to all medical entomologists: rat to rat by fleas, but not by physical factors without them.

1906—Ricketts (1871–1910): Tick transmission of Rocky Mountain spotted fever and first promotion of guinea pigs as test animals. Liston had earlier used them to trap plague-infected fleas in Bombay.

1907—Mackie (1875–1944): Relapsing fever spirochaetes transmitted by lice. Tick transmission of "human tick fever" in tropical Africa, first recorded by Livingstone in 1857, had been independently confirmed in 1904–1905 by Dutton and Todd among others.

1908—Chagas (1879–1934): Trypanosomes, cause of disease later named for him, transmitted by bugs, *Panstrongylus megistus.*

1909—Doerr, Franz, and Taussig: Sand fly fever agent transmitted by bites of *Phlebotomus papatasi;* and later through their eggs by Moshkovsky et al in 1937.

1909—Nicolle (1866–1936), Comte and Conseil: Typhus rickettsiae transmitted by lice.

1913—Townsend (1863–1944): Incrimination of *Phlebotomus verrucarum* as the vector of *Bartonella* organisms, cause of Oroya fever in Peru; confirmed in 1929 by Noguchi and Shannon.

1913—Leiper (1881–1969): Development of *Loa loa* filariae in *Chrysops;* additional details given by Klein in 1915.

1913—Mitzmain (1882–1941): Mechanical transmission of surra in Oriental livestock by horseflies, *Tabanus striatus.*

1917—Davies and Weldon; Bruce, Byam, and Bacot: Transmission of trench fever rickettsiae by lice.

1917—Montgomery: Nairobi sheep disease, first virus transmitted by ticks, and in 1931 first shown to be transovarial by Daubney and Hudson, who also reported first arthropod-borne zoonosis, Rift Valley fever, in man and livestock.

1928—Sharp: Development of human filariid worms, *Acanthocheilonema perstans,* in biting midges, *Culicoides austeni.*

1930—Theiler (1899–): First use of white mice in yellow fever studies, which promoted whole field of arbovirology.

1933—Kelser (1892–1952): Transmission of western equine encephalitis virus by *Aedes aegypti.*

It is difficult to assign a definite date of discovery for many important findings. The leishmaniases were associated on epidemiological and experimental grounds with insects almost since the discovery of the parasite by Leishman in 1903, but the conclusive proof was not forthcoming until the completion of transmission experiments of kala azar by Swaminath, Shortt, and Anderson in 1942, and of Oriental sore by Adler and Ber in 1941.

Other disease agents have been shown to be arthropod-borne, both biologically and mechanically, and a vast amount of detail has been added to our knowledge of those listed above. There can be no period in medical entomology to equal the drama and excitement of the initial golden age of discovery, all within recent memory.

Steinhaus (1946), who provides many pertinent references, aptly states:

"Only through a thorough understanding of the general microbiology of insects and ticks can we acquire a clear picture of the manner in which these arthropods transmit disease-producing organisms."

Developments in Pertinent Systematics

The early taxonomists.—Among the insects first described by Linnaeus (Chapter V), it is natural that domestic pests, *Culex pipiens, Musca domestica, Aedes aegypti, Pulex irritans, Pediculus humanus, Cimex lectularius,* should be included in his initial system of binary nomenclature (the doubts are inconsequential, now, that the presently known *A. aegypti* could have been the original one found in Egypt). Additional genera and species of the blood-sucking and parasitic arthropods were established by Latreille, De Geer, Fabricius, Blanchard, Meigen, Say, and other pioneer entomologists of the Old and New Worlds. Stimuli to systematic work was the eventual proof that domestic and blood-sucking pests may also be vectors of disease. The colonial powers, with their vast tropical holdings at the beginning of this century were especially concerned, and the British Museum quite early commissioned F. V. Theobald to monograph the mosquitoes of the world, the first volume of which appeared in 1901 and the fourth and last in 1910. Howard, Dyar, and Knab produced another monumental, four-volume monograph of the American mosquitoes in 1912–1917, and Dyar's widely used *Mosquitoes of the Americas* appeared in 1925. One should acknowledge the pioneer studies of the Brazilian mosquitoes by Goeldi and Peryassu. In France, Blanchard's monograph of the mosquitoes appeared in 1905.

Modern influences.—These and burgeoning pioneer studies of other parasitic groups, for which space precludes review here, presaged the ultimate realization of the necessity for the specific identification by specialists of faunal elements implicated in disease-ridden areas. It soon became recognized that it was insufficient simply to refer to a given disease as carried by an *Anopheles* or a *Culex* mosquito, a rat flea or a dog tick, for example.

Unlike World War I, the mobility in World War II was a great stimulus to work on the taxonomy of medically important arthropods. Entomologists were assigned to epidemiological mosquito survey and control units, and many of these young men took full advantage of their presence in foreign lands to build large collections of insects which accelerated subsequent studies, notably in the British and U. S. National Museums, of pertinent faunas from all over the world.

The significant influence of new systematic concepts, based on biological species (Mayr) and dynamic populations (as developed in previous chapters), on medico-veterinary entomological taxonomy will not be augmented here. One impressive example is worth recalling. In Europe, the epidemiology of malaria remained an enigma because in many areas there were dense populations of *Anopheles maculipennis* but no disease. The literature during the 1920s abounded with references to anophelism without malaria. The great malariologists of Europe (Missiroli, Swellengrebel, de Buck, Roubaud, Hackett, and others) sought answers in the behavior of local mosquito populations in their various habitats. In 1926, Falleroni made the important discovery that "*maculipennis*" females could lay two kinds of eggs. Evidence continued to accumulate on the biological differences between these populations, and culminated in the cross-breeding experiments of Bates and the cytotaxonomic studies of Frizzi. It then became clear that instead of one species, there were seven species and subspecies, some of which were malaria vectors because they were attracted to man as a source of food, while others preferred to feed on nonsusceptible animals.

Genetics is playing an increasingly important part in sorting out factors in similar species complexes that became evident in part through puzzling vagaries in vectorship.

Progress in Epidemiology

Malaria.—Following the discovery in 1898 by Ross that bird malaria was transmitted by *Culex* mosquitoes, and by Grassi, and Sambon and Lowe, that human malaria was transmitted by *Anopheles* mosquitoes, an important step forward was the demonstration by Watson in Malaya, and by Christophers in India, that not all *Anopheles* mosquitoes were involved as vectors. In each region only one or a few species were sufficiently attracted to man to insure transmission of the parasites from one person to another. This led to the principle of species sanitation. In all malarious areas of the world, malariologists began sorting out the dangerous malaria vectors. Their breeding habitats were delineated, so that it now was possible to concentrate resources available for malaria control against each specific vector. Early control consisted of application of larvicides to breeding places, or destruction of these habitats by drainage, filling, water level management, or other engineering and environmental sanitation procedures.

Notable pioneering examples of the feasibility of malaria control were Gorgas in Panama and Watson in Malaya. No longer was it necessary to consider the tropics the "white man's grave." Immigrants, as well as natives, could live in the warm regions of the world in good health so long as they were protected from insect-borne diseases, especially malaria. Such protection was economically feasible for larger communities, but in rural areas and in small, scattered communities malaria retained its devastating impact on man's health and economic well-being.

It would be difficult to exaggerate the impact of DDT on public health. Application of a small amount of this insecticide as a residual deposit two or three times a year to adult resting places brought about a fantastic reduction in malaria and certain other diseases as well. Pilot projects in certain countries, carried out in cooperation with the World Health Organization, were so encouraging that in 1954 the objective of the WHO-sponsored programs was raised from control to regional eradication. Meanwhile malaria had been eradicated from the United States.

In spite of the general success of residual insecticides for malaria control, problem areas became manifest. In southeast Asia, it was found that elusive outdoor-biting members of the *Anopheles leucosphyrus* group were able to maintain high rates of transmission and were not affected by residual deposits in houses. In certain parts of western Central America, malaria transmission by *A. albimanus* has not been brought to a complete halt because of outdoor biting, the apparent ability of the mosquitoes to enter and leave treated dwellings without acquiring a lethal dose of the poison, insecticide resistance, and the migrating habits of the laboring population. Vast problems continue to exist in Africa which have not been solved with residual insecticides.

It appears, therefore, that we must return to the principle of species sanitation in those areas where residual insecticides are not as effective as envisioned. This means more thorough investigations of the regional epidemiological factors that permit the mosquitoes to maintain transmission.

Plague.—Few bacterial diseases are customarily arthropod-borne, but none has had a more sinister history of periodic devastation of mankind than *Bacillus* (now *Yersinia*) *pestis.* Of course, in the very early days, plague was a general term for any epidemic disease, and became identified with the dread Black Death later in the Middle Ages. Wu Lien-Teh reasoned in 1926 that, from a hypothetical origin in the high Asiatic Plateau region, plague was spread from Hong Kong and other Far East ports and has been with mankind from time immemorial. When, as later chronicles tell us, die-offs of rats preceded human epidemics of bubonic plague in major cities of Asia and the Old World, the stage was set for the modern discoveries not only of the obligatory role of rat fleas in initiating epidemics in rats and people, but also the part of certain flea species in interepidemic maintenance as well. While fleas are considered short-lived off their hosts, infected prairie dog fleas have been recovered from burrows several months after extermination of the host colony (Kartman, 1970).

Justice cannot be done in this limited review to the many contributions to facets of the complex plague problem. The literature must be consulted as suggested below. It was 3 years after Kitasato and Yersin independently in 1894 described the bacilli, that Ogata in Formosa confirmed suspicions that rat fleas were carriers by infecting mice with suspensions of fleas from plague rats. This was followed closely in 1898 by Simond's similar demonstration of infected flea feces, and his often overlooked proof that healthy rats in cages over flea-free infected rats remained healthy. This anticipated the much more convincing (including similar) tests in 1902–1903 by Gauthier and Rauboud, and in 1906 and following years, of the renowned Indian Plague Research Commission as reviewed by Hirst in 1953. In their cage experiments, exposure by contact or to contaminated soil caused no infection of healthy animals until fleas were introduced on plague-infected rats in adjoining cages.

But this was far from the whole epidemiological story; for a very readable historical account the reader is referred to Hirst (1953). In 1905 Liston had reported the rich growth of bacilli in fleas or their heavily contaminated feces (ejected while feeding), and 4 years later, Bacot and Martin made memorable observations which included transmission by regurgitation from bacilli-blocked fleas during attempted feeding. Highly fatal, contagious pneumonic-type plague has long been known; a pertinent, isolated, self-limited incident in 1919 near Oakland, California, is revealing. A ground squirrel hunter contracted primary bubonic plague with secondary pneumonia in 3 to 4 days which resulted in 14 primary pneumonic cases and 13 deaths among contacts of relatives and friends!

Filariasis.—The complete life cycle of filaria parasites was not variously reported in its entirety until long after the initial discovery by Manson. The separation of *Wuchereria bancrofti* into the periodic and aperiodic forms, the former transmitted by night-biting *Culex, Anopheles,* and some *Aedes,* and the latter by day-biting *Aedes (Stegomyia) polynesiensis,* was paralleled by a similar recognition of periodic and subperiodic strains of *Brugia malayi.* A most interesting discovery was that of Edeson and Wharton that the subperiodic form exists as a zoonosis in the swamp forests of Malaya, with monkeys as hosts as well as man. This form has also been found in this type of habitat along the west coast of Palawan, Republic of the Philippines. The vectors are several species of swamp-inhabiting *Mansonia.* The periodic form, which occurs through wide areas of Asia, is transmitted by some *Anopheles* as well as by *Mansonia;* it seems to have become completely adapted to man as its human host.

Leishmaniasis.—Experimental transmission of the *Leishmania* parasite through *Phlebotomus* was completed only after many years of study by a number of investigators. Epidemiological evidence pointed to the abundant man-biting *P. argentipes* as the vector of kala azar in northeast India and Assam. Furthermore, Oriental sore was associated with *P. sergenti* in the Mediterranean and Mideast areas. Not until 1928 did Adler and Theodor demonstrate *Phlebotomus* with natural infections in Jericho. They also were able to infect these flies experimentally, and to transmit the parasites in vitro through membranes. Oriental sore could be caused in human volunteers by inoculation of the parasites from infected sand flies into the skin. What was not possible, however, was a simple further step, to produce the lesion by allowing laboratory-infected flies to bite human volunteers.

Meanwhile, in India, Shortt, Swaminath, Smith, and others were experiencing the same frustrations. Infected flies could be found in nature, they could be experimentally infected, but no infection could be made to occur by the bites of experimentally infected flies.

The break came in 1940. Up to this time the *P. argentipes* used in the transmission experiments had been given only blood meals. In an attempt to enhance their longevity, Smith, Hadler, and Ahmed began to feed them a maintenance diet of soaked raisins. Not only did this result in a higher survival rate, but the flies developed massive infections with the flagellates, which moved forward through the alimentary tract and into the mouth parts. Experimental transmission of *L. donovani* to hamsters by the bites of these flies was effected readily and in 1942 Swaminath, Shortt, and Anderson reported experimental infection of 5 out of 5 human volunteers by the bites of such infected *P. argentipes.*

At about this time, Adler and Ber (1941) reported that they were able to cause Oriental sore by the bites of *P. papatasi* which had been given a maintenance diet of 3 parts of 2.7% saline and 1 part inactivated defibrinated rabbit blood.

Proof that certain species of *Phlebotomus* were vectors explained much of the epidemiology of the leishmaniases. Furthermore, in much of the world there are zoonoses. The domestic dog was considered to be an important reservoir for kala azar in the Mediterranean, as it is for muco-cutaneous leishmaniasis in South America. In 1960–1962, Shekhanov and Suvorova, and Dergachova and Dolmatova described endemic foci in Central Asia, where burrowing rodents are reservoirs for *L. tropica,* and species of *Phlebotomus* living in these burrows act as vectors among the reservoirs as well as to people. Hoogstraal and his associates have conducted an extensive investigation of kala azar in the Sudan, where the Nile grass rat, *Arvicanthus,* is the reservoir, and the vector, *P. orientalis,* causes infections in people who go into the acacia grove enzootic foci.

The American *L. brasiliense* exists in many different strains, and in fact several specific names have been given to them. In Panama, the spiney rat is the reservoir, and further work no doubt will reveal additional wild animal reservoirs. There are several species of *Phlebotomus* associated with these reservoirs and with man, and it is among these that the vectors will be found. Partial evidence against some of them is already at hand and further work is underway to clarify a somewhat confusing situation.

Arboviruses.—The early work on yellow fever led to the hypothesis that this disease could persist endemically only in large centers of population in the tropics. The reason behind this was that man was thought to be the only vertebrate host, and domestic *A. aegypti* the only vector. As a patient had a viremia high enough to infect the mosquito only for 2 to 3 days early in the infection and then either died or developed solid immunity, there seemed to be needed a continuous source of virus to the mosquito in the form of a steady succession of human, possibly subclinical cases. Furthermore, there would have to be a year-round high density of *A. aegypti.* Such conditions were met only in large cities of the tropics.

Gorgas, in Havana and Panama, demonstrated the feasibility of eradicating yellow fever by reducing the *A. aegypti* population. The Rockefeller Foundation, with the cooperation of certain South American governments, set out to eradicate yellow fever by applying Gorgas' anti-aegypti measures to all of the urban endemic centers. All went well; yellow fever disappeared from these cities (for a good historic account, the reader is referred to Strode, 1951).

However, in 1932 a disturbing event occurred; this was an outbreak in the Valle do Chanaan, Brazil, in a rural community where *A. aegypti* could not be the vector. Meanwhile there had been much progress in laboratory studies. Reed had said that the causative agent could pass through bacterial filters. That the agent indeed was a filterable virus was proven by Bauer and Mahaffy in 1930 after the important discovery that Indian rhesus monkeys (and later white mice) were susceptible for intensified laboratory studies. However, filterability of mosquito virus was not demonstrated until after the deleterious effect of saline diluents was discovered. Passage of the virus in brains of white mice and chick embryo cultures produced the modified 17D strain (Theiler) which has become a widely-used vaccine. The mouse neutralization test for immunity to yellow fever was also devised in 1931 by Theiler. It now was possible to determine the distribution of yellow fever throughout suspected areas of endemicity.

Not only *A. aegypti,* but many other species of mosquitoes, especially those belonging to the subgenus *Stegomyia* in tropical Africa and the genus *Haemagogus* in the American tropics were shown to be highly susceptible and capable of transmission (Philip, 1930). In poorly adapted species, such as *Eretmopodites chrysogaster,* the extrinsic incubation period (i.e. in mosquitoes) was peculiarly prolonged, or even suppressed in the crab-hole *Aedes,* though the virus persisted for their caged lives. That quantitative factors were possibly involved was suggested by Hudson and Philip to explain the longer periods also required in *A. aegypti* fed early and late in the fastigium of fever in the donor—a subject not known to have been investigated as yet in other mosquito-borne viruses. *A. aegypti* were even infected 12 to 24 hours before febrile rise but transmission was not accomplished by Philip either transovarially (as early supposed and as in sand fly fever) nor mechanically by interrupted feeding.

It now also became possible to search for the existence of yellow fever not only in people but also in wild animals, and to suspect mosquitoes other than *A. aegypti* as vectors. The extensive efforts by the Rockefeller Foundation laboratories showed that yellow fever was widely endemic as a zoonosis in monkeys over most of the rain-forested areas of South America, and also in the tropical forest belt of Central Africa. Waves of viremia, which became known as jungle yellow fever, spread through these monkey populations, with arboreal mosquitoes serving as vectors and possibly also as reservoirs during seasons when the monkey populations are reduced and mosquito breeding at a low level.

In the Americas, prime suspects as vectors of jungle fever were species of the genus *Haemagogus.* In attempts to prove this, workers at the Rockefeller Foundation Laboratory at Villavicencio, Colombia made extensive efforts to isolate the virus from wild-caught *Haemagogus.* However, they encountered difficulty in obtaining enough mosquitoes for these tests. One day M. Boshell, watching woodcutters at work, noted that when the tree fell there was an increase in the number of *Haemagogus* biting. He immediately grasped the implication that these mosquitoes normally spent most of their time in the forest canopy. Platforms were built at various heights in the jungle trees, and on them large numbers of *Haemagogus* were captured, from which yellow fever virus was readily isolated. In Africa, similar research by the Rockefeller Foundation and later British Medical Laboratory in Entebbe also resulted in the isolation of virus from a forest canopy mosquito, *Aedes (S.) africanus.*

In South America the disease occurs in people whose activities take them into the forest, and it has been said that it is an occupational disease of woodcutters. In Africa, *A. africanus* does not come down from the forest canopy to bite people. Monkeys do descend, however, especially to raid banana plantations. Here they encounter another mosquito, *A. (S.) simpsoni,* which transmits the virus from monkeys to man. The proof, years earlier, by Philip in Nigeria that both could readily infect experimental monkeys had seemed academic at the time.

The 17D and the French mouse brain vaccines have protected millions of people from yellow fever in endemic areas. We no longer fear the devastating urban epidemics. Research on yellow fever is thus at a low ebb. However, the disease exhibits phenomena which are poorly understood. One is its ability to survive the dry season, and Trapido and Galindo have suggested that a sabethine mosquito, *Sabethes chloropterus,* which can survive the nonbreeding dry season as adults, can serve as a reservoir as well as a vector. In parts of east Africa, the disease persists where the monkey populations are very low. Indeed, how does it survive where there are normal monkey populations, which either die or become immune as the epidemic wave passes through them?

In the Rockefeller laboratories, before advent of the above vaccine, five Rockefeller field investigators died of laboratory infections (Strode).

Dengue.—There continues to be considerable interest in this long-known mosquito-borne disease. Although evidence of transmission was obtained in 1902 by Graham and confirmed by a series of investigators from Bancroft in 1906 to Simmons, St. John, and Reynolds in 1931, it was not possible to isolate the virus from wild-caught mosquitoes until 1956, when Hammon and his team did so from *A. aegypti* during an outbreak in Manila. Meanwhile Sabin had reported in 1954 the existence of two types of dengue virus, one isolated from U.S. troops in Sicily, the other from Australian troops in New Guinea. The Manila epidemic was marked by hemorrhagic manifestations in young children, with about a 10% fatality rate. Hammon isolated two additional types from these cases, dengue 3 and 4. In 1958, dengue broke out in Bangkok; again isolations of this virus as well as of chikungunya virus were made from *A. aegypti.* A number of isolations of all these strains have been made subsequently in India, Thailand, and Vietnam. Improvements in techniques, especially the use of tissue cultures, have facilitated such isolations and differentiation in the laboratory.

Except for the hemorrhagic disease, which occurs in young children during the course of *A. aegypti*-transmitted epidemics, dengue is seldom fatal, though it may cause severe illness. Future trends in dengue research will therefore include the development of modified strains which can be used for vaccines. A diffuse outbreak was observed by Rosen et al in 1954 in Tahiti among the residents of scattered, semirural communities. *A. (S.) polynesiensis* was suspected of being the vector, and the ability of this species to transmit the virus was proven by him and colleagues. *A. scutellaris* is also alleged to be a vector in New Guinea. Continuing problems concern comparison of the vector efficiency of the various species of the *A. scutellaris* group. It is further questioned if there is a monkey or other animal reservoir that can keep the virus alive during the long periods of time between the large human epidemics.

Other Arboviruses.—In 1933 and 1937 outbreaks of viral sleeping sickness occurred in St. Louis. A similar disease had been recognized in Japan for some years before this; an epidemic in 1924 caused about 4000 deaths there. Again there was a recrudescence in 1934 in St. Louis. In 1930 an epidemic of neuropathic disease of equines occurred in California; in 1933 outbreaks of a similar disease took place along the eastern seaboard and through the midwestern states. A virus was suspected. In 1931 Meyer, Haring, and Howitt isolated one from horses in California, and Giltner and Shahan, two years later, from horses in New Jersey; the virus of the St. Louis encephalitis was isolated in 1933 by Muckenfuss, Armstrong, and McCordoc, and that of Japanese encephalitis in 1934 by Hayashi. These viruses have been named according to the area from which they first were found; i.e. western equine encephalomyelitis (WEE), eastern equine encephalomyelitis (EEE), St. Louis encephalitis (SLE), and Japanese B. encephalitis (JBE) or Japanese encephalitis (JE). Suspicion arose as to the similarity of the causative agents of encephalitis in man and in horses, and in 1938 Schoening, Giltner, and Shahan produced an infection in a horse with virus obtained from the brain of a child who had died of encephalitis, while Cox and Philip infected horses in 1941 with SLE after finding this type as probable cause of an equine outbreak in Colorado.

Epidemiological evidence pointed to an insect vector of the outbreaks in both horses and animals, and the existence of a reservoir host which could produce an asymptomatic viremia. Such a reservoir was found for WEE by Hammon and his associates during the 1941 outbreak in the Yakima Valley of Washington. It was the domestic fowl. Meanwhile, in 1938 Tyzzer, Sellards, and Bennett isolated EEE virus from ring-necked pheasants and Fothergill, LeRoy, and Dingle did so from pigeons. Subsequent work showed that many species of wild birds are capable of producing viremias of high titers for periods of 3 or 4 days.

The first attempt to transmit the virus was that of Kelser in 1933. He infected guinea pigs with WEE, allowed *A. aegypti* to bite them, and after 6 to 18 days fed these mosquitoes on other guinea pigs and a horse, which subsequently showed evidence of having been infected. Mosquito transmission of several other encephalitis viruses was subsequently demonstrated, among which was West Nile virus passed between young hamsters by *A. albopictus* in 1943 by Philip and Smadel.

That mosquitoes were natural hosts for the viruses was demonstrated by Ten Broeck and Merrill (1935–1936) in studies with *A. aegypti.* Attempts to find virus in wild mosquitoes were unsuccessful at first. However, Mitamura et al (1937–1939) reported finding *Culex pipiens pallens* and *C. tritaeniorhynchus* infected with JE. In 1941 Hamman and his co-workers caught *C. tarsalis* naturally infected with SLE and WEE. In the following years many isolations were made of WEE from *C. tarsalis,* as well as other species which play a secondary role, if any, as vectors. EEE was not found in wild mosquitoes until 1948, when Howitt et al found it in *Mansonia perturbans.* Diligent searching by workers at the Communicable Disease Center in Georgia resulted in a number of isolations from the swamp-dwelling *Culiseta melanura.*

The St. Louis virus was first isolated from *C. tarsalis* by Hammon et al in 1941 and later in 1957 from *C. quinquefasciatus* by Ranzenhofer et al.

In this brief review it is impossible to discuss the tremendous amount of work that has taken place in recent years on the arboviruses and the startling number, approaching 300 entities catalogued by 1972, has already been mentioned. These have been isolated from various mammalian and avian hosts, but especially from mosquitoes and some other biting flies, and from ticks. Man usually is infected accidentally and seldom if ever serves as a source of virus to the vectors as he does for urban yellow fever and classical louse-borne typhus.

With these basic facts at hand, there has been a lessening of interest in searching for more viruses, their vertebrate hosts, and vectors. No doubt many more exist. Some may be responsible for human febrile illness; others may never be noticed. Nevertheless, even with the best known of these viruses, there are gaps in our knowledge, such as how they survive during interepidemic seasons, and what is involved in the spread of EEE from the wild swamp biocenosis to urban communities. Thomas, Rush, and others recently furnished a suggestive lead by infecting mosquitoes on garter snakes with long WEE viremia which could be exacerbated on revival of winter-dormant infected snakes. For a more adequate review, the reader is referred to Horsfall & Tamm (1970).

The Rickettsial diseases.—Good condensed histories of these are also provided by Horsfall & Tamm (1970). In his well-known book *Rats, Lice and History,* Zinsser reiterated that pestilence was of greater historical importance than armies in bringing about momentous events in man's political power struggles. He believed that a significant reason for the decline of the Roman empire was the difficulty in maintaining such a large and complex social and political organization in the absence of our modern sanitary technology. There was no defense against the devastating outbreaks of disease which repeatedly swept through the civilized world. To quote Zinsser: ". . . soldiers have rarely won wars. They more often mop up after the barrage of epidemics. And typhus, with its brothers and sisters,—plague, cholera, typhoid, dysentery—has decided more campaigns than Caesar, Hannibal, Napoleon, and all the inspector generals of history."

The actual causative agents for these early epidemics, though called by such terms as red and black death, are obscure. Zinsser, from his examination of the clinical histories, concluded that Hippocrates (born 460 B.C.) described typhus. In 430 B.C., Athens was devastated by an epidemic which Zinsser hesitated to attribute to typhus. In 1566, when Maximillian II of Germany was campaigning against the Turks in Hungary, an epidemic caused him to abandon this operation. This according to Zinsser, was typhus, and the returning soldiers carried the infection throughout Europe, leaving behind them local epidemics in cities and towns. He believed that this was the event in history by which typhus was introduced into Europe, and which also led to a digression from a rat-flea-man cycle to the man-louse-man cycle. From what we now know of the host relationships of louse-borne *Rickettsia prowazeki* and flea-borne *R. typhi,* this seems unlikley. Although *R. prowazeki* may well have evolved from a primordial rodent-inhabiting flea-borne rickettsia, it is a different species than modern *R. typhi,* and though often hypothesized, evidence is controversial (though accepted by Steinhaus, 1946) that the latter can readily transform into *R. prowazeki*; the two undoubtedly stemmed from a remote common ancestor. The possibility of an animal reservoir for *R. prowazeki,* which could in part explain its interepidemic survival, has received some added credence since 1951 with reports of puzzling involvement of livestock in Ethiopia and Egypt by Gutfreund, Imam, and others, plus isolation from their ticks in Ethiopia (see review by Philip, 1968).

It is obvious that the discoveries of the causative agents of typhus and related diseases, and of their transmission by acarids and insects, were among the most significant advances in the field of medical entomology as were those of distinctly animal rickettsioses, particularly in Africa, such as heartwater (*Cowdria ruminantium*).

In 1884 Murchison described the rapid spread of typhus among people who were in contact with patients and even to those who handled clothing or other objects belonging to these patients (so-called fomites).

The demonstration in 1909–1910 by Nicolle, Anderson, and Goldberger, and Ricketts and Wilder that monkeys could be infected by inoculation of blood from human cases confirmed the presence of a pathogen in the blood, and the vectorship by the human body louse was proven during these studies. The nature of this organism was not known for certain, although in Hamburg, da Rocha-Lima in 1916 described *Rickettsia prowazeki* from louse guts which he claimed to be the causative agent, thus memorializing two laboratory victims of the disease. Though da Rocha-Lima bitterly decried in 1951 the credit he thought was over-extended to these two martyrs (". . . the findings of Ricketts were not . . . confirmed as a discovery"), Philip (1954) replied that prior recognition of the etiologic agents, named or unnamed, has not lessened the illustrious contributions of all three of these pioneers. The organisms Ricketts and Wilder compared to the noncultivable, Giemsa-staining diploccoid bodies of spotted fever can hardly be doubted as referring to the respective etiologic agents in spite of Ricketts' understandable reluctance to name them so early in the studies. Subsequent workers have confirmed da Rocha-Lima's incrimination of this organism. Wolbach, Todd, and Palfrey in 1922, found the microorganisms within the cells of certain tissues in man and guinea pigs, as well as in the epithelial cells of the gut of the louse of what at that time, they considered the agents of Mexican or New World typhus as contrasted with Old World or classic typhus. They named their agent *R. typhi* and this name has been retained in consequence for the murine typhus agent though their studies included lice rather than the then unknown flea vectors. Sporadic murine cases were likely among their human subjects. A list of synonyms, including *R. mooseri,* is discussed by Philip in *Bergey's Manual of Determinative Bacteriology,* 7th ed., (1957).

The demonstration that the infectious agent was in the peripheral blood led to the obvious hypothesis that a bloodsucking insect was a possible vector. Furthermore, the many observations over the years that outbreaks of typhus occurred during times of stress when lousiness was conspicuous pointed to the louse as the most logical candidate. Prowazek, in 1913, infected a monkey by injection of a crushed louse that had fed 2 days earlier on a typhus patient. A year later, Sergent, Foley, and Vialette also transmitted the infection to a monkey as well as a human volunteer by inoculation of crushed lice. There was conjecture among the earlier workers as to whether the lice transmitted by their bites, but in 1922 Atkin and Bacot showed that the bite was not infectious. The details of the intracellular development of the parasite in the gut of the louse were supplied by the studies of da Rocha-Lima and Wolbach, Todd, and Palfrey.

On epidemiological grounds, Maxcy recognized in 1929 that sporadic cases of typhus occurring in the United States were not the same as the classical European typhus, although the latter also occurred in the United States as sporadic cases among immigrants. These were recrudescences of classical typhus without lice, and were called Brills' disease. The causative organism was found in rats and fleas in 1931 by Dyer, Mooser, and others and the details of the transmission of *R. typhi (mooseri)* through fleas were established by the studies of Mooser and Castaneda a year later.

Rocky Mountain spotted fever was recognized as a distinct disease entity by practicing physicians in the northwestern United States from about 1870. It was described in 1899 by Maxcy in the Portland Oregon Medical Sentinel as "an independent specific disease" which was not included in any of the available textbooks. Wilson and Chowning of the University of Minnesota carried out clinical and pathological studies in 1902 and 1903 in western Montana. They suggested that ground squirrels could be reservoirs and that ticks could be vectors as believed by many local ranchers. In 1906 H. T. Ricketts showed that the infection could be caused in guinea pigs by inoculation of blood from patients, that interrupted feeding of female and male ticks could result in transmission, and that ticks could be found with natural infections. Although Ricketts generally is credited with having worked with *Dermacentor andersoni,* he did not learn until later that two Boise physicians, McCalla and Brereton, had preceded him by a year in infecting a volunteer with a tick from a patient, which he characteristically acknowledges. In 1951, Philip and Kohls noted that at least some of his work was done with the one-host winter tick, *D. albipictus*. Ricketts described the organisms in tick tissues, as well as their transovarial transmission from mother to offspring, and stage-to-stage survival. This work was confirmed and extended in 1919 by Wolbach who named the organism *Dermacentroxenus* (=*Rickettsia*) *rickettsi.*

Details of the host relationships and other aspects of the epidemiology of Rocky Mountain spotted fever have been supplied through a long series of studies by the Rocky Mountain Laboratory, of the US Public Health Service, in Hamilton, Montana.

In 1931 Rumreich, Dyer, and Badger reported the disease from the eastern United States, where *D. variabilis* was shown to be the vector. As early as 1929, a virulent typhus was recognized by Travassos in Sao Paulo, Brasil; it later was shown to be identified with Rocky Mountain spotted fever. The disease was discovered for the first time by Calero, Nunez, and Silva in Panama in 1950. With the earlier findings of this disease in Mexico by Bustamante and colleagues in 1946, it became obvious that it occurs throughout North, Central, and parts of South America where climatic conditions favor the survival of the several species of tick vectors. The disease is thus more appropriately known as American rather than Rocky Mountain spotted fever.

Control

The ultimate practical goals of medico-veterinary entomology are the interruption of transmission of arthropod-borne diseases of man and principally his domestic animals, or of the wild ones that impinge on his economy. What made the field so fascinating was that this involved studies of all the pathobiotic factors which allowed diseases to exist, including the identities of the vectors (whether obligatory or coincidental) and their habits. Suppression could then devolve on attacking some phase or phases in the life cycle most vulnerable to alteration or suppression. One of the earliest of these selective operations was obviously against females of most blood-sucking

Diptera in attempts to alter their harborages, to trap them or latterly, to induce sterile mating.

The advent of the powerful residual insecticides brought an abrupt change in conceptual control. Where there was a problem, crews merely moved in with spray and blanketed the area or domiciles with the pesticides. As noted above, this was so successful against malaria that worldwide eradication was envisioned (but, of course, not the controversial, environmental side effects that ensued).

The chronicles of DDT have been recorded many times and may be briefly reiterated here only for the historical record. Ultimate wide application of DDT to control of agricultural pests has already been discussed in previous chapters. Othmar Zeidler, a Viennese pharmacist who was a chemistry student in Strasbourg, synthesized DDT in 1874 on an empirical basis. In the late 1930s, Paul M. Mueller of the Geigy Company in Switzerland discovered the unique insecticidal properties of this compound, for which he was awarded a Nobel Prize in 1948. Now this was at the time that a search was under way for more effective louse control with the impending onset of World War II, which conjured specters of an ancient military scourge—typhus.

The experience with another louse-borne pathogen, trench fever, which although not fatal caused much morbidity amongst more static troops in World War I, had not been forgotten, while ancillary benefits could accrue by suppression of louse-borne relapsing fever in endemic areas. The military authorities were seeking better methods of controlling human lice; especially desirable was a chemical which could be applied to the clothing and body, with minimum hazard, as well as prevent reinfestation for long periods, which the old Serbian steam barrels of WWI could not do. A louse powder called MYL was initially formulated with short-lasting pyrethrum as the killing agent. Then a sample of DDT was furnished to the US Department of Agriculture by the Geigy Company in 1942 and its superiority was quickly evident. After demonstrations of its effectiveness in delousing civilians in Egypt and Algiers, many millions of people, soldiers and civilians alike, were dusted with 10% DDT powder in a carrier. Typhus was again appearing in various combat sectors. One of the great preventive medical accomplishments in WWII, was the abortion of a potential epidemic in Naples in the winter of 1943–1944 by strenuous delousing, particularly among local residents and refugees, and, to a limited extent, by vaccination.

The unique insecticidal properties of DDT and its low cost of manufacture led to a rapidly expanding usage for the control of other insect pests as well as disease vectors. Progress in development of control of diverse disease-bearing pests, such as ticks and mites, cannot adequately be reviewed in this history, but the ravages of malaria may be used to exemplify historic complexities of one such problem.

It is difficult for those of us living at present in the United States to appreciate how devastating malaria can be. We become complacent about this disease which was not uncommon with us at one time. Geddes Smith noted (1941) that no one knew how much malaria there was in the world, but that the total number of cases appeared to be astronomical, including an estimated 2,000,000 per year in the southern United States. With the changes taking place in environmental practices, the disease disappeared first from the upper Mississippi Basin and later from the southern states as well. In the South, this process was hastened by antimalarial programs conducted by local and federal agencies.

By 1952 the elimination of indigenous malaria in the United States was judged to have been accomplished and the National Malaria Society was officially disbanded. DDT was being used during the latter part of the antimalarial campaign, but it appears doubtful whether it contributed more than an acceleration of a malarial decline which began many years before.

The United States, as well as northern Europe, was a marginal area where persistence of malaria was so precarious that the economic and social changes accompanying the population growth may have contributed in a large measure to the elimination of the disease. The situation in tropical countries is far different. An estimate was published in 1968 by the World Health Organization that prior to the use of DDT, 1,692,000 people lived in malarious areas of the world. Perhaps we can obtain a better appreciation of the disease if we refer to a few observations by some experienced malariologists of a generation ago. J. A. Sinton wrote when he was Director of the Malaria Survey of India: "India is like Prometheus bound by chains of apathy to the rock of expediency and financial stringency, while the vulture of malaria devours the vitals of her people." From his analysis of mortality data available at the time, Sinton concluded: ". . . in a year of ordinary malarial prevalence, such as 1931, about one million persons died in British India alone as a *direct* result of this disease." Far greater mortality rates were recorded during epidemics. However, the direct influence of malaria on mortality is only part of the story. People whose health is undermined by malaria succumb more easily to other diseases, and Sinton estimated that at least another million people were killed each year in India indirectly by malaria. We should note also that young children are frequent victims; records given by Sinton indicate that 40 to 70% of the deaths due directly to malaria in endemic zones occurred in children under 10 years of age. The indirect effect, such as malnutrition of infants whose malaria-infected mothers are unable to lactate, as well as lowered resistance to the highly prevalent parasitic and viral infections, greatly increased the mortality rates in this age group. Other side effects of malaria, the misery, suffering, and inability to work, leading in marginal societies to starvation, are immeasurable. It appears quite probable that malaria, as well as yellow fever, was introduced into the New World by the early European explorers and colonists, and outbreaks of these diseases established the evil reputations achieved by many of these colonial settlements. One such community was Porto Bello, in Panama, which was described as "an open grave ready to swallow all who resorted there" (Haring, in Simmons). During the construction of the Panama railroad, from 1850 to 1855, according to Robinson, the death rate was not more than 40%, but the hospitals were always filled, and "sulphate of quinine became a prime necessity, almost an article of diet." Tomes said in 1855 that "no one who resided over two months in Aspinwall escaped fever."

The above are only a few examples of the many accounts of the devastating effects malaria has had upon the health of people living in tropical and subtropical countries. They may, however, help us to appreciate why malaria was described by Leathers as "one of the greatest scourges inflicted upon humanity," by Manson as "directly and indirectly it is the principal cause of mortality and death in the tropics and sub-tropics," and by Osler as the "greatest destroyer of humanity."

Following his work in Havana on the control of the yellow fever mosquito, W. C. Gorgas was placed in charge of the sanitation of the Panama Canal Zone, and proved that it was possible to control malaria as well as yellow fever by eliminating mosquito breeding habitats or by larviciding. In an address to the Royal Geographical Society in 1942, Sir Malcolm Watson said: "Of Panama, you all know the great genius of Surgeon General Gorgas. The French failed because they had not discovered how to control malaria and yellow fever. . . . In 1915 the Panama Canal represented the greatest engineering and medical triumph the world had seen." This statement is not only a tribute to Gorgas, but reminds us again of the gravity of the malaria problem as it existed, not only in Panama, but throughout the world's tropical and subtropical regions.

Meanwhile, another discovery was made by Watson himself. In Malaya, of some 30 species of *Anopheles,* only 3 were found to be important malaria vectors, and by concentrating the resources available in an attack on these 3 in their special breeding places, malaria control became practical. Darling in Panama, Christophers and Sinton in India, and other pioneer malariologists began to sort out the really dangerous species. Thus, was established the principle of species sanitation previously alluded to. In each problem area, one identified the dangerous vectors, discovered their breeding places, and then, by one means or another, reduced the density of the vector population by attacking the larvae. This was accomplished through the use of larvicides, especially paris green (developed in Marshall Barber's laboratory about 1920 in Nashville, Tenn.); destruction of the breeding habitats by drainage, filling, impounding, canalizing streams, and so-called reclamation of marshes; removal of aquatic and emergent vegetation from the margins of lakes and impoundments; and periodic flushing of the breeding places by manipulation of the water level.

As in Panama, malaria control measures now became feasible in those communities where people were willing and able to pay for it. These measures were expensive, and could be applied only to larger centers of population. Furthermore, control was not always completely satisfactory. In a banquet address delivered at the Fourth International Congress of Tropical Medicine and Malaria in 1948, Dr. Lewis W. Hackett said, perhaps not entirely facetiously, "We larviciders were pretty smart and our strategy was not to fight nature but to collaborate with her. We would usually manage to start operations immediately after a big epidemic and ride to success on the ebbing tide. Our motto was: 'God is our co-pilot'." The great seedbed of infection continued to be the scattered rural populations of the tropics and subtropics. The general pessimism about further progress against the disease in the vast tropical areas of Africa, Asia, and the Americas was stated in 1941 by Dr. Mark Boyd: ". . . there are many regions in the world where any attempt to control this disease based on the application of available measures is beyond local resources. Malaria will likely continue to be endemic in such regions until cheaper control methods are devised." No one realized it at the time, but this cheaper method was already at hand; it was DDT.

A new approach to malaria control was the frequent spraying of houses with an insecticide in order to kill those specimens which had fed the previous night, and which now were resting in dark corners. B. De Meillon and G. A. P. Ross in Africa, and P. F. Russell and others in India, showed that malaria transmission indeed could be markedly reduced by this method; however, it was effective only if each house was sprayed once or twice a week, and this meant the use of a lot of insecticide and many spray crews. Russell and Knipe (1940) commented: ". . . the cost of the above method, although relatively not expensive, was too great for the average village in this area to pay for by yearly taxation." What was needed was an insecticide which could be applied to the walls of houses, and which would kill mosquitoes for several months at least. Also this material should be nontoxic to people and domestic animals; it should not smell bad or be irritating.

When DDT became available, entomologists demonstrated that not only did it protect people from lice, but that any insect which rested for a short time on a wall or other surface to which a thin coating of DDT had been applied would die. Furthermore, the effectiveness was maintained for months. The results were astounding; from everywhere came reports of the practical disappearance of this disease. In Italy, the renowned malariologist, Missiroli, made a fateful decision which he described as follows: "Taking as a point of departure the fact that the control of adult insects is less expensive than larval control, I drew up a plan aimed at eradicating malaria from Italy within a period of five years. I set forth this program in a lecture held on January 20, 1946 at the Istituto Superiore di Sanita. It was heard with benevolent astonishment . . . I suggested, as an integral part of this new approach, that all prophylactic measures employed so far be dropped: larval control, mechanical protection, quinine prophylaxis and the great land reclamation scheme." The success of this program is manifest in the fact that in 1949 and 1950, for the first time in 2,000 years, not a single death due to malaria occurred in Italy, Sicily, or Sardinia.

In view of the encouraging results of several pilot control programs, in 1948 the World Health Organization established a policy of aiding those governments desiring to organize control programs. These were so successful that in 1954 the objective of the antimalarial campaigns became nothing less than eradication. The United States invested large sums of money in these campaigns through its Agency for International Development.

Raising the sights to eradication may have been too ambitious because this accentuated a feeling of pessimism when it became obvious that the goal was still beyond reality. Although malaria practically disappeared from most problem areas, eradication means elimination of the parasite itself. This has proved to be impossible in many areas for several reasons. There may be difficulties in spraying every house in remote areas. Some outdoor biting may take place, especially when people remain out of doors after sundown. Some populations of *Anopheles* have become resistant to DDT. In discussing the impact of DDT resistance on malaria control, Busvine and Pal say: "The effect ranges from an inconvenience to an apparently insuperable obstacle." However, WHO's J. W. Wright tells us that in only 1% of the areas in which DDT has been used extensively has resistance become so severe that this insecticide no longer is useful. In spite of these and other problems, the success of the antimalarial program can be seen from information provided in 1968 by the World Health Assembly. Of the approximately 1.7 billion persons inhabiting the original malarious areas, by 1967 over 1.3 billion were living in areas where eradication was in progress or had already been achieved. Thus, this scourge had been removed from over two-thirds of the people, who, a few years ago, were living under the risk of infection. The WHO Expert Committee on Malaria in its Thirteenth Report states: "To have progressed this far and to have brought a sustained measure of well-being to a total of 953 million people, more than one quarter of the world's population, is an international achievement without parallel in the provision of public health service." Again we must emphasize that this had been possible only because of the unique insecticidal properties of DDT.

A combination of several events brought about a disillusionment as to the continued effectiveness of the chlorinated hydrocarbons. One was the development of resistance on the part of many pest species, and another, which is now the subject of so much concern and controversy, is the consequences of continued accumulation of the slowly degradable compounds in the environment and their present and future danger to fish and wildlife, and indeed to man himself.

The pressures to ban DDT have been exerted by various conservationist groups. The latest trend, therefore, in the control of disease-bearing arthropods is the search for alternate methods which will permit us to place less and less reliance on chemical insecticides.

What has been considered to be a most significant advance was the eradication of the screwworm fly from Curacao and the southeastern United States by the liberation of males sterilized by irradiation. This method has been successful in certain areas also against fruit flies. The male mosquito, however, is not nearly as robust an insect as the screwworm fly, and experimental efforts to reduce natural populations of *Aedes aegypti* and *Anopheles quadrimaculatus* by liberation of sterilized males have not given very promising results. In addition to the short life span of the male mosquitoes, treated males seem to be less competitive than the local males for the wild females. Treatment with chemosterilants seems to have less effect on competitiveness; furthermore, both males as well as females are rendered sterile. However, here again we are dealing with powerful and dangerous chemicals which can not be applied to the environment without strict precautions.

The emphasis in the search for new methods for the control of medically important arthropods, in addition to the sterile male technique, include, under the heading of genetic manipulation, lethal genes, behavioral modifiers, abnormal sex ratios, hybrid sterility, cytoplasmic incompatibility, and translocation semisterility. Some of these hypotheses appear to be rather remote from immediate practical application; however, others have actually progressed from laboratory experiments to field trials. In one of these tests in Okpo, a small town near Rangoon, Laven liberated males of a strain of *C. quinquefasciatus* (=*C. fatigans*) which were cytoplasmically incompatible with the local population of this species. Over a period of several weeks, the proportion of sterile egg batches in the breeding places steadily increased, while that of the fertile ones decreased until, at the end of the observation period, no more fertile eggs could be found.

A second field trial involved the use of sterile hybrid males of the *Anopheles gambiae* complex. These males were liberated in a small village near Bobo-Dioulasso, Upper Volta. Unfortunately, as Davidson and colleagues reported in 1970, they appeared to have little effect on the natural population of *A. gambiae*.

Other possible biological methods include the use of predators and parasites, and competitive population displacement. Eventually, some of these may lead to effective control which would permit us to set aside those insecticides which are environmental hazards. Meanwhile, there continues to be an essential requirement for medico-veterinary entomologists to have a thorough understanding of the epizootiologic perspectives of arthropod-borne diseases, and of the life cycles and natural histories of the vectors, so that the most effective combination of chemical and biological methods can be applied which will control a given disease without unacceptable side effects.

In closing, we have had to forego recounting historic, unique, and often ingenious developments or attempts to control vectors of many other arthropod-borne maladies, such as fly breeding in barnyard and pasture manure and advocacy of bat rookeries to control local pest mosquitoes or of woolly sheep drives through tick-infested brush, dissemination of insect and fungus parasites, violence engendered by enforcement of past stock dipping and quarantine regulations, personal protection with specific vaccines or repellents (certain African natives were reputed to annoint their bodies with freshly "harvested" cattle urine as a repellent!), and reduction of fatal hazards with appropriate antibiotics. Nor have we had space to review such important laboratory tools as refinements in use of experimental animals and tissue cultures. These all should have had a place in an expanded history of our subject.

LITERATURE CITED

Bruce, D. 1896. The tsetse disease or nagana. *Agr. J. Cape Town,* 9:358–90. Also: further report on the tsetse fly disease or nagana in Zululand. London: Harrison & Sons. 69 pp.

Hirst, L. F. 1953. *The Conquest of Plague*. Oxford: Clarendon. 478 pp.

Hoeppli, R. 1959. *Parasites and Parasitic Infections in Early Medicine and Science*. Singapore: Univ. Malaya Press. 526 pp.

Horsfall, F. L. Jr., Tamm, I. 1970. *Viral and Rickettsial Infections of Man*. Philadelphia: Lippincott. 4th ed. 1282 pp.

Howard, L. O. 1930. A history of applied entomology. *Smithson. Misc. Coll.,* Vol. 84, publ. no. 3065. 545 pp.

Kartman, L. 1970. Historical and ecological observations on plague in the United States. *Trop. Geog. Med.* 22: 257–75

Manson, P. 1878. On the development of *Filaria sanguinis hominis,* and on the mosquito considered as a nurse. *Nature* Mar. 28, 1878. 439 pp.

Newsholme, A. 1927. *Evolution of Preventive Medicine*. Baltimore: Williams & Wilkins. 226 pp.

Philip, C. B. 1968. A review of growing evidence that domestic animals may be involved in cycles of rickettsial zoonoses. *Zentrbl. Bakteriol. Parasitenk. Infek. Hyg.* 1 Orig., 206: 343–53

Reed, W., Carroll, J., Agramonte, A. 1901. Experimental yellow fever. *Ann. Meet. Assoc. Am. Phys., 16th* (see US Senate Doc. 822, Yellow Fever, 1911:110–30)

Ricketts, H. T. 1906. The study of "Rocky Mountain spotted fever" (tick fever) by means of animal inoculations *J. Am. Med. Assoc.* 47:33–42

Rosen, H. R. 1929. The discovery of insect transmission of pathogenic microorganisms. *Science* 70:355

Smith, T., Kilborne, F. L. 1893. Investigations into the nature, causation and prevention of Texas or southern cattle fever. *US Dep. Agr. Bur. Ann. Ind. Bull.* No. 1. 301 pp. (Also in USDA Bur. Ann. Ind., 8th and 9th Ann. Reports for years 1891–1892. pp. 177–304.)

Steinhaus, E. A. 1946. *Insect Microbiology*. Ithaca, NY: Comstock. 763 pp.

Sternberg, G. 1884. *Malaria and Malarial Parasites*. New York: William Wood. 329 pp.

Strode, G. K. 1951. *Yellow Fever*. New York: McGraw-Hill. 710 pp.

FOREST ENTOMOLOGY

F. Schwerdtfeger
Department of Forest Zoology, University of Göttingen
Göttingen, Germany

Survey and Arrangement

The scientific study of insects that live in the forest goes back to antiquity. It is not, however, until the existence of such insects is seen in relation to a forestry practiced by man that we can speak of forest entomology. For, "by forest insects we do not mean all insects living in the forest but only those which have influence on the thriving and the utility of those wood plants with which the forester is concerned" (J. T. C. Ratzeburg, 1839).

Forestry first came into being in central Europe when the originally ample wood supplies started to vanish and the fear of timber shortage gained ground. This development began in the seventeenth century but occurred mainly in the eighteenth century. In the second half of the eighteenth century, a gradually increasing number of publications on forest entomology appeared, almost always caused by the striking mass occurrence of an injurious insect. This phase may be regarded as the early period of forest entomology.

It was the work of J. T. C. Ratzeburg who "has frequently been called the father of forest entomology" (S. A. Graham, 1929), that terminated the early period and, at the same time, was the beginning of the classical period of forest entomology. In the second third of the nineteenth century, he compiled the knowledge of this time, increased it to an extent that is almost unbelievable and thus laid a solid foundation on which his successors could base the structure of forest entomology.

This further development was at first a continuation of what Ratzeburg had done, namely, the continued and more complete listing of those insects relevant in forestry and the gathering of facts, especially as to their taxonomy, morphology, and bionomics, and to their injurious or useful effects. We can call this stage the period of stock-taking or inventory research. Its results have been summarized in a number of textbooks and manuals.

At the beginning of the twentieth century, the emphasis in forest entomological research shifted from the gathering of facts to the discovering of correlations and causes. People tried to find out how the insect is influenced by its environment, how the populations of certain insects increase to outbreak proportions, and how different insect populations in a stand affect one another. In other words, the autecology, demecology, and synecology of forest insects moved into the foreground of scholarly interest. This phase in the history of our field of study may be called the ecological phase or the period of causality research. Its results serve as valuable aids in developing economical methods of protecting forests from injury by insects.

This European development of forest entomology so far encompasses more than two centuries. On the other continents the scientific study of forest insects, following the rise of controlled forestry, began considerably later; for instance, in North America in the second half of the nineteenth century (S. A. Graham, 1929), in India at the beginning of the twentieth century (J. C. M. Gardner, 1939), and in Africa and other parts of the world in our day. Everywhere, the main features of the development of forest entomology are the same as they were in Europe: the demand for forest entomological knowledge is aroused by the occasional mass appearance of injurious insects; one begins with the stock-taking of forest insects and investigates their taxonomy, morphology, bionomics, and their economic significance. Then the ecological aspects of the matter are studied, especially for the purpose of finding means of predicting future injuries in time and preventing them most effectively.

This basic correspondence in the development of the regional forest entomologies yields the principle of arrangement according to which this history of forest entomology is to be presented here. After the early period in Europe and the work of Ratzeburg, the periods of inventory research and causality research, and the working out of methods for the timely recognition and prevention of insect injuries will be successively described. The arrangement of the material according to critera derived from the history of the field appears to be appropriate for another reason: the development not being synchronous in different parts of the world—mainly Europe and North America—a chronographic presentation would have to deal with the different regions separately, thus leading to repetition.

In our procedure we must accept two things: since the same stage of development of forest entomology is reached at different times in different regions, in one chapter dealing with a certain stage of development scientists and achievements will be mentioned whose chronographic dates can differ by some 50 or 100 years. Moreover, even in one region there is no clean cut between the different historical stages; it is only the emphasis that has shifted and if, for example, the main interest is absorbed by demecological research in a certain period, it does not mean that in the same period less significant and therefore neglected forest insects cannot be studied morphologically or bionomically.

Concise historical presentations of forest entomology are those by O. Nüsslin and L. Rhumbler (1922, 1927), who describe only the European development, however, and by S. A. Graham (1929, 1952), with a survey of the development in Europe and North America.

Early European Period

The first forest entomological publications that deserve the name appeared in Germany. They were called forth (and this is typical of the beginnings of forest entomological observations and research everywhere, as I have mentioned) by extraordinary outbreaks of certain insect species. It is typical, too, that the authors of the early period in most cases were neither foresters nor zoologists but scholars of other branches of knowledge, especially theologians and medical men who happened to witness the phenomenon of the mass occurrence of an insect and became interested in it. In those days most foresters lacked the training necessary to go into such a phenomenon scientifically, and the zoologists were mainly bookworms who were little concerned with outdoor biology.

The beginning of forest entomological literature is marked by *Nachricht von einer Raupe, so etliche Jahre her an manchen Orten in Sachsen vielen Schaden gethan hat (Report on a Caterpillar which did Great Damage in Some Places in Saxony Several Years Ago),* a thin book which appeared at Regensburg in 1752. The author was the Regensburg preacher and doctor of divinity and worldly wisdom, J. C. Schäffer, who became known as an entomologist by a number of noteworthy publications. On a journey he happened to go to a region where caterpillars were doing enormous damage in forests and gardens. Trying to get more information from the residents he was confronted with utter ignorance and thus was stimulated to go into the phenomenon more deeply. Cause of the damage was the gypsy moth *Porthetria (Lymantria) dispar.* Schäffer described correctly and in a detailed manner the exterior of the caterpillar, its genesis from the egg, its development, feeding activities, pupation, and the metamorphosis of the pupa into a moth. A colored copper plate engraving offers clear illustrations of all developmental stages. Schäffer's views as to the causes of the outbreak seem to be almost modern: he mentions favorable weather conditions, failure of natural enemies, and, especially, suitable food. By means of a numerical example he shows how, due to the great fertility of the moths, a vast multitude of caterpillars can spring from small quantities of larvae within a few years. The figures take into account the sex ratio of the moths and the average natural mortality which is rated as one third of the progeny per generation.

In the following years more publications on forest entomological matters appeared but much time elapsed before a treatise of similar quality was presented. The publications of the following decades were mainly attempts by the editors of forest magazines and compendiums to impart to foresters what was known about forest insects. Thus, the oldest forest magazine, the *Allgemeine Oeconomische Forst-Magazin,* which was edited by J. F. Stahl and first appeared in 1763, included in its first year's volume *Einige Beobachtungen von den Insecten wilder Bäume (Some Observations on the Insects of Uncultivated Trees)* by an anonymous author in which insect species, damages, and gall deformations on different species of forest trees were described, according to observations by C. Linné. There were also occasional treatises by foresters who had had experience with some kind of insect injury and described it.

The impulse for a real further development in forest entomology resulted from a forest catastrophe of great proportions and was caused by the bark beetle *Ips typographus*. It had done increasing damage especially in the spruce forests of the Harz Mountains. About the year 1772, a real disaster began which lasted more than three decades and destroyed the greater part of the spruce stands. Faced with this catastrophe, the Electoral Chamber in Hanover took action which was decisive for the development of forest entomology: they turned to the scientists and in 1782 commissioned the Royal Society of Science in Göttingen to call upon experts to find out how the *Wurmtrocknis,* as the bark beetle pest was called, could be stopped. Thus, for the first time and from the highest authorities, a forest entomological problem was placed for discussion before a wide forum. The effect of this public appeal was the publication of numerous scholarly and less scholarly treatises which dealt with the bionomics of the bark beetle, its damage, and the control measures which seemed to be effective. Two books are outstanding among the bulk of the publications: *Abhandlung über die Wurmtroknis (Treatise on the Worm Dryness)* 1787, by J. F. Gmelin, the famous Göttingen professor of medicine, and the work with the title *Ueber einige Insektenarten, welche den Fichten vorzüglich schädlich sind, und über die Wurmtrockniß der Fichtenwälder der Harzes (On some Insect Species which are Injurious Mainly to Spruce, and on the Worm Dryness in the Spruce Forests of the Harz Mountains)* which appeared in 1794. Its author was C. H. von Sierstorpff, Master of the Hunt of the Duke of Brunswick-Lueneburg. The two books share a high standard of presentation by providing evidence of outstanding knowledge. On the other hand, they are basically different. The scholar, Gmelin, drew a broad picture of the bark beetle pest; he considered all literature published up to his day, carefully collected the accessible records, and cautiously balanced the opinions of his informants. The immediate connection with the natural phenomenon and observations of his own are missing, however. We find them with the practical man, von Sierstorpff, who laid aside the existing literature which he thought inconsiderable, relied on his own experience and drew an excellent picture of the life and the injurious effects of the bark beetle. While Gmelin cautiously examined possible control measures, von Sierstorpff offered practical instructions as to how the pest could be checked.

Extensive damage caused by Lepidoptera in conifer forests gave rise to further publications. Two important ones must be mentioned: *Ueber den Raupenfraß und Windbruch in den Königl. Preuß. Forsten in den Jahren 1791 bis 1794 (On Caterpillar Injury and Windfall in the Royal Prussian Forests in the Years 1791 to 1794)* by C. W. Hennert, printed in 1797, and *Geschichte der kleinen Fichtenraupe, oder der Larve von der Phalaena Monacha Linn. (History of the Small Spruce Caterpillar or the Larva of*

Joh. Friedr. Gmelin's

der Weltweisheit und Arzneikunst Doctors, der leztern öffentlichen ordentlichen Lehrers zu Göttingen, der Römisch-Kayserlichen Akademie der Naturforscher, der Churmainzischen zu Erfurt, der Königlich Grosbritannischen Gesellschaft der Wissenschaften zu Göttingen, und der physikalischen zu Zürch Mitglieds

Abhandlung

über die

Wurmtroknis.

Leipzig,

im Verlag der Crusiusischen Buchhandlung,

1787.

FIGURE 1. Title page of J. F. Gmelin's *Abhandlung über die Wurmtroknis.*

P. M. L.) by J. H. Jördens which was published in 1798. Again, none of the authors was either a forester or an entomologist. Hennert had followed a military career, became supervisor of palace building, had also to do with forest surveys, and was thus confronted with forestry problems. Ratzeburg called him a genius. His book describes with scientific thoroughness

Figure 2. C. W. Hennert (1739–1800).

the experiences he had gained at an outbreak of the pine tent caterpillar, *Dendrolimus pini,* in the pine forests of the Brandenburg Marches. Jördens was a physician and wrote many medical publications. Because of a mass occurrence of the nun moth, *Lymantria monacha,* in Thuringia and Saxony he was stimulated to observe and study this injurious insect. H. Weidner has recently reported on Jördens' life and work and on forest entomological publications by other contemporary authors.

By the first of the nineteenth century a considerable foundation of forest entomological knowledge had become established, compiled during approximately 50 years. It was documented by numerous publications, each of which had been provoked by a certain event, and had a single forest insect as its subject. To put this isolated knowledge together to form an overall structure of forest entomology was attempted by one of the most famous forestry experts of his day, J. M. Bechstein. Originally he was a theologian but, in following his passion for hunting and nature, he had acquired the forestry knowledge of his time and, in Thuringia, founded a private institution for the teaching of forestry and hunting which later became a public one. As a writer on forestry he was extremely prolific. In 1798 he published the first section of a *Naturgeschichte der schädlichen Waldinsekten (Natural History of the Injurious Forest Insects).* With this first section the matter rested. Bechstein had probably seen that forest entomology required special knowledge he lacked. The man who had this special knowledge, at least to a certain degree and was willing to place it at Bechstein's disposal, was G. L. Scharfenberg, a parson who pursued entomology and collected butterflies. In 1804 and 1805 the two men edited the *Vollständige Naturgeschichte der schädlichen Forstinsekten. (Complete Natural History of the Injurious Forest Insects),* the first reference book of forest entomology that deserves the name. The three-volume work with its 1042 pages of text and 13 copper plates was well organized. After a general introduction which dealt with the morphology and bionomics of insects and gave a short survey of the entomological literature, there was a special introduction to the natural history of forest insects, and then in systematic order the description of those species that were injurious to the forest. An appendix presented the beneficial insects or those deserving protection. The work is captivating by its ample and well-arranged material, but on closer inspection shows serious shortcomings. Both authors lacked the closer personal acquaintance with forest insects. Thus, very many species were described without regard to their actual economic importance. The existing forest entomological literature is considered in quite an inadequate manner. In 1818 Bechstein published a textbook of forest entomology with the title *Forstinsectologie* which was part of an encyclopedia of forestry and hunting. In spite of its considerably smaller size of 551 pages, it is more substantial than the earlier reference book. The number of species presented is limited to those that seem to be important; they are evaluated as being more or less injurious, the literature has been better organized and the emphasis shifted toward the description of habits, harmfulness, and possible control measures.

Julius Theodor Christian Ratzeburg

So far, all forest entomological publications had come from men who were not professionally concerned with forest insects. The first to dedicate himself completely to forest entomology, having pursued other interests previously, was J. T. C. Ratzeburg. He was the first forest entomologist.

Ratzeburg was born in Berlin in 1801. His father was a professor at the school of veterinary science in Berlin and manager of the Royal Pharmacy. The son also studied pharmacy first, then medicine, and habilitated at the University of Berlin in 1828. With J. F. Brandt, he jointly published a two volume *Medicinische Zoologie,* 1829–1833, a compilatory work of good but not extraordinary standard. In 1830, he was appointed teacher of sciences at the newly established forest school at Eberswalde, 30 miles northeast of Berlin. There, a forester and a mathematician were teaching with him. One of the subjects he taught was entomology. He soon realized its significance in forestry as insect damage, mostly of lesser extent, was frequent in the outlying areas surrounding Eberswalde. The reference book by Bechstein-Scharfenberg that he consulted for information and the other

FIGURE 3. J. T. C. Ratzeburg (1801–1871).

summary literature proved to be insufficient. Thus, in 1835, Ratzeburg began the preliminary studies for a work that was to win him worldwide fame.

He proceeded along three different lines: first, he collected and examined everything that had been written on the subject up to this time; for a scientist a natural course to follow but one which had been used only rarely by his predecessors. Second, Ratzeburg tried to make use of the knowledge that existed with foresters and naturalists. At his suggestion, the Ministry called upon all forest officials to communicate to him their observations of and experiences with forest insects. An extremely active correspondence ensued which crossed the Prussian borders and included foreign countries. Thus, he inquired in Holland, Petersburg, and London in order to formulate a judgment on the destructiveness of the ship-timber beetle, *Lymexylon navale;* further information came from Courland, Sweden, and Switzerland. Third, and mainly, Ratzeburg endeavored to widen his personal experience in order to be able to confirm or reject that of other authorities. He was out in the forest almost every day; he travelled and made observations outdoors and on insects kept in cages.

The result of this work was the three volume publication of 813 quarto pages and 53 plates *Die Forst-Insecten oder Abbildung und Beschreibung der in den Wäldern Preußens und der Nachbarstaaten als schädlich oder nützlich bekannt gewordenen Insecten (The Forest Insects or Illustrations and Descriptions of those Insects which have Become Known as Either Harmful or Beneficial in the Forests of Prussia and the Neighboring States).* The first volume, dealing with Coleoptera, appeared in 1837, and only two years later in a considerably enlarged second edition. The second and third volumes, containing Lepidoptera and the other forest insects, respectively, were published in 1840 and 1844. Every higher forest official in Prussia received a copy at the expense of the treasury, which was quite an extraordinary thing to happen when one considers the pronounced thriftiness of the Prussian budget. With satisfaction, Ratzeburg wrote in the preface to the third volume that the work "has been acquired also by teachers at the universities and schools, by clergymen, gardeners, economists, and even by military men, and has crossed not only the borders of Germany but also those of Europe." In 1929, the American, S. A. Graham evaluated *Die Forst-Insecten* as "a monumental work which even today has never been surpassed in excellence and scope," and in 1931, the Russian W. N. Stark, called it "the fundamental work of modern forest entomology."

The descriptions, as far as their structure is concerned, followed the form that seemed natural and had been used by previous authors: after general introductions, the forest insects are presented, one by one, according to their morphology, bionomics, economic importance, and, where necessary, the possibilities for their control. The exceptional quality of the work lies in the thoroughness and carefulness of the text, the critical evaluation of facts already known, and the immense multiplicity of new material. Within it one finds results and views which were far advanced and which were taken up and even considered new several decades later. One example is found in the description of the nun moth, *Lymantria monacha.* Cage experiments are mentioned in which the development of an individual from the egg to the adult at temperatures of 12–15, 14–17, 16–19, and 19–22°R is measured to last a total of 204, 149, 117, and 98 days, respectively; thus (in 1840), the temperature dependence of the metabolism of a poikilothermic animal is shown clearly and with exact figures. The fact was not stated again with equal precision until done so by E. D. Sanderson in 1910 and H. Blunck and L. M. Peairs, both in 1914. Specifically for the nun moth, the same result was not found until 1935, almost a century later, by W. Zwölfer who obviously did not know of the data provided by Ratzeburg. Special mention must be made of the clear, scientifically faultless plates, part of which are lithographed, others in copper and colored by hand.

For students and others for whom the large work was too bulky and costly, Ratzeburg wrote a shorter guide, *Die Waldverderber und ihre Feinde (The Forest Destroyers and their Enemies).* In addition to the insects, the description of which, however, occupied most of the space, it also dealt with injurious birds and mammals. The book appeared in 1841 in a first edition, was edited by Ratzeburg himself to its sixth edition in 1869, and, after his death, by J. F. Judeich in a seventh edition. Every edition was en-

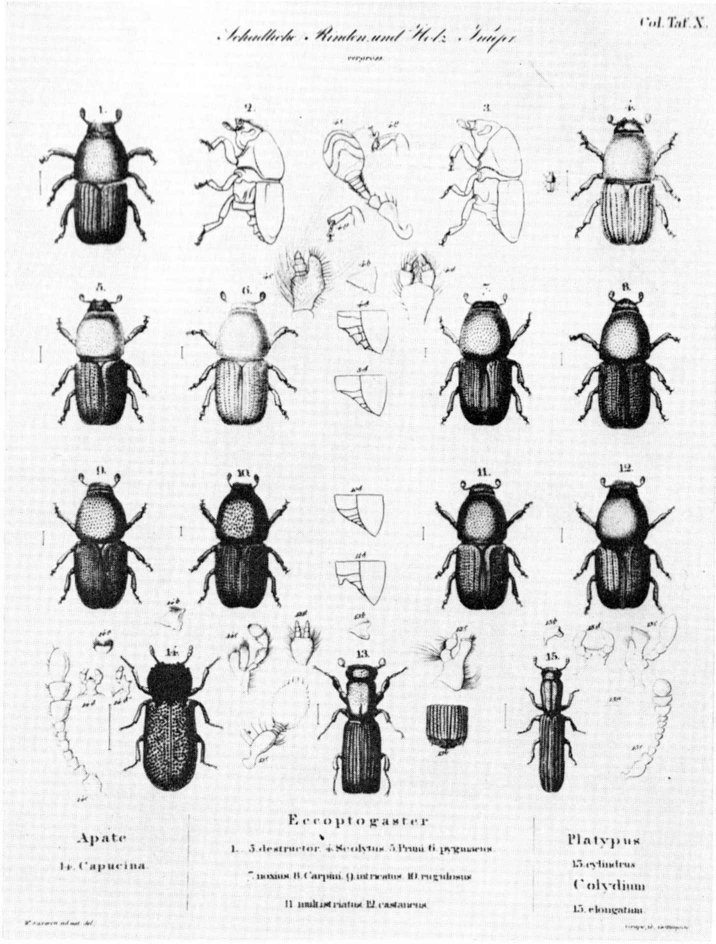

FIGURE 4. Copper plate from Ratzeburg's *Die Forst-Insecten.* Original size 27x22 cm.

riched by new observations and experiences. This guidebook was also translated into French. Due to its extremely wide circulation, it contributed to its author's fame even more than the magnificent *Forst-Insecten.*

Of the numerous shorter treatises and writings by Ratzeburg, none can be mentioned here. Two several-volume works must, however, be given attention: *Die Ichneumonen der Forst-Insecten (The Ichneumonids of the Forest Insects)* in three volumes which appeared in 1844, 1848, and 1852, respectively, and contain the descriptions of about 1000 species of Ichneumonidae, and *Die Waldverderbniss (Forest Destruction)* in two volumes

FIGURE 5. Copper plate from Ratzeburg's *Die Forst-Insecten*. Detail.

and printed in 1866 and 1868. The latter work was looked upon by Ratzeburg as the counterpart of his *Forst-Insecten* as he had, by this time, made the trees and the injuries done to them by animals, mainly insects, the center of attention.

Ratzeburg died in 1871. His lifelong work opened new corridors in all fields of forest entomology. Therefore, we shall find his name again and again as the text proceeds.

Period of Inventory Research

As far as central Europe was concerned, Ratzeburg's work had, in the main, settled the question of which insects are of economic importance in forestry, how they live, what kind of damage they cause, and by which means man could fight them. Thus, a period in the development of forest entomology was coming to an end which elsewhere was in its beginnings or had not even yet begun: the period of stock-taking or inventory research.

Some German forest entomologists living at the same time or later than Ratzeburg, made it their task to fill gaps and made additions to his work. The publications of H. Nördlinger which he called *Nachträge zu Ratzeburgs Forstinsekten" (Supplements of Ratzeburg's Forest Insects)* are characteristic of this trend; they started in 1848 and were collectively edited in a second edition in 1880. B. Altum, Ratzeburg's successor as a teacher at the forest school at Eberswalde, also made numerous contributions covering almost all groups of forest insects. Others tried to widen the knowledge of certain insect groups by studying them intensively. Thus, the monographs on sawflies and horntails by T. Hartig, in 1837, and on bark beetles by W. J. Eichhoff in 1881 came to be written.

In France, E. Perris studied the Coleoptera of *Pinus maritima,* especially the buprestids, cerambycids, curculionids, and scolytids. His principal work, *Histoire des Insectes du Pin maritime,* appeared in 10 parts between 1851 and 1870. He was first in having trees felled every month in order to trace the infestation, and the development of the beetles during the course of the year.

At the beginning of the twentieth century the investigation of forest insects made rapid progress in other European countries as well. The outstanding forest entomologists, who contributed essentially to the inventory research of their countries until the middle of the century and later, must be mentioned here: A. Barbey in Switzerland and France, F. A. Wachtl and W. Sedlaczek in Austria, J. Komárek and A. Pfeffer in Czechoslovakia, G. Cecconi in Italy, G. Ceballos and J. A. Torrent in Spain, C. M. Baeta Neves in Portugal, A. T. Gillanders and R. N. Chrystal in England, J. E. V. Boas in Denmark, I. Tragardh and V. Butovitsch in Sweden, U. Saalas in Finland, M. Nunberg and W. Koehler in Poland, M. Rimsky-Korsakow and W. N. Stark in Russia. Some of them published national compendiums of forest entomology which will be listed in a later section.

In North America, forest entomology developed quite late compared to the other branches of applied entomology. This is not surprising when one considers that in the nineteenth century there were plenty of forests and wood supplies seemed to be inexhaustible. There was no regulated forestry and tree injuries caused by insects passed unnoticed. Injuries done to shade trees received more attention, and thus, in North America, initially tree entomology arose rather than forest entomology. Observations on insects which are injurious to trees were made in the reports on *Noxious, Beneficial and Other Insects of the State of New York* by A. Fitch from 1856 to 1870 and in the *Treatise on Some of the Insects Injurious to Vegetation* published by T. Harris in 1886. At about the same time, there were contributions to tree entomology by B. Walsh, C. V. Riley, J. H. Comstock, and S. A. Forbes, who were mainly concerned with agricultural entomology, however. The first compendium on North American forest insects bearing the title *Insects Injurious to Forest and Shade Trees* was published by A. S. Packard in 1881 and 1890; it is a compilatory summary of known facts, and quotes many of the more important articles by earlier authors.

The first real forest entomologist in North America was A. D. Hopkins. K. Escherich in 1913 called him the North American Ratzeburg. He originally worked in agricultural entomology and was induced to turn to forest entomology by a mass occurrence of the bark beetle *Dendroctonus frontalis* in 1891, which destroyed as much as 80% of the trees in some forests in West Virginia. He was made head of the Division of Forest Insect Investigations, established in 1902, in the Bureau of Entomology and Plant Quarantine of the US Department of Agriculture. He initiated a period of intensive bark beetle research as Eichhoff had done before him in Europe. In numerous short publications and extensive monographs which came out between 1893 and 1921, he, to a large extent, clarified the taxonomy and bionomics of the North American bark beetles. His work was continued and supplemented by M. W. Blackman in the US and J. M. Swaine, the originator of forest entomology in Canada. The results of these and other scientists were collated by W. J. Chamberlin in the monograph *The Bark and Timber Beetles of North America* (1939).

In the first half of the twentieth century forest entomological research in North America gained a wide basis through the establishment of numerous forest insect laboratories. At some universities forest entomological studies were also carried out. Through intensive research work the questions of which insects were injurious in North American forests and how they lived were essentially settled about the middle of the century. Documents of these successful studies are S. A. Graham's *Principles of Forest Entomology,* which appeared in several editions, and the two substantial compendiums *Insect Enemies of Eastern Forests* by F. C. Craighead (1950) and *Insect Enemies of Western Forests* by F. P. Keen (1952).

On other continents the period of stock-taking is still going on. The publications by C. F. C. Beeson (1941) and R. N. Mathur (1962) in India, by L. G. E. Kalshoven (1953) in Indonesia, by K. M. Moore (1957–1963) in Australia, by G. B. Rawlings (1958) in New Zealand, and S. J. Curry (1965) and K. W. Brown (1965) in East-Africa may serve as examples.

Period of Causality Research

With the increase in forest entomological knowledge the period of inventory research gradually merged into and was finally almost replaced by the period of causality research. The new research was characterized by analytical procedure which aimed at revealing the influence of environmental factors and of the properties of populations affecting their dynamics. Research was done along two lines: on one hand the effects of certain factors and properties were generally investigated, on the other hand, concrete cases of the fluctuation of a population and its dependence on the ruling conditions were studied. The end of this analytical procedure is synthesis, which is to yield a complete picture of causes and effects in the population dynamics of forest insects.

Analysis of Relevant Environmental Factors

As early as 1752 Schäffer had named weather, enemies, and food as environmental factors that determine the injurious occurrence of a forest insect. These three groups of factors became the preferred objects of research from the second quarter of the twentieth century onwards.

In Europe this research was stimulated and furthered by the German K. Escherich. After studying medicine and zoology he dedicated himself to entomology. He went to the US in 1912 in order to get to know aspects of study and methods of applied entomology that were already far advanced there, especially on the agricultural sector. As professor at the University of Munich he later gave enormous impulse to forest entomology. His work *Die Forstinsekten Mitteleuropas (The Forest Insects of Central Europe)* planned to consist of 5 volumes, of which, however, only 4 appeared from

1914 to 1942, is the twentieth century counterpart of Ratzeburg's *Forst-Insecten*. He was an inspiring personality who guided a whole generation of scientists. Many of the studies that will be mentioned below are due to his initiative.

By statistical comparisons the Austrian E. Zederbauer had found in 1911 that outbreaks of the nun moth *Lymantria monacha* and other forest insects occur only in regions of a certain climate. With the same method F. Eckstein (1923) and W. Berwig (1926), both disciples of Escherich, found correlations between the weather of certain years and the occurrence of outbreaks of the pine looper *Bupalus piniarius* and the pine noctuid *Panolis flammea*. Since that time an author writing on the mass occurrence of an insect almost always attempts to correlate the phenomenon with the meteorological data of the corresponding time. A new course was followed by W. G. Wellington in Canada (1954, 1957) when he examined the potential connections between population dynamics and atmospheric circulation processes instead of static climatological and meteorological data.

Deeper insight into the effect of weather factors on insects was gained by experiment. The data on the influence of temperature on the speed of development of the nun moth, which were given by Ratzeburg in 1840, have already been mentioned. Other early figures on the speed of development at different temperatures were published by Regener (1865) as to the egg, larval, and pupal stages of the pine tent caterpillar *Dendrolimus pini*. The first observer to combine two weather factors, temperature and air humidity, in extensive experiments was C. Hennings in 1907; he observed the beginning and the duration of the different developmental stages of the bark beetle *Ips typographus* in cultures that were kept at four different temperatures and two humidities. Besides the well-known strong dependence on temperature he found a weaker but nevertheless distinct influence of air humidity. Stimulated by V. E. Shelford's publications, especially the one on the codling moth in 1927, W. Zwölfer included the mortality of the developmental stages in his studies. His subjects were the pine noctuid *Panolis flammea* (1931) and nun moth *Lymantria monacha* (1934 and 1935), the stages of which were kept at different combinations of temperature and humidity. The results were given the form of development duration curves and mortality diagrams. Since the dependence of insect life on temperature and humidity was so clearly shown and expressed by figures and curves in these investigations and others that were carried out in the same way, a temporary overevaluation of weather factors set in: they were considered as determining the population dynamics of insects while other factors were evaluated too low or neglected completely.

One of these factors is the occurrence of natural enemies of the insects. In *Die Ichneumonen der Forst-Insecten* (1844–1852) Ratzeburg had already thoroughly studied parasites; he did not think, however, that they were of great importance in the natural control of injurious insects, because in his opinion they prefer hosts which are already ill. That this is not true was made evident, for instance, by L. O. Howard and W. F. Fiske in 1911, when they reported on the successful biological control of the gypsy moth which had been brought to North America from Europe: the unhindered reproduction and spread of the insect in the New World was stopped by also importing the parasites which had continuously been keeping it down in Europe. An illustration by Escherich (1913), according to which only 36 of 100 caterpillars of the pine noctuid produced adults and the rest produced parasites, turned out to be of lasting impressiveness. Since then the literature on parasites, complexes of parasites, and parasitizing percentages has become immense. The influence of predators is much harder to determine than the effectiveness of parasites, documented by the percentage of parasitized individuals. By examining birds' stomachs and by feeding experiments S. A. Forbes (1880) and G. Rörig (1903) showed what quantities of insects birds can eat. Their results give no clue, however, what percentage of an insect population is eaten by birds. These kinds of studies, which give information about the significance of birds in the natural control of forest insects, have only recently started to be carried out, e.g. by P. B. Dowden et al in the US (1953), V. Tichý and J. Kudler in Czechoslovakia (1962), L. J. Mook in Canada (1963), and W. Altenkirch in Germany (1965–1968). The survey by C. H. Buckner 1966 must be mentioned. Attention must also be called to the publication on *Calosoma sycophanta* by A. F. Burgess (1911), as an early and exemplary work on a predatory forest insect. Pioneers in the field of diseases of forest insects were R. Hartig and T. Bail who in 1869 reported about fungal epizootics on insects injurious to the forests. T. Bail (1870) and C. von Tubeuf (1893) carried out valuable investigations into the occurrence and importance of the fungus *Empusa aulicae* which infests the caterpillars of *Panolis flammea* and repeatedly brought its outbreaks to a sudden stop. There are more recent surveys of fungal diseases of insects by E. Müller-Kögler (1965) and M. F. Madelin (1966), who do not, however, restrict themselves to those species occurring in the forest.

It was comparatively late that the importance of food as an influential environmental factor in the development of forest insects was subjected to experimental research. It can be studied most easily with polyphagous species by offering them different kinds of food which they also accept under natural conditions. H. Maercks (1935), H. Sattler (1939), and A. Mayer (1940) fed caterpillars of *Lymantria monacha* with numerous forest plants they normally eat; the survival rate differed widely and ranged from 0% to 100% depending on the food plant. That a specific host can have a variable trophic valence or infesting disposition for the monophagous insect also is a well-known fact; this was more closely defined by F. Kangas (in 1948), as far as the so-called secondary insects are concerned, which can only live in or on trees weakened by other causes. E. Merker (1955) used the osmotic value of pressed bark sap as a quantitatively measurable characteristic of the vitality of a tree; F. Schwerdtfeger (1948), J. P. Vité and J. A. Rudinski (1960) found the intensity of the oleoresin exudation pressure to be decisive in a tree's stronger or weaker resistance against the infestation by secondary bark beetles. The beetles find susceptible trees by means of olfactory attractants which emanate either from the trees themselves (Kangas, 1968) or are emitted as pheromones by individuals which have already settled (*Symposium on Volatile Attractants,* 1970). That the so-called secondary forest insects can successfully infest trees in a certain physiological condition only has long been known. Astonishing, however, was the realization that primary insects which live on completely healthy trees, as was thought, also depend in their development on the changing physiological state of those trees, i.e. on a variable disposition of the host. This was realized when H. Büttner (1956) and H. Oldiges (1958) reared larvae of Lepidoptera and Tenthredinidae on conifers fertilized with different substances or not fertilized at all: the mortality of the larvae, the weights of the resulting pupae, and the egg production of the adults coming from these pupae proved to be different. Further investigation confirmed these results with other forest insects too (Stark, 1965). W. Schwenke (1961) attributed the effect to a shifting of the protein-sacharine relation in the assimilating organs of the host tree, which view, however, is considered incorrect by D. Otto (1970). The causality between the nourishment and the water supply of a host on the one hand and its trophic valence or disposition for phytophagous insects on the other hand obviously needs further investigation. A. J. Paramoniv (1953) and D. F. Rudnew (1963–1964) go so far as to regard the factor of food, i.e. the physiological state of the host, as determining the population dynamics of all forest insects.

An environmental factor, e.g. a parasite, can only become effective if it meets the animal in question, in this case the potential host. This seems to be a truism. W. Thalenhorst, however, showed in 1951 that coincidence, i.e. the effective temporal and local meeting of two or more partners of an ecosystem, can be a quite complex process and in many cases will be decisive for the future of a population. Just one example: the caterpillars of the oak tortricid *Tortrix viridana* emerge from the eggs at springtime when the oak buds start to open. In order to develop further the caterpillars must enter buds which are at a certain stage of opening. Depending on the spring weather the buds may not have reached that stage or be exactly at that stage or even beyond it at the time of the hatching of the caterpillars and the latter will either die to a large extent or develop further correspondingly.

Analysis of Population Characteristics

W. Zwölfer (1934) carried out his above-mentioned investigations, in which he examined the influence of temperature and humidity on the development and the mortality of the nun moth, in two consecutive years, 1931 and 1932, using caterpillars of the same origin: at all temperature levels the mortality was higher in 1932 than in 1931. E. Janisch (1938) simultaneously reared caterpillars of the same species, which came from two locations that were far apart, feeding them different plants: with all food plants the one origin had a higher survival rate than the other. Thus was shown, that the effects of environmental factors are different depending on the average physiological state, i.e. the constitution of the insect population, and that this constitution can be locally different and changeable in time. That not only the abiotic and trophic factors but also the third group of environmental influences, the natural enemies, depend on the changeable constitution of the insect as far as their effectiveness is concerned, can be clearly seen in the fluctuating ratio of the larch sawfly *Pristiphora erichsonii* and the parasitic ichneumonid *Mesoleius tenthredinis:* the parasite, imported from England to North America in 1910 for the biological control of the sawfly, at first caused a heavy decrease of the injurious insect; since 1940, however, it has lost its effectiveness almost completely in some regions, because more and more of the host larvae have developed the capability of incapsulating the parasite's eggs, thus rendering them innocuous (Muldrew, 1953).

W. G. Wellington (1957, 1960) first demonstrated on the western tent caterpillar *Malacosoma pluviale* that the behavior of the animals can also be individually different within a population and is subject to change in time. These differences in behavior are often connected with the insects' physiological condition and especially with changes in population density.

The importance of density for the development of forest insect populations was emphasized by H. Eidmann in 1937. He spoke of overpopulation phenomena which cause the abundance of a population to decrease automatically when a certain level is reached. On the basis of his observations Eidmann mentioned as overpopulation phenomena the increased competition for food, a decrease of fertility, the rising susceptibility for diseases, etc. Thus trains of thought had been formulated by a forest entomologist which at the same time and later were developed in demecology, especially by A. J. Nicholson in Australia (1933, 1954) and A. Milne in England (1957, 1962).

Analysis of Population Development

The above-mentioned studies, which are but examples of the total literature, had yielded important results as to which environment factors, coincidence conditions, and population characteristics are causally responsible for the population development of forest insects and consequently for the occurrence or nonoccurrence of the injuries caused by them. In order to realize which factors and processes actually determine the development of a population in concrete cases F. Schwerdtfeger (1932, 1934) studied the qualitative and quantitative population changes of the pine noctuid *Panolis flammea* during periods of one year each and presented the data in the form of tables. They were the first presentations to grant a short and precise insight into the abundance dynamics of a forest insect and their causes. They were reinvented, as it were, by R. F. Morris and C. A. Miller (1954) in their investigations on the spruce budworm *Choristoneura fumiferana,* and were named life tables; they are regarded as a very important aim and have aided in forest entomological studies ever since (Harcourt, 1969).

The insight into the causalities of population dynamics of forest insects was deepened when people started to analyze the development of certain populations through several generations and to put to work teams of scientists to whom specialized tasks had been assigned. G. Wellenstein was the first to direct this kind of team work for several years when an outbreak of the nun moth *Lymantria monacha* was going on in East Prussia from 1933 to 1937; in the center of the infested area a camp with living quarters and laboratories was set up; there up to 37 people were doing observations and experimental research for four summers. The result was a monograph published in 1942 which was sensational at the time: in 19 contributions, 8 authors presented the results of their studies of the bionomics and population dynamics of the nun moth and its enemies, and their experiences as far as survey, prognosis, and control was concerned. Since then several long-term team work investigations of that kind were carried out or started with increasingly improved approaches and methods; the key factor analysis may serve as an example (Morris, 1959; Varley-Gradwell, 1960). Their number, however, remained limited since they require a great amount of planning and organization as well as personnel and money. The following investigations must be mentioned: R. F. Morris and his colleagues on the spruce budworm *Choristoneura fumiferana* in Canada from 1945 to 1959, whose results were published in 1963; the studies on the oak tortricid *Tortrix viridana* and the winter moth *Operophthera brumata* by G. C. Varley in England, which were begun in 1949 and 1950, respectively, and have not been finished yet (Varley-Gradwell; 1956, 1963); P. Bovey in Switzerland on the grey larch budmoth *Zeiraphera diniana* (Bovey, 1958; Auer, 1961; Baltensweiler, 1964); and H. Klomp (1966) on the pine looper *Bupalus piniarius* in the Netherlands. In these projects methods of automatic data acquisition and computer technique are being used to an ever-increasing extent (Watt; 1962, 1966).

While the above-mentioned investigations are concerned with insects which at times occur in masses and can be very injurious in such cases, W. Thalenhorst has been studying since 1950 the population dynamics of different sawfly species which feed on the spruce in Germany but have never been observed in great density. From a comparison between the results gained on species which always remain latent and species with a heavily fluctuating density important contributions towards our knowledge of population dynamics can be expected (Thalenhorst; 1958, 1968).

Synthesis

Long-term team investigations have rapidly widened our knowledge of the cause and effect complex underlying the occurrence of injurious forest insects. These successes are to a large extent due to model-analysis methods which were first used in Canada and the US. The principle of these methods is that the action of single and also combined factors can be represented in a definite and biologically meaningful mathematical expression, the model. The second stage is to test the working of the model as a whole by comparison with actually observed data. Examples are the above-mentioned publications by Morris (1963), Klomp (1966), and Auer (1968).

Through a reasonable combination of the results of analysis we come to a synthesis which gives us a somewhat complete picture, even if there are some gaps left, of what is happening in an insect population. In the cases of those forest insects on which long-term investigations have been or are being done we already have overall pictures of the population-dynamic processes, which certainly call for additions and perhaps corrections but can serve as a basis for the forester's decision on how to prevent or control insect injuries. In the near future it will be the main task of forest entomology to combine faultless analyses to syntheses of the population-dynamic cause and effect complexes of all economically important insects.

In a historical survey of the methods and results of forest entomology its manifold contributions towards a general theory of population dynamics must at least be briefly mentioned. In view of the astonishingly high parasitization of forest insects which K. Escherich found in his cage experiments, he put forth in 1914 the opinion that parasites played the decisive role in the natural control of insects. H. Eidmann (1937) regarded overpopulation phenomena and D. F. Rudnew (1963–1964), the quantity and quality of food as the determining factors in abundance dynamics of forest insects. J. M. Franz (1950) pointed out the great importance of the constitution of a population. F. Schwerdtfeger (1968) tried to find a common denominator of these theories and others that did not originate from forest entomologists, in his integrated theory of abundance dynamics in animal populations.

Utilization of the Results of Causality Research

It is the signification of causality research to utilize its results for the real aim of forest entomology as an applied science, i.e. to offer the forester suitable methods of saving his economic object, the forest, from insect injury. Such methods are, however, not only a result of the rise of causality research but were to a large extent devised and used on an empirical basis long before.

There are principally two practicable ways of preventing insect injuries in forests. One can in the long run create conditions which will hinder or even make impossible the injurious mass occurrence of insects; these methods are summarized as forest hygiene. Or in the event of an outbreak of an insect suitable means are applied for a short time in order to stifle the pest at birth or put an end to it quickly; the complete system of methods used in this manner is called forest therapy. In some cases it will be necessary to have the insect population continuously surveyed and to have prognoses of its dynamics in order to be able to take quick steps in the face of danger.

Forest Hygiene

The catastrophic bark beetle pests which struck Germany in the second half of the eighteenth century and, as I mentioned, gave rise to scientific publications like the excellent books by Gmelin (1787) and von Sierstorpff (1794), made people realize one basic fact, i.e. that the increase of important bark beetle species is caused and accelerated by an ample supply of newly felled trunks or sickly or dying trees where they find suitable breeding places. From this realization the principle of clean forestry was derived which central European foresters have observed ever since; it designates the postulate that all fallen or felled trees or sickly ones should be processed and their bark removed as soon as possible, so that they cannot become hotbeds of bark beetle reproduction.

Another forest hygienic postulate originated in Germany as well: in the course of the nineteenth century, economic thinking encouraged the tendency of planting pure conifer forests in the place of preceding deciduous or mixed stands. It soon turned out that these monocultures suffered much more from injurious insects than did the former forests that were appropriate to their habitat. Quite early there were people who warned, and K. Escherich, in particular, never tired of pointing out, that conifer monocultures were exposed to risks and that deciduous or mixed forests were much less easily affected by crises. These warnings were supported by comparative studies which were carried out by F. Schneider (1939) and A. D. Voûte (1946) in mixed primeval forests and monocultures in the tropics. More recent studies are listed in the reviews by S. A. Graham (1956) and H. Francke-Grosmann (1963). E. Schimitschek (1969) calls the preservation of autochthonous forest vegetation and its restoration wherever it has been destroyed the main task of forest hygiene.

From other forest hygienic measures (cf. the survey in Schwerdtfeger, 1970) the following must be mentioned: strengthening of the trees through good alimentation, especially through fertilization (Merker, 1958; Schwenke, 1961), and the permanent settlement of polyphagous enemies of the injurious insects, especially birds (von Berlepsch, 1899; Henze, 1943) and ants (Schulz, 1924; Gösswald, 1951; Otto, 1966). Under certain conditions these measures have proved to be effective, but in many cases they cannot be recommended for economical reasons.

A special kind of long-term measure directed against specific insect species, however, is the importation of the enemies of injurious insects. This is being used especially in North America, where numerous injurious insects were imported without their enemies from Europe since the middle of the nineteenth century, so that there were no hindrances to their spread and increase in their new home. The later importation of those predators and parasites which keep them down in their original home aims at setting up the same natural checks in the infested region. This is not a specifically forest entomological method, but it has been used successfully against forest insects among others. The first campaign of this kind began in Massachusetts in 1905 under the direction of L. O. Howard; in time almost all enemies of the gypsy moth *Porthetria (Lymantria) dispar* and the brown-tail moth *Nygmia phaeorrhoea (Euproctis chrysorrhoea)*, which were not native pests, were imported; this undertaking, which stretched over 25 years, had the effect that the two Lepidoptera are no longer injurious everywhere all the time, but become injurious at intervals and in some areas as they do in Europe (Howard-Fiske, 1911; Burgess, 1932). Mention must be made of the successful importations of enemies of the larch sawfly *Pristiphora erichsonii* (Muldrew, 1955), the European larch casebearer *Coleophora laricella* (Graham, 1956), and the European spruce sawfly *Diprion hercyniae* (Finlayson-Finlayson, 1958).

Forest Insect Survey

If a short-term control measure, e.g. the spraying of an insecticide, is to have its full effect, the injurious occurrence of the insect must be recognized in time. This timely recognition can only be secured by continuously surveying the occurrence of injurious insects.

The continuous survey of significant forest insects was regulated in the Prussian state forests as early as in the second half of the nineteenth century. In the records of Prussian forest boards the author found annual figures on the density of pine insects which go back to the year 1881. The method used was the same as it is now: the insect stages hibernating in the litter are collected from an area of a certain shape and size, and counted; this is done every year at the beginning of winter in selected pine stands and the figures are entered on forms designed for the purpose. In other countries a forest insect survey was established much later, in most cases from the second quarter of this century onwards. In many countries there is none at all so far.

The methods in use today differ according to their aims, the intensity of forestry, and geographical and financial conditions. The simplest form is the registration of injuries on the basis of annual or more frequent reports which is, e.g. the procedure in the Netherlands as described by J. Luitjes and A. D. Voûte (1958). In countries with extensive and inaccessible forests the airplane is used in the recognition of injuries, as is done in the US (Bongberg, 1958) and in Canada (McGugan, 1958). Both methods have the disadvantage that frequently the mass occurrence of an insect will not be recognized until it is already in full operation and injuries have been done. A timely recognition in all cases can only be ensured by the permanent checking of population densities, as was described above for some pine insects. The procedure to be chosen will differ according to the bionomics of the insect and its stages; the methods used in North America are listed in R. F. Morris (1960) and B. H. Wilford (1960), those of central Europe in F. Schwerdtfeger (1970).

Prognosis

The prognosis of the occurrence of an injurious insect is in the majority of cases based on surveillance. Its task is to find out when and where an insect will cause enough damage so that control measures will be necessary.

We can distinguish long-term prognosis which predicts an event a long time, perhaps a year or more, before it occurs, and short-term prognosis which aims at pointing out for the coming weeks or months the trend of a gradation which is already going on.

Long-term prognosis is relatively easy and safe with those insects (their number is comparatively small, however) which increase and decrease in density in regular intervals, i.e. have a cyclic fluctuation. A good example is the grey larch budmoth *Zeiraphera griseana,* which in the Alps reaches a maximum of its cycle every 6 to 8 years (Baltensweiler, 1964). Long-term prognosis is much more difficult and less reliable with species of an irregular, acyclic fluctuation. On the basis of well-known and hypothetical correlations between the weather and the occurrence of outbreaks, attempts were made to predict the dynamics of an insect population by means of meterological data. W. C. Wellington (1952) demonstrated that there are outbreaks of the spruce budworm *Choristoneura fumiferana* in those years when a decreasing number of cyclones cross the respective regions and the influence of dry air masses is more predominant than usual; on the other hand increases of the forest tent caterpillar *Malacosoma disstria* are in preparation when the weather is mainly determined by maritime masses of air and cyclones. Apart from these cases the development of long-term prognosis in forest entomology, as far as it will be possible at all, is still in its beginnings.

Short-term prognosis is better developed, in many cases up to perfection. In those cases where the development of an insect is directly influenced by the weather, meterological data can be consulted. Thus in northwest Germany according to B. Ohnesorge (1961) injuries by the spruce aphid *Liosomaphis abietina* must be expected in spring when in the preceding winter the medium temperature of December through March was above 2.5°C, the medium temperature of the coldest month above 0.5°C, and the minimum temperature during the period above −14°C. In general, short-term prognosis is based on the checking of the population density. Whenever high or increasing densities of an injurious insect are found, e.g. of the pupae of the pine noctuid in hibernation, further characteristics of the population structure must be analyzed in order to predict the extent of caterpillar injury in spring; these characteristics are in particular the sexual index, the size or weight of the female pupae as an indicator of the egg production of the adults, and their state of health. Ratzeburg demonstrated in 1856 that the parasitization of caterpillars and pupae increases every year during an outbreak, and he gave directions as to how parasitization can be examined on caterpillars of *Dendrolimus pini.* In order to judge whether the density of those individuals which are capable of development, and will most probably cause damage, must be regarded as economically serious, F. Schwerdtfeger (1932) introduced the term of critical figure: it designates the density which forecasts heavy injuries and calls for control. Since then critical figures have been worked out for a considerable number of significant forest insects. A survey of critical figures and their application in central Europe, and more generally of the methods of prognosis used there, was given by F. Schwerdtfeger in 1970. In North America the US Forest Service published directions for the evaluation of forest insect outbreaks; a review confined primarily to evaluation of population trends was published by F. B. Knight in 1967.

Control

Even the earliest authors of forest entomology recommend or describe control measures which we call mechanical, chemical, and biological methods today. Hennert (1797) reported that during the great outbreak of *Dendrolimus pini* in the Brandenburg Marches adults, eggs, and pupae of the insect were collected in remarkable quantities and ditches were dug to catch the caterpillars migrating on the ground. There were attempts to kill the caterpillars by burning sulfur and spraying tobacco lye; in order to destroy the larvae hibernating in the litter, pigs were driven into the woods, and it was suggested that the predatory beetle *Calosoma sycophanta* be bred artificially and put out into the forest.

Until the first quarter of the twentieth century almost exclusively mechanical methods were put into practice more or less successfully. The following ones are most frequently used: collecting the cockchafer *Melolontha* sp., often intricately organized and carried out on large areas, the successfulness of which was subjected to critical observations by Ratzeburg as early as 1839; digging ditches to catch the weevil *Hylobius abietis;* attracting and destroying bark beetles by means of trap trees; fixing tar-rings and later tanglefoot-rings on trunks in order to intercept the caterpillars of *Dendrolimus pini* on their way up the trees. All these methods and numerous others were being used at the time when Ratzeburg's *Forst-Insecten* appeared. K. Eckstein compiled them in a book in 1904.

As far as chemical methods are concerned, in the nineteenth century sulfur and tobacco lye, which Hennert had already mentioned in 1797, and, moreover, lime-wash and soapy water, were recommended again and again and were also used on a small scale in a few places, especially in nurseries. The first synthetic insecticide was ortho-dinitrocresol-potassium proposed as Antinonnin for the control of the nun moth by C. O. Harz and W. von Miller in 1892. With a slightly altered formula it was used on a relatively large scale for the control of different forest insects in central Europe in the 1930s. When the proposal was first made it was premature, for there was no machinery to spread the substance over large areas and in high forest stands. It was not until the airplane had been developed that chemical control, which had already gained a firm footing in agriculture and fruit growing, could be extended to forest protection as well; it soon became the most widely used method. In 1912 the Prussian forester Alfred Zimmermann had patented his idea of spreading poisonous substances from an aircraft in order to exterminate the nun moth and other forest pests. The idea was not realized, the patent was forgotten. It was in the US in 1921 that C. R. Neillie and J. S. Houser brought about the first instance of pest control by means of an airplane: lead arsenate was dusted over a small catalpa stand which was infested by the catalpa sphinx *Ceratomia catalpae;* the result was surprisingly good (Uphof, 1923). In 1924 a similar experiment was made in Switzerland, and in 1925 there were the first large-scale campaigns in Switzerland and in Germany. Since then chemical control and the aviochemical method in particular has been further developed and has become an indispensable requisite in forest insect control, in spite of its indisputable disadvantages, especially in respect to the endangering of the biocenosis (Balch, 1958). Its development during the last 50 years is characterized by the change from dusting to spraying, the substitution of organic-synthetic contact insecticides for arsenical stomach poisons, the increasing predominance of DDT and the realization of its disadvantages, the growing use of helicopters instead of airplanes, and the construction of efficient ground machinery.

As a biological control the driving of pigs into the forest was practiced from the eighteenth century until this century (Wendt, 1930). By this means the number of insects in the litter, especially the pupae of *Bupalus piniarius* and *Panolis flammea,* could be more or less heavily reduced. The use of parasitic insects was recommended and practiced in two ways at the beginning of the nineteenth century: first in the form of so-called Raupenzwinger, i.e. caterpillar cages where caterpillars were kept in great quantities and their parasites were expected to show mass reproduction, and second, by the moving of naturally parasitized hosts to parts of the forest where parasitization was still low. Ratzeburg critically studied both methods in 1840, found the former to be ineffective and recommended the second. The first use of parasites artificially reproduced in mass breeding against a forest insect took place in Germany in 1933: during an outbreak of *Panolis flammea* masses of *Trichogramma* sp. were put out on an area of 25 acres; there was no effect at all (Wellenstein, 1934). In the following years experiments with other parasites and also predatory insects against different forest pests were carried out; they were always made on small areas and some of them were surprisingly successful; for technical and economical reasons there has never been a practical application on a large scale so far. The prospects of microbial control of forest insects seem to be more favorable. The experiments with and the practical use of *Bacillus thuringiensis* in particular, which were started at the beginning of the 1960s, have yielded encouraging results at least with some insect species (Krieg, 1967). After the early attempts of Kloeck (1925) and J. Ružička (1925) to spread the polyhedrosis of the nun moth artificially were slightly encouraging, the application of certain viruses now seems to be promising too (surveys in Tanada, 1959; Franz, 1961; Beirne, 1962; and Schwerdtfeger, 1970).

Textbooks and Reference Works

Some of the more important summary works of forest entomology published in different countries are listed in the literature cited. Besides textbooks and reference works which deal exclusively with forest entomology, those books pertaining to forest zoology and forest protection which treat forest insects extensively, will also be mentioned.

Literature Cited

Altum, B. 1874. *Forstzoologie. III. Insecten*

Anderson, R. F. 1960. *Forest and Shade Tree Entomology*

Barbey, A. 1913. *Traité d'Entomologie Forestière*. 2nd ed. 1925

Bechstein, J. M., Scharfenberg, G. L. 1804–1805. *Vollständige Naturgeschichte der schädlichen Forstinsekten*. 3 vols.

Bechstein, J. M. 1818. *Forstinsectologie*

Beeson, C. F. C. 1941. *The Ecology and Control of the Forest Insects of India and the Neighbouring Countries*

Boas, J. E. V. 1923. *Dansk Forstzoologi*

Cecconi, G. 1924. *Manuale di Entomologia Forestale*

Chamberlin, W. J. 1924. *Forest Entomology*

Chamberlin, W. J. 1931. *An Introduction to Forest Entomology*

Chrystal, R. N. 1937. *Insects of the British Woodlands*. 2nd ed. 1948

Craighead, F. C. 1950. *Insect Enemies of Eastern Forests*

de Koning, M. 1922. *Boschbescherming*

Dingler, M. 1927. *Schutz gegen Tiere*. 1st vol. of Hess-Beck, Forstschutz. 5th ed.

Doane, R. W., Van Dyke, E. C., Chamberlin, W. J., Burke, H. E. 1936. *Forest Insects*

Döbner, E. P. 1862. *Handbuch der Zoologie*. 2nd vol. Wirbellose Tiere.

Eckstein, K. 1897. *Forstliche Zoologie*

Escherich, K. 1914–1942. *Die Forstinsekten Mitteleuropas*. 4 vols.

Felt, E. P. 1905–1906. *Insects Affecting Park and Woodland Trees*

Felt, E. P. 1924. *Manual of Tree and Shrub Insects*

Froggatt, W. W. 1923. *Forest Insects of Australia*

Gäbler, H. 1955. *Forstschutz gegen Tiere*

Gillanders, A. T. 1908. *Forest Entomology*. 2nd ed. 1912

Graham, K. 1963. *Concepts of Forest Entomology*

Graham, S. A. 1929. *Principles of Forest Entomology*. 3rd ed. 1952

Györfi, J. 1957. *Erdészeti Rovartan*

Haber, A., Nunberg, M. 1956. *Zoologia dle Leśników*

Henschel, G. 1861. *Leitfaden zur leichteren Bestimmung der schädlichen Forst-Insekten*. 3rd ed. entitled *Die schädlichen Forst- und Obstbauminsekten*. 1895

Herrick, G. W. 1935. *Insect Enemies of Shade-Trees*

Hess, R. 1876–1878. *Der Forstschutz*. 3rd ed. 1898

Holmgren, A. E. 1867. *De för träd och buskar nyttiga och skadliga Insekterna*

Judeich, J. F., Nitsche, H. 1895. *Lehrbuch der Mitteleuropäischen Forstinsektenkunde*. 2 vols.

Kauschinger, G. 1846. *Die Lehre vom Waldschutz*. 7th ed. by H. von Fürst. 1912

Keen, F. P. 1952. *Insect Enemies of Western Forests*

Kollar, V. 1837. *Naturgeschichte der schädlichen Insekten in Beziehung auf Landwirthschaft und Forstcultur*

Kovačevič, Z. 1956. *Primijenjena Entomologija. III. Šumski Stetnici*

Nördlinger, H. 1884. *Lehrbuch des Forstschutzes*

Nüsslin, O. 1904. *Leitfaden der Forstinsektenkunde*. 2nd ed. 1912

Nüsslin, O., Rhumbler, L. 1922. *Forstinsektenkunde* (3rd ed. of Nüsslin, 1904). 4th ed. 1927

Packard, A. S. 1881. *Insects Injurious to Forest and Shade Trees*

Packard, A. S. 1890. *Insects Injurious*

to Forest and Shade Trees
Pawlowsky, E. N., Stackelberg, A. A., Eds. 1955. *Injurious Forest Animals.* (russ.)
Peirson, H. B. 1927. *Manual of Forest Insects*
Pfeffer, A. 1961. *Ochrana Lesů*
Ratzeburg, J. T. C. 1837–1844. *Die Forst-Insecten.* 3 vols.
Ratzeburg, J. T. C. 1841. *Die Waldverderber und ihre Feinde.* 6th ed. 1869
Ratzeburg, J. T. C. 1866–1868. *Die Waldverderbniss.* 2 vols.
Rimsky-Korsakow, M. N. 1935. *Forest Entomology.* 2nd ed. 1938. (russ.)
Ritzema Bos, J. 1891. *Tierische Schädlinge und Nützlinge für Ackerbau, Viehzucht, Wald- und Gartenbau*
Schimitschek, E. 1944. *Forstinsekten der Türkei und ihre Umwelt*
Schwerdtfeger, F. 1944. *Die Waldkrankheiten.* 3rd ed. 1970
Stark, N. 1931. *Enemies of the Forest.* (russ.)
Stark, W. N. 1931. *Injurious Forest Insects.* (russ.)
Stefanow, D. 1956. *Forest Protection.* (bulg.)
Taschenberg, E. L. 1874. *Forstwirthschaftliche Insekten-Kunde*
Tragårdh, I. 1914. *Sveriges skogsinsekter.* 2nd ed. 1939
Živojinovič, S. 1948. *Šumarska Entomologija*
Živojinovič, S. 1958. *Zaštita Šuma*

HISTORY OF APICULTURE[1]

GORDON F. TOWNSEND
University of Guelph, Guelph, Canada

and

EVA CRANE
Bee Research Association, Chalfont St. Peter, Gerrards Cross, Bucks., England

Honey bees now live in all parts of the world except the extreme polar regions, but this was not always so. Until the sixteenth century they were confined to the Old World, where they had evolved and were widely distributed long before man appeared on the earth. Primitive man learned to get honey by robbing the bees' nests in hollow trees or rock crevices; a painting made in a rock shelter in the mountains of eastern Spain in Mesolithic times, probably about 7000 B.C., survives to show us how this was done (Figure 1). Bee hunting is still carried out in various parts of the world, and honey can still be a lifesaving food for primitive peoples in times of famine.

The twentieth century has seen rapid changes take place in the beekeeping industry. This has been the case particularly in the Western Hemisphere and Australia, where abundant sources of nectar have led to rapid technological advances. In these areas it is not uncommon for one producer to handle more than 50–100 tons of honey per year. The honey bee has become important not only as a producer of honey and wax and as a pollinator, but also as a research animal in biological studies.

EARLY HISTORY

Beekeeping proper started when man learned to safeguard the future of the colonies of bees he found in hollow tree trunks or elsewhere by a certain amount of care and supervision. Gradually, separate hives came to

[1] The authors found great difficulty in bringing together in such short space an adequate recognition of the part played by various individuals and countries in a subject which covers both an industry and a science. The first part of the summary (by E. C.) gives a world picture. This is followed by the history of the development as an industry (by G. F. T.). Undue emphasis may appear to be placed on American developments in this section but it was in North America where apiculture first became a recognized industry. The last sections deal with the more fundamental disciplines and only the more decisive steps in the view of the author (G. F. T.) are followed in chronological order.

FIGURE 1. Rock painting depicting honey gathering discovered in the Cuevas de la Araña near Bicorp in Valencia, Spain.

be used as substitutes for the natural dwellings of bees; for convenience and safety they were collected together in an apiary. Hive construction depended on what local materials were at hand and on the local skills of the various communities. It was an inevitable development in any region populated by honey bees as soon as man advanced from hunting and collecting food to producing it in a settled existence.

In the great forests of Europe, the earliest hive probably was a log from a fallen tree in which wild honey bees had nested. The log would be separated by chipping away the rest of the tree with axe and adze, a technique used throughout the Stone Age. Cork and other types of bark were also made into hives and, later, planks cut from tree trunks were used.

The earliest centers of culture were in the Middle East, in hot, dry open country which was not forested. The first hives there were probably pots in which swarms happened to settle. Pottery vessels were made during most of the Neolithic period, from perhaps 5000 B.C. onwards, and water pots are still used as hives in some Mediterranean lands. In ancient Egypt and adjoining regions, pipe hives (long tubes piled horizontally) made of clay and other materials were used.

In agricultural communities, techniques were developed for making containers of basketwork as well as pottery, and these baskets were also used to house bees. Baskets have changed little through the ages, and baskets of coiled straw are made today in the same way as before 500 B.C. The bone awl, essentially the same as that of a Mesolithic basket maker, was used for making skeps for bees as recently as within the last decade in a Yorkshire dale in England. Woven baskets came later, and were made of various materials such as pliable hazel twigs; examples made between 3000 and 2000 B.C. have been found in Egypt. Wickerwork hives still linger on in a few parts of Europe.

All these primitive hives fulfilled certain necessary functions: they protected the bees and their combs from wind, rain, and extremes of heat or cold; their flight entrances were small enough for the bees to guard; and there was some other opening through which the beekeeper could get at the honey and wax which constituted his harvest. Wood, bark, and clay were themselves weatherproof; straw and wicker hives were generally protected with an additional cover, and wicker hives were often plastered with mud and cow dung.

Primitive hives were usually small, because the beekeeper wanted to encourage swarms to populate his empty hives. Primitive beekeeping consisted of little more than providing the hives, and killing the bees (for instance by plunging the hive into boiling water) to get the honey and wax. In ancient Egypt, smoke was used to drive bees from their hive, and by ancient Roman times bees were fed. At some time in the Middle Ages, beekeepers devised a form of protection to wear when handling their hives.

Until the sixteenth century, a significant age for the honey bee, the beekeeper's calendar remained virtually unchanged; in early summer he caught and hived the swarms which issued; in late summer he killed the bees in most of his hives, cut out the combs, and strained the honey from the wax; and in the fall, if necessary, he provided food in the remaining hives, which he overwintered. Burning sulfur was commonly used for killing bees.

Little was understood as to what went on inside the hive, for the events there could not be seen. It was not realized that the large "king" bee was in fact a female, the mother of the other bees in the hive, nor were the sexes of the workers and drones understood, let alone the facts of mating between queen and drone. It was not known that the bees themselves secreted the wax with which they built comb, nor that their visits to flowers had anything to do with the formation of seeds and fruits.

Beekeeping — 1500 to 1851

Three separate streams of events, each of great significance in the history of bees and beekeeping, were set in motion in the sixteenth century and led to Langstroth's discovery of the bee space in 1851. The bee space is the space (approximately one-fourth inch) through which a bee will pass but not seal one surface to another with comb or propolis. This fundamental advance by Langstroth made it possible to construct beehives with removable frames and thus control the development of a colony of bees. First, scientific and technical developments enabled beekeepers to understand the fundamental facts of the life cycle and biology of their bees; second, and coupled with the first, there were developments in beekeeping methods which gave the beekeepers slightly more control over their bees, as well as greater opportunities for observing the bees inside the hive; and third, the honey bees themselves spread over two new continents, from one of which was to come the greatest single advance in the science and craft of beekeeping.

Developments in beekeeping technique.—Between 1500 and 1851 there were many attempts to devise ways of taking the honey from hives without killing the bees. For instance, several colonies were united together in a single hive for overwintering, instead of killing all but one colony. The uniting was done by "driving" the bees; their own hive was inverted, and an empty hive placed on it with the two open ends in contact but held apart at an angle; the sides of the inverted hive were drummed, causing the bees to leave it by running up into the empty hive. Several other colonies were driven into the same hive, where the queens would fight it out until only one remained. The process of driving had been known in the Middle Ages, but had not been favored then.

Alternatively, where fairly large hives were used, like those made of logs or cork, the lower third of all the combs might be cut out with a specially shaped knife; the remainder constituted a permanent brood nest, and each year the honeycomb was built afresh by the bees and cut out by the beekeeper. With the small straw skeps this was not possible, and various types of extensions were added at the top, over a hole left in the top of the skep. The extension might be a smaller skep (a cap) or a glass jar (a bell). The bees stored honey, but they did not breed there: it was a true honey super. Alternatively an eke was placed under the skep; this was a straw cylinder a few inches high which formed an extension of the skep downwards. All these extensions could be removed complete with honey combs without disturbing the brood nest.

Hives made of wooden boards were also used. Initially these were simple boxes, but various elaborations were devised later, and collateral hives had boxes at the side for honey storage.

Throughout these centuries the minds of beekeepers in the most progressive areas were constantly occupied with the problem of getting more control over the bees and their activities, and of learning what was going on inside the hive. It is difficult for us now, with the problem solved, to enter into the minds of these experimenting beekeepers, who struggled so long and so unsuccessfully to find a way of getting combs they could easily remove from the hive. Observation windows in the hive walls were easy enough to make but did not show much of what was going on inside. There is a tantalizing entry in Samuel Pepys' Diary (1665): "After dinner to Mr. Evelyn's; he being abroad, we walked in his garden, and a lovely noble ground he hath indeed. And among other rarities, a hive of bees, so as being hived in glass, you may see the bees making their honey and combs mighty pleasantly."

Réaumur tells us that the Italian astronomer Maraldi found single-comb observation hives in the garden of the French Royal Observatory in Paris in 1687. Huber's leaf hive in Switzerland came over a century later; it consisted of a number of frames hinged together at one side like the leaves of a book, and the bees built combs in the frames. It was invaluable for his observations, but was an observation hive only and quite unsuited to practical beekeeping.

Between 1650 and 1850 many hives with top bars and frames were invented, but after these two centuries of effort there was still failure on the fundamental point: whatever bars or frames were used, the bees attached their comb to the walls of the hive as well, and the combs could, therefore, only be removed from the hive by cutting them out. Only two of the many inventions need be referred to here. About 1806 a Ukrainian beekeeper Peter Prokopovich produced the first movable-comb hive to be used on a commercial scale (he kept up to 10,000 colonies). This hive had three vertical compartments, the top one having wooden frames with notched bee passages in the end bars. The frames were removed from the back of the hive, but as the bees attached the frames to the hive walls with comb or propolis, this was not at all easy. The second invention was so fundamental, and made so early, that it might well have altered the whole history of beekeeping if it had been more widely known and understood.

Woven basket hives were in early use in Greece, and at some stage (we shall probably never know when) some beekeepers started using them with the open end uppermost. The open end was covered with wood which, by the seventeenth century at any rate, was cut into "bars" about 1½ inches wide. Each bar was made slightly convex on its under side, and the bees attached their combs along the ridges so formed, i.e. one along the under side of each bar. What distinguished this hive from all the other "bar hives" was the fact that it was wider at the top than at the bottom and, presumably because of this slope of the hive walls, the bees did not attach their combs to them. News of these hives reached England in 1682, when Sir George Wheler described them in his book *A Journey into Greece*. He reported that in spring the number of occupied hives was doubled by removing half the combs from each hive and placing them in an empty one. The Greek beekeepers had in fact produced a workable movable-comb hive (Figure 2). Aristotle's account of the life of the bee makes it seem possible that he used one of these hives when writing his *Natural History*.

Wheler's report had a considerable influence on hive development in England and elsewhere, but the crucial step, which would have given the desired movable-comb wooden hive, was never taken. It seems to the authors that if one of the many bar hives had been wider at the top than at the bottom, the bees would have made their own bee space between the combs and the hive wall, and modern beekeeping might have started 150 years earlier than it did.

In most parts of Central and Eastern Africa bees are still kept in bark or log cylinders in a somewhat similar way to the early primitive beekeeping of Europe (Figure 3). Attempts have been made to introduce modern methods and European-type bees, both without success. For some as yet undetermined reason (possibly disease) the European bees do not survive in Africa.

FIGURE 2. Greek basket hive, possibly the first movable comb hive.

Modern methods have not been successful mainly because of the excessive cost of equipment, but also because the methods of management are completely foreign to traditional practices. By stepping back in time to modifications of the Greek bar hive, some success in the improvement of beekeeping practice is being accomplished particularly in Rhodesia and Kenya.

The spread of honey bees over the world.—We must now leave this story of the beekeeper's unsuccessful attempts to invent the hive they needed, to follow the adventures of the bees themselves during the same two and a half centuries. The honey bee belonged to the Old World: Europe, Africa, and Asia. Prior to 1500 there were no honey bees in the New World: the Ameri-

FIGURE 3. Present day beehives in Tanzania, East Africa.

cas, Australia, and New Zealand. But, like the dog, the honey bee has accompanied man on most of his major migrations, and the early settlers in each part of the New World took hives of bees with them. Records of the establishment of honey bees in America are scanty enough, but it seems likely that there were importations into South, Central, and North America in the 1500s or 1600s. In 1622 honey bees were reported in Virginia. In 1640, the town of Newbury, Massachusetts established a municipal apiary. The expert put in charge became the town's first pauper! Honey bees were not introduced to the west coast of North America until the 1800s. In 1809, bees were imported to Alaska from Russia by a Russian monk and in 1830 they were carried to California from Alaska. No one knows whether these bees survived, but in 1856 grey-banded bees were found wild in California and these could very likely have been descendants of those originally imported from Alaska. Honey bees were then taken north from California to British Columbia in Canada. The first honey bees were landed in Australia at Sydney in 1822 and W. C. Cotton took the first consignment to New Zealand (from England) in 1842. It is thus little more than 100 years since honey bees, members of the genus *Apis,* have lived in all five continents.

Beekeeping — 1851 to 1900

By 1851 the honey bee had completed its colonization of almost the whole world, the major unconquered territory (Siberia) not being occupied until the land itself was settled during the present century. The most progressive beekeepers knew enough about their bees to do great things with them, but this was prevented because, despite all their efforts, they still had no suitable hive.

The step which changed this was made in 1851 by Lorenzo Lorraine Langstroth, an American born in Philadelphia, and living there at the time. Langstroth had shown a rather unusual interest in insects as a child, and this was revived when, as a young pastor in Andover, Massachusetts, he visited a friend who kept bees and saw a glass globe filled with honey in the comb. Before he returned home he bought two colonies of bees in box hives. He soon also acquired a Huber leaf hive and obtained various books on bees, including Huber's *Letters* and Edward Bevan's *The Honey-bee* (1838). He used the bar hive with a shallow super described by Bevan, and improved it by deepening the grooves on which the bars rested, leaving about ⅜ inch between the cover and the bars (this is the origin of our present top bee space). He found that this facilitated the removal of the cover board on which the glasses rested. The key development, which cuts the history of beekeeping into two halves, was made in the fall of 1851, and we have Langstroth's own words to describe it:

> Pondering, as I had so often done before, how I could get rid of the disagreeable necessity of cutting the attachments of the combs from the walls of the hives, and rejecting, for obvious reasons the plan of uprights, close fitting (or nearly so) to these walls, the almost self-evident idea of using the same bee space as

in the shallow chambers came into my mind, and in a moment the suspended movable frames, kept at a suitable distance from each other and the case containing them, came into being. Seeing by intuition, as it were, the end from the beginning, I could scarcely refrain from shouting out my "Eureka" in the open streets.

Langstroth's intuition was justified: the bees did in fact respect the bee space left between the hive and the frames in which the combs were built; they did not build comb across the space, and the frames were, therefore, truly movable.

The movable-frame itself was in common use in the United States by 1861. It was introduced into England in 1862, and the writing of Charles Dadant in the French and Italian journals, starting in 1869, brought about its introduction into Europe; it soon spread to other countries, each of which used variants, built on the same basic principle.

With this discovery modern beekeeping began, and development in the next half-century was in the nature of an explosion compared with the slow and halting progress of the centuries before. The use of movable frames led directly to the invention by Johannes Mehring in Germany of beeswax foundation, a sheet of wax pressed into the form of the bases of hexagonal cells. This saved beeswax and ensured that the bees built regular worker comb in the frames. Langstroth was thinking only of the brood chamber when he devised his frames; he used glasses above the crown board for the honey. But it was soon realized that if the honey chamber or super was furnished with similar frames, these could easily be removed when filled with honey. If means could then be found to extract the honey without destroying the comb, the comb in its frame could then be used again. This led to the invention of the centrifugal honey extractor in Austria in 1865 by Major F. Hruschka and possibly in France a few years earlier. The perfection of the queen excluder by Abbé Collin of France in 1865 enabled the beekeeper to keep the queen, and hence the brood, out of the honey chamber. By using the bee escape, produced in 1891 by E. C. Porter in the United States, he could get the honey chamber free from bees before he removed the frames of honey. In 1870 Quinby invented the modern bee smoker, which was improved considerably by Bingham. J. S. Harbinson of California is credited with originating the first comb honey section in 1857. This was followed by many other developments for the handling of comb honey. Perhaps the crucial moment in the production of comb honey was the year 1873 when Harbinson shipped a carload of section honey to Chicago.

With the general adoption of comb honey sections, the tendency was more and more towards small hives to force the bees to work in a small space furnished by the sections. This, naturally, led to problems with swarming. In 1892 G. W. Demaree described a system for swarm control in the *American Bee Journal;* this method, in modified forms, is used by a great number of beekeepers even to the present date. During the period much discussion was taking place, in America and also in Europe, over the proper size for a hive of bees, particularly through the writings of Charles Dadant in the French bee journals. The argument on one side was Dadant's recommendation of a larger hive to provide greater room for brood rearing (and thus stronger and healthier colonies) as against Doolittle's suggestion of five- and six-frame hives with little room for the storage of pollen and honey. In the late 1800s and early 1900s when this controversy was taking place, those advocates of the small hive soon found their crops diminishing. This trend continued until, unknowingly, by the use of the Demaree system of swarm control, they were again increasing the amount of brood comb available and their crops started to again increase. By the early 1920s, after more than one-half a century of trial, error, argument, and experiment, only three types of hives were in common use in North America. The Langstroth hive in both the eight- and ten-frame sizes had remained popular; they now are used with two brood chambers, thus giving the bees plenty of space. In addition to these, the modified Dadant hive has replaced the various kinds of equipment used by the large-hive men of older times.

The *American Bee Journal* was first published in January 1861 and its editor was Samuel Wagner who was destined to have a profound influence on the new industry of honey production. The first issue of *Gleanings in Bee Culture* appeared in January 1873; its founder and first editor was A. I. Root. It has remained in this family throughout its entire history. Although many bee journals have come and gone since that period, these two have continued to the present day. Root had a remarkable characteristic of wanting to experiment in many areas, with the result that he played an active part in many of the early developments. One of his contributions was an almost perfect foundation mill. In 1879, he advocated the sale of bees by the pound, thus making the first step towards the package bee business which was to develop on a very large scale in the next century. While partial success was achieved during these early periods, the shipping of package bees did not develop to any great extent, due to a lack of demand.

Beekeeping After 1900

It was during the period from 1900 to the mid-1950s that almost every modern beekeeping method and many of the modern pieces of beekeeping equipment were devised. Many of these developments took place in North America and spread rapidly to other areas of the world, further refinements being made on them. In most cases, the techniques were developed by the beekeepers themselves, but they were supported strongly by government-backed research designed to assist in producing practical results.

For such developments to take place, incentive was necessary and, fortunately, three incentives soon appeared. At the turn of the century, Maeterlinck published *The Life of the Honey Bee.* Though this book appears to be written with little practical experience, it aroused a worldwide interest in bees and was published in many languages. Then World War I (1914–1918) created a shortage of sugar. Finally, sweet clover (*Melilotus*) was widely used as a forage crop, particularly in the plains areas of the western United States and Canada. It was not long until such phrases as "an acre of wheat or a hive of bees" appeared in many journals. Thus, on the American continent was born a small industry which was destined to lead the way in apiculture from a practical point of view for years to come. With a rising young industry in its midst, the various governments began to realize that support was necessary at several levels. Possibly the first appointment in a regulatory or extension role was that of Morley Pettit in Ontario, Canada, as Provincial Apiarist in 1909.

Several appointments in the Bureau of Entomology in the United States Department of Agriculture were to play a leading part in the development of beekeeping, both nationally and internationally. The first of these was E. F. Phillips who joined the Bureau in 1905 and in 1907 was appointed apicultural expert. Phillips was a very strong character and had an ability to persuade others to listen to his views. While he had little practical experience in beekeeping, he was able to recognize the needs of the industry. He stressed the urgent need for an understanding of bee diseases, for fundamental knowledge on anatomy and embryology of the honey bee, and on wintering problems; and he immediately went about trying to see that these problems were solved. In 1907 he had White appointed to look after diseases. In 1908 J. A. Nelson joined him to carry out the program on embryology; in 1909 R. E. Snodgrass joined his group and in 1911 published his first work on the anatomy of the honey bee. In 1912 Demuth joined him and carried out the first extensive work on wintering of bees. Phillips received many requests to speak on the application of their research. In 1915 he published his first basic text on honey bees, which was to be used as a reference for many years to come. In 1924 Phillips became disenchanted with the administration duties of the position and left to become Professor of Apiculture at Cornell University. About the same time, he began to travel very widely; he was perhaps the first American interested in beekeeping to make an extended trip throughout Russia and many of the other Eastern European countries.

During Phillips' years at Cornell, he attracted graduate students from throughout the world who, in years to come, filled many senior positions in the United States and elsewhere. It was very fortunate that the American beekeeping industry had such a forthright speaker during this period of development.

Fortunately, Phillips was followed in his USDA program in 1924 by James Hambleton. A quiet-spoken man of excellent administrative ability, he was able to build a good research organization in many regions of the United States. Hambleton became known, not for his direct apicultural work, but for his ideas of depth and substance which flowed constantly to his staff. He traveled widely in different countries and became well known at international meetings. He was a strong supporter of the use of the honey bee for pollination and encouraged Farrar, Todd, and others in their behavior studies.

Farrar's work centered around colony behavior as related to management, and during the thirty or more years in which he was publishing in this field, he strongly influenced the manner in which American beekeepers handled their colonies. He particularly emphasized the influence of colony population in honey production and was perhaps the first to put emphasis on the two-queen system which has been used rather widely in recent years. He also pointed out very strongly the need for pollen supplements or pollen in the brood-rearing program of the bees, a factor which beekeepers in the past had sometimes tried to reduce. The work of these three men, and the work which they instigated, soon had an effect throughout the world and provided the base for much of the practical beekeeping being carried on at the present time.

It was during this same period that the use and recognition of the honey bee reached its height as a pollinator of farm crops. The fact that the pollen which the bees collect is the male "seed" of the flower, which fertilizes the ovum, was discovered in England in 1750 by Arthur Dobbs. He also observed that honey bees gathered pollen from only one kind of flower on each flight and suggested that disastrous cross-fertilization would result if this were not so. The part played by bees in fertilizing flowers was established quite clearly by Sprengel in 1793. Even in ancient times, measures were taken which insured that the fig tree was pollinated, even though the growers may not have known exactly what they were doing. During the early part of the century, much emphasis was placed on the use of the honey bee in fruit crops, particularly apples. This culminated in a report by Brittain from Canada, published in 1933, on apple pollination studies in the Annapolis Valley.

Improvements in techniques have been introduced since that time, such as the use of hand-collected pollen and of special hive inserts to distribute this pollen with the aid of honey bees, but the basic principles as described by Brittain still stand.

With the encouragement of Hambleton and the behavior studies of Vansell and Todd, the use of the honey bee in legume crops reached its ultimate in the mid-1950s when Charles Reed of California organized a large-scale company for the pollination of alfalfa. At the height of its operation, it was using over 80,000 colonies of bees to pollinate 40,000 acres of alfalfa in one county in California. Many groups have entered the field since that time, and the use of the honey bee on a migratory basis for the pollination of both fruit and legume crops has now become an established practice. The latest major publication in this field is *Insect Pollination of Crops* by J. B. Free of England in 1970 which brings together the known requirements for the use of honey bees and other insects in the pollination of crops, and the necessary requirements for each crop.

During this period the developments in European countries centered around fundamental research. Many research units were established after World War II, and scientific publications began to appear in increasing numbers. Fortunately, in 1949 the Bee Research Association was established, originally by British beekeepers. The need for a sound abstracting system was recognized from the start, and the Association's journal, *Apicultural Abstracts,* soon received the support of the Commonwealth Agricultural Bureaux. The BRA now has a file of more than twenty years of published abstracts, all cross-indexed on a detailed subject classification and placed on a computer at the University of Guelph in Canada for subject and author retrieval.

The history of the accomplishments in fundamental research can best be outlined under the various disciplines.

Honey Bee Physiology and Behavior

Even in the time of Aristotle, men marveled at the social organization of a colony of bees. We now know that behavior and physiology of the honey bee, particularly with respect to glandular secretions, are closely related, but much detail is still missing.

Swammerdam was perhaps the first to give a detailed description of the life history of the bee, but his findings were not published until 1737. Francois Huber, the blind scientist working with his wife and assistant Francis Burnens in Switzerland, is often regarded as the one who laid the foundation of our scientific knowledge of the life history of the honey bee. He confirmed many observations of earlier writers, and refuted others, and made many new contributions of his own. His *New Observations on the Bee,* published in 1792, included such subjects as: ventilation of the hive by fanning; rearing queens from eggs and larvae destined originally to be workers; fertilization of the queen in flight; and the essentiality of pollen for brood rearing.

Perhaps the most widely known advances in the apiculture field have been studies of bee behavior. It was the work of Karl von Frisch of Austria which aroused the world's interest and created the impetus for much subsequent research. He was not, of course, the first to notice some of these characteristics of the honey bee. Artistotle, in his *Historia Animalium,* stated that each bee is followed by several others in the field. Spitzner of Germany in 1788 noticed the dance of the bees and associated it with bees going to the field. In 1923 Park of the United States noticed both the round dance and the tail-wagging dance; Buttel-Reepen of Germany suggested in 1915 that bees communicate by sounds. Frisch started his work on bees about 1919 and in 1920 made his first report on the round dance.

In 1826 Müller of Germany first suggested that the insect eye perceives the stimuli from all the sensory areas as a mosaic. Cheshire of England described in 1886 the number of facets in the eyes of the worker, queen, and drone. Hess of Germany demonstrated in 1920 that bees are very sensitive to ultraviolet light. In 1949 Frisch demonstrated that bees can detect polarized light; and in 1950 Autrum and Stumpf suggested that their analysis of polarized light was made possible through the ommatidium.

Frisch's work continued on through the second world war, after which a series of papers was published which aroused worldwide attention. There have been so many workers in this field who have contributed important phases of the program that it is impossible to give proper credit to all. Many of the contributors were associated with Frisch either as students or co-workers. Ribbands' book *The Behaviour and Social Life of Honeybees,* published in 1953, brought much of this work together and acted as a base for expansion into many other areas of honey bee behavior.

In 1855 Dönhoff in Germany noted that bees stretched out their tongues when the antennae were touched with honey, but it remained for Minnich in 1932 in the United States to locate the taste receptors on the antennae and legs of the honey bee. In 1838 Lefebvre of France located the sense organs for smell on the antennae.

Many researchers have attempted to solve the problem of how bees can produce queens and workers from the same kind of egg. In 1888 von Planta of Germany published the results of his analysis of the food of the larval honey bee. These led him to the conclusion that during the first three days of larval life young larvae are fed on the same food whether they are intended to be workers or queens, but that on the fourth day the food given to the larvae destined to become workers is changed, and pollen and nectar are added. Thus, a belief which has continued through to the present day has developed: that any larva fed on the young larval food (royal jelly) throughout its life will become a queen. While von Planta's analyses have since been proven inaccurate, his conclusions are essentially correct.

The only satisfactory way of solving this problem was to rear young larvae in an incubator and eventually to use artificial diets. Rhein of Germany was possibly the first to rear bee larvae in the laboratory in 1933; but he obtained only workers from larvae fed on royal jelly. It remained for Weaver of the United States, in 1955, to rear larvae to adult queens in the incubator on fresh royal jelly. He was, however, not able to repeat the tests on stored royal jelly, and concluded that the queen-determining substance was very labile. In 1959 Smith of Canada reared larvae to queens on stored royal jelly; although all larvae received the same treatment, out of 25% emerging as adults only one third were queen-like, the rest being intermediate or worker-like; so if there were a queen-determining substance in royal jelly, it cannot be as labile as was formerly thought. Haydak of the United States proposed in 1943 that mass feeding of queens was at least one of the factors involved, since workers were placed on a reduced diet after the third day. As yet there has been no success in rearing the larvae to workers or queens on an artificial diet.

Rembold, Weaver, and Shuel and Dixon have prepared theories as to how the differentiation takes place. Each has some supporting evidence but no conclusive answer.

Of recent interest is the development of information concerning chemical substances secreted by honey bees and their relationship to social behavior. In 1942 Gertrud Hess concluded from her experiments that some material substance produced by the queen and circulated in the food exchange controlled ovary development in worker bees. Little attention was paid to these results till almost simultaneously in 1954, when deGroot and Voogd of Holland, Butler of England, and Pain of France, published on this subject, all confirming the work of Hess. Butler showed that a material was secreted by the queen bee which, in some way, inhibited the colony from building queen cells. He also suggested that this may inhibit the development of laying workers, and he called this material queen substance, a name which has stayed with it. At the same time, deGroot and Voogd extracted a material which inhibited the development of ovaries in worker bees, thus supporting Hess' and Butler's theory of control of laying workers. Publications from these three countries almost simultaneously aroused a tremendous interest in a new area of research. Butler, in summarizing his own and other work, suggested that a queen honey bee secretes a chemical compound which is taken from her body by worker bees, is passed from bee to bee, and controls not only the development of ovaries in workers but also the production of queen cells in general. Many other workers, including Gary of the United States and Simpson of England, have played an active part in the development of this field. This work was soon followed by the characterization of these compounds by Callow and Barker of England, and Barbier of France. Karlson of Germany called these chemical messenger substances pheromones in 1959. Since that time, several workers have been busy trying to identify other compounds secreted by other glands in the honey bee. Of particular interest is the work of Boch and Shearer of Canada relating to geranic and nerolic acid and citral which is secreted by the Nasonov gland and is attractive to foraging bees; thus it provides one means of marking good forage areas. These same workers have also identified alarm substances such as 2-heptanone in the mandibular glands of honey bees of certain ages, and isoamyl acetate in the sting. Hügel of France and Taber of the United States, in 1962 and 1963, respectively, presented evidence that plants in their pollen exhibit substances attractive to honey bees. This was confirmed by Boch and Lepage in 1968, and two compounds were tentatively identified. As information concerning the social behavior of honey bees slowly unfolds, it may prove yet to be one of the most useful developmental areas yet studied in the field of apiculture.

Embryology and Anatomy of the Honey Bee

The first recorded information on the embryology of the honey bee is that which Weismann of Germany included in a letter which he sent to the Russian investigator Elias Mecznikow, which was published in 1866. This was followed shortly afterward by articles by Otto Bütschli of Germany and Kowalevski, the Russian embryologist, both of which were well illustrated and contained many sections relating to the embryology of the honey bee. Kowalevski was perhaps the first to make use of microtome sectioning of the honey bee egg. In 1884 an Italian, Battista Grassi, published a very complete account of the embryology of the honey bee using the microtome method. In Germany, Nachtsheim published in 1913 a rather extensive article on the fertilization and maturation of the egg of the honey bee. He reported that the drone has only 16 chromosomes and the female bees twice this number, and that queens and workers originate from the same eggs. The work relating to chromosome number was not confirmed until 1948 by Sanderson and Hall in Scotland.

The first description of the queen bee as a female which laid eggs was published in Spain in 1586 by Luis Méndez de Torres. In 1609 Charles Butler in England mentioned in his famous *Feminine Monarchie* that drones were male bees. This was followed by Richard Remnant's *Discourse or Historie of Bees,* in which it was stated that the worker bees were females. Meanwhile, in 1652 an Italian, Prince Cesi, had published the first drawings of bees made under a microscope.

In 1811 Ramdohr (cited by Cheshire in 1886) announced the discovery of a pair of salivary glands in the thorax of the honey bee, and two other pairs of glands were found in the head by Meckel in 1846. These were later described by Cheshire as the pharyngeal glands. It was Fischer in Germany who first noted in 1871 that these glands were used for the production of brood food. The thoracic glands were described by Schiemenz in 1883, and the postcerebral glands by Cheshire in 1886. The mandibular glands were described by Meinert in 1860 as a pair of oval sacs which open at the base of the mandibles. Many other glands have been identified, including the wax glands by Butler in his *Feminine Monarchie* (later described by Dreyling of Germany in 1903) and scent glands described by Zoubarev in 1883, based on dissections by Nasonov.

Valuable publications in this area were Nelson's *Embryology of the Honeybee,* published in 1915, and Snodgrass's *Anatomy of the Honeybee,* published in 1925 and revised in 1956.

One significant addition to the embryology of the honey bee since Nelson's publication is that of Woyke in Poland, who in 1966 demonstrated that diploid drones may be found in the embryonic stages and, by laboratory rearing, as adults.

Queen Rearing and Bee Breeding

The fact that bees could raise a queen from worker eggs or very young larvae was published in Germany in 1568 by Nickel Jacob, but the primary facts about the mating of the queen with the drone were not recorded until 1771, when Anton Janscha in Slovenia published an account of the event. It was Langstroth's comments on Huber's plan of increasing the number of colonies (by dividing the colony into two parts and permitting the bees in the queenless section to rear another queen) which led to the beginning of commercial queen rearing. Langstroth recommended the forcing of swarms to produce an abundance of queen cells which could then be removed and used to start new colonies. Increasing the number of colonies by forming nuclei appears to have been the result of experiments by

Langstroth in the United States and Dzierzon of Germany, working independently. Many articles were published by various beekeepers on methods of producing queens; G. M. Doolittle of the United States summarized the best of the various suggestions and, with his own innovations, recommended a system which, with only minor alterations, is used to the present day.

With the problem of commercial queen production under way, the next problem was shipment of the queens. The first queens were probably shipped by Robinson to Langstroth within the United States in 1863. Many types of cages for shipping queens were developed; Frank Benton of the United States used a small block of wood with most of its interior cut away and a small supply of sugar candy put in it. This type of cage, with small variations, has remained in use since that time.

With some of the problems of transporting queens solved, attempts were made to import to America better breeding stock from Europe. Many unsuccessful and partially successful attempts were made. Until the development of the queen cage, most shipments were made in colonies or boxes containing combs. Most attempts failed; the first successful transatlantic shipment was made in 1870, by Grimm of the United States, who brought queens from Italy.

From 1872 onward Charles Dadant imported large numbers of queens annually and is generally credited with finally solving the problem of transatlantic shipping of queen bees.

In 1879 D. A. Jones of Canada and Frank Benton of the United States went to Cyprus and Palestine in a quest for honey bee breeding stock. Jones returned to Canada with a large number of queens, leaving Benton behind to continue the work. This program continued for a number of years, the bees being mainly established on islands in Georgian Bay in Canada. In 1923 restrictions were placed on the importation of queens into the United States and Canada to avoid further bee diseases, particularly acarine. Further importations were not carried out until Smith of Canada in 1961 and 1962 was able to transport immature stages satisfactorily, under controlled conditions, thus avoiding the transfer of acarine mites on adult bees.

With the problems of production and stock availability attended to, attempts were made during this same period to overcome the problems of breeding. Any program to breed a better honey bee must include some method of controlled mating, but queen bees will not mate within the hive. They mate in flight with drones which may originate at some distance from the queen's hive. To this problem are added the genetic complications of parthenogenesis and haploid drones.

The first recorded attempt to control the mating of the queen appears to be that of Réaumur of France, who in 1740 confined the queen and drones together in a glass dish. It was not until the work of Janscha in 1771 that it was realized that mating took place outside the hive. Many attempts were made to control the mating of queens during the next hundred years, all without success. The methods used were confinement of queens by Demaree,

Kramer, and David; tethering the queens by Demaree in 1881 and Chuck in 1882; squeezing fluid from the drones onto the queen larvae by Lee in 1884; and squeezing seminal fluids of the drone onto the vulva of the queen by McLain in 1887. Interest was first shown in isolated mating stations in Switzerland at the turn of the century. Many other European countries followed suit. Possibly the first attempt at instrumental insemination was that of Huber towards the end of the eighteenth century when he unsuccessfully tried to introduce semen into the vagina of the queen by means of a hair pencil. Subsequent developments mentioned have all been made in the United States. The use of a syringe for instrumental insemination was introduced by McLain in 1886. Prior to 1915 Bishop had attempted to inseminate queen bees in several manners, but soon realized that more information concerning anatomy was necessary. In 1920, he published detailed studies on the male and female reproductive organs. He mentioned the valve fold in the queen's vagina, but apparently did not recognize its significance in instrumental insemination.

The development of practical instrumental insemination is generally recognized as beginning with Watson's demonstration in 1926. In 1930 carbon dioxide was used as an anaesthetic for the queen by Laidlaw, at the suggestion of Hambleton. In 1933 Laidlaw, using a hand method developed by Quinn of the United States where partial insemination had been obtained, was able to show that the difficulty was the blockage of the vagina caused by the valve fold. This discovery led the way for the rapid development of modern instruments for instrumental insemination.

In 1945 Mackensen was able to show that two treatments of carbon dioxide induced queens to lay, whether they were inseminated or not. Present instruments for insemination of queen bees are quite successful, thanks to the untiring efforts especially of Mackensen, Roberts, Nolan, and Laidlaw. With the availability of breeding stock and the controlled methods of mating now available, the whole future program of improved breeding still lies ahead.

Bee Disease and Poisoning

Bee diseases have plagued the beekeeper since pre-Biblical times and have been partially responsible in some areas for the fact that beekeeping has not developed into an industry. It is only in recent years that satisfactory controls for most of the diseases have been developed.

There are early references to diseases in the writing of Aristotle and Pliny, but the descriptions are rather vague. In 1717 Schirach of Germany gave one of the first classifications of bee diseases, mentioning (among other diseases) dysentery and foulbrood. Schirach and other writers following him all considered the foulbrood to be of more than one form. Dzierzon of Germany described in 1882 two forms of foulbrood in his *Rational Beekeeping* which agree closely with what have been known for some decades as European and American foulbrood.

Early writers attributed bee diseases to many causes, blaming improper food, the queen, etc. In 1884 Cheshire of England drew attention to the fact that at least one disease was due to a bacillus. White of the United States first described clearly in 1907 the cause of American foulbrood as *Bacillus larvae* and, five years later, named *Bacillus pluton* as a primary cause of European foulbrood. This was not confirmed until 1956 by Bailey in England who renamed it *Streptococcus pluton.*

Zander in Germany was the first to make a proper study of diseases of adult bees. In 1907 he indicated the existence of two such diseases, both giving rise to dysentery. One form he considered to be noninfectious and comparatively harmless, and attributed it to physical or food conditions. The second form Zander described as being contagious and caused by a protozoan *Nosema apis,* which was named by him. (Nosema disease was first discovered by Dönhoff in Germany who reported the presence of spores in the intestine of bees in 1857. This discovery was almost forgotten until 1909 when Zander found the parasites in the wall of the intestine of the infected bees.) Following the reports of nosema disease in Germany, it was observed in other European countries and shown to exist in the United States by White in 1914. Little attention was paid to it, however, until the early 1940s when Farrar showed in Wisconsin what heavy losses occurred when colonies were kept in greenhouses during the winter, or were wintered outside without packing, and fed stimulants in spring such as pollen supplements.

Acarine disease was first recognized by Rennie in Scotland in 1921, when he was examining the so-called Isle of Wight disease. The mite responsible was named *Acarapis woodi.* The disease appears to be largely confined to certain areas in Europe and Asia, but a few recent cases have been reported in South America.

Many attempts were made through the years to control the various diseases of bees, but it was not until the era of antibiotics that much headway was made. Nickel Jacob of Germany gave, in 1568, the first recorded description of a treatment for brood diseases; these recommendations lasted until the use of drugs was adopted in 1944. Two Americans, Haseman and Childers, recommended sodium sulfathiazole for the control of American foulbrood in 1944. This was followed shortly afterwards by recommendations for the use of terramycin for the same organism and later by streptomycin for European foulbrood. Large numbers of antiprotozoan compounds were tried for the control of nosema disease, but it was not until 1952 that Katznelson and Jamieson in Canada pointed out that fumagillin, produced by the mould *Aspergillus fumigatus,* could control this disease. Serious attempts have been made over several years to produce strains of bees resistant to the various diseases, first by O. W. Park and later by Rothenbuhler, in the United States. While they were able to produce strains showing resistance, these have not as yet been used commercially with success.

The first recorded instance of the possible effect of poisons on honeybees appeared in 1681 in *Systema Agricultural* by Worlidge of England, relating to the control of ants. Powdered arsenic mixed with honey was placed in a box with holes punched in it and hung in a tree. The specific instructions concerning bees read "Make not the holes so large that a bee may enter lest it destroy them."

Orchard-spraying experiments began about 1878, after the success obtained with Paris green against the potato beetle. Since that time, beekeepers have periodically sustained heavy losses from spray poisons. The problem reached such proportions in the 1950s that beekeeping associations in different countries were able to persuade governments to introduce legislation to assist in the control of bee poisoning. The first record of such legislation is, however, "An Act for the Further Protection of Bees" introduced in Ontario, Canada, in 1892. This Act provided fines for anyone spraying blooming fruit trees with Paris green or any other such poison.

Until the mid-1940s arsenic continued to be the main poison causing honey bee losses. Since that time, many others have entered the picture. But it is now possible for the beekeeper's colonies to coexist with most of the insecticides, if properly applied. Of outstanding concern is the drift from aerial applications, which in some areas has resulted in losses in the thousands of colonies of bees. The problem has been greatest, perhaps, in California; there Anderson and Atkins have been testing most of the newly introduced pesticides since 1950 for their toxic effects on honey bees.

Literature Cited

Butler, C. G. 1954. *The World of the Honeybee*. London: Collins

Chauvin, R. 1968. *Traité de Biologie de l'Abeille* (5 vols.). Paris: Masson et Cie

Cheshire, Frank R. 1886. *Bees and Beekeeping*. Vol. 1. London: Upcott Gill

Eckert, J. E., Shaw, F. R. 1960. *Beekeeping*. New York: MacMillan

Free, J. B. 1970. *Insect Pollination of Crops*. London: Academic

Grout, Roy A. 1963. *The Hive and the Honey Bee*. Hamilton, Ill.: Dadant & Sons

Huber, Francois. 1814. New observations on bees. *Am. Bee J.* (Transl. C. P. Dadant, 1926, Hamilton, Ill.)

Nelson, James Allen 1915. *The Embryology of the Honey Bee*. Princeton, NJ: Princeton Univ. Press

Pellett, Frank C. 1938. *History of American Beekeeping*. Ames, Iowa: Collegiate Press

Phillips, M. G. 1967. *The Bee Man*. Ithaca, NY: Cornell Univ. Press

Ribbands, R. 1953. *The Behaviour and Social Life of Honeybees*. London: Bee Assoc.

Snodgrass, R. E. 1956. *Anatomy of the Honey Bee*. Ithaca: Comstock

Townsend, G. F., Shuel, R. W. 1962. Some recent advances in apicultural research. *Ann. Rev. Entomol.* 7:481–500

GENETICS—THE LONG STORY[1]

SPENCER W. BROWN[2]
Department of Genetics, University of California
Berkeley, California

INTRODUCTION

Genetics is a conceptual science owing its identity to a common interest in mechanisms of both transmission and expression of the heritable entities. Entomology owes its coherence to a common interest in a group of organisms, the insects, and taxonomy thus seems likely to remain the core science of insect biology.

Entomologists and geneticists have cooperated most closely in the study of evolution. With some possible exceptions, recent systems of taxonomy have been constructed to parallel presumed phylogenies as closely as possible. Since evolutionary change is dependent on heritable variation, genetics has had the responsibility for analyzing evolutionary mechanisms which result first in changes in gene frequencies within groups of mutually interacting individuals and ultimately in all of life. The association of genetics and entomology in the cytotaxonomy, cytogenetics, and population genetics of insects has been highly productive, responsible for the major portion of what is known from nature about the evolutionary process.

The science of ecology has the awesome responsibility of analyzing all the numerous external factors which influence the development and reproduction of organisms. Natural selection adds another component to the constants of short-term ecological dynamics thereby transforming them, in part, into long-term vectors driving and directing the evolutionary process. Neither the organism nor environment remains constant; the complex or system evolves as a whole. Undoubtedly, the most dramatic of all such changes was the pre-Cambrian shift from a reducing atmosphere to an oxidizing atmosphere which set the stage for life as we know it.

The immense complexity of the interactions between the environment and any one species of a complex association of organisms is the reason there are so few clear-cut examples of natural selection. Complex processes are not

[1] This paper is dedicated to Professor Curt Stern on his seventieth birthday in appreciation of both his historical contributions to genetics and his contributions on the history of genetics.

[2] The author's interest in entomological aspects of genetics comes largely from work with coccid chromosome systems, made possible by grants from the National Science Foundation, currently NSF GB–8196:2.

usually amenable to simple analysis. It is thus not surprising that some of the best examples of natural selection are those in which man has introduced a fairly simple change in the environment, and the response has been relatively easy to analyze: industrial melanism of insects and resistance to insecticides. Undoubtedly, insects will be equally important in providing examples of more subtle, complex situations for which the investigators will need to chart out changing spiderwebs of cause and effect involving all environmental, developmental, and genetic components of the system.

It is therefore fortunate that for several decades *Drosophila melanogaster* was the queen bee of genetics. Although work with other organisms, such as fungi, worms, marine invertebrates, algae, mammals, higher plants, and other insects was begun before or during the heyday of *Drosophila,* almost all significant basic concepts in transmission genetics were either first developed by *Drosophila* workers or conspicuously verified by them. In fact, the influence of *Drosophila* genetics has been so great that the analysis of genetic systems of other organisms has occasionally been hindered by attempts to fit these systems to the *Drosophila* pattern. Nonetheless, the comprehensive work with *Drosophila* has provided the entomologist with most of the spectrum of basic concepts in transmission genetics tailor-made to fit insect biology. Continued work with *Drosophila* developmental genetics promises to do much the same for gene action in highly differentiated multicellular organisms.

An attempt will be made in this brief historical sketch to contrast the amazing progress of the science of genetics itself with the retention of much older ideas by the vast majority of people.

Synopsis of the phases of genetics.—The history of genetics can be divided rather easily into four major phases: the prescientific, the scientific pre-Mendelian, the Mendelian, and the molecular. The first phase began in prehistory with man's first awareness of resemblance between parent and offspring and continued through many thousands of years.

Although Koelreuter's work on plant hybrids, published in 1761, may be taken as the start of the second phase (44) he had many important predecessors. There had also been others who had experimentally hybridized plants. However, both in the scope of his procedures, which included inter se pollination of first generation hybrids and backcrossing them to their parents, and in his detailed records of the results, Koelreuter established the field of plant hybridization as an exact discipline.

In the year 1900 Mendel's principles were rediscovered and Mendel's own work rescued from obscurity. The period between Koelreuter's work and 1900 had been one of intense interest and activity in many fields of biology related to heredity. For Charles Darwin, hereditary transmission of variation was the only possible basis for evolution, providing the raw material on which natural selection could act. Other biologists were peering through microscopes at cells and chromosomes, applying new scientific methods to plant breeding, or adapting statistical techniques to the analysis of inheritance. When the Mendelian revolution occurred, the results of some of the preceding work were immediately seen to conform to Mendelian principles. In other areas, bitter battles were waged before the older work could be brought into harmony with the new Mendelism and the proper benefits derived from the combination. Such conflicts were fairly well resolved by 1930.

The fourth phase has been the most dramatic explosion of knowledge in the history of biology, and has led to the recognition of the encodement of hereditary factors in DNA and their eventual expression via RNA and protein. Although the beginnings can be traced to early roots, molecular genetics is largely the child of the third quarter of the present century. The identification of the primary informational molecules has led in turn to analysis of the profound, subtle, and complex systems of molecular events occurring within organisms and is directly pertinent to the dynamic processes of ecology and evolution.

The history of genetics is noteworthy for major lapses or dead periods. Koelreuter, Mendel's chief predecessor in experimental plant hybridizations, was not appreciated for almost a century, until about the time of Mendel himself. While Koelreuter's work was at last being recognized, Mendel, in turn, was being neglected. When Mendel began to be appreciated, Garrod's insightful analysis of human biochemical genetics was disregarded. Such instances demonstrate how preconceptions rigidly held by biologists themselves have grossly retarded the development of genetics.

Prescientific Concepts of Heredity

There has apparently never been an attempt to garner the numerous and conflicting prescientific ideas about the obvious transmission of traits from parent to offspring. Many of the ideas on heredity merge with ethical, moral, political, economic, religious, and mythical concepts in a most confusing manner, and if presented at all, the ideas on heredity have usually been buried in a mass of historical narration or ethnographic detail. A brief and admittedly incomplete listing of some of these concepts must suffice to illustrate both the strengths and the weaknesses of this living legacy.

Like begets like.—This concept was sufficiently valid to have enabled the Neolithic peoples of both the Old World and the New to create the requisite economic basis for civilization, surplus food. In spite of all the attendant superstition, the Neolithic farmer understood quite well the necessity of selecting his seed and breeding stock with care (45).

Hybridization is bad.—This belief is an obvious corollary of the basic tenet, like begets like, since outcrossing with unselected types would destroy the results of patient selection.

Hybrids can occur between disparate forms.—This belief disappeared from science in the nineteenth century but has not yet been eradicated from the general populace.

Men of our tribe are more worthy than those of other tribes, castes, religions, and so on.—A vast array of boundaries occur within and between human societies. Such group structures have several genetic consequences. Gene flow across the boundary will be inhibited because marriages are either permitted or encouraged only within the group. If the status of the two groups differs markedly, the gene flow is apt to be a unilateral funneling from males of the group with the higher status to females of the disadvantaged group.

Incest taboo.—This most nearly universal of all human cultural attributes is in marked contrast with the just-mentioned affinity of relatives for each other. The taboo sets an inner boundary to gene flow by prohibiting father-daughter, mother-son, and brother-sister intercourse. Although there have been continued and recent attempts to base the origin of the taboo on the deleterious effects of inbreeding (e.g. 20) it seems likely that the taboo is an especially prominent facet of an adaptive complex of behavioral traits important for success in hunting (1, 41, 42).

Prescientific Explanations of Inheritance

Attempts to account for the hereditary process were distinct from the set of beliefs just outlined, and neither challenged them nor to any great extent supplemented them. There are, of course, no records of Neolithic concepts of inheritance, but again, there is evidence from contemporary, preliterate peoples and from the earliest written records which gives at least some ideas about the earliest thoughts on the mechanisms of heredity.

For example, the Azande of the Eastern Sudan, an agricultural and hunting people numbering about two million, have concepts about heredity strongly reminiscent of those of some of the classic Greek philosophers (10).

An Early Record

The present versions of the earliest events recorded in the Old Testament can be traced to accounts composed in the kingdoms of Judah and Israel during the eighth and ninth century B.C. In essays on the history of genetics, attention has been paid to the well-known Biblical story of Jacob's genetic manipulations to increase the numbers of speckled, spotted, and ring-streaked animals in his father-in-law's flock, which indicates some of the beliefs prevalent in the early civilizations in the Fertile Crescent. At first glance, this tale indicates belief in the efficacy of maternal impressions in altering the characteristics of the offspring; in the importance of the mother's capacity, whether from nature or nurture, to bear and feed their young; and in the transmission of characteristics from sires to young. The only real entry in the account is that of the transmission of paternal traits, but this entry must be immediately discounted since the culture itself was strongly patriarchal, and the sire's traits would be the first to be noticed and admired in his progeny.

The Classic Period

Two opposing schools of thought which emerged in the Greek classic period have been aptly characterized by Furley (12). The idealists, or Aristotelians, subscribed to a "continuous theory of matter, the supremacy of final causes, the finite universe, and the uniqueness and eternity of our world." The mechanists or atomists, whose leader was Democritus, believed in "the atomic theory of matter, mechanical causation, the infinity of the universe, the plurality of worlds, and the transcience of our world."

The two parties conformed to their principles in their thinking about heredity. The atomists apparently adopted the Hippocratic belief that particles from all parts of the body were transmitted via the semen and were responsible for the development of the offspring. Darwin, who held similar views, named the process pangenesis.

Aristotle (*Generation of Animals* I: vii and viii; 27) enjoyed demolishing such beliefs, and noted that ancestral traits may appear even though the ancestors themselves could have supplied no part of the semen. He also noted (loc cit IV: iii) that the concept of a large number of different sorts of determiners contributed by both the male and the female was somewhat better; since such determiners could not exist in reality, they must exist as potentialities. But potentiality was precisely what Aristotle himself had in mind in claiming that the male contributes only a dynamic influence which shapes the matter provided by the female. Thus the amorphous semen could not contain the invisible particles envisioned by atomists. Presumably his own immaterial force would not have been expected to be visible.

The last great exponent of the atomist school was the Roman, Lucretius, who lived shortly before the Christian era. In his major work, *De Rerum Natura,* Lucretius forcefully expressed the atomist point of view (43), and gave potentially equal significance to the two parental contributions which might, however, vary in any one conception. He realized that ancestral traits could be transmitted unexpressed before appearing anew and in different combinations.

The Pre-Mendelian Scientific Period

The era of Christianity brought with it an increased concern with mysticism and a decline in the interest in natural science. The emphasis on the supernatural appeared in the writings of Galen (second century A.D.) and continued to manifest itself in a variety of ways. After the fall of Rome, the legacy of civilization was preserved outside central Europe in Ireland, Constantinople, Syria, and Persia. The Arabs, spurred on to cultural triumphs by their new religion, Islam, were impressed by the learning of antiquity and developed it further. The Arabs are credited with the introduction of the experimental method, first of importance in chemistry and only much later in the analysis of inheritance. Modern science did not begin until Copernicus (1473–1543) remodeled the solar system, and Harvey (1578–1657) performed the same service for the human circulatory system.

The development of information on reproduction and heredity was especially slow and tortuous. Koelreuter (1733–1806) was the first to attempt a scientific study of heredity. Extension much beyond Koelreuter's contribution was blocked by the misapprehension that the biological entity recognized as a species should somehow appear as a heritable entity. Thus, it was not until the nineteenth century that the sophistication of the Greek atomists was attained, and not until the twentieth that it was superseded with the belated recognition of Mendel's work. The present section begins with the laborious steps which preceded Koelreuter: epigenesis, preformation, and epigenesis again.

Precursors.—Harvey (1578–1657) was the great biological revolutionary of the Renaissance. After his work on the circulation of the blood, future anatomists could no more return to prior error than could the followers of Copernicus and Galileo continue to believe in the Ptolemaic universe. In his seventies, Harvey published an account of the generation of animals. He believed all the higher animals, including man, produced eggs, even though these could not be detected. Development of the egg proceeded by a process of epigenesis—the realization, in the developing form of the new individual, of the potential responsible for its start.

Harvey's idea that the new individual was somehow or other responsible for its own development left open the possibility of spontaneous generation and of such horrifying events as the birth of rats and cats from women. Epigenesis was countered by the preformationism of the greatest of the early insect anatomists, Swammerdam (1637–1680) who overturned, in a few short years of study, centuries old beliefs on the generation of lowly creatures. These did not arise from debris, but, as he demonstrated in his own work on insect metamorphosis, the preformed elements only enlarge during development while the old slough off. Swammerdam's demonstrations gave to ontogeny the precise aspects of orderly process, explained why offspring must resemble their parents, and ruled out spontaneous generation. It was thus a clear advance over earlier concepts.

Most historians of biology like to shy away from the subsequent period when preformation was pushed to ridiculous extremes. Some microscopists saw a preformed little man inside the head of the spermatozoon. Bonnet (1720–1793) saw embryos within embryos of the parthenogenetic, summer generations of aphids, and developed his famous theory that the original female of the species contained the germs for all future offspring. However, an earthworm can regenerate from a small piece of itself, or a crab grow a new claw; Bonnet realized that germs for development must be present in the body as well as the egg. Strict preformationism was thereby weakened to include concepts of predetermination and preorganization which thus foreshadowed much later concepts of development. Preformationists were divided into parties; the animalculists saw the new individual in the sperm; the ovists saw it in the egg. Leeuwenhoek (1632–1723), the famous Dutch microscopist, had observed a cross between a white doe rabbit and a grey buck. The offspring were all grey. The male parent, as Aristotle had claimed, was preponderant.

Later modification and rebuttals of the preformation theory included Buffon's famous hypothesis that there were germs of life everywhere—thus could new species come into being. Others, including Linnaeus, believed the influence of the maternal parent governed the formation of certain parts of the offspring's anatomy and the paternal parent determined the rest, the "two-layer" theory.

What has perhaps been so irritating about this phase of biological history is that common sense seems to have been left behind. Instead of a fusion of the ideas of epigenesis and preformationism into a composite at least equal in stature to the concepts of the Greeks, epigenesis was rejected and preformationism proliferated absurdities. The two concepts foreshadow our present knowledge of the realization of developmental potentialities based on gene action. A thorough study of the intellectual history of the eighteenth century may yet show the interactions among personalities, politics, and religion responsible for this major lapse. Eggs and sperm were both known, at least in frogs and fishes, and the first phases of fertilization had been observed. It had been realized from ancient times that parental characteristics were often quite mixed, mingled, and melded in the offspring. Even Aristotle did not deny biparental hereditary determination, but explained transmission of maternal characteristics as a lessening of the prepotency of those of the male. Why, then did no one venture to put it all together?

During the middle of the eighteenth century, Caspar Wolff (1733–1794) resurrected epigenesis. He could see no structure in the undeveloped egg, and believed that the new individual, driven by an essential spark of life, grew its own parts from food supplied to it. His work was largely overlooked, and preformationism hung on into the nineteenth century.

Meanwhile, interest in gardening, herbs, and plant collecting had started a line of activity which was eventually to reveal the laws of hereditary transmission. Although sex in plants had been recognized since ancient horticultural practices with fig and date palm pollinations, the annual practice was devoted solely to the production of fruit rather than to plant improvement.

Camerarius (1665–1721) was apparently the first to systematize and test the information on sex in plants, and concluded that the pollen is the male, the ovary the female element. Early in his career Linnaeus (1707–1778) was stimulated by a report of Camerarius' work to study the neglected vital organs of plants, the pistils and stamens, which he used as the basis for his taxonomic system. He himself made several demonstrations of the necessity of pollination for seed production and artificially produced a species cross of goat's beard (*Tragopogon*). Linnaeus began his career as a firm believer in special creation, but changed remarkably as experience with blurred species boundaries and plant hybrids led to serious doubts of the immutability of species. Others during the seventeenth and eighteenth century had also been speculating on sex in plants and certain well-known hybrids were produced, but little true experimental work had been done (see Roberts, 30, Chap. 3).

Koelreuter.—The first person to make a systematic study of plant hybridization was Joseph Gottlieb Koelreuter (1733–1806). He was the first true antecedent of Mendel, a debt Mendel himself recognized, and, like Mendel, was almost completely neglected by his contemporaries.

Koelreuter performed hundreds of hybridization experiments involving over a hundred different species of plants. His first hybrid, *Nicotiana rustica* × *N. paniculata,* grew amazingly well but proved to be sterile, at least in the first trials. He thus confirmed his belief in the stability of the species and refuted the claims of the Linnaean school that new species could arise directly from hybridization. Because reciprocal crosses gave essentially the same results, intermediate between the two parents, Koelreuter's results refuted both preformationism and such concepts as the two-layer theory of hereditary determination. The results with *Nicotiana, Dianthus,* and *Mirabilis* showed a remarkable difference between the uniformity of the hybrids (F_1) produced by crossing two parental species and the considerable variation of the offspring of the F_1 hybrids, the F_2, among which the various characteristics reflected one or the other original species, the original hybrid, or were intermediate, and all in a confusing and bewildering array of different combinations of characteristics. By pollinating the original hybrid with pollen from one of the parental species, and continuing this process through several generations, Koelreuter was able to "change" one species into another.

To entomologists, Koelreuter is especially noteworthy for his systematic studies on insect pollination of plants with showy flowers such as iris and waterlily. He performed the correct experiment by exposing the flowers to the wind but excluding the insects and found no seeds were set. Koelreuter also inserted glass tubes into flowers to draw out the nectar that attracted the bees and other insects.

Christian Konrad Sprengl (1750–1816) enlarged upon Koelreuter's work on insect pollination. He recognized the relationship between flower form and the pollinating insects and discovered that in many bisexual (perfect) flowers, the stamens and pistils do not mature at the same time (dichogamy) and the flowers are therefore dependent upon insects. Such observations showed that cross pollination between individual flowers and plants was much greater than anyone had previously believed. The value of Sprengl's work was not recognized until Darwin cited it in the *Origin of Species* as additional support, in the mutual dependence of flowers and insects, for his concept of natural selection.

Maupertuis.—We are indebted to Glass (14, 15) for his appreciative résumés of the life and contributions of the remarkable Pierre Louis de Maupertuis (1698–1759) whose ideas were so far ahead of his time that they had little impact on most of his contemporaries. Although some of his concepts influenced others whose ideas were perpetuated, Buffon, Diderot, and Bonnet, Maupertuis' approach to the problem of inheritance through mathematical analysis of transmission of simple traits was not seen again until the work of Gregor Mendel, over a hundred years later. For a human family in which polydactyly appeared in four successive generations, Maupertuis calculated that the odds were $8 \times 10^{12}:1$ against such a pedigree occurring by chance.

The Nineteenth Century

In most histories of biology and genetics, usually two or three of the several different lines or fields of interest in heredity are omitted or seriously distorted in accounts of the nineteenth century. It will certainly not be possible to rectify this situation in the present brief sketch, but at least the several fields can be enumerated and their significance noted. The various areas seem to have been fairly well isolated, though not completely so. Although there is at present a considerable inertial lag in the transfer of information from one field to another, there were, in addition, profound conceptual barriers which slowed or prevented transfer of information during the nineteenth century as well as hindering its initial production.

There were eight fields in which at least some of the participants had a pronounced interest in the mechanisms or results of hereditary transmission: 1. plant and animal breeding, 2. medicine, 3. plant hybridization, 4. cytology, 5. biometry, 6. evolution, 7. chemistry, and 8. anthropology. It has been the work of the twentieth century to fuse all these varied lines of interest in heredity to make the tremendous body of knowledge which is modern genetics. The process of fusion is still incomplete, and the reasons are the more recent counterparts of the nineteenth century conceptual barriers.

Plant and animal breeding.—During the nineteenth century plant breeders began to follow in earnest the ancient practices of animal breeders of controlled mating. Both groups began to keep better records and demonstrations of the heritable nature of many traits became available for use in producing more desirable strains.

The best account of this period in plant breeding is that of Roberts (30) who gave considerable attention to the work of the practical breeders. The most striking aspect of a century's work was, however, as Roberts (l.c., p. 283) pointed out that nobody except Mendel had attempted to look carefully at the second, variable hybrid generation to determine "whether, beneath this apparent disorder, there might not be concealed some law." The facts used by Mendel as the basis for his conceptual analysis were the expression of the dominant trait in the first generation hybrid and the segregation or separation of the dominant and recessive traits in the progeny in ratios specific for the selfed hybrid and for its backcross to the recessive parent.

In 1799, Knight, an English horticulturalist reported dominance of the grey over the white color of the seed coat in the garden pea. In 1823, he noted the backcross of the hybrid plant to the white-seeded parent gave two types of plants, those with white and those with grey seeds. (The seed coat is, as Knight recognized, a tissue of the maternal plant and would not show the results of hybridization until the seeds resulting from the cross pollination were grown to maturity.) Goss, during the 1820s noted dominance, this time immediate, of white over blue cotyledons in the garden pea; in the next generation, a mixture of the two types were produced, often in the same pod. Goss separated the blues from the whites; the blues bred true, the whites often produced further mixtures. Such Mendelian facts as these continued to be collected by the plant breeders from work with peas, melons, lupines, and wheat throughout the nineteenth century. In 1901 Spillman published extensive data on first and second generation crosses in wheat. In many of the second generation progenies, ratios approximating the expected 3:1 were obtained. Spillman realized quantitative laws were somehow involved but did not go further.

The old rule of like begetting like was still being applied in selection for desirable agronomic characteristics, but it was more cogently applied—the procedures were stricter. In France, de Vilmorin produced superior varieties of sugar beet by testing the sugar content of the individual beets, and sowing separately the seeds from each of the superior plants.

Stud books and other records began to be kept more accurately and extensively by the animal breeders, but progenies were small and populations also small outside pure breeds within which variation was apt to be frowned upon. Late in the nineteenth century, however, Galton was able to test his "law of ancestral heredity" from data on coat color in basset hounds for which the pedigrees extended over twenty years.

Medicine.—The situation in medicine was analogous to that in agriculture; in both fields there was a growing awareness of the significance of heredity, more and better information, but no real insight into the mechanism of inheritance. Hemophilia, which posed a serious threat to the religious practice of circumcision, had been recognized since the Babylonian Talmud (fifth or sixth century A.D.) as transmissible to males through the maternal lineage. Maupertuis' study of polydactyly has already been mentioned, and the abnormality was also investigated in the eighteenth century by Réaumur. At the close of the eighteenth century, the hereditary nature of red-green color blindness was recognized. During the nineteenth century other abnormalities, including chorea—Huntington's study, and certain types of deafness—the work of Alexander Graham Bell, were added to the roster. The importance of heredity became increasingly obvious, but its basic mechanism remained elusive (35).

Plant hybridization.—There was no sharp distinction between the hybridizers and the breeders except the former were more apt to pursue theoretical goals and attempt to generalize. Except for Mendel's work, most of the hybridization experiments of the nineteenth century had as their goal, as did the earlier work of Koelreuter, the elucidation of the nature of the species. Since the species was believed to be the fundamental unit, it was expected that species would somehow also manifest themselves as heritable entities, and indeed the first time the term "factor" was used (Gaertner, 1849; also Wichura, 1865) it referred to the entire contribution to the hybrid of one of the parental species (30).

Again, with the exception of Mendel and the late nineteenth century hybridizers who were to discover and confirm his neglected work, the contributions of the nineteenth century plant hybridizers can be characterized as a continuation and extension of Koelreuter's work. The vast amount of information collected indicated over and over again the intermediacy of the first generation hybrid and the variability of the second generation.

A continuum is often recognized leading from Camerarius through Koelreuter and Gaertner to Mendel. Although Mendel studied Gaertner's compendium carefully, there were in it few additions of real value to Koelreuter's work. But both Naudin and Naegeli are of considerable interest. Naudin was also concerned with species but in some of his crosses, the species differed in a single striking attribute. Naudin perceived that random union of pollen and eggs, each of which contained one or the other of the two parental specific essences, would lead to the three types of plants observed in the next generation, as may be here symbolized: A♀ with A♂, and B♀ with B♂ to give the two parental species, and A♀ with B♂ and the reciprocal to reconstitute the hybrid.

Naegeli was an outstanding botanist with significant contributions in cytology, morphology, and systematics. His interest in hybridization was largely theoretical; in a series of papers published in 1865–1867, previous work on plant hybridization was codified and summarized with the species maintaining its primacy. It seems likely that at least part of the considerable difficulties resulting for Mendel from his relationship with Naegeli stemmed from Naegeli's very considerable authority as a plant scientist and a vested interest which shielded him from new ideas.

Cytology.—Except for the work of Mendel, the closest approach to the true nature of inheritance came during the nineteenth century from a study of cells and their behavior. The workers in this area were not handicapped by the species problem and progress was, comparatively speaking, remarkably smooth.

Early in the century, improvements in the microscope made it possible to study the cell in some detail. Although cells had certainly been observed earlier, the work of Schleiden and Schwann in the late 1830s established cells as fundamental units of both plants and animals. In the 1840s Unger,

Naegeli, and others argued for the formation of new cells by division, and in 1855, Virchow expounded his famous dictum, all cells come from cells. About three decades later, enough facts had been collected to enable Strasburger and Flemming to make a similar claim about plant and animal nuclei: all nuclei come from nuclei.

The fusion of two nuclei on fertilization was first seen in animals by Hertwig in 1875 and in plants by Schmitz, 1879, in the filamentous green alga, *Spirogyra.* In 1866, Haeckel had suggested the nucleus was the organ of inheritance, and the events of syngamy seemed to confirm this concept.

The mitotic cycle was analyzed in greater detail. In 1882 Flemming discovered the longitudinal splitting of the chromosomes, and shortly thereafter van Beneden and Heuser observed the two halves moving to opposite poles. In 1883, Roux claimed that the longitudinal split could have meaning only if the chromosome was differentiated along its length; the chromosome behaved as though qualities rather than quantities were being transmitted to the daughter cell. This was the first statement of one of the great unifying concepts of biology, the chromosome theory of inheritance. But a vast amount of work remained before the theory was experimentally demonstrated.

In 1883 van Beneden discovered that the eggs and sperms each had half the number of chromosomes found in the body cells, and believed that the others were somehow lost in the process of gametogenesis. In 1887, Weismann made the startling prediction that a regular process would be found, either a reduction division, or as had been previously suggested, an elimination of whole chromosomes. The doubling of the chromosome number at syngamy would thus be reversed. Weismann's prediction was a challenge to cytologists who, in the early 1890s, discovered pairs of chromosomes during meiosis and began the modern, accurate description of this fundamental process.

Thus by 1895, there was a picture of the chromosomes as the bearers of hereditary determinants; they kept in step with the major features of the life cycle, doubling in number on fertilization, dividing in an orderly fashion during development, and reducing their number prior to gamete formation. In 1892, Ruckert had suggested that each of the meiotic pairs of chromosomes consisted of a maternal and paternal element and that the nodes between them (now, chiasmata) represented points of exchange of material between the two parental chromosomes.

Although cytologists had made excellent progress and showed valid insight into the significance of their observations, the lack of experiment led to the construction of elaborate theories. In 1884, Naegeli proposed a hereditary substance, idioplasm, which was not confined to the nucleus but extended from one cell to another throughout the organism. The suggestion of a specific hereditary substance continued to have appeal after Naegeli's elaborations became obsolete. Five years later, De Vries advanced intracellular pangenesis: a complete set of hereditary particles was maintained in each cell nucleus, but additional particles circulated between the nucleus and the cytoplasm. This is an amazing prevision of the maintenance of DNA in the chromosomes of the nucleus while RNA copies move from it to the cytoplasm.

In addition to his prediction of reduction division, Weismann made three other proposals. The concept of the germ line or germ plasm was originally expressed by Galton in the term "stirp". Weismann's elaboration of the idea helped to rule out both transmission of acquired characteristics and pangenesis. Observations have repeatedly confirmed that lines of reproductive cells are established very early in embryogenesis so that the body of the animal can be looked upon as a side branch of the sexual life cycle. Weismann did not fare so well with two of his other ideas. His elaborate theory of chromosome structure cannot be cited in detail; suffice it to say that he proposed a disintegration of chromosomal units during ontogeny to release the necessary hereditary determiners to the appropriate cells. He also pictured germinal selection in which the cells struggled for survival during ontogeny. Embryologists were quick to point out, however, that no such struggle manifested itself during early development; each cell occupied its allotted place without turmoil.

Although no immediate further progress would have been likely from the observational methods of the cytologists, their work, especially the analysis of meiosis, paved the way for an acceptance of Mendelism. Shortly after the rediscovery of Mendel's principles, his factors were to find a ready home in the chromosomes.

Biometry.—Charles Darwin's cousin, Francis Galton, began modern biometry by comparing measurements of children with that of their parents. Statistical methods, especially the Gaussian curve of errors, were borrowed from the astronomers to interpret the data. On the average, children tended to be closer to the mean than were the parents. The process of variation seemed to be a sporadic production of divergent individuals followed by a regression toward the average. Galton was influenced by such observations to develop his ancestral law of inheritance in which the parents were assumed to contribute a half, the grandparents a fourth, etc, to each characteristic of the individual. As a statistical tool, the ancestral law worked well also with discontinuous variation as Galton was able to show with coat color of basset hounds (30).

The observed regression of offspring toward the population average led Galton to discount the assumption of evolutionists and biometricians that continuous variation was of paramount importance in evolution. Galton was also apparently looking for a way out of the trap of blending inheritance. He proposed to Darwin a scheme in which various numbers of parental particles determined the characteristic. For two such particles, one from each parent, Galton obtained the Mendelian 1:2:1 ratio, and proposed a method for estimating the actual number of particles controlling a given character by counting the number of progeny in each class. But neither Galton nor Darwin pursued the subject further.

Galton's associates, chiefly Karl Pearson, continued to improve on the statistical methods but remained confirmed Darwinians, convinced of the significance of continuous variation. A bitter battle broke out between the Pearson group and William Bateson who agreed with Galton's views on discontinuous variation. Bateson was one of the early champions of the Mendelian principles, but the biometricians were not readily convinced.

Evolution.—It is a truism that traits which are not transmissible to the offspring can have no meaning for evolutionary change. Thus, evolutionists have always had a special interest in the mechanisms of inheritance. The necessity of explaining the evolutionary process has all too often determined the thinking about heredity as though the problem to be solved were the type of hereditary mechanism which would sustain evolutionary change. Heredity thus has no rights of its own.

The major evolutionist of the early nineteenth century, Lamarck, had a simple concept of the inheritance of acquired characteristics: a one-eyed race of men could be obtained by mating with each other individuals whose left eyes had been removed at birth. To obtain the changes that must have occurred during the course of evolution, Lamarck added to the simple concept of transmission of acquired characteristics the idea that those body parts strengthened through use would be enhanced in subsequent generations and the unused parts eventually lost. Lamarck has been subjected to much quite unjustified ridicule; he intuitively perceived what is only now becoming worked into the pattern of evolutionary thought, that a change will have no selective value unless the creature's repertoire of behavior includes the capacity to exploit the change. Thus, if the ancestors of the giraffe had been unwilling or unable to reach up for their food, there would have been no selective advantage in increasing height and the ecological niche of the high browser would have been forever closed to their descendants. Nordenskiold (25, p. 322) quoted, "It is not the organs—that is to say, the form and character of the animal's bodily parts—that have given rise to its habits and peculiar properties, but, on the contrary, it is its habits and manner of life and the conditions in which its ancestors lived that has in the course of time fashioned its bodily form, its organs, and its qualities." The most dramatic example of Lamarckian evolutionary, *not* his genetic, thinking has been in the interpretation of the ancestry of man; the increasing size of the brain would have had little selective value without the behavioral changes resulting from using and making tools. "The tool-using, ground-living, hunting way of life created the large human brain rather than a larger-brained man discovering certain new ways of life." (Washburn & Lancaster, 1968).

Roberts (30) has paid special attention to the later phases of Darwin's career during the time when he was corresponding with Naudin and had begun to be influenced by Galton. Darwin had developed a concept of pangenesis in which gemmules were pictured as moving from various parts of the body to the germ cells. He later proposed an almost Mendelian behavior for his gemmules in hybrid crosses. Although Galton had pointed out to him that limiting the number of gemmules to two would give a 1:2:1 ratio in the offspring of a hybrid, Darwin did not use this concept but contented himself with noting that segregation and combination of gemmules could give some of the results observed in plant hybridization experiments. Finally, Galton's concept that different numbers of gemmules might be specific to different characters would have given Darwin the means for including gradual evolutionary change (numerous gemmules) and the behavior of nonblending characters (few gemmules) in the same hereditary scheme (26).

It is not generally recognized that Darwin came closer to finding the fundamental laws of heredity than any other nineteenth century investigator except Mendel and Mendel's late nineteenth century discoverers. Considering his heroic contributions in so many other fields (13), and his long-standing bias against the evolutionary significance of the type of character necessary for a Mendelian analysis, it is remarkable that Darwin came as close as he did.

The evolutionists themselves may not have been the ones to succeed in uncovering the secrets of inheritance, but the obvious significance of heredity to evolution, and the overwhelming impact of Darwinian evolution on the development of biology during the last decades of the nineteenth century gave an immense impetus to the analysis of heredity (6).

Chemistry.—The fact is usually forgotten that DNA first achieved recognition as the hereditary substance in the nineteenth century. In 1871 Miescher isolated nucleoprotein from fish sperm and pus cells and Altmann, in 1889, split it into its two components, nucleic acid and protein. Chromatin appeared to be a combination of the two substances, and it was recognized that nucleic acid was responsible for its basophilic staining reactions. In the first edition, 1896, of his classic work on the cell, Wilson recognized nucleic acid as the hereditary material of the chromosome, but this view became obsolete during the twentieth century when inaccurate evidence indicated that nucleic acid disappeared from nuclei during important phases of the life cycle. Realization of the possible significance of DNA did not begin again until the fifth decade of the twentieth century, and then developed slowly (36).

Anthropology.—At the turn of the nineteenth century, Blumenbach laid the foundations of physical anthropology. By his day, intense interest in human racial differences had been awakened by voyages of discovery and colonization of the New World. There has been a continuing, popular interest in the results of racial mixture and the hereditary basis of physical, intellectual, and emotional attributes.

Gregor Mendel

The career of Mendel and the fate of his work have been pretty much of an enigma for twentieth century biologists. However, recent insights into the significance of his education and the preparation for his definitive work with peas have gone a great way toward clarifying the obscurity. If the term molecular is used in its most general sense, then Mendel was the first physicist to undertake molecular genetics. He turned away from the species as an entity in inheritance and sought to analyze the nature of the hereditary transmission of the simplest, definite differences he could find, not the complex accumulation of differences between species but the mundane contrasts between smooth or wrinkled and yellow or green seeds, and tall or dwarf plants, an attempt much like Galileo's interest in the way dropped things fall.

The most important years in Mendel's intellectual development were those he spent first at the University of Vienna and later as a demonstrator at a physics institute in Vienna. A considerable debt is due Olby (26) for his careful assessment of the significance of this period in Mendel's life. Important aspects of Mendel's work seem to be traceable to the influences he experienced in Vienna.

At the university, Mendel studied plant science with Unger whose radical ideas on evolution led to bitter attacks from the clergy. A chief reason for Mendel's attempt to follow hereditary transmission was the light such work might possibly shed on evolution (22).

Also at the university, Mendel studied physics with two men who looked for mathematical solutions to the problems they wished to solve. One of these was Doppler, noted for the Doppler effect, the change in wavelength attributable to the relative motion of the emitting body. Mendel is well known, of course, for his statistical formulations, 3:1, 1:1, and so on, but it is not so often pointed out that he followed out the mathematical consequences of various aspects of his system, such as the percentage of homozygosity in succeeding generations of self-pollination.

As a physics demonstrator, Mendel became acquainted with the physicist's method of isolating describable aspects of a system and working out means for their analysis. Doppler's description of a simple relationship did not depend on the nature of the waves. Many nineteenth century biologists, faced with an analogous situation would have overlooked their "Doppler effects" in a vain struggle to define or characterize the entire system.

The tragedy of Mendel's career was his association with Naegeli who had just completed an exhaustive study of speciation in *Hieracium,* an apomictic composite. Naegeli failed to grasp the significance of Mendel's work and suggested he study *Hieracium.* The frequent occurrence of apomixis, not yet understood, led, of course, to disastrous results for any possible confirmation of the work with peas. Mendel did not look further for an understanding colleague, and his work remained unnoticed by the scientific community for 34 years; by then, Mendel was dead (see also 28).

Several other sources of Mendel's inspiration have been suggested, such as experiments with mice which Mendel was afraid to publish. However, Mendel was seriously and openly concerned with human heredity (35). Of special interest to entomologists is another suggestion that Mendel may have heard of Dzierzon's work with bees: first, that males were produced from unfertilized eggs; and second, that Italian and German drones were produced in a 1:1 ratio by hybrid queens (26). Most likely, Sturtevant (37) was right in remarking that Mendel's own words, that work with flowers came first, should be believed.

Modern Genetics

The short story of modern genetics began just 73 years ago with the confirmation of Mendel's work by de Vries (7), Tschermak (38), and Correns (4). During the next two decades, the various strands of nineteenth century interest began to be woven together to make a remarkably strong foundation for all subsequent work. Two groups soon saw the applicability of the new principles to practical problems. The precise Mendelian rules gave plant and animal breeders an exact method for incorporating new and desirable traits into otherwise valuable breeds. Physicians realized that transmission of certain human traits, including metabolic disorders, conformed to Mendelian expectations. In 1909, Garrod published his classic work on human biochemical genetics, *Inborn Errors of Metabolism.* The subsequent decline of human genetics prior to its present renaissance is amply detailed by Ludmerer (21).

Anthropology, the other field concerned with man, had to wait several decades for the development of population genetics. Then the utilization of the nonabnormal, Mendelizing traits of the blood groups enabled a genetic analysis of human populations. Likewise, the further development of the chemical basis of heredity was postponed until the 1940s when the first new insight into the possible significance of DNA came from Avery's isolation of DNA as the transforming principle of *Pneumococcus.*

Of the remaining fields, cytology fitted hand and glove with the new Mendelism; biometry and evolution were the scene of bitter conflict; and Mendel's own field, plant hybridization, eventually became a specialized area of plant cytogenetics and cytotaxonomy.

The biometricians and evolutionists who saw in continuous variation the only likely basis for evolution rejected the new approach derived from analysis of disjunct characteristics, Mendel's tall versus short pea plants, and so on. The leading contributor among the biometricians, Karl Pearson, continued to refine various statistical tools that later were used to rationalize the differences between the biometrical and Mendelian schools. R. A. Fisher published the first such attempt in 1918 and his 1930 book on *The Genetical Basis of Natural Selection* eventually eliminated the conflict between the two approaches. Theoretical contributions of immense significance came also from J. B. S. Haldane and Sewall Wright. The earlier, considerable insights of Weinberg were, however, often overlooked and his work repeated later by Fisher and Wright (34). The study of evolution, which had been quite confused, was made respectable biology again by Dobzhansky's *Genetics and the Origin of Species,* first published in 1937. In it, evidence from a variety of approaches was combined in a clear, consistent account. Meanwhile, way back in 1908, both Hardy and Weinberg had published their independent derivations of the famous rule on the maintenance of the proportion of alleles in mixed populations (33). Development of this area from Darwin to the synthesis of Mendelism, Darwinism, and biometry has been analyzed in detail in a book by Provine (29).

The major line of development began with the repetition of Mendel's experiments with other plants in 1900 and was almost immediately fused with cytology. In papers of 1902 and 1903, Sutton showed that the behavior of the chromosomes in meiosis in the grasshopper was exactly that expected if the chromosomes carried the Mendelizing factors. Shortly thereafter, Boveri's observations on postfertilization abnormalities of cell division in the sea urchin indicated that each chromosome controlled or influenced the realization of a specific set of developmental potentialities. The next few years saw the identification of the sex chromosomes and the different, heteropycnotic state in which they so often occur. Both had been reported from a study of meiosis and spermiogenesis in an hemipteran insect by Henking in 1891, but without realization of their significance. However, some prominent geneticists were not willing to accept the chromosome as the home of the hereditary factors without further proof.

Drosophila and the chromosome theory.—An entomologist, C. W. Woodworth, was apparently the first to cultivate *Drosophila* in the laboratory and W. E. Castle was the first geneticist to appreciate its value for genetic experiments.

In the hands of T. H. Morgan and a small group of young collaborators at Columbia University, *Drosophila* was used as the basis for amplifying Mendelian genetics and giving it its present form. Although linkage had previously been reported in plants, the first significant contribution from Morgan's laboratory was the report, in 1910, of the criss-cross nature of sex linkage in *Drosophila.* With his collaborators he began the large-scale task of analyzing the physical relationships of the factors with each other and soon found that a linear linkage map satisfactorily summarized the data.

Several recent books (3, 9, 37) outline the history of modern transmission genetics in considerable detail, and the basic facts and concepts are familiar to most biologists. It should be emphasized, however, that the most significant contribution of the Morgan school was the standard of excellence established for experimental work in genetics. C. B. Bridges' experimental verification of the chromosome theory of inheritance was a striking validation of the approach of the Morgan group. Bridges obtained XXY female *Drosophila* from spontaneous nondisjunction (XX egg plus Y sperm). These were recognizable if both Xs carried a recessive gene derivable only from the mother. Bridges was able to establish a sex-linked criss-cross transmission in the reverse direction: the Y chromosome from the mother, and the X from the father. The expected extra Y chromosomes were demonstrated cytologically. There was no doubt then that the chromosomes carried the hereditary factors and most of the skeptics were convinced.

Refined experimental procedures were the key to H. J. Muller's analysis of mutation in *Drosophila* which preceded by several years the demonstration that X rays spectacularly increase the frequency of mutations and chromosome rearrangements. Other outstanding examples of highly refined experimental work were Stern's proof of somatic crossing over, the demonstration that genetic crossing over is accomplished by a physical exchange (Stern, *Drosophila*; Creighton and McClintock, maize), McClintock's analysis of the behavior of chromosome rearrangements in meiosis, of the behavior of broken chromosome ends and of ring chromosomes during the life cycle, and Sturtevant's demonstration of the duplicate nature of the *Bar* factor in *Drosophila*. The results of these experiments are given in most textbooks on genetics but the how and why of the experiment are all too often omitted.

Entomology and genetics.—The most important theoretical contributions to entomology from the developing science of genetics have been in the area of population dynamics and evolution. The results of this work have largely determined current concepts of speciation, natural selection, and population dynamics and are providing a key to the immensely difficult area of the ecological basis of the observed processes. In the United States, Sturtevant and Dobzhansky initiated the study of inversions in chromosome III of *Drosophila pseudoobscura* in natural populations. Because the inversions were overlapping it was possible from the banding patterns of the giant salivary chromosomes to construct a phylogenetic tree from them (abcdef ⟷ abedcf ⟷ aebdcf). Each locality had specific frequencies of those inversions present and changes in frequency of specific chromosomes were found to be closely correlated with geographic distribution. Thus, an exceedingly refined adaptation was demonstrated for the hereditary materials themselves. In England, E. B. Ford followed the distribution of wing patterns in butterflies and showed again a precise adaptation. An especially entertaining story is that of the work of Ford and his colleagues on the selective value of industrial melanism in butterflies. Although birds and butterflies had been carefully observed for decades by countless naturalists in Britain, no one had reported predation by birds of these particular butterflies until the work on natural selection began (11).

Insects have been of considerable interest to cytogeneticists because of the unorthodox nature of some of the chromosome systems. In *Drosophila,* itself, crossing over usually does not occur in the males of most of the species. Other examples pose much more serious challenges to any simple explanation of chromosome behavior during the life cycle. In *Sciara,* the zygotes have three X chromosomes; developing females eliminate one of these from the soma and the germ line; developing males eliminate two from the soma but only one from the germ line. Meiosis in the male is highly unusual. The paternal chromosomes are eliminated and both chromatids of the maternal X chromosome are included in the one sperm cell formed by each meiotic event (23). A vast array of cytological tricks are played by the coccids; in one system, an entire set of chromosomes becomes heterochromatic in the early development of the male (16). Such systems are important in showing that the chromosomes must somehow or other be "informed" of what to do.

More recently, the work of Beermann (2) on tissue to tissue differences in puffing patterns in the giant chromosomes of *Chironomus* was a direct demonstration of the way in which genes may be turned on or turned off during development. Since puffs were later seen to be sites of active synthesis of RNA, the door had been opened to intensive analysis of the control of gene action in eukaryotes.

In more practical areas, genetics has become significant in three fields, insecticide resistance, biological control, and improving the strains of two beneficial insects, the honey bee and the silkworm.

Control.—The story of insecticide resistance is a mirror image to that of disease resistance in plants. In plants, the breeder introduces a resistant factor to prevent damage by a natural enemy, usually a fungus of some sort. In insects, a resistant factor is introduced by nature through mutation, and prevents damage by an insecticide. The genetic bases are parallel; resistance may be due to a varying number of genes, or to dominant or recessive genes (5).

The failure of insecticides in many situations has led to a considerably increased emphasis on biological control. Although there were spectacular earlier successes, some situations may become so complex that mistakes in alpha taxonomy occur. With increased sophistication has come an awareness, however, that the insect selected for control of another insect must be adapted to the specific ecotype of the host and to the general ambience. Since adaptation requires an array of specific factors, or even requires a gene pool from which the appropriate combination can be drawn, the problem becomes exceedingly complex very quickly. The obvious, simplest approach is to secure the control agent from the home territory of the pest. But even here, the genetic problems of the number of individuals to be collected and the inevitable screening of genotypes by laboratory rearing before field release intrude themselves. A recent and highly stimulating review by van den Bosch (39) outlines these and other current problems in biological control and a recent, more popular article describes the devastating financial loss resulting from the use of pesticides (40).

Eventually it should be realized that all such problems involving two organisms are essentially warfare between two genetic systems, and that man may often be able to determine the outcome by proper manipulation of one of the genetic systems. A model case is the experiment by Lerner and co-workers with two *Tribolium* species. Selection of appropriate stocks determined which of the two species would win out in competition for survival under controlled laboratory conditions (32).

From the time of H. J. Muller's demonstration that shortwave radiation induces mutations, it has been generally realized that increasing dosage to the gametes of one parent increases a class of zygotes called dominant lethal. The damage induced in one gamete is so extreme that development does not occur even though the other gamete is quite normal. The first experiments with the goal of putting the dominant lethal effect to practical use were apparently those of R. C. Bushland with screw-worms. After irradiation the sterile male produced sperms capable of fertilizing the eggs, but the larvae did not hatch. Results of a series of experiments indicated that the method would be applicable to control if certain conditions obtained, and Curaçao, an island off the coast of Venezuela, was chosen for a large-scale test. The results were incredibly good; after five months the screw-worm was completely eradicated. Three consecutive papers in an issue of the *Journal of Economic Entomology* carried a summary of this work, *Atomic Radiation for Insect Control,* the first by Knipling (19). Since the success on Curaçao, much more extensive schemes of screw-worm control have been put in operation. The method will not work except in quite specific situations, but for these it provides an elegant solution.

Bees and silkworms.—Since remote ages, bees have been the most fascinating of all insects because of their honey, their hives, and their behavior. The intimate details of their life cycle, genetic system, and evolution are proving equally interesting, especially those concepts which help to account for their behavior in hive building, food collection, food storage, swarming, communication, defense, and relationships between the members of the hive, especially the altruism shown their fertile sisters by the sterile workers. Male bees are haploid, but otherwise the genetic system seems quite orthodox. However, breeding experiments are not typical since all crosses must be made by artificial insemination. In pigment changes, eye color mutations in bees seem to be similar to some of those known for *Drosophila* and Lepidoptera. The high frequency with which gynandromorphs occur in some strains of bees indicates an interesting lability in early development and genetic markers have been used to determine the origin of the haploid sector. Behavior genetics is of special interest. Resistance to American foulbrood can be achieved by hygienic behavior which is apparently determined by a recessive factor for uncapping cells with dead or diseased larvae and another recessive for their removal.

Breeding bees for high production and other desirable characteristics also involves considerable attention to behavior. Bees with preference for alfalfa can be selected, and probably such preferences can also be obtained for a wide variety of other pollens. Stinging behavior is apparently under quite complex genetic control. Crosses between the ferocious African bee and the Italian bee have been analyzed for several components of stinging behavior; the F_1 hybrid is intermediate or like the Italian parent in most respects but takes a much longer time to become fierce! Selection for honey production and disease resistance have been successful but special problems arise since inbreeding leads to brood inviability because of homozygous sex alleles. A similar problem appears in obtaining inbred lines to be crossed for superior hybrids. Unfortunately, genetic markers cannot be used to maintain heterozygosity at the sex locus during an inbreeding program because none of the known markers is linked with it.

Progress in bee genetics and evolutionary analysis are summarized in reviews by Kerr & Laidlaw (17) and Rothenbuhler, Kulinčevič & Kerr (31). Undoubtedly bee genetics, not only of the honey bee, but of other species of social and solitary bees, has a great future, especially in the field of behavior genetics.

Unlike the honey bee, the silkworm is easy to breed in the laboratory and its genetics has been quite thoroughly explored. Again, an unusual genetic system is encountered because the female, not the male, is heterogametic. Considerable attention was paid to the biochemical genetics of the silkworm, particularly in regard to pigmentation, and interesting cases of maternal inheritance of pigmentation have been analyzed (18). Unlike the work with bees where the various aspects have fitted together in a coherent picture, that with the Lepidoptera has been quite divided between the exploration of formal, biochemical, and developmental genetics with the silkworm and of population and ecological genetics and evolution with quite different species.

Conclusions

Over two thousand years were required before an appreciable improvement occurred in the atomist theory of inheritance. Hippocrates or Democritus would have had little difficulty in perceiving that Mendel had brought order to much the same sort of hereditary particles which they envisioned. And it is quite likely that they would have found in the chromosome theory and the concept of the germ line a verification of their insistence on the importance of physical particles in inheritance, if not of their origin.

About a hundred years after Mendel's work was published, the next great advance in genetics showed that the particles of inheritance were stretches of DNA precisely encoded with information controlling development and metabolism.

It is likely that the main work of the next few decades will be attempts to integrate, on the ultimate ecological foundation, the considerable store of knowledge already accumulated. Like every other aspect of the organism, the genetic system must be geared to environmental requirements. But the genetic system is also the basis for evolution. Thus, one of the foreseeable foci of integration is the analysis of the evolution of the ability to evolve an ability to exploit an environmental system from which there is no direct feedback of information to the evolving organism (24). With their large numbers and multiplicity of adaptive forms, insects will be of increasing importance in reaching the final goal of understanding living organisms in terms of living organisms. As Dobzhansky has frequently remarked and Monod has emphasized, evolution is by nature a process of chance, of opportunity. Thus, changes in living organisms can never be reduced, in a narrow sense, to deterministic molecular changes. Evolution of the molecular basis of organic existence not only follows the same laws as those which have been seen to govern changes in any other aspect of the organism, but now, as Monod so rightly insists, such changes can be much more precisely defined in molecular terms.

The fuller understanding of the ecological foundations of evolution and the genetic mechanisms responsible for change will have profound practical significance. In nature, adaptation usually occurs as the result of the slow accumulation of small changes. In this way, adaptive complexes can be built up from a series of uncorrelated characteristics. Unless several such characteristics are determined by one or a few genes, similar adaptive complexes are difficult to achieve in breeding programs. Since insects will always be significant pests and no single control method will be a panacea (39, 40), the establishment of on-going control programs for long as well as short term efforts would be highly desirable. If investment in long term programs can be encouraged, then long term genetic remodeling, via selection, would be feasible. Species could be domesticated and reconstituted, behaviorally and otherwise, to fit the breeder's preconceived picture of an ideal biological control agent. Likewise, more and better types of resistance could be built into the animals and plants under attack.

The history of genetics is noteworthy for the long eclipses between significant findings and their recognition by the scientific community. Maupertuis, Camerarius, Koelreuter, Mendel, Garrod, and Weinberg were all neglected. One of the main tasks of the next few decades will be an attempt to analyze systems as a whole and to integrate the resultant information for both theoretical and practical use. It is to be hoped that the next eclipse of genetics will not be its omission from programs where it is vitally needed.

ACKNOWLEDGMENTS

Suggestions on source material were kindly provided by Dr. H. Sharat Chandra, Professor Melvin M. Green, Professor I. Michael Lerner, Professor Curt Stern, and Professor Sherwood L. Washburn. Professor John W. Beardsley, Mrs. Roberta M. Brown, and Dr. Chandra read parts of the manuscript. Thanks are due Miss Jerilyn Hirshberg for bibliographic assistance and Mrs. Merlene Wood for manuscript preparation.

LITERATURE CITED

1. Aberle, D. et al 1963. The incest taboo and the mating patterns of animals. *Am. Anthropol.* 65:253–65
2. Beermann, W. 1956. Chromomerenkonstanz und spezifische Modifikationen der Chromosomenstruktur in der Entwicklung und Organdifferenzierung von *Chironomus tetans. Chromosoma* 5: 139–98
3. Carlson, E. A. 1966. *The Gene: A Critical History.* Philadelphia: Saunders
4. Correns, C. 1900. G. Mendel's law concerning the behavior of progeny of varietal hybrids. *Genetics* 35(5/2):33–41; 1950. Transl. L. Piternick
5. Crow, J. F. 1957. Genetics of insect resistance to chemicals. *Ann. Rev. Entomol.* 2:227–46
6. Crow, J. F. 1959. Darwin's influence on the study of genetics and the origin of life. *Biol. Contrib., Univ. Texas, Austin* Pub. No. 5914; pp. 49–55
7. de Vries, H. 1900. Concerning the law of segregation of hybrids. *Genetics* 35(5/2):30–32; 1950. Transl. A. Hannah
8. Dobzhansky, Th. 1937. *Genetics and the Origin of Species.* New York: Columbia Univ. Press
9. Dunn, L. C. 1965. *A Short History of Genetics.* New York: McGraw-Hill
10. Evans-Pritchard, E. E. 1932. Heredity and gestation, as the Azande see them. *Sociologus* 8: 400–14
11. Ford, E. B. 1964. *Ecological Genetics.* London: Methuen
12. Furley, D. J. 1967. *Two Studies in the Greek Atomists.* Princeton, N. J.: Princeton Univ. Press
13. Ghiselin, M. T. 1969. *The Triumph of the Darwinian Method.* Berkeley: Univ. California Press
14. Glass, B. 1947. Maupertuis and the beginning of genetics. *Quart. Rev. Biol.* 22:196–210
15. Glass, B. 1960. Eighteenth-century concepts of the origin of species. *Proc. Am. Phil. Soc.* 104:227–234
16. Hughes-Schrader, S. 1948. Cytology of the coccids (Coccoidea-Homoptera). *Advan. Genetics* 2: 127–203
17. Kerr, W. E., Laidlaw, H. H. Jr. 1956. General genetics of bees. *Advan. Genet.* 8:109–53
18. Kikkawa, H. 1953. Biochemical genetics of *Bombyx mori* (silkworm). *Advan. Genet.* 5:107–40
19. Knipling, E. F. 1955. Possibilities of insect control or eradication through the use of sexually sterile males. *J. Econ. Entomol.* 48: 459–62
20. Lindzey, G. 1967. Some remarks concerning incest, the incest taboo, and psychoanalytic theory. *Am. Psychol.* 22:1051–59
21. Ludmerer, K. M. 1972. *Genetics and American Society.* Baltimore, Md.: Johns Hopkins Univ. Press
22. Mendel, G. 1866. Experiments in plant hybridization. *Roy. Hort. Soc. London,* transl. of original in *Verh. naturf. Ver. in Brunn,* Abhandlungen, iv:3–47, 1865. Reprinted separately, Harvard Univ. Press, Cambridge, Mass., 1956
23. Metz, C. W. 1938. Chromosome behavior, inheritance, and sex determination in *Sciara. Am. Natur.* 72:485–520
24. Monod, J. 1970. *Le Hazard et la Nécessité.* Paris: Edition du Seuil
25. Nordenskiöld, E. 1935. *The History of Biology.* New York: Tudor Publ. Co. New ed.
26. Olby, R. C. 1966. *The Origins of Mendelism.* New York: Schocken Books, Inc.
27. Peck, A. L. 1943. *Aristotle: Generation of Animals.* London· Wm. Heinemann, Ltd.
28. Piternick, Leonie K., Piternick, G. 1950. Transl. Gregor Mendel's letters to Carl Naegeli 1866–1873. *Genetics* 35(5/2):1–29
29. Provine, W. B. 1971. *The Origins of Theoretical Population Genetics.* Chicago: Univ. Chicago Press
30. Roberts, H. F. 1927. *Plant Hybridization before Mendel.* Princeton, N. J.: Princeton Univ. Press
31. Rothenbuhler, K. C., Kulinčevič, J. M., Kerr, W. E. 1968. Bee genetics. *Ann. Rev. Genet.* 2: 413–38
32. Sokoloff, A. 1966. The genetics of *Tribolium* and related species. *Advan. Genet.* Suppl. 1

33. Stern, C. 1943. The Hardy-Weinberg law. *Science* 97:137–38
34. Stern, C. 1962. Wilhelm Weinberg. *Genetics,* 47:1–5
35. Stern, C. 1965. Mendel and human genetics. *Proc. Am. Phil. Soc.* 109:216–26
36. Stern, C. 1970. The continuity of genetics. *Daedalus,* Fall, 1970; *Proc. Am. Acad. Arts Sci.* 99(4): 882–908
37. Sturtevant, A. H. 1965. *A History of Genetics.* New York: Harper & Row
38. Tschermak, E. 1900. Concerning artificial crossing in *Pisum sativum. Genetics* 35(5/2):42–47; 1950. Transl. A. Hannah
39. van den Bosch, R. 1971. Biological control of insects. *Ann. Rev. Entomol.* 2:45–66
40. van den Bosch, R. 1972. The cost of poisons. *Environment* 14(7, Sept.):18–31
41. Washburn, S. L., Lancaster, C. S. 1968. The evolution of hunting. In *Man the Hunter,* ed. R. B. Lee, I. DeVore, 293–303. Chicago: Aldine
42. Washburn, S. L., Howell, F. C. 1960. Human evolution and culture. In *Evolution After Darwin II. The Evolution of Man,* ed. S. Tax, 33–56. Chicago: Univ. Chicago Press
43. Winspear, A. D. 1963. *Lucretius and Scientific Thought.* Montreal: Harvest House, Ltd.
44. Zirkle, C. 1935. *The Beginnings of Plant Hybridization.* Philadelphia: Univ. Pennsylvania Press
45. Zirkle, C. 1951. The knowledge of heredity before 1900. In *Genetics in the 20th Century,* ed. L. C. Dunn, 33–57. New York: Macmillan

A HISTORY OF BIOLOGICAL CONTROL

K. S. Hagen and J. M. Franz
Division of Biological Control, Department of Entomological Sciences
University of California, Berkeley
and
Institut für Biologische Schädlingsbekämpfung, Biologische Bundesanstalt
für Land und Forstwirtschaft, Darmstadt, Germany

Tracing the history of biological control parallels describing the development of applied entomology and applied ecology. As the knowledge of entomology, particularly entomophagy, advanced, the use of natural enemies arose, and controlling populations of pest insects using predators, parasites, and later pathogens demonstrated the importance of natural enemies as regulators of insect population abundance. In recent years, the use of selective pesticides in place of broad spectrum pesticides to rectify weak predator-prey ratios, gave rise to the integrated control concept. Capitalizing on natural biological control, integrated control underlined the importance of the biological control components in agroecosystems. More recently some have broadened the dimensions of biological control to include the autocidal techniques such as the sterile male and genetic intraspecific manipulations of populations.

The emergence of the concept of suppressing pest insects by using their natural enemies arose from observations, made by naturalists and agriculturists in Asia, Arabia, and Europe, of one insect eating another. As early as the ninth century in China, predaceous ants were used against citrus pest insects. By the early eighteenth century, the larger predators such as birds, the showy, large arboreal ground beetles, and the conspicuous lady beetles were appreciated as natural control agents and some local transferring of insect predators was done in Europe to curb insect outbreaks. The European anatomists and microscopists in the early 1700s observed and later interpreted the habits of ichneumonid larvae as parasites that killed their host caterpillars. The existence of food chains was realized in the middle 1700s. As the diversity of Ichneumonidae in European forests came to light when hundreds of species were described by nationalist-systematists during the early and middle 1800s, the idea arose that each species of phytophagous insect possesses its own parasite complex. By the 1850s in both the Old and New Worlds, the complexity of factors operating in natural populations was appreciated and gave rise to the ecosystem concept.

It was not until the 1830s that insect pathogens, first the fungi later the protozoa and bacteria, were identified as causing diseases in insects. Viruses were so identified at the beginning of the twentieth century. It was in the 1870s that the use of diseases was first advocated to control pest insects (see chapter by Cameron).

As the knowledge of entomophagy increased there was international transport of birds in the late 1700s, insects in the 1800s, and fish in the early 1900s in attempts to control pest insects. Also in the 1860s phytophagous insects were first employed to control weeds. It was realized that many serious pests were of exotic origin and that in endemic habitats these same insects usually were not considered pests. The first and perhaps most spectacular success of applied biological control was the introduction of the vedalia beetle from Australia into California in 1888 to control the cottony cushion scale. The scale pest which was killing millions of citrus trees was brought under complete control within two years following the vedalia beetle introduction.

The outstanding success of the vedalia became quickly known in all major agricultural centers of the world and led to many introductions of parasitic and predaceous insects particularly coccinellids. But then in a few years, after many reckless insect importations had been made, it was realized that biological control was not a panacea. However, there were during the early 1900s a few additional outstanding biological control successes, particularly in Hawaii and California. Furthermore, in these years the technique of mass culture of entomophagous insects and their periodic colonization in the field was developed in California.

During the period between 1920 and 1940 plant quarantine regulations were enacted and quarantine facilities for processing imported natural enemies were constructed in certain biological control centers. Biological control research became more sophisticated and screening for hyperparasites became imperative. Research on the culture, biology, and systematics of entomophagous insects was greatly intensified. A number of insect and weed pests were brought under complete biological control by new natural enemy introductions. So impressed were the biological control workers in the early 1900s with the suppression of certain insect and weed pests after the establishment of introduced natural enemies that regulation of insect populations in nature was viewed as being mainly the result of the action of biotic agents.

Important ecological concepts of population regulation germinated from the hypotheses proposed by the biological control researchers in the first 50 years of the twentieth century. The interrelationship of a natural enemy with its phytophagous host was looked upon as population symbiosis. The negative feedback reaction of the host-parasite relationship gave birth to the density dependence concept and is considered the main ecological basis of biological control. Another school of insect ecologists during this period considered the physical environmental factors as being the main cause of insect population regulation. The monitoring of both insect populations and the physical environment, coupled with studies of closed laboratory population systems, led to a blending of the philosophies of both schools and an integration of ideas occurred in the middle 1900s. It is recognized that the physical environment and the density of an insect population governs the velocity and the degree to which biotic forces have effect (see chapter by Andrewartha and Birch).

Through the late 1940s into the 1960s chemical control of insect pests with persistent organic insecticides was so spectacular and successful that the biological control approach received little support. In fact, it was considered passe except for a few research centers. Entomological research, in general, also suffered a setback for it was thought that little had to be known about the ecology or biology of insects to bring about insect reduction with the new chemicals. Blind applications at any time easily wiped out insects. But the intensity of the suppression of insects to extremely low densites was a powerful selective force to which many insects reacted and pesticide resistance became a wide-scale reality. The appearance of secondary pests particularly tetranychid mites and coccids and the problem of insecticide residues stimulated renewed interest in alternative methods of pest control. The biological control approach was again seriously considered around the world as one of these alternatives. The stage was also set in the 1950s for the development of integrated control which involved the blending of natural and applied biological control with the use of selective pesticides which spared natural enemies. Recently integrated control systems of even greater complexity have been developed. But all embrace an important biological control component.

The search for other alternate methods of pest control gave rise to the use of the intraspecific sterile male and possible other genetic manipulation techniques. Under the broad concept of biological control, the use of organism against organism, the autocidal techniques are considered under biological control today. Biological control usually requires international cooperation and in recent years international organizations have been formed which should further advance the field.

The above description of the evolution of biological control was largely based on the following references: Anon. (2), Askew (5), Bodenheimer (16), Carson (19a), Clausen (21, 23, 24), Compere (27), Day (31), DeBach (33, 34), Doutt (40, 41), Essig (48), Flanders (55), Fleschner (58), Franz (60), Howard (75, 76), Howard & Fiske (77), Huffaker (79, 81), Kirby & Spence (89), Knipling (90, 91), Köllar (93), Liu (99), Marchal (105), Nicholson (116), Ordish (121), Rabb & Guthrie (125), Silvestri (137, 138), Smith (142, 144–146, 148), Steinhaus (152), Stern et al (153), Swan (154), Sweetman (155), Telenga (158), Thompson (160, 162), Turnbull & Chant (169), Usinger (171), van den Bosch (172), van den Bosch & Messenger (173), Whitten (176), Woodworth (181), Zaslavskii & Sugonyaev (184).

AUSTRALIAN REGION

Australian entomologists have pioneered in many areas of biological control particularly in the field of weed control, ecology, population dynamics, and systematics of Hymenoptera. Australia has been a source of insect pests as well as a reservoir of important pest natural enemies which have been exported around the world. Fortunately because of its isolation many cosmopolitan insect pests have not reached Australia, but some old world pests were accidentally imported.

Biological Control of Insect and Mite Pests

Wilson (178) reviewed the development of biological control in Australia and New Guinea, and this is the source of much of what is included in the following comments. In 1901, Western Australia appointed George Compere as Entomologist to the Department of Agriculture. Compere was an avid supporter of biological control and was responsible for many introductions of natural enemies into Australia up to 1910 when he returned to California. As Wilson pointed out many of the insects introduced in these early days were unidentified and the work was poorly documented.

W. W. Froggat (1858–1937) a pioneer entomologist in Australia contributed greatly to Australian economic entomology. He was aware of the benefits of natural enemies in the control of insects, but like L. O. Howard, he was not overly enthusiastic about the biological control approach. He discounted many claims made by the California school after visiting California in 1907 (75). R. J. Tillyard (1881–1937) came to Australia from New Zealand where he had been very active in successfully introducing natural enemies against exotic pests. He was appointed in 1928 to take charge of the entomological work of the Commonwealth in the new capital center at Canberra. Biological control of weeds was Tillyard's particular interest in economic entomology and he encouraged research in this area. A. J. Nicholson (1895–1969) succeded Tillyard at Canberra. His interest in biological control arose through his studies on insect mimicry. He was struck by the powerful force that predators must have had upon the natural selection in the creation of mimics. His thoughts turned to the balance of nature and its processes. Subsequently many highly significant contributions to ecology were made. His contributions to biological control were more theoretical than applied yet Nicholson was aware of the influence of pesticides on beneficial insects (117).

Most of the biological control research in Australia has been undertaken by the Commonwealth and state authorities. The earliest introduction of a natural enemy was the coccinellid *Harmonia conformis* from eastern Australia into western Australia in 1896. There have been many more such intentional movements of eastern Australian natural enemies to the west. In all the years from 1896 to 1960, Wilson (178) concluded that useful results against Australian insect pests had been obtained in 12 out of 26 biological control attempts using natural enemies of exotic origin. Of 4 cosmopolitan pests some control was achieved against 3 of them. Of 6 native pests only 1 was considered as having some success. The latter was an introduction of *Gambusia affinis* in 1926 and with the help of some native predaceous fish some mosquito control was obtained.

Biological Control of Weeds

The most outstanding biological control project in Australia has been the great reduction of the prickly pear, *Opuntia* spp., by insects. This success is to weed control what the vedalia is to insect control. About 60 million acres of land were infested mainly in Queensland by the prickly pear in 1925.

The first *Opuntia* enemy was an introduction in 1787 of a cochineal insect from Argentina for dye production, but it is not known if it became established, according to Wilson. In 1903, a *Dactylopus* sp. from India was introduced purposely for destroying the cactus, and it became established. In 1913 more insects were introduced, but between 1921 and 1935 the main effort was made and 48 exotic insects were introduced from various parts of the world. *Cactoblastis cactorum,* a phyticid moth, was among the group of insects imported in 1925. By 1928 it became quite apparent that this species would bring about the destruction of the two main *Opuntia* spp. Dodd (37) fully described the campaign against the prickly pear in Queensland. Even today in areas of Queensland where once *Opuntia* thrived, the children of ranchers some 40 years later know the name *Cactoblastis* and what it has done.

Although not as spectacular, the weed St. Johnswort has been reduced in some areas particularly by imported European beetles. The chrysomelids introduced into California from Australia have spectacularly cleared pastures of the weed.

New Zealand and New Guinea

Predators and pollinators were introduced into New Zealand in the 1890s and since then there has been interest, but of varying intensity, in the biological approach to control of insect pests and weeds (63). W. M. Maskell (1849–1899), educated in England, came to New Zealand and was one of the first resident entomologists. He described some insects in New Zealand including the cottony cushion scale, *Icerya purchasi,* in 1878. Maskell provided the important fact to early California orchardists that *I. purchasi* was present in New Zealand. This information, plus that of Crawford's in Australia, narrowed the territory down for foreign exploration carved out by Koebele in 1888 when the famous vedalia was found in Australia and sent to California (48, 75).

T. W. Kirk (1890s), the first government entomologist, was responsible for the introduction of many beneficial insects into New Zealand including the vedalia beetle against *I. icerya* and *Rhizobius ventralis* against *Eriococcus coriaceus* from Australia and bumble bees for pollination from England. R. J. Tillyard who was in charge of Cawthron Research Institute at Nelson from its inception in 1920 was interested in biological control. He imported *Aphelinus mali* from the United States against *Eriosoma* with success. He was also interested in the control of weeds (75). Tillyard, together with D. Miller in the early 1920s, developed a rather intensive biological control research. This lasted into the late 1930s, by which time the Entomology Division of D.S.I.R. was involved in conjunction with Cawthron Institute entomological staff as the Entomological Research Station. There have been 10 highly successful insect pest control projects where natural enemies have been imported; also a number of introduced pests have come under control by endemic parasitic Hymenoptera without human intervention (63). Biological control of weeds has received considerable attention through the years but without much success. St. Johnswort has been controlled in some areas.

In New Guinea successes of biological control with imported natural enemies have been reported against one cosmopolitan pest (of two attempted), the sweet potato hawk moth by the giant toad, and against three native pests (of five attempted), coconut grasshoppers, coconut leaf miner, and mosquitoes (178).

ETHIOPIAN REGION

About one half of the mainland countries of the Ethiopian region have practiced biological control, but only a few countries have consistently been active in this field. Some associated islands in both the Atlantic and Indian Oceans have considered biological control as the main approach to solving pest problems. Greathead (69) has recently reviewed the biological control attempts in Africa. Africa has been a source of some important natural enemies exported around the world, but it is surprising that for its size and the diversity of its biota that this region has not been the main source of agricultural pests to other lands; thus it has not been a great reservoir to tap for natural enemies. The *Oryctes* spp. of Africa and the giant African snail has attracted many explorers to Africa seeking natural enemies of these two exported pests.

Indian Ocean Islands

The island Mauritius has been particularly active through the years and was the location of the first international movement of a natural enemy against an insect pest. The Indian myna bird, *Acridotheres tristis,* was imported from India by Compte de Maudave in 1762 to combat the red locust, *Nomadacris septemfasia.* This introduction is credited with reducing the locust outbreaks (112).

According to Greathead (69) the first government entomologist D. d'Emmerez de Charmoy attempted solving pest problems using natural enemies and succeeded in controlling an *Oryctes* sp. after establishing a *Scolia* sp. which was one of 64 species introduced between 1915–1939 against the scarab. During this period *Aspidiotus destructor* was controlled by imported coccinellids, and *Opuntia* spp. were strikingly reduced by various *Dactylopius* spp.

Biological control attempts in Madagascar began in 1914 with the introduction of *Gambusia* against *Anopheles* larvae and *Dactylopius* against

Opuntia spp. in 1925. From 1949 natural enemies were imported against six agricultural insect pests, the giant African snail, and against *Anopheles* larvae; two of the insect pests were controlled and a predaceous fish was established (69).

Western Africa

Other than in Ghana very little has been done using the classical biological control approach. In Ghana, the pest insects of cocoa have received the most attention, with the endemic mealybugs which are vectors of swollen shoot virus being the main target. Twenty-nine imported natural enemies were released against the mealybugs from the late 1940s into the 1950s but no effective control was obtained (69).

Eastern Africa

Kenya, Tanzania, and Uganda have often turned to the biological control method to solve their pest problems. Beginning in the early 1920s through the 1930s about 40 exotic coffee mealybug natural enemies were imported from abroad, but it was not until the late 1930s when a proper identification was made of the coffee mealybug, *Planococcus kenyae,* that the origin was narrowed by Le Pelley (95) to Uganda, NW Tanzania, and the Congo. A search here yielded two encyrtid parasites, *Anagyrus* spp, and by 1940, widespread control was obtained. Later several coccinellids from the endemic center of the mealybug were introduced and established. Greathead (69) concluded that the natural spread of the nearby African parasites was restricted since they were confined to an ecological "island" and could not spread naturally to the coffee growing areas. Melville (107a) estimated that over 20 million dollars had been saved at a cost of less than 70,000 dollars expended for the applied research. Of 21 other agricultural pests that were subjected to biological control attempts, 6 were successful.

Southern Central Africa

Since 1892, with the introduction of the vedalia against the cottony cushion scale, South Africa has consistently used the biological control approach against many of their pests. A similar history of reactions by some entomologists occurred in South Africa after the vedalia success, as it did in the U.S. C. P. Lounsbury (100), the first government entomologist, deplored, as did L. O. Howard, the reckless introductions of many coccinellids against an assortment of pests which mainly all failed, and Lounsbury was greatly relieved when the craze for importation of ladybirds died down. It was not until the 1920s that more serious attempts were made with imported biological control agents in South Africa and some successes were achieved. *Aphelinus mali* sent from the U.S. controlled *Eriosoma lanigerum.* Later this parasite was sent from South Africa and successfully established in many of the other African countries.

One of the most outstanding projects was the importation of an egg parasite, *Patasson nitens,* found in Australia by F. G. C. Tooke in 1926 and released against the eucalyptus snout beetle, *Gonipterus scutellatus,* in South Africa. By 1935, the weevil was controlled in most areas (164). The parasite has been sent with success from South Africa to many countries in Africa and to its islands where the pest occurs.

The biological control of weeds has been a major area of research in South Africa since 1913. The program against various *Opuntia* spp. has been emphasized. The introduction of various *Dactylopius* spp. has been the main effective natural enemy used. *Cryptolaemus montrouzieri,* introduced successfully from Australia against the citrus mealybug as well as an endemic coccinellid, moved over in some areas to feed on the *Dactylopius.* This had reduced the impact of these coccids on the suppression of the *Opuntia* (69).

The first full-time biological control entomologist in South Africa was G. C. Ullyett who began his work in 1936 and who was a pioneer in many areas of biological control including the development of mathematical models of parasite-host systems. South Africa has also supported additional biological control workers who have contributed extensively to the knowledge of the entomological fauna of South Africa as well as to that of the Ethiopian Region.

NEARCTIC REGION

It is not surprising that practical application of biological control evolved most rapidly in the New World particularly in North America. Canada pioneered mainly in attacking forest pests while in the U.S. the early economic entomologists were concerned about controlling agricultural insect pests. Immigrating farmers and missionaries from the Old World brought with them seeds, fruits, and cuttings of many agricultural plants. Inadvertently the overwintering eggs as well as other diapausing stages of spider mites, aphids, scales, Lepidoptera, and Diptera were also introduced along with the plant material.

United States

As the new crops developed so did the uninvited guests, and very few natural enemies arrived along with the pests. The pest explosions in fruit trees and grain increased in magnitude through the years as land was further cleared and as the tilled fields coalesced. To aid the farmers, various state entomologists were appointed first in the East with Asa Fitch (1809–1879) in 1854 at New York; followed by the appointment of Benjamin D. Walsh (1808–1870) in the Midwest at Illinois in 1866 and C. V. Riley (1843–1895) at Missouri in 1868. Finally in the Far West in California (1881) Matthew Cooke (1829–1887) was named head of a horticultural commission to combat insect pests. In the meantime, the federal government had formed special agricultural commissions directed to handle insect problems, and in 1863 the first U.S. entomologist was appointed (48, 75, 77, 104).

The outbreaks of the wheat midge *Sitodiplosis mosellana* in the East stirred Fitch to request parasites of the midge from England, for he realized that the midge occurred in the Old World and possessed parasites there. This approach was also tried by C. J. S. Bethune of Canada in 1864, for the midge was also prevalent in northern wheat fields. Walsh also sought support to introduce the parasites of the midge from governmental assemblies. In describing his failures to provoke interest in importing natural enemies, Walsh wrote in the *Practical Entomologist,* that the introduction of exotic natural enemies will be the control method used by future generations. Walsh pointed out that over one half the insect pest species attacking crops at that time were of foreign origin. However, just a few years later, Walsh's protege, Riley, accomplished the first international movement of entomophagous mites and insects. While at Missouri Riley in 1873 aided in the export of a predatory mite *Tyroglyphus phylloxerae* to France for the control of the grape phyloxera which had come from the New World. The first importation of an insect parasite into the U.S. was also engineered by Riley in 1883, when he successfully established *Apanteles glomeratus* against the cabbage butterfly.

In California, the diversity and expanse of new crops grew rapidly and these crops were being plagued by many insect pests most of which were of foreign origin like the crops they attacked. Cooke (29) a self-trained entomologist did not assign much importance to natural enemies. However, his contemporaries, mostly orchardists who served on Cooke's commission to fight insect pests, Felix Gillet, S. F. Chapin, and Elwood Cooper advocated natural enemies. Gillet suggested in 1881 that "stations" be established in the state to culture parasites by the millions to release against insect pests.

The introduction of the vedalia.—In the campaign against the cottony cushion scale, Cooke, Chapin, and Cooper were appointed in 1882 to a special committee of the Horticultural Board to consider ways of controlling the cottony-cushion scale, *Icerya purchasi,* which was beginning to play havoc in the citrus groves of southern California. Riley suspected that the scale was of Australian origin since he was aware of W. G. Klee's correspondence with W. M. Maskell in New Zealand and F. Crawford in Australia. Maskell wrote that it did originate in Australia and that it was being called the "Australian Bug" by the people in the region of Cape of Good Hope where the scale had invaded.

Riley was invited to speak at a convention of fruit growers meeting in California in 1887. He outlined a plan of attack aimed at controlling the scale pest, including the desirability of introducing from Australia parasites which were holding the scale in check. Albert Koebele (1852–1924) and D. W. Coquillett (1856–1911) were Riley's two field agents in California. Money was obtained to support Koebele's search for natural enemies in Australia in 1888. Koebele quickly found a dipterous parasite *Cryptochaetum iceryae* and a small lady beetle attacking the scale in the Adelaide area. The lady beetle was first described in the genus *Vedalia,* and today it is known as *Rodolia cardinalis.*

A total of 129 vedalia beetles were sent to California between November 1888 and January 1889 by Koebele to Coquillett who placed the beetles on a small infested citrus tree enclosed in a mesh tent. By early April the scales were gone on the caged tree and the beetles were allowed to invade the surrounding citrus trees. By early June over 10,000 beetles were collected and distributed in various southern California groves. By October 1889, the vedalia had cleared the scale pest from hundreds of acres of citrus, and the shipment of oranges from Los Angeles county went from 700 to 2000 railway cars. *Cryptochaetum* also became established and today is an important enemy of the scale along the California coast (39).

The news of the success of the vedalia campaign reached around the world, and the countries that had been invaded by *I. purchasi* requested the vedalia from California. In most of these new release sites, the predator repeated its spectacular performance. Through the years and even today new introductions and reintroductions of the vedalia occur.

The use of DDT to control citrus thrips in the 1940s in California quickly demonstrated that the cottony cushion scale was still a potential pest for it flared into damaging numbers. As the vedalia was being killed by the DDT, this clearly pointed to a way of determining the importance of natural enemies through the use of insecticides that selectively kill parasites or predators but not the pest, and DeBach (32) labeled this technique as the insecticidal check method. The California citrus growers after having the cottony-cushion scale resurge from DDT treatments for thrips gladly paid one dollar apiece for a vedalia beetle.

Ladybird beetle era.—Koebele returned from Australia as a hero in 1889, and between the USDA and California, Koebele was contracted to return to Australia to search for more natural enemies. He was so impressed with the potential of lady beetles as predators that he collected and sent 46 species from Australia to California on his second trip in 1891–1892; but only four species became established. One of these *Cryptolaemus montrouzieri* played an important role in the development of the mass culture and periodic colonization a few years later.

The discovery of huge overwintering aggregations of *Hippodamia convergens* in the Sierra Nevada led the state entomologists to collect the beetles by the ton and to transport them to the valleys for aphid control (19). The findings of Davidson (30) indicated no value from the releases because of dispersal.

L. O. Howard (1857–1950) who had succeeded Riley as bureau chief was critical of the California entomological schemes. He was concerned about the reckless handling of importations and would have stopped the state from importing insects if it were not for the appointment of H. S. Smith in 1913 in charge of importations for California. It was Howard who brought integrity to the use of natural enemies through his research and selection of workers.

At the University of California at Berkeley the first entomology professor appointed was C. W. Woodworth in 1891. He was influenced by S. A. Forbes at Illinois, whom he assisted, and learned the concept of the ecosystem in nature. Woodworth (181) was apparently first to state that regulation of populations is through agents that change as the population density changes.

Gypsy moth campaign.—The biological control approach during this campaign developed basic ecological concepts, new biological information of entomophagous insects, systematics of tachinids, ichneumonids, and chalcidoids, and by no means least of all, the development of men who became leaders and teachers.

Howard & Fiske (77) described the evolution of the gypsy moth project since the introduction of the moth into Massachusetts. A detailed report by Forbush & Fernald (59) described the introduction and spread of the gypsy moth. According to Clausen (23) the importation of natural enemies from Europe and Japan from 1905 to 1914, renewed in 1922–1923, involved 40 species of parasites and predators being released in eastern U.S. Of these, 9 parasites and 2 predators have become established. Up to 1927 a total of about 92 million parasites had been released. Clausen (23) believed that the combined effect of the natural enemies has given appreciable control.

The greatest contribution to come from the gypsy moth work was the concept that "a natural balance can only be maintained through the generation of facultative agencies, which effect the destruction of a greater proportionate number of individuals as the insect in question increases in abundance." Smith (145) agreed with Howard and Fiske but proposed that their facultative agencies be called "density dependent factors". Furthermore the entomologists who were influenced by the research conducted in the gypsy moth laboratory at Melrose Highlands in Massachusetts in the early 1900s were Howard, Chief of the Bureau of Entomology, Fiske, in charge of the gypsy moth laboratory, and the following scientific investigators: A. F. Burgess in charge of ground beetles, H. S. Smith and P. H. Timberlake assigned to work on the biology of parasitic Hymenoptera, and C. H. T. Townsend and W. R. Thompson assigned to work on the biology of the dipterous parasites.

California Biological Control

Harry S. Smith (1883–1957) was appointed in 1913 superintendent of the state insectary at Sacramento, after leaving the USDA gypsy moth laboratory in Massachusetts. In 1923, the biological control work of the State of California was transferred to the University of California's Citrus Experiment Station at Riverside and Smith retained the responsibility of importation and screening of natural enemies into the State.

At Riverside a new quarantine laboratory-insectary facility was constructed in 1929. In 1945 Smith organized another biological control laboratory at Albany which was part of the state-wide Department of Biological

Control of the University of California. At the Albany laboratory and insectary Smith directed a biological control campaign against the oriental fruit moth and against the Klamath weed, *Hypericum perforatum,* with C. B. Huffaker and J. K. Holloway as the investigators. Smith also organized a laboratory of insect pathology with E. A. Steinhaus in charge.

Smith attempted to build a staff of entomologists that would quantify field evaluations of established and endemic natural enemies, that would delve into population ecology, that appreciated systematics, and that would investigate biology and physiology of entomophagous insects. Smith pursued all these areas himself. He was first to propose the term "biological control" (142) and interpreted Howard and Fiske's population regulation concept, proposing terms of density-dependence such as density dependent, density independent, and inverse density dependent mortality factors (144). He (143) also defined multiple and superparasitism. "Prof. Harry" known to his colleagues and students left an indelible mark in the advancement of entomology and ecology, and wove these two fields together in developing a science of applied ecology called biological control.

The contributions made by the Department of Biological Control in the Smith era were many, including complete biological control of several serious insect pests and development of mass culture techniques of both hosts and natural enemies, new ecological and biological relationships were discovered, the first attack on weeds in the U.S. using insects was initiated, and an insect pathology laboratory was formed.

C. P. Clausen former head of the biological control investigations for the USDA became chairman of the University of California's Department of Biological Control in 1951 and retired in 1959. During this period biological control of aphids was initiated, coccid enemy importations continued, and spider mite enemies were given much attention; it was the dawn of integrated control in California.

Biological control of arthropods.—The outstanding biological control achieved by using imported natural enemies in California was the complete control of several coccids, the citrophilus mealybug, the black scale along the coast of California, and the nigra scale. Various *Aphytis* spp. were imported in the late 1940s from many different countries and released against several diaspine scales, olive parlatoria, fig scale, and California red scale; these gave good control and when they were supplemented by other parasites or replaced by other strains of *Aphytis* economical control eventually was achieved, indicating again the value of multiple releases (33, 35). Imported *Aphytis maculicornis* alone greatly reduced *Parlatoria oleae* on olives in California, but the introduction of another aphelinid *Coccophagoides utilis* added enough mortality of the scale to make this another case of complete biological control in groves not receiving insecticides (84).

Beginning in the 1950s biological control was attempted against various aphids and much success has been achieved. The aphids still offer much opportunity for biological control. Like in the *Aphytis* the recent work with strains of aphid parasites points out again that geographical areas to be searched for natural enemies should have a similar climatic picture to the area of the proposed introduction (172).

Biological control of weeds.—Cooperation between J. K. Holloway of the USDA and C. B. Huffaker achieved phenomenal control of Klamath weed in California and later in other western states from imported species of *Chrysolina* beetles from Australia (74). Huffaker (80) reviewed the status of the biological control of weeds.

Mass culture and periodic colonization.—*Cryptolaemus montrouzieri* showed promise in controlling the citrus mealybug, but each winter the beetles only survived in the extreme southern coastal part of California. Smith & Armitage (147) developed a mass culture technique for the beetles utilizing mealybugs which in turn were cultured on potato sprouts. Periodic colonizations of the beetles were made each spring in the citrus orchards to control various mealybugs. Thus, *C. montrouzieri* was one of the first insect natural enemies to be manipulated in this manner. Both county and citrus grower groups undertook to build insectaries and release *Cryptolaemus* in citrus orchards by the millions each year. There were 15 organizations in the 1920s in California engaged in beetle production (23). Even today there are associations of citrus growers that not only mass culture *Cryptolaemus* but also mass produce other parasites of scales and mealybugs (51); thus what Gillet proposed in the late 1800s became a reality. Furthermore, there are today an increasing number of private commercial insectaries that sell various natural enemies such as *Trichogramma* and *Chrysopa carnea* for use against various pests.

The mass culture and periodic colonization techniques comparable to the magnitude of *Cryptolaemus* production were developed for *Trichogramma* spp. by Flanders (52), for *Macrocentrus ancylivorus* (50), and for *Aphytis* spp. against California red scale (33, 35, 54). *Chrysopa carnea* larvae were mass cultured on coddled tuber worm larvae (49) and eggs were obtained by artificial diet (71, 72); *Chrysopa* eggs were placed in pear orchards for mealybug control (42). The first mass culture techniques of fruit flies on artificial diets and their parasites were developed by Finney and Hagen under Smith's direction in 1950 in Hawaii.

The biological control of spider mites began to receive much attention in the late 1940s with mite outbreaks associated with use of persistent hydrocarbon pesticides. The importance of phytoseiid mites was demonstrated by Huffaker & Kennett (82, 83) by inoculating new strawberry plantings with both the prey-mite and predator-mite. Control was achieved for three years while only in the third year in normal practice would the typhlodromid/prey ratio be effective.

Removal of mite predators by hand picking showed the importance of predators controlling avocado brown mites (57).

Augmentation of natural enemies in the field.—The conservation, augmentation, and manipulation of native and established natural enemies is an area in biological control receiving much attention recently, and in an area of increasing monocultures, the need of manipulating parasites and predators becomes greater. A review of the various efforts up to 1962 is in DeBach (33).

Supervised control.—The routine monitoring of pest populations and their natural enemies in fields and prescribing to the grower what, if any, control is needed is called supervised control. This was first developed by Ray F. Smith of the Division of Entomology at Berkeley in 1946, and was based on work done by Michelbacher & Smith (109) on the role of a key endemic parasite, *Apanteles medicaginis,* of the alfalfa caterpillar. The establishment of an economic injury level of the caterpillar, coupled with an easy caterpillar and parasite sampling technique permitted predicting whether or not damage by the caterpillar would occur one or two weeks later, and appropriate action was recommended (149). This often prevented unnecessary insecticide applications. Supervised control evolved into integrated control procedures.

Biology.—The host parasite relationship involving autoparasitism or adelpho-parasitism was discovered by S. E. Flanders in 1937. The knowledge of such interrelationships has made establishment of certain complex parasitic Hymenoptera possible (53).

USDA Biological Control Since 1910

Besides the gypsy moth and brown tail moth there are about 10,000 species of insects and mites in the United States, that have some degree of importance as pests. Today about 700 could be considered important pests and of these pests 212 are of foreign origin (131). It is the 212 exotic pests that biological control workers aim to contain using the classical method of importing their enemies. After 80 years some 128 beneficial species have been successfully imported. Thus, there is much to be done in the classical biological control approach in the U.S., not to mention the possibilities of augmentation and manipulation of endemic natural enemies.

After the USDA gypsy moth and brown tail moth program ended, biological control work was carried out by several independent divisions concerned with different insect pests. During 1911 to 1913 parasites of the alfalfa weevil *Hypera postica* and *H. punctata* found in Italy were imported. Among the parasites that became established in Utah and later (1933) in California was *Bathyplectes curculionis.* In coastal valleys of California this parasite does a spectacular job but in the Central Valley parasitism drops off, being restricted by climate. In Utah and Colorado today it is not as effective.

The European corn borer discovered in Massachusetts in 1917 invoked biological control attempts in 1919 when a laboratory was set up in France in which H. L. Parker and W. R. Thompson (1887–1972) screened for natural enemies; later the Orient was explored. This large-scale program was concluded in 1938. Of 24 species of parasites imported, 6 became established, but each of these parasites became rather limited in its distribution and rarely became abundant enough to effect control (7).

The next significant effort was the attack on the Japanese beetle. Foreign exploration took place in the Orient from 1920 until 1933 mainly by C. P. Clausen, and 15 species of parasites were introduced, 5 became established. It was difficult to determine the impact of these 5 natural enemies, but 2 *Bacillus* spp. inadvertently introduced (milky disease), killed the grubs, and exerted good control (23).

The European earwig invaded Portland, Oregon, about 1924. A search in Europe revealed two tachinids which were imported. *Bigonicheta spinipennis* became established in the northwestern states, and the city of Portland cultured and released the tachinids for years (23). In the 1930s Flanders introduced *B. spinipennis* into the San Francisco Bay Area. The tachinid was recovered from many locations along the central California coast in 1962. Although the parasitism rarely exceeds 18% there seems to have been a general suppression of the earwigs in the San Francisco Bay Area (K. S. Hagen, unpublished data).

Under C. L. Marlatt *Chilocorus similis* was introduced from China for the San Jose scale. It became established temporarily in Georgia. A *Chilocorus* similar to *similis* has been recovered in southern California. The arrival of the oriental fruit moth in eastern U.S. became a target for H. W. Allen, J. K. Holloway, and G. J. Haeussler. Of 20 species of parasites colonized, only 2 became established and they were not very effective. The unique development in this project was the move of endemic *Macrocentrus ancylivorus* on to the oriental fruit moth and its development as an effective parasite of the introduced pest (23).

In 1934, a Division of Foreign Parasite Introduction was organized with responsibility for foreign investigations and for quarantine handling of imported material. The screened natural enemies were turned over to various field stations or state cooperators. There were also entomologists in France and Japan who searched for parasites. After World War II, the Division of Foreign Parasite Introduction, headed by C. P. Clausen, was reduced in size with entomologists located at five stations: Beltsville, Maryland; Moorestown, New Jersey; Albany, California; Paris, France; and Mexico City, Mexico (131). During this period in the U.S. the use of insecticides had reduced most biological control research except in California. The small biological control group of the USDA in 1951 was placed under the Insect Identification and Parasite Introduction Branch of Entomology Research Division.

Clausen (23) summarized the biological control of insect pests of agricultural crops in the continental United States over a 60-year period. Among the 95 species of insects that were established 81 were parasitic in habit and 14 predaceous. Approximately 390 species were imported and released in varying numbers, but failed to establish themselves. Dowden (43) made a similar summary as Clausen's but listed the 88 parasites and predators that were liberated against 33 forest insect pests from the turn of the century up to 1960. Of the 44 species that became established, 36 were parasites and 8 predators.

During the past 80 years entomologists of the USDA, University of California, and Hawaii have imported at least 420 beneficial species and have attempted to colonize them. Of these, 128 are now established in the continental United States, and 60 insect pests and 10 weeds have been the target of this work. The result of these introductions is complete or substantial control of at least 15 insect pests and 2 weeds is involved (131).

Biological control of weeds.—The first weed for which insects were imported was the Klamath weed. This importation was extremely successful in the western states (74).

Periodic releases and augmentation.—The first large-scale movement of parasites took place in 1907 when Hunter of the University of Kansas collected millions of green bug, *Schizaphis graminum,* parasitized by *Aphidius testaceipes* early in the season in southern Kansas and transferred them into northern Kansas where the aphid population had not yet exploded. Success is claimed using this method by Hunter (85). Mass releases of *Trichogramma* have been carried out since the days of Howard and Fiske. This technique received a great boost when Flanders (52) developed a simple mass culture technique. Since that time *Trichogramma* have been used against many pest Lepidoptera. The control achieved is questionable. (See DeBach 33.)

Canada

Biological control in Canada has been consistently supported since the early 1930s, and many basic contributions have been made to the theory and practice of biological control. While research and support of biological control dwindled in the 1940s and 1950s in most of the world at the height of the persistent pesticide period, Canada retained a vigorous effort in many phases involving biological control. The biological control programs conducted in Canada are nicely summarized in Technical Communications No. 1 and 4 published by the Commonwealth Agricultural Bureau in 1962 and 1971, and by Turnbull & Chant (169). Recently LeRoux (96) described the biological control attempts on pome fruits.

Biological control in Canada has largely been directed at forest pests, but the earliest attempts were made against agricultural pests, pioneered by amateur entomologists such as Reverend C. T. S. Bethune who in 1864 suggested that parasites be imported into Canada and W. Saunders who in 1882 made the first introduction of an entomophagous insect into Canada (a *Trichogramma* sp. released against the imported currantworm). James Fletcher, the first Dominion Entomologist (1884–1908) and C. G. Hewitt from 1910 made additional and successful introductions. Both Fletcher and Hewitt were given assistance and encouragement by Howard. By 1919, biological control became an established method of control in Canada (6, 64, 169).

In 1915 a small Natural Control Investigations Laboratory was built at Fredericton where J. D. Tothil, A. B. Baird, A. E. Dustan, and 7 other entomologists worked. The fall webworm, tent caterpillars, and spruce budworm and later the oyster shell scale, the cherry tortrix, and the larch sawfly were some of the target insects that natural enemies were being introduced via the Fredericton laboratory to control.

The function of the Fredericton laboratory was transferred in 1923 to Ontario. Here parasites were introduced against the greenhouse whitefly, the European earwig, and the ornamental fruit moth. By 1929 the biological control center was located at Belleville and was renamed the Dominion Parasite Laboratory.

According to Baird (6) the year 1933 saw the beginning of a new era in biological control with an all-out effort directed by J. M. Swaine to curb a serious infestation in eastern North America by the European spruce sawfly, *Diprion hercyniae*. It was a joint effort between Canada, Farnham House Laboratory, and the United States Bureau of Entomology. A 40-room insectary was constructed at Belleville in 1936 with quarantine facilities. Parasites were imported and cultured by the millions and distributed throughout Canada. The campaign was very successful in that the introduced parasites together with a virus disease fortuitously introduced and spread by the parasite brought the European spruce sawfly under control (6).

In 1940, the Farnham House Laboratory in England was closed and the Superintendent, W. R. Thompson, was transferred to Belleville. During the war years Canada provided the facilities and services to the Farnham House service which was reorganized as the Commonwealth Institute of Biological Control.

The Belleville laboratories expanded in 1955 to one of the largest biological control centers in the world, and was renamed as the Entomology Research Institute for Biological Control and was part of the Research Branch, Canada Department of Agriculture. At this time Baird as head of Belleville was succeeded by Brian Beirne. There were about 40 workers associated with Belleville at this time. In the period 1910–1956, nearly a billion individuals of some 220 species of insect parasites and predators were released against 68 species of insect pests; 4 phytophagous insects were released against a weed. More than 50 beneficial species became established, and at least 20 species were important in the control of 26 pest species (6).

Biological control of insect pests of crops and ornamentals.—From 1882 up to 1959 biological control was attempted against 23 agricultural pest insects releasing 86 natural enemies, mainly parasitic Hymenoptera (102).

From 1959 to 1968 biological control was attempted against 39 agricultural pest situations. Intentional introductions of parasites and/or predators were made against 11 species with 5 becoming established but only 2 with any marked success (88).

Biological control of forest insect pests.—Balch (8) in the Forest Biology Division sums up biological control results as positive. Between 1910 and 1958 releases of natural enemies were directed against 36 pests. Of these 16 were of foreign origin whereas 20 were endemic to North America. The pests included 24 defoliators, 4 scale or sucking insects, 4 needle or leaf miners, and 4 bark beetles. A total of 104 species was brought to Canada. These species consisted of 89 parasites and 15 predators. Of the species released, 35 were recovered. Of these, 12 species were widely established and exerted a measurable degree of control; 10 species were established but their control value was doubtful. The other 12 species were established but not evaluated (101).

Biological control of weeds.—The first weeds attacked in Canada using insects were the Klamath weed in 1951 and toadflax in 1957 (102). By 1968, six candidate weed pests were targets of insects released against them. The Klamath weed control is considered a complete control, and also complete control of the tansy ragwort was achieved by the cinnabar moth. Only slight reduction of toadflax has been accomplished by an introduced noctuid. Two thistles, as target weeds, have been only slightly reduced by introduced insects (72b).

Integrated control.—One of the first demonstrations of integrated control was in the apple orchards of Nova Scotia. The control strategy was engineered by A. D. Pickett (123a). Stopping all sprays in 1943 in blocks of apples he was able to demonstrate the power of natural biological control. When forced to treat for apple scab or codling moth he used proven selected pesticides. This pest control strategy has become extensive and is practiced in other agricultural areas in Canada (103) and in the world.

MEXICO

Biological control in Mexico is a relatively new approach to insect control. The most outstanding project was one initiated against the citrus blackfly, *Aleurocanthus woglumi.* This pest is of Asiatic origin and initial biological control against this pest began in the West Indies and natural enemies were first introduced into Cuba. The discovery of the pest in Mexico was made in 1935 and the blackfly spread rapidly. In 1943 a project was undertaken by the Mexican Department of Agriculture and the USDA for the importation of *Eretmocerus serius,* and it became established and controlled the pest in some areas. A further search for parasites was made in tropical Asia and 4 additional parasites were found and were established in Mexico. *Amitus hesperidum* became the most effective parasite. In different areas of Mexico different parasite species dominated (23). During the late 1950s a newly organized Departmento de Control Biologico of the Defensa Agricola carried out an extensive and successful program of colonization with the parasites (24, 56, 136).

NEOTROPICAL REGION

The attempts to use natural enemies in the South and Central American countries are scattered and uneven. After the vedalia success in California there were shipments of the vedalia made to citrus growing areas and later *Aphelinus mali* was shipped from the USA against the woolly apple aphid. Only in recent years have there been Departments of Biological Control organized. Bennett (11) reviewed the biological control of sugar cane borers in both the West Indies and South America; some tachinid introductions have provided significant control.

South America

Argentina.—The first studies of natural enemies of insect pests were made in 1897 when L. Bruner of the University of Nebraska and the French entomologist J. Künekel d'Herculais studied the possibility of combating the white peach scale by means of natural enemies. There were five attempts made to establish introduced natural enemies up to 1932; three were successful, among them *Prospaltella berlesei* against the white peach scale in 1908, *A. mali* controlled the woolly apple aphid by 1922, and *Rodolia cardinalis* against the cottony cushion scale in 1932 (33, 75).

Brazil.—Economic entomologists in Brazil seem to have relied mainly on the use of pesticides to control agricultural pests, but some workers are concerned about insecticide resistance and the impact of pesticides on natural enemies (65, 94, 107, 114). Ciociola (20) and Gomes (65) presented a history of biological control in Brazil.

There have been relatively few natural enemies imported into Brazil but there have been some outstanding successes. *Aphelinus mali* introduced in 1923 gave substantial control of the woolly apple aphid, and *P. berlesei* has achieved complete control of the white peach scale in 1921 (33). Gomes (65) discussed the introduction of *Prorops nasuta* from Africa in 1934 to control the coffee berry borer. The parasite became established, but did not reduce the pest density satisfactorily. *Tetrastichus giffardianus* was established against *Ceratitus capitata* in 1937 but with poor results (5). Three species of tachinids have been released against the sugar cane borer and are still being used (61, 66, 150). Oliveira (120) reported relative success using *Bruchobius latisceps* against *Acanthoscelides obsoletus.*

Chile.—A history of biological control in Chile has been written by Graf-Marin & Cortes-Peña (68) and by Gonzalez & Rojas (67). Interestingly the first natural enemies to be introduced into Chile were made by a private citizen T. Schneider in 1903 when he introduced *Hippodamia convergens* and *Rhizobius ventralis* from California. Two pteromalid parasites of *Scolytus rugulosus* were imported in 1915 and apparently they did not become established. Like the rest of South America, Chile imported *A. mali* against *E. lanigerum* with success in 1922, and in 1929 when invaded by *I. purchasi,* the vedalia and *Cryptochaetum iceryae* were introduced in 1931. The *Rodolia* not only completely controlled *I. purchasi* but also *I. palmeri.* Five chalcidoid parasites were imported against *S. oleae* in 1933 with no success until *Metaphycus helvolus* was introduced from Peru in 1946 with partial control resulting (86). Both the citrus mealybug and the longtailed mealybug were brought under substantial control by a complex of parasites and predators imported from California during the 1930s. Graf-Marin & Cortes-Peña (68) include a bibliography of 104 references dealing with biological control and systematics of parasitic Hymenoptera. Caltagirone (18) lists the entomophagous insects and their hosts found in Chile.

Colombia.—Biological control in Colombia seems to have been practiced at levels comparable to Venezuela. In 1933, *A. mali* was introduced from the U.S. and complete control of *Eriosoma lanigerum* was obtained (33). Recent attempts to control *Diatraea saccharalis* have involved the introduction and mass release of the Peruvian race of *Paratheresia claripalpis* which has a shorter life cycle than the native race (11).

Ecuador.—Substantial control of *Icerya montserratensis* was obtained from the introduction of *Rodolia cardinalis* from the U.S. in 1941 (33).

Guyana.—According to Myers (113, 114) egg masses of *Diatraea* moths were collected and the egg parasites *Trichogramma* and *Telenomus alecto* that emerged were used for release or for a stock culture for mass rearing on *Sitotroga* eggs, one of the first examples of mass culture of a *Trichogramma* on a factitious host. This was overshadowed by the introduction of the tachinid *Metagonistylum minense,* commonly known as the amazon fly, from Brazil (26). The importance of floral pollen being available for the adult parasitic wasps was stressed by Myers (113, 114).

Peru.—In 1909, C. H. T. Townsend, the famous dipterist, who had worked for the Federal Bureau of Entomology was appointed Entomologist and Director of Entomological Stations of Peru. After 4 years, he returned to the United States and worked in the Gypsy Moth Parasite Laboratory for 4 years; thence to Sao Paulo, Brazil for 3 years, returning to Peru where he worked on cotton insects, and in 1926 became Chief Entomologist of Agricultural Experiment Station of Lima. The cotton growers association of the Canete Valley brought W. E. Hinds from the Louisiana Experiment Station to Peru in 1925, and Hinds suggested that the dusting of arsenicals be done by aircraft which was done in 1926. G. N. Wolcott and Townsend, according to Howard (75), had much impact on economic entomology in South America.

Johannes E. Wille who became chief entomologist of Peru wrote a book in 1943 and a second revised edition in 1952 on agricultural entomology of Peru. He emphasized the importance of natural enemies of agricultural pests. In 1956, Wille (177) reviewed biological control in Peru and the following successes were noted. The white scale on cotton was controlled by natural enemies introduced in 1904. The woolly apple aphid was controlled by *A. mali,* and *I. purchasi* by *Rodolia.* The black scale was controlled by three imported parasites from the U.S. Peru is one of the few areas in the world where *Hippodamia convergens* has been introduced and established, and now has spread into Chile. The culture and release of the native tachinid *Paratheresia claripalpis* gave good control of *Diatraea saccharalis.* Wille's studies of predation in cotton fields revealed the value of predaceous anthocorids and mirids coming from corn plants that were planted in cotton fields. The corn also acted as a trap plant for *Argyrotaenia sphaleropa* which can attack cotton but the parasites were so effective on controlling the pest in corn that cotton is hardly attacked. Hambleton (72a) also pointed out the diversity and importance of natural enemies in cotton.

There is a long dynamic history concerning pest control in Peruvian cotton varying from no pesticides to the most intensive use of pesticides on cotton in the world and finally evolving into an integrated control program where selective pesticides are used variably depending upon insect abundance (10).

Uruguay.—According to DeBach (33) complete biological control of the cottony cushion scale by *Rodolia* occurred in Uruguay in 1916, and complete control of the woolly apple aphid by *A. mali* was achieved in 1921 and complete control of the white peach scale by *Prospaltella berlesei* in 1912.

Venezuela.—It appears that little has been published on biological control attempts in Venezuela. According to DeBach (33) both *A. mali* and *Rodolia cardinalis* had been introduced in 1941 and the vedalia completely controlled *I. purchasi. Aphelinus mali* has given partial control of the woolly apple aphid. Myers (114) describes agricultural ecology; and Guaglumi (70) lists the insects and their hosts. H. E. Box used the cane borer parasites *Metagonistylum minense* and *Paratheresia claripalpis* from Trinidad, Mexico, Venezuela, and Peru. The geographical populations of the latter were crossed and the resulting hybrid from Venezuela × Trinidad parents appeared to be more vigorous and possibly better suited to Venezuela (24). Although *P. claripalpis* showed initial promise, *M. minense* subsequently proved to be the effective parasite (17).

Central America

Only in recent years have there been any recorded biological control attempts in Central America. The most outstanding case is the complete biological control of the citrus blackfly, *Aleurocanthus woglumi,* by *Eretmocerus serius* in Costa Rica and Panama in 1932. The parasite originally collected in Malaya was established in Cuba first and later distributed widely. In recent years there has been an increasing interest in biological control and integrated control.

West Indies

Considerably more effort has gone into utilizing biological control in the West Indies for pest control than has apparently been done in South America. Myers (113, 114) and Simmonds (139) have reviewed the biological control attempts in the West Indies and British West Indies, respectively.

Myers and Box were the first workers that were active in biological control in the Caribbean in 1930. It was not until 1946 when the Commonwealth Institute of Biological Control opened a station in Trinidad that biological control was again considered; sugar cane has received much attention since it is the most important economic crop in the British West Indies. The *Diatraea* spp. are the most important pests of sugar cane, and Myers and Box (1931–1938) used tachinids to control *D. saccharalis* on several islands. Later releases of the same and other species of tachinids on other islands became successful. In Barbados, *Trichogramma* have been mass cultured from 1931 to at least 1956 and released mainly against *D. saccharalis.* Recently Alam, Bennett & Carl (1) reported successful control of this borer by *Apanteles flavipes* and *Lixophaga diatraeea.* The coconut white fly has been satisfactorily controlled by *Prospaltella* and a coccinellid. A histerid beetle controls the banana weevil in poorly kept plantations (139).

Puerto Rico

There have been at least 4 outstanding successes in biological control in Puerto Rico. Wolcott (179) reports 2 cases where natural enemies were so effective they disappeared from lack of prey after reducing serious pest populations. The introduced coccinellid *Chilororus cacti* proved so effective in totally eliminating the scale insects *Pseudalacaspis pentagona* and *Asterolecanium pustulans* from Puerto Rico for 12 to 14 years that no beetles survived to control new infestations. The giant Surinam toad was so effective in eliminating white grubs *Phyllophaga portoricensis* and *P. vandinei* from cane fields for 15 years that no toads survived to control new white grub infestations.

The introduction of *Cryptolaemus montrouzieri* and *Rodolia cardinalis* have proven a great success against mealybugs and *I. purchasi,* respectively. A review of the effectiveness of parasites and cultural practices on the control of the sugarcane borer is given by Wolcott in *Insects of Puerto Rico.* He was reluctant to go to broad-scale use of insecticides on sugarcane fields in deference to relying on natural control.

Cuba

A most outstanding biological control success was the introduction of a parasite *Eretmocerus serius* and a small coccinellid predator *Catana clauseni* collected in the Malayan region. The pest was the citrus blackfly, *Aleurocanthus woglumi,* which invaded the West Indian Islands in 1913 and became a serious pest of citrus and other trees. Full economic control was accomplished mainly by the parasite (24).

Bermuda

Bennett & Hughes (12) reviewed biological control of insect pests in Bermuda, and these authors conducted many attempts of biological control themselves. Of 15 biological control projects attempted in Bermuda the most successful has been the one directed at *I. purchasi*. *Rodolia cardinalis* alone did not control the cottony cushion scale in colder parts of the island, but with the introduction of *Cryptochaetum iceryae* the scale pest is no longer a problem any place on the island. Several other projects gave worthwhile results: the oleander scale, the palmetto scale, the potato tuber worm, and mealybugs. Predatory snails have been introduced to use against an introduced phytophagous snail and a passerine bird has been introduced to prey upon tree lizards which prey heavily upon beneficial insects (12, 141).

ISLANDS OF OCEANIA

Of all the biogeograhical regions of the world, the islands of Oceania can probably claim to be the sites of the greatest number of outstanding biological control successes. In the Hawaiian Islands alone there have been at least 24 cases of biological control of pest insects by imported entomophagous insects, and Fiji has had 6 cases (33).

Hawaiian Islands

The response of the Hawaiian entomologists to an invasion of an alien pest is nearly automatic. Usually there is no question about what type of control will be used against the pest for an explorer-collector is on his way to the country suspected as being the origin of the pest. The 24 cases of successful biological control of insect pests include 2 complete, 10 substantial, and 12 partial. This record is slightly better than recorded from California: 2 complete, 9 substantial, and 8 partial control results (33). As Doutt (41) pointed out, the successes achieved are the result of having a permanent organizational structure maintained solely to conduct biological control and not because of their salubrious climates or because they are ecological islands as some ecologists have suggested. Both Hawaii and Cailfornia have had permanent organized biological control units since the early 1900s.

The history of entomology in Hawaii is nearly the history of biological control in Hawaii. Timberlake (163) described a history of biological control in Hawaii and Pemberton (123) wrote a history of entomology in Hawaii which essentially is a description of the history of biological control.

In 1893, Albert Koebele, famous for discovering the vedalia beetle, was hired by the Hawaiian Government to control insect pests. The vedalia beetle had been introduced in 1890 and had controlled the scale. When Koebele arrived in Hawaii he first secured colonies of *Cryptolaemus montrouzieri* from California. They originally were collected in Australia by Koebele and sent to California. Koebele travelled extensively searching for entomophagous insects in the Orient and Australia from 1894 to 1896 and among the natural enemies sent to Hawaii 6 coccinellids and 2 hymenopterous parasites became established. There is no complete record of the total number of natural enemies sent to Hawaii in these early days. Between 1890 and 1895, 14 species of lady beetles became common in Hawaii and at least 7 other species were temporarily established but disappeared. This period 1890–1910 as in Australia, California, and South Africa was a lady beetle era and not many records were kept nor detailed evaluation of releases made.

In 1900, the sugar cane was being attacked by the sugarcane leafhopper. The leafhopper had invaded Hawaii from Australia. By 1904, the sugar cane industry was suffering enormous losses and was threatened with complete destruction by this insect. The native natural enemies could not cope with the leafhopper. Because of this insect a Division of Entomology was established in 1904 in the Hawaiian Sugar Planters Experiment Station (HSPA) with R. E. L. Perkins in charge. Other members of the staff were Koebele, G. W. Kirkaldy, F. W. Terry, O. H. Swezey, and Frederick Muir. Later D. T. Fullaway, F. X. Williams, and H. S. Osborn joined the HSPA staff. These entomologists are synonymous with biological control in Hawaii and produced many successes. They left a heritage of experience and contributed to the formation of biological control as a practical science (55, 75, 123, 163).

From explorations of Koebele, Perkins, and Muir, egg parasites were found in Australia and established against the sugarcane leafhopper and considerable control of the leafhopper was achieved in many areas of Hawaii. It was not until 1920 from Muir's trip to Australia that a predaceous mirid, *Tytthus mundulus,* was discovered and sent to Hawaii. This mirid proved so effective that by 1923 the leafhopper populations were definitely on the wane in all plantations, and in a few years the leafhopper became scarce. This is a classic case in the remarkable biological control of an important economic insect pest.

A series of insect pests in sugar cane were all brought under control by introduced natural enemies and sugar cane even today is rarely treated with insecticides. So efficient was the biological control research that the entomology staff of the HSPA has dwindled to only two or three today. There are no serious insect problems in sugar cane today in Hawaii.

The Territorial Board of Agriculture, now the State Department of Agriculture, was responsible for many successes in biological control. They had a permanent staff member for over twenty years with Noel Krauss as the foreign explorer for natural enemies. Today, as before, the first approach considered to control a new pest is the biological control approach.

Biological control of insect pests.—Among the 24 cases of biological control of insect pests the use of *Gambusia* for mosquito control was first suggested and carried out in Hawaii. One of the largest entomological campaigns organized against an insect pest was one during the late 1940s into the 1950s. The target pest was the oriental fruit fly, *Dacus dorsalis*. The cooperation of about 50 entomologists of State Departments of Agriculture of Hawaii and California, the USDA, the HSPA, the Pineapple Research Institute, and the Universities of Hawaii and California combined their efforts under the direction of Walter Carter. *Dacus dorsalis* was suppressed, biological control was one direction taken, and this is a case considered under substantial biological control (25, 115).

Biological control of weeds.—Hawaii has led the world in the control of noxious plants through biological control methods and was one of the first countries to undertake such work (123). Lantana, Pamakani, and *Opuntia* weeds have been reduced on ranges and pasture lands once unusable. They are now relatively free of the weeds because of introduced insects.

Biological control of snails.—The giant African snail came to Hawaii in 1936, and gradually spread to most of the Islands. F. X. Williams in 1947 sought natural enemies in East Africa. Large carabid beetles were introduced but it is the carnivorous snails that are now thought to give better results. *Gonaxis kibweziensis* and *Euglandina rosea* appear to be the most important natural enemies introduced.

Fiji Islands

Three remarkable cases of biological control of insect pests were achieved in the Fiji Islands: the coconut moth, the coconut scale, and the leaf-mining beetle. These insect pests were attacked by well-organized plans made by outstanding entomologists.

Coconut moth.—*Levuana iridescens* was a severe menace to the coconut industry of Fiji. A committee was appointed in Fiji in 1924 to investigate the possibilities of controlling the coconut moth. The committee appointed J. D. Tothill as Director of the Levuana Campaign along with R. W. Paine and T. H. C. Taylor and the work began in 1925. In little more than a year the pest was completely suppressed by a tachinid fly, *Ptychomyia remota,* which was introduced from Malaya (165). An important aspect in this project is that the parasite *P. remota* is a normal parasite of another genus and species. As Clausen (24) said in the history of biological control, instances of successful utilization of parasites from another host genus are exceedingly rare.

Coconut scale.—In 1926, the Levuana Committee in Fiji decided to extend its activities mainly in the direction of attempting the biological control of serious coconut pests. The first of these other pests to be investigated was the coconut scale, *Aspidiotus destructor,* which was effectively controlled in 1928 by the introduction of a coccinellid, *Cryptognatha nodiceps,* (156). Taylor (156) also introduced parasites against the coconut scale but they were ineffective. On this basis, he claimed that biological control workers should seek predators of diaspine scales instead of parasites. By today, however, the parasite successes over scales far outnumber the predator (24).

Coconut leaf-mining beetle.—In 1929, Tothill was appointed Director of Agriculture of Uganda and from that time on, until 1934, the entomological work of the Committee in Fiji was continued by Taylor and Paine. The third pest in Fiji they attacked was *Promecotheca reichei.* Taylor went to Java to investigate in detail the parasites found by Paine to be associated with other species of *Promecotheca* in 1932. *P. reichei* is indigenous to Fiji and has its own parasites in Fiji but the invasion of a predaceous mite nullified the beetle parasites and the leaf-mining beetle populations exploded. Taylor outlined the requisites a parasite should have in order to fit into the pest beetle's phenology. In Java he found the parasite that fulfilled the requirements outlined and it was the eulophid *Pleurotropis parvulus.* The parasite was introduced, cultured, and released, and in one year complete economic control was achieved (157).

Taylor followed Tothill to Uganda in 1935 and in 1953 joined the Anti-Locust Research Centre. Taylor's critical and outspoken views contributed much to biological control. He expressed the view that biological control seldom works in continental areas and that the method of biological control was not compatible with the use of chemicals. On both scores it appears he was wrong, but it was his thinking and pronouncements that forced biological control workers to consider his concepts.

Coconut spike moth.—*Tirathaba trichogramma* was the next pest to be attacked, and Paine in the middle 1930s was the main worker (24). Parasites were introduced into Fiji; *Apanteles tirathabae* and *Erycia basifulva* were the important natural enemies. The moth pest causes premature nutfall and this condition was checked but control was not complete.

Trust Territory and Guam

The Trust Territory of the Pacific Islands comprises the Micronesian Island groups known as the Carolines, Marianas, and Marshalls, with the exception of Guam. In 1947 the Pacific Science Board of the National Academy of Sciences, National Research Council appointed the Invertebrates Consultants Committee for the Pacific to plan and direct the entomological work in these islands. One of the main objectives of this Committee

has been the introduction into the islands of beneficial insects to combat insect pests and weeds, using the classical biological control approach (62). During the first 8 years up to 1956 this committee had introduced 43 beneficial insects and 1 insect disease against 20 insect pests and 1 weed. Of these, 12 are known to have become established and several of the important pests brought under control by the introduced natural enemies (62).

The coconut rhinocerus beetle, *Oryctes rhinoceros,* was one of main targets for biological control and in view of this many parasites and a disease have been released. The one parasite that stands out is *Scolia ruficornis* from Zanzibar. It now is present on many islands in the Pacific and is exerting some control of *Oryctes* (62, 78).

In Micronesia the fluted scale, *Icerya aegyptiaca,* is being controlled by a *Rodolia pumila* which was suspected to have been accidentally introduced into various islands where the Japanese were moving *R. cardinalis* into islands to control *I. purchasi.* Beardsley (9) discusses the history of movements of natural enemies of the fluted scales in Micronesia.

ORIENTAL REGION

Some areas in the Oriental Region have been active for many years while in countries like India and Pakistan they are only in recent years beginning to introduce natural enemies; although there was some movement of *Dactylopius* in the middle 1800s for control of *Opuntia,* and it was also introduced into Ceylon (168). The Commonwealth Institute of Biological Control has stations in India and Pakistan where many local entomologists are on these staffs. The aim of these stations is mainly to search and export natural enemies. Therefore, much is being learned about the pests in these countries, and the local people are being trained in biological control techniques. Rao et al (125a) should be consulted for biological control activity carried out in Asia.

Malaysia

The biological control and integrated control approaches are being used extensively on the pests of oil palms in Malaysia (81, 180). B. J. Wood has documented the increase in pest problems from use of certain insecticides and the timing of insecticides, for he concluded that natural biological control of many insects is prevalent, and it is very simple to create pest problems. G. R. Conway (28) calls for the ecological approach to pest control in the tropical environment. He has observed man-created insect pest outbreaks by using insecticides indiscriminately.

Asia

Although not a biogeographical region, the biological control researchers have a common front in attacking rice insects. This is clearly seen particularly in the attack strategy against the rice stem borers.

Nishida (119) discusses the ecology of rice stem borers in relation to the rice paddy ecosystem, and points out that this ecosystem is the most important and extensive agricultural ecosystem in Southeast Asia. He describes the ecosystem and traces the native home of *Chilo suppressalis* and *Tryporyza incertulus.* Nishida (119) concludes his analysis by saying:

> There are two schools of thought concerning the control of rice pests in Asia. One school believes that rice should be grown in an entomologically sterile environment, and the other believes that rice should be grown in a biologically balanced environment so that phytophagous pest insects are present but are maintained at a subeconomic level. The ideal method is growing rice in an entomologically sterile environment, but certainly it is not the most economically practical means with a low value crop like rice.
>
> Since it is not economically practical to grow rice in an entomologically sterile environment, it becomes the responsibility of all those concerned with pest control of rice in Asia to realize the importance of the conservation of the rice ecosystem and to take all measures necessary to prevent the abuse of insecticides. It is not too late to take action for there are still vast virgin rice areas untouched by modern insecticides for thousands of years.

Indeed the rice agroecosystem is complex and the number of natural enemies diverse. The reviews of Nickel (118) and Yasumatsu (183, 183a) on the natural enemies of rice stem borers clearly attests to the complexity that exists with just this one group of pests. Bess (14), Nickel (118), and Yasumatsu (182) discuss the feasibility of biological control of rice stem borers.

Japan

A review of biological control of insect pests in Japan was made by Watanabe (175). Of eleven insect pests where biological control was attempted, five were successful. The first attempt was in 1911 when the introduction of the vedalia was made against *I. purchasi* resulting in a great success. The spiny black fly, a pest in citrus orchards, became difficult to find after the introduction of *Prospaltella smithi* from South China in 1922. The ruby scale, a pest of tea and other plants in Japan, came under successful control when *Anicetus beneficus* was found in northern Japan by Yasumatsu in 1948 and released in southern Japan. *Aphelinus mali* introduced in the 1920s controlled the woolly apple aphid until certain insecticides were used. Periodic releases of a native egg parasite, *Anaphes nipponicus,* successfully controlled the rice leaf beetle in rice fields but farmers did not adopt the method (175).

Sasaki (133) was the first to observe the microtype egg deposition on leaves by a tachinid which caused the Uji disease of silkworms. Later his son described the complete life cycle of the involved tachinid, *Blepharipoda zebina.*

PALAEARCTIC REGION

In comparing the contribution of biologists living in various parts of the earth to the development of biological control, the remarkable element stemming from the Palaearctic zone seems to have been the stimulation, the emergence of ideas leading to practical results usually elsewhere, mainly in the Nearctic Region. Most types of modern biological control in the sense of manipulation of organisms to reduce pest populations have been suggested first in the Old World, but the most important and popular successes have been achieved later on in other parts of the world. A more detailed look at the matter, however, will reveal not only its biological and social background, but also so many exceptions to this rule that generalizations become difficult. It will be the purpose of this part of the review to sketch roughly the local and rather independent emergence of various types of biological control in the Palaearctic and its influence on the approach to pest control in other biogeographical regions.

The base for the practical utilization of potential beneficial organisms was some deeper knowledge of living beings and their interrelations. Different examples of parasitism with insects as hosts and as parasites were described and illustrated already in the seventeenth century (3, 111, 128). In 1701, Leuwenhoek (97) definitely stated that hymenopterous larvae are parasites and kill their hosts; in 1706, Vallisnieri (166—1730) showed the nature of insect parasitism, and Réaumur (127) in his famous memoirs (1734–1742) devoted a whole chapter and some good plates to parasitic larvae on caterpillars. He also probably was the first to clearly suggest the use of entomophagous insects, namely the aphidivorous fly (the lacewing) to keep a greenhouse free of aphids. Linnaeus in his essay on pest insect control in orchards (published 1763 under the pseudonym C. N. Nelin) proposed biological control by means of collecting and releasing *Calosoma sycophanta,* other carabids, *Coccinella septempunctata, Chrysopa,* and Aphidiidae (171). He also mentions the first use of snails (*Cepea nemoralis*) to reduce the growth of moss and "tree lice" on apple trees, as this was done by the mayor of Lund (Linnaeus, 98; quotation by H. Sachtleben, 130). In the same time, *Picromerus bidens,* the predacious pentatomid, was already proposed for the reduction of bed bugs (*Cimex lectularius*) (22, p. 587).

In the first half of the nineteenth century, suggestions on the use of entomophagous insects were frequently made, e.g. in 1800 by Erasmus Darwin (after Riley, 129) and by Kirby & Spence (89) in England; Hartig (72c) in Germany suggested the concentration and confinement of catepillars, e.g. of *Dendrolimus pini* in special places ("Raupenzwingern") in order to create foci for the activity of parasites after their emergence from the host. These suggestions indicated that the general attitude had matured for the next step: the actual experiment and demonstration of the practical value of such suggestions. The incipient activity in the Palaearctic which gradually developed during the subsequent decennia shall now be briefly reviewed. Only some of the older examples will be selected as a more recent review on the situation in Europe is available (60).

Utilization of Native Arthropods

The first practical experiments in the field were carried out by Boisgiraud in France near Poitiers, around 1840 (87). He released collected *Calosoma sycophanta* against gypsy moth larvae (*Porthetria dispar*) on poplars and used *Staphylinus olens* successfully (and *Carabus auratus* without success) against earwigs (*Forficula auricularia*) in his garden. Obviously stimulated by these experiments, the Milan "Societa d'incorregiamento delle arti e mestieri" offered a golden medal for the development of this method of pest control. The paper submitted by A. Villa (174) describes a remarkable reduction of several garden pests (*Cetonia, Pieris, Forficula*) by the release of Carabidae and Staphylinidae in his garden, compared with surrounding fields. Although the conclusions were challenged by Bassi and heavily disputed later on (166), the experiments served as activators for more trials. It was gradually realized that the destruction of pest insects should spare beneficial species, and various propositions as well as actual experiments were made in the nineteenth century. In France, the removal of infested apple buds as control measure of the apple blossom weevil (*Anthonomus pomoron*) was refined by caging the buds and releasing into the orchards parasites after their emergence. Experiments using one million infested buds led to the release of about 250,000 parasites; after the second year of this treatment, the isolated orchard was said to have been free of pests for the subsequent ten years (36).

In Germany, Ratzeburg had early recognized the importance of parasitic Hymenoptera and other entomophagous insects in the forest. Although he was skeptical concerning the use of them, he described (126) an experimental transfer of heavily parasitized larvae of *Dendrolimus pini* to a more marginal area of the outbreak zone with the subsequent remarkable increase of parasitization by *"Microgaster"* probably *Apanteles nemorum*. Ratzeburg recommended also the colonization of the red forest ant (*Formica rufa*-group). This colonial predator had sometimes caused reduced defoliation by caterpillars around their natural colonies ("green islands") in otherwise heavily defoliated pine and spruce forests. The protection of forest ant colonies was enforced by law in some parts of Germany for several centuries. The artificial foundation, however, of new colonies in various forests is of more recent age. Experiments of this type began in Germany at the end of the last century (110, pp. 33–34), continued in the 1920s (134), and have now spread to most Palaearctic countries faced with forest insect problems. Some positive results have been reported mainly against larvae of a few Lepidoptera and Hymenoptera (sawflies) (review cf. 122).

P. Marchal in his excellent review "Utilisation des Insectes auxiliaries entomophages" (105) described other early examples of deliberate protection, furthering, or transfer of entomophagous insects. They concern parasites of the wheat blossom midge (*Contarinia tritici*), of the grape tortricid (*Eupoecilia ambiguella,* syn. *Clysia ambiguella*), and of several scale insects in Italy and France.

The mass production and release of native beneficial insects seems to be another example of a special type of biological control that got started in the Palaearctic and was then transferred to and further developed in North America. The idea of mass cultivating ("farming") the native egg parasite *Trichogramma evanescens* was put forward first by Enock in England (45). The earliest field experiments recorded took place in Russia where (according to Howard, 75, p. 302). A. F. Radet(e)zky and N. N. Troitsky in Tashkent started to work in 1913 with egg parasites of the genus *Trichogramma.* A species, called in those days *T. semblidis* (also: *Pentarthron semblidis*), was used against the codling moth (*Laspeyresia pomonella*) after mass rearing on other hosts. I. A. Portchinsky (124) recommended *Phalera bucephala* as substitute host for production in winter. Further older Russian literature on *Trichogramma* has been listed by Hase (73). This tradition was summarized later on by N. F. Meyer who published several papers on biological control (108) and by Telenga (158). Also nowadays, huge areas of beets, vegetables, and orchards are treated regularly in the USSR by release of *Trichogramma* against noctuids and codling moth, respectively.

It is very probable that H. S. Smith knew of these experiments when he suggested to S. E. Flanders in 1925 to study the possibility of mass rearing of *Trichogramma minutum* for release against the codling moth (52). This contributed to the great technical progress in the economic mass production of this and other parasites in North America. It may be added also that the idea of spraying host eggs parasitized by *Trichogramma* to the plant leaf surface originated in Europe (135) and was improved, later on, in North America (159).

Utilization of Entomophagous Arthropods from Other Faunal Regions

It is well known that the real breakthrough of this type of biological control was the introduction of the coccinellid *Rodolia cardinalis* from Australia to California in 1888. This way of repairing self-induced damages had, quite naturally, its greatest economic justification in such faunal zones where many pest arthropods had been unintentionally imported previously. Such importations were usually a consequence of former migrations of parts of the human population. Therefore, the social and religious background of such important emigrations as they occurred in Europe in the eighteenth and nineteenth centuries should be considered as the ultimate causes of the transfer of pest insects from the Palaearctic to the Neararctic and other faunal regions. The economic necessity to combat such imported pests initiated the early ideas and activities. Asa Fitch, the first State appointed entomologist of the United States, published also the first suggestion to import beneficial insects against introduced pests in 1861. He wrote to Curtis in England to obtain parasites of the wheat midge (*Sitodiplosis mosellana*) apparently without success (quoted after Howard, 75, p. 47–48). The first actual experiment took place in France where Planchon tried in 1873 to colonize the predatory mite *Tyroglyphus phylloxerae* against the grape phylloxera

(*Phylloxera vitifoliae*) imported from North America. This experiment was not successful. The material and, very probably, a good deal of the idea stemmed from C. V. Riley. This great American entomologist was born in London in 1843 and he studied in England, France, and Germany and emigrated to the U.S., transferring, thereby, the scientific background to a faunal region where the importation of beneficial arthropods was urgently needed and could well develop. He initiated the introduction from England and successful acclimatization in the U.S. (Washington, D.C.) of the braconid *Apanteles glomeratus* to reduce the number of imported cabbage worms (*Pieris rapae*) in 1884 (23). Another forerunner was the introduction of *Coccinella undecimpunctata* from England to New Zealand in 1874. After the great success against the cottony cushion scale (*Icerya purchasi*) in California, mentioned above and initiated again by Riley, a series of experiments of this type followed (cf. list given by Trouvelot, 167). In the Palaearctic, the early phase of the introduction type of biocontrol was merely a repetition of what had already been tested elsewhere. Examples are the colonization of *Rodolia cardinalis* in Egypt against *Icerya aegyptiaca* (1892); in Portugal (1898), in Italy (1902), and in France (1912) against *I. purchasi* (105, 106); the colonization of *Aphelinus mali,* parasite of the woolly apple aphid (*Eriosoma lanigerum*), in France (1920) and subsequently in most Mediterranean countries, in central, eastern, and even partially in northern Europe (130). The limited efficiency of this parasite, as of most entomophagous insects in climatically suboptimal regions, was clearly demonstrated in the Palaearctic. This enforced early tests in the direction of integrated control because insecticides had to assist in the suppression of pests without destroying beneficial species (see below).

The first case where the introduction of entomophagous insects originated in the Palaearctic region and reduced successfully an imported pest was the importation and acclimitization of *Prospaltella berlesei* from North America to suppress the white peach scale (*Pseudaulacaspis pentagona*) on mulberry in Italy. A. Berlese obtained the parasite from L. O. Howard in 1906 and succeeded in reducing completely the scale insect in most parts of Italy, after artificial distribution, until 1914. The heated scientific discussion between Berlese and his famous successor at the Portici laboratory, F. Silvestri, on the theoretical and practical superiority of parasites or predators, of single or multi-species introductions is also part of the history of biological control (for reviews see Howard, 75, pp. 254–56, Sachtleben, 130, pp. 72–74).

Another example of autochthonous Palaearctic biological control by introduction, but of a special type, is the case of *Encarsia formosa* (Eulophidae), the parasite of the greenhouse whitefly (*Trialeurodes vaporariorum*), a famous pest on tomatoes, cucumbers, and other vegetables (151). *E. formosa* was detected in greenhouses of southern England in 1926, mass bred, and 287,000 of them distributed to English greenhouses in 1928. From there it was shipped to Canada (1928), New Zealand (1933), Australia (1934), and other regions of the world. Quite recently, the same parasite is being sold by a Swiss firm for use in cucumber greenhouses, in combination with a predatory mite attacking *Tetranychus urticae* in the same environment. The predator (*Phytoseiulus persimilis*), originating from Chile, has first been investigated in Germany by Dosse (38). It was subsequently distributed almost all over the world and may serve as another of the few examples of biological control by importation that started in the Palaearctic.

The limited importance of introduced beneficial arthropods for Eurasia is compensated for by the intensity of the export of parasites and predators (rarely of pathogens) from this region. This activity was initiated by American entomologists looking for natural enemies of the gypsy moth (*Porthetria dispar*) after its introduction to Massachusetts in 1869. Beginning in 1905 and ably directed by Howard, they stimulated studies of parasites and predators of all developmental instars in many European countries and shipped them alive to the receiving station at Saugus, Massachusetts. A similar study of the brown tail moth (*Euproctis chrysorrhoea*) and its natural enemies was included (77). The cooperation with European entomologists was most productive and was extended to Japan in 1908. The headquarters were in Budapest from 1926 to 1935. This is not the place to describe the scope and the results of this huge experiment. It was the first large-scale operation of this type and served as instructive model for both American and European entomologists. Further actions of the same type followed later concerning, e.g. the European corn borer (*Ostrinia nubilalis*). This operation started in 1919 in France and later covered the whole Palaearctic including Japan, Taiwan, and other parts of eastern Asia. It yielded over 5 million beneficial insects covering over 100 species (for reviews cf. Sachtleben, 130, pp. 64–66).

The western Palaearctic parallel of this type of study of entomophagous arthropods in their native country and subsequent export of the promising species to the newly invaded zone got started at the Farnham House Laboratory in England by the former Imperial Bureau of Entomology. Under the direction of W. R. Thompson (1928–1958) and F. J. Simmonds (since 1958) it became the well-known Commonwealth Institute of Biological Control (CIBC). Its headquarters remained at Farnham Royal until 1940 when, during World War II, work in Europe became impossible. The headquarters were shifted to Belleville, Ontario, Canada, later to Ottawa, and, in 1962, to Curepe, Trinidad (140, 161). A European Station was set up in 1948, first at Feldmeilen near Zurich, and since 1958 at Delémont, also in Switzerland. Other stations are situated in the West Indies, South America, India, Pakistan, and East and West Africa. The work of the CIBC has broadened from the original limited task of shipping beneficial insects to Commonwealth countries to that of a scientific institution which deals with practical as well as with fundamental projects. The study of general ecology of various pests and their natural enemies and the research into the potential value of certain beneficial species allows the careful selection of most promising candidates for introduction to other countries beyond the limits of the British Commonwealth. In addition, a world catalogue of parasites and predators is being prepared together with a technical bulletin and summaries on the present status of biological control in various parts of the world. Thus, this institute represents one of the most effective approaches to biological control that originated in the Palaearctic and is still an integral part of this faunal region.

Utilization of Vertebrates

The use of domesticated animals for the local reduction of pest insects has a long tradition in Europe. Driving groups of semidomesticated pigs into the forest was a widely practiced method to control such species that live or hibernate in the litter like the white grubs (*Melolontha* sp.), pupae of *Bupalus piniarius, Panolis flammea,* or cocoons of sawflies (for details see Eckstein, 44, pp. 152–56). Ratzeburg (126, p. 176) recommended it and earlier descriptions are known from the end of the eighteenth century (e.g. Esper, 47, p. 350). The utilization of pigs was undoubtedly inspired by the natural activity of the wild boar which is still considered to be of great value in reducing focal populations of pest insects like *Acantholyda nemoralis* in the huge forests of Poland (92).

The most common use of domestic fowl for the reduction of field insects like *Blitophaga undata* (Silphidae) on beet was combined with special equipment, called the "Huhnerwagen" (fowl's carriage) in Germany. It was built for transportation of fowl (for illustrations cf. Eckstein, 44, p. 202) and its effect was said to be satisfactory at local outbreaks (15). This type of biocontrol failed, however, against unpalatable prey like the Colorado potato beetle (*Leptinotarsa decemlineata*).

A typical Palaearctic sport, the old tradition of falconry, should be also mentioned here because it has been used to keep away herons from fishponds and crows from crop fields (170). Its modern version is the recent utilization of falcons to keep airfields clear of sea gulls in Canada, another example of the transfer of techniques from the Old to the New World.

Of all known techniques to utilize wild vertebrates for biological control only the use of native, mainly of insectivorous, birds should be mentioned because of its history in some European countries. Birds cause frequently such strong emotional responses in many people that the discussion as to the usefulness of some of them continued for over a century. Here, only the beginning of an active utilization of insectivorous birds shall be reported. The provision of nesting boxes in orchards and forests has been recommended in Germany since the early nineteenth century. Impractical models were gradually replaced by others built after the shape of natural woodpecker's holes. A great impetus to the further spread of active bird colonization in northern Central Europe as a way of biological control of pest insects was the work of the Baron von Berlepsch who, among other activities, founded the first "Vogelschutzwarte" (bird protection station) at Seebach in Thuringia which was officially acknowledged by the Government in 1909. (For further details of the early history of this facet of biocontrol see v. Berlepsch, 13.) The tradition continues up to the present time: in Germany, the provision of nesting boxes (which need some sort of official approval) is very popular on many state and private forests, parks, and orchards.

SUMMATION

The biological control method is here to stay and it is being recognized as one of the most economical and effective methods we have to control arthropod pests and range weeds. The classical method of importing the pest's original natural enemies which are often left behind is usually the most economical approach to pest control. The greatest biological control successes have been of this type. Importation of new natural enemies alone on a worldwide basis has resulted in the permanent control of 120 pest species. Of these 42 pest species were completely controlled, 48 represent cases of substantial control, and 30 of partial control (34). DeBach (34) brings out an important point that the same pest species that have been subsequently controlled in other countries should each be added as cases of success. Thus, there have been 253 successful projects in the world.

Sailer (131) calculated that since there are 212 insect pests of foreign origin in the USA and each phytophagous species has at least 3 more or less host specific natural enemies then there will be more than 600 beneficial insects available in the endemic areas of the 212 pest species, but only 128 natural enemies have been introduced in 80 years.

The monocultures that are annually tilled or annual crops grown in relatively sterile environments of no trees, no bushes, no weeds, no rocks is not the place to embark on the classical method of biological control, but it is here that there is also a great future for other biological control approaches. But the technology of mass culture of beneficial insects must be advanced; there are a few recent breakthroughs in the direct mass culture of predators on artificial diets. With inexpensive insects, releasing predators and parasites should become competitive with the costs of insecticides. The use of the mass culture and periodic release technique does not offer permanent control such as a success with classical biological control; however, the control achieved with periodic releases is usually sustained over a longer period of time than the control obtained by use of a pesticide and, of course, there is no problem of chemical residues.

The provisioning of missing requisites in the environment for natural enemies is just dawning. Artifical diets used in the field have been shown to increase the effectiveness of certain predators.

The contributions made by biological control workers to ecology, biology, systematics, and physiology have greatly advanced entomology. The biological control worker of the future must be attuned to all the above fields or form teams that will cover all areas of entomology to be successful. The role of natural enemies in the regulation of arthropod populations is of such importance that every insect control campaign automatically should consider integrated control, and among the methods to be considered in controlling a pest, biological control should be one of the first.

ACKNOWLEDGMENTS

We would like to acknowledge the assistance of Dr. F. D. Bennett, Dr. L. E. Caltagirone, and Mr. R. L. Tassan.

LITERATURE CITED

1. Alam, M. M., Bennett, F. D., Carl, K. P. 1971. Biological control of *Diatraea saccharalis* (F.) in Barbados by *Apanteles flavipes* Cam. and *Lixophaga diatraeae* T. T. *Entomophaga* 16(2):151–58
2. Anon. 1969. Principles of plant and animal pest control. *Insect-Pest Management and Control,* Vol. 3. Washington, D.C.: Nat. Acad. Sci., 1965. 508 pp.
3. Aldrovandi, U. 1602. *De Animalibus Insectis.* Bologna
4. Askew, R. R. 1971. *Parasitic Insects.* New York: Am. Elsevier, 316 pp.
5. Autuori, M. 1942. Dados a respeito de *Tetrasticus giffardianus* Silv., parasita de *Ceratitis capitata* Wied. *Sao Paulo, Arquivos do Instituto Biologico* 13:149–62
6. Baird, A. B. 1956. Biological control, pp. 363–67. In: Entomology in Canada up to 1956. A review of developments and accomplishments, ed. R. Glen. *Can. Entomol.* 88(6):290–371
7. Baker, W. A., Bradley, W. G., Clark, C. A. 1949. Biological control of the European corn borer in the United States. *US Dep. Agr. Tech. Bull. No. 983.* 185 pp.
8. Balch, R. E. 1960. The approach to biological control in forest entomology. *Can. Entomol.* 42(4): 297–310
9. Beardsley, J. W. Jr. 1955. Fluted scales and their biological control in the United States Administered Micronesia. *Proc. Hawaii. Entomol. Soc.* 15(3): 391–99
10. Beingolea, O. 1963. Empleo de Insecticides Organicos en el Peru y Pasibilidades de Reducirlo por medio del Control Integrado. *Rev. Per. Est. Agr.* 5(1)
11. Bennett, F. D. 1971. Current status of biological control of the small moth borers of sugar cane *Diatraea* spp. *Entomophaga* 16(1):111–24
12. Bennett, F. D., Hughes, I. W. 1959. Biological Control of insect pests in Bermuda. *Bull. Entomol. Res.* 50:423–36
13. Berlepsch, H. Frh. v. 1929. *Der Gesamte Vogelschutz.* Verlag J. Newman, 12th Aufl., Neudamm
14. Bess, H. A. 1967. Feasibility and problems of chemical control and biological control of rice stem borers (Research on the natural enemies of rice stem borers). *Mushi* 39(Suppl.):45–50
15. Blunck, H., Görnitz, K. 1925. Lebensweise und Bekämpfung der Rübenaaskäfer. *Arb. Biol. Reichsanst.* 12:42
16. Bodenheimer, F. S. 1958. Animal ecology today. *Uitgeverij Dr. W. Junk, Den Haag.* 276 pp.
17. Box, H. E. 1956. The biological control of moth-borers (*Diatraea*) in Venezuela. Battle against Venezuela's cane borer. *Sugar,* N.Y. 51(5):25–27, 30, 45; (6):34–36; 57(7):30–44
18. Caltagirone, L. 1957. Insectos entomofagos y sus huespedes anotados para Chile. *Apart. Agr. Technica. Año* 17(1):16–48
19. Carnes, E. K. 1912. Collecting ladybirds (Coccinellidae) by the ton. *Mon. Bull. State Comm. Hort.* 1(3):71–81
19a. Carson, R. 1962. *Silent Spring.* Boston: Houghton Mifflin, 368 pp.
20. Ciociola, A. I. 1972. Some aspects of non-chemical insect control in Brazil. Submitted for publication
21. Clausen, C. P. 1936. Insect parasitism and biological control. *Ann. Entomol. Soc.* 20:201–23
22. Clausen, C. P. 1940. *Entomophagous Insects.* New York: McGraw-Hill, 688 pp.
23. Clausen, C. P. 1956. Biological control of insect pests in the continental United States. *US Dep. Agr. Tech. Bull. No. 1139.* 151 pp.
24. Clausen, C. P. 1958. Biological control of insect pests. *Ann. Rev. Entomol.* 3:291–310
25. Clausen, C. P., Clancy, D. C., Chock, Q. C. 1965. Biological control of the oriental fruit fly (*Dacus dorsalis* Hendel) and other fruit flies in Hawaii. *US Dep. Agr. Tech. Bull. 1322.* 102 pp.

26. Cleare, L. D. 1939. The Amazon fly (*Metagonistylum minense* Towns.) in British Guiana. *Bull. Entomol. Res.* 30:85–102
27. Compere, H. 1961. The red scale and its insect enemies. *Hilgardia* 31(7):173–278
28. Conway, G. R. 1968. Crop pest control and resource conservation in tropical South East Asia. *International Union of Conservation, Nature and Natural Resources,* Morges, Switzerland, ed. L. M. Talbot, M. H. Talbot, pp. 159–63
29. Cooke, M. 1883. *Injurious insects of the Orchard, Vineyard.* H. S. Crocker and Co. Calif. 472 pp.
30. Davidson, W. M. 1924. Observations and experiments on the dispersion of the convergent ladybeetle (*Hippodamia convergens Guern.*) in California. *Trans. Am. Entomol. Soc.* 50: 163–75
31. Day, P. R. 1972. Crop resistance to pests and pathogens. In *Pest Control Strategies for the Future.* Washington, D.C.: Nat. Acad. Sci., 376 pp.
32. DeBach, P. 1946. An insecticidal check method for measuring the efficacy of entomophagous insects. *J. Econ. Entomol.* 39: 695–97
33. DeBach, P., Ed. 1964. *Biological Control of Insect Pests and Weeds.* New York: Reinhold, 884 pp.
34. DeBach, P. 1972. The use of imported natural enemies in insect pest management ecology. *Proc. Tall Timbers Conf. Ecol. Animal Control by Habitat Management, Tallahassee, Florida, 1971.* 3:211–33
35. DeBach, P., Rosen, D., Kennett, C. E. 1971. Biological control of coccids by introduced natural enemies. In *Biological Control,* ed. C. B. Huffaker, pp. 165–94. New York: Plenum
36. Decaux, F. 1899. Destruction rationellé des insectes qui attaquent les arbres fruitiers, par l'emploisimultane des insecticides, des insectes auxilaires, et par la protection et l'élévage de leurs ennemis naturels: les parasites. *J. Soc. Hort. France* 22: 158–84
37. Dodd, A. P. 1940. The biological campaign against prickly-pear. *Commonw. Prickly Pear Bd., Brisbane.* 177 pp.
38. Dosse, G. 1959. Der Einfluss von Temperatur und Nährung auf verschiedene Raubmilbenarten und Hinweise auf die Möglichkeit einer biologischen Bekämpfung von Spinnmilben in Gewächshäusern. *Versh. Int. Pfl. schutz Kongr. 4th. Hamburg,* 1957. 1:929–32
39. Doutt, R. L. 1958. Vice, Virtue and the Vedalia. *Bull. Entomol. Soc. Am.* 4:119–23
40. Doutt, R. L. 1964. The historical development of biological control. In *Biological Control of Insect Pests and Weeds,* ed. P. DeBach, pp. 21–42. New York: Reinhold
41. Doutt, R. L. 1972. Biological Control: Parasites and Predators. In *Pest Control Strategies for the Future.* Washington, D.C.: Nat. Acad. Sci., 376 pp.
42. Doutt, R. L., Hagen, K. S. 1950. Biological control measures applied against *Pseudococcus maritimus* on pears. *J. Econ. Entomol.* 43(1):94–96
43. Dowden, P. B. 1962. Parasites and predators of forest insects liberated in the United States through 1960. *US Dep. Agr. Handbk. No. 226*: 70 pp:
44. Eckstein, K. 1915. *Die Technik des Forstschutzes gegen Tiere.* Berlin: Verlag P. Parey, 2nd Aufl. 254 pp.
45. Enock, F. 1895. Remarks on *Trichogramma evanescens* Westw. recorded by the Secretary of the South London Entomology and Natural History Society. *The Entomol.* 28:283
46. Escherich, K. 1914. *Die Forstinsekten Mitteleuropas.* Berlin: Verlag P. Parey, Band 1. 432 pp.
47. Esper, E.H. Ch. 1786. *Die Schmetterlinge in Abbildungen nach der Natur mit Beschreibungen. 4. Teil, 1.* Erlangen
48. Essig, E. O. 1931. *A History of Entomology.* New York: Macmillan, 1029 pp.
49. Finney, G. L. 1950. Mass-culturing *Chrysopa californica* to obtain eggs for field distribution. *J. Econ. Entomol.* 43(1):97–100

50. Finney, G. L., Flanders, S. E., Smith, H. S. 1947. Mass culture of *Macrocentrus ancylivorus* and its host, the potato tuber moth. *Hilgardia* 17:437–83

51. Fisher, T. W. 1963. Mass culture of *Cryptolaemus* and *Leptomastix*—natural enemies of citrus mealybug. *Calif. Agr. Exp. Sta. Bull. 797,* 39 pp.

52. Flanders, S. E. 1930. Mass production of egg parasites of the genus *Trichogramma. Hilgardia* 4(16):465–99

53. Flanders, S. E. 1937. Ovipositional instincts and developmental sex differences in the genus *Coccophagus. Univ. Calif. Pub. Entomol.* 6(15):401–22

54. Flanders, S. E. 1951. Mass culture of California red scale and its golden chalcid parasites. *Hilgardia* 21(1):1–42

55. Flanders, S. E. 1955. The organization of biological control and its historical development. *Meded. Landbouwhogesch., Gent.* 20: 257–70

56. Flanders, S. E. 1969. Herbert D. Smith's observations on citrus blackfly parasites in India and Mexico and the correlated circumstances. *Can. Entomol.* 101 (5):467–80

57. Fleschner, C. A. 1958. Field approach to population studies of tetranychid mites on citrus and avocado in California. *Proc. Int. Congr. Entomol., 10th, Montreal, 1956* 2:669–74.

58. Fleschner, C. A. 1959. Biological control of insect pests. *Science* 129(3348):537–44

59. Forbush, E. H., Fernald, C. H. 1896. *The Gypsy Moth.* Boston: Wright & Potter, 495 pp.

60. Franz, J. M. 1961. Biological control of pest insects in Europe. *Ann. Rev. Entomol.* 6:183–200

61. Gallo, D. 1952. Contribução para o controle biológica da broca da cana de açúcar. *Anais da Escola Superior de Agricultura "Luiz de Queiroz"* 9:136–42

62. Gardner, T. R. 1958. Biological control of insect and plant pests in the Trust Territory and Guam. *Proc. Int. Congr. Entomol., 10th, Montreal* 4: 465–69

63. Given, B. B. 1967. Biological control of weeds and insect pests in New Zealand. *Mushi* 39(Suppl.): 17–22

64. Glen, R. 1962. Preface, pp. ix–x. In: A review of the biological control attempts against insects and weeds in Canada. *Commonw. Arg. Bur. Farnham Royal, Bucks., England. Tech. Communicat. No. 2*

65. Gomes, J. 1962. Histórico do combate biológico no Brasil. *Rio de Janeiro, Boletim do Instituto de Ecologia e Experimentacao Agricolas* 21:89–97

66. Goncalves, C. R. 1962. Perspectivas de combate biológico à principais pragas das plantas cultivadas na baixada fluminense. *Rio de Janeiro, Boletin do Instituto de Ecologia e Experimentacao Agricólas* 21: 65–78

67. Gonzalez, R. H., Rojas, S. 1966. Estudio analitico del control biologicao de plagas agricolas en Chile. *Agr. Tecnica* 26(4): 133–47

68. Graf-M., A., Cortes-Peña, R. 1940. Introduction de hiperparasitas en Chile: Resumen de las importaciones hechas y de sus resultados. *Proc. Pac. Sci. Congr. Pac. Sci. Assoc., 6th.* 4:351–57

69. Greathead, D. J. 1971. A review of biological control in the Ethiopian region. *Tech. Comm. Commonw. Inst. Biol. Control 5,* 162 pp.

70. Guagliumi, P. 1966. Insetti e Aracnidi dello Piante comuni del Venezuela Segnalati nel Periodo 1938–1963. *Relazioni e Monogr. Agrar. Subtrop. e Trop. N. Ser. N. 86,* 391 pp.

71. Hagen, K. S. 1950. Fecundity of *Chrysopa californica* as affected by synthetic foods. *J. Econ. Entomol.* 43:101–4

72. Hagen, K. S., Tassan, R. L. 1970. The influence of food Wheast® and related *Saccharomyces fragilis* yeast products on the fecundity of *Chrysopa carnea. Can. Entomol.* 102(7):806–11

72a. Hambleton, E. J. 1944. *Heliothis virescens* as a pest of cotton, with notes on host plants in Peru. *J. Econ. Entomol.* 37(7): 660–66

72b. Harris, P. 1971. Biological control of weeds in Canada, 1959–1968. Current approaches to biological control of weeds. *Commonw. Inst. Biol. Contr. Tech. Comm. No. 4,* pp. 67–76

72c. Hartig, G. L. 1827. *Anleitung zur Vertigung oder Verminderung der Kienraupen.* Berlin

73. Hase, A. 1925. Beiträge zu Lebensgeschichte der Schlupfwespe *Trichogramma evenescens* Westwood. *Arb. Biol.: Reichsanst.* 14:171–224

74. Holloway, J. K., Huffaker, C. B. 1951. The role of *Chrysolina gemellata* in the biological control of Klamath weed. *J. Econ. Entomol.* 44:244–47

75. Howard, L. O. 1930. A history of applied entomology. *Smithsonian Misc. Collect., Publ. 3065.* 84, 564 pp.

76. Howard, L. O. 1933. *Fighting the Insects. The Story of an Entomologist.* New York: Macmillan, 333 pp.

77. Howard, L. O., Fiske, W. 1911. The importation into the United States of the parasites of the gypsy moth and the brown-tail moth. *US Dep. Agr. Bur. Entomol., Bull. 91*

78. Hoyt, C. P., Catley, A. 1967. Current research on the biological control of *Oryctes. Mushi* 39 (Suppl.):3–8

79. Huffaker, C. B. 1958. Experimental studies on predation: Factors and predator-prey oscillations. *Hilgardia* 27:343–83

80. Huffaker, C. B. 1959. Biological control of weeds with insects. *Ann. Rev. Entomol.* 4:251–76

81. Huffaker, C. B. Ed. 1971. *Biological Control.* New York: Plenum, 511 pp.

82. Huffaker, C. B., Kennett, C. F. 1953. Developments toward biological control of cyclamen mite on strawberries in California. *J. Econ. Entomol.* 46(5):802–12

83. Huffaker, C. B., Kennett, C. E. 1956. Experimental studies on predation: Predation and cyclamen-mite populations on strawberries in California. *Hilgardia* 26(4):191–222

84. Huffaker, C. B., Kennett, C. E. 1966. Studies of two parasites of olive scale, *Parlatoria oleae* (Colvée). IV. Biological control of *Parlatoria oleae* (Colvée) through the compensatory action of two introduced parasites. *Hilgardia* 37(9):283–335

85. Hunter, S. J. 1909. The green bug and its enemies. *Bull. Univ. Kan.* 9(2):1–163.

86. Isla Marco, R. 1959. Notes on the biological control of pests of agriculture in Chile. *FAO Plant Prot. Bull.* 8:25–30

87. Joly, N. 1842. Notice sur les ravages que la *Liparis dispar* (*Bombyx dispar* Latr.) a exercés aux environs de Toulouse, suivie de quelques réflexions sur un nouveau moyen de détruire certains insectes nuisibles. *Rev. Zool. Soc. Cuvierienne,* pp. 115–19

88. Kelleher, J. S. 1971. Current approaches to biological control of agricultural insect pests. In Biological control of agricultural insects in Canada, 1959–1968. *Commonw. Agr. Bur. Tech. Communicat. No. 4*:1–10

89. Kirby, W., Spence, W. 1815. *An Introduction to Entomology.* London: Longman Brown, Green and Longmans, 285 pp.

90. Knipling, E. F. 1955. Possibilities of insect control or eradication through the use of sexually sterile males. *J. Econ. Entomol.* 48(4):459–62

91. Knipling, E. F. 1972. Sterilization and other genetic techniques. In *Pest Control Strategies for the Future.* Washington, D.C.: Nat. Acad. Sci., 376 pp.

92. Koehler, W. 1954. (The prospect of the role of wild boar in biological combating of injurious insects.) *Roczn. Nauk Lesn., Warszawa, 4, Pr. Nr. 111,* 125–140. In Polish

93. Köllar, V. 1840. *A Treatise on Insects Injurious to Gardeners, Foresters, and Farmers.* London: William Smith, 113, Fleet Street. 377 pp. (Translated from the German, and illustrated by engraving, by J. and M. Loudon). William Smith, London.

94. Lepage, H. 1943. A "Escama Vermelha" dos citros em Sao Paulo *Aonidiella aurantii* (Maskell). (The citrus red scale in the State of Sao Paulo). *Argui-*

vas Instit. Biol., Sao Paulo 14: 311–30

95. Le Pelley, R. H. 1948. The biological control of the common mealybug on coffee and native crops in Kenya. *Rep. Commonw. Entomol. Conf., 5th, London, 1948.* pp. 65–67

96. LeRoux, E. J. 1971. Biological control attempts on pome fruit (apple and pear) in North America, 1860–1970. *Can. Entomol.* 103:963–74

97. Leuwenhoek, A. von 1701. Part of a letter concerning excrescences growing on willow leaves, etc. *Phil. Trans. Roy. Soc. Lond.* 22:786–99

98. Linnaeus, C. 1756. *Reisen durch das Königreich Schweden.* Band 1. Leipzig

99. Liu, G. 1939. Some extracts from the history of entomology in China. *Psyche* 46:23–28

100. Lounsbury, C. P. 1940. The pioneer period of economic entomology in South Africa. *J. Entomol. Soc. S. Africa* 3:9–29

101. McGugan, B. M., Coppel, H. C. 1962. A review of the biological control attempts against insects and weeds in Canada. Part II—Biological control of forest insects, 1910–1958. *Tech. Comm. Commonw. Inst. Biol. Control, Trinidad* 2:35–216

102. McLeod, J. H. 1962. A review of the biological control attempts against insects and weeds in Canada. Part I—Biological control of pests of crops, fruit trees, ornamentals, and weeds in Canada up to 1959. *Tech. Comm. Commonw. Inst. Biol. Control, Trinidad* 2:1–34

103. MacLellan, C. R., Proverbs, M. D., Jaques, R. P. 1971. *Carpocapsa pomonella* (L.), codling moth (Lepidoptera: Olethreutidae). 12–15. In *Biological control programmes against insects and weeds in Canada 1959–1968. Tech. Comm. 4,* England: Commonw. Agr. Bur., 266 pp.

104. Mallis, A. 1971. *American Entomologists.* New Jersey: Rutgers Univ. Press, 549 pp.

105. Marchal, P. 1907. Utilisation des insectes auxiliaires entomophages dans al lutte contre les insectes nuisibles à l'agriculture. *Ann. Inst. Nat. Agron., 2 Sér.* 6:281–354

106. Marchal, P. 1913. L'*Icerya purchasi* en France et l'acclimatation de son ennemie d'origine australienne le *Novius cardinalis. Ann Serv. Epiph.* 1:13–26

107. Mariconi, F.A.M. 1963. Situacão atual da entomologia econômica Bresileira. *Piracicaba, Revista de Agricultura.* 38(4):171–76

107a. Melville, A. R. 1959. The place of biological control in the modern science of entomology. *Kenya Coffee* 24:81–85

108. Meyer, N. F. 1931. *Biologische Bekämpfungsmethode gegen schädliche Insekten.* Moskau, Leningrad. (Russ).

109. Michelbacher, A. E., Smith, R. F. 1943. Some natural factors limiting the abundance of the alfalfa butterfly. *Hilgardia* 15: 369–97

110. Mitteldorpf. 1872. *Die Vertilgung der Kienraupe durch Teerringe.* Berlin

111. Mouffet, T. 1634. *Insectorum sive Minimorum Animalium Theatrum.* London

112. Moutia, L. A., Mamet, R. 1946. A review of twenty-five years of economic entomology in the Island of Mauritius. *Bull. Entomol. Res.* 36:439–72

113. Myers, J. G. 1931. A preliminary report on an investigation into the biological control of West Indian insect pests. *Imp. Inst. Entomol., Farnham House Lab., Empire Marketing Bd. Publ. 42,* 173 pp.

114. Myers, J. G. 1935. Second report on an investigation into the biological control of West Indian insect pests. *Bull. Entomol. Res.* 26(2):181–252

115. Newell, I. M., Haramoto, F. H. 1968. Biotic factors influencing populations of *Dacus dorsalis* in Hawaii.

116. Nicholson, A. J. 1933. The balance of animal populations. *J. Anim. Ecol.* 2(1):132–78

117. Nicholson, A. J. 1940. Indirect effects of spray practice on pest populations. *Proc. Int. Congr. Entomol., 7th, 1938,* pp. 3022–28

118. Nickel, J. L. 1964. Biological con-

trol of rice stem borers: A feasibility study. *Int. Rice Res. Inst. Tech. Bull. 2.* 111 pp.
119. Nishida, T. 1967. On the ecology of stem borers in Asia. (Research on the natural enemies of rice stem borers). *Mushi* 39(Suppl.):23–31
120. Oliveira, M. A. 1948. Contribuicao ao estudo do combate biologico do Gorgulho do Feijao (*Acanthoscelides obsoletus* Say) pelo *Bruchobius latisceps* Ash. *Pelotas, Agros* 1(14):258–68
121. Ordish, G. 1967. *Biological Methods in Crop Pest Control.* London: Constable, 242 pp.
122. Otto, D. 1967. Die Bedeutung der Formica-Völker für die Dezimierung der wichtigsten Schädinsekten. Ein Literaturbericht. *Waldhygiene* 7:65–90
123. Pemberton, C. E. 1964. Highlights in the history of entomology in Hawaii 1778–1963. *Pac. Insects* 6(4):689–729
123a. Pickett, A. D., Putman, W. L., LeRoux, E. J. 1958. Progress in harmonizing biological and chemical control of orchard pests in eastern Canada. *Proc. Int. Congr. Entomol., 10th, Montreal, 1956* 3:169–74
124. Portchinsky, I. A. 1913. *Phalera bucephala L.* and its importance for the artificial breeding of *Pentarthron (Oophthera) semblidis* in winter). *Mem. Bur. Entomol. Sci. Commit., Central Board Land Admin., Agr.,* 10: 1–16 (Russ.).
125. Rabb, R. L., Guthrie, F. E., Eds. 1970. *Concepts of pest management.* Raleigh: N. Carolina St. Univ. 242 pp.
125a. Rao, V. P. et al 1972. A review of the biological control of insects and other pests in South-East Asia and the Pacific Region. *Commonw. Inst. Biol. Contr. Tech. Comm. No. 6,* In press
126. Ratzeburg, J. Th. Ch. 1840. *Die Forst-Insecten.* 2. Berlin
127. Réaumur, R. A. F(erchauld) de 1734–1742. *Memoires pour servir a l'histoire des Insectes.* Paris.
128. Redi, F. 1668. *Esperienze intorno alla generazione degli insetti.* Firenze.
129. Riley, W. A. 1931. Erasmus Darwin and the biologic control of insects. *Science* 73:475–76
130. Sachtleben, H. 1941. Biologische Bekämpfungsmassnahmen. In Sorauer: *Handbuch der Pflanzenkrankheiten,* 6, 2. Halbbd., 1–120, P. Parey, Berlin.
131. Sailer, R. I. 1972. Foreign exploration and importation of exotic arthropod parasites and predators. *Proc. Bio. Control Plant, Insect and Dis. Pests and Weeds Inst., Mississippi.* In press
132. Santis, L., Esquivel, L. 1966. Tercera lista de Himeopteras parasitos y predatores de los insectos de la Republica Argentina. *Rev. Mus. La Plata (n. ser.) Sec. Zool.* 9:47–215
133. Sasaki, C. 1886. On the life history of *Ugimyia sericaria* Rondani. *Tokyo J. Col. Sci.* I:1–46
134. Schulz (Wirschkowitz). 1924. Künstliche Vermehrung der Ameise. *Dtsch. Forstwirt,* Nr. 88.
135. Schutte, F., Franz, J. F. 1961. Untersuchungen zur Apfelwicklerbekämpfung (*Carpocapsa pomonella* (L.) mit Hilfe von *Trichogramma embryophagum* Hartig. *Entomophaga* 6:237–47
136. Shaw, J. G., Lopez, F., Chambers, D. L. 1970. A review of research done with the Mexican fruit fly and citrus blackfly in Mexico by the Entomology Research Division. *Bull. Entomol. Soc. Am.* 16(4).186–93
137. Silvestri, F. 1909. A survey of the actual state of agricultural entomology in the United States of North America. Extracts from *Reprint Bull. Soc. Ital. Agr.* 14(8). In *The Hawaiian Forester and Agriculturist.* 6: 287–336. Transl. from Italian by J. Rosenstein.
138. Silvestri, F. 1932. The biological control of insect and weed pests. *South-Eastern Agr. Coll. J.* 30: 87–96
139. Simmonds, F. J. 1958. Recent work on biological control in the British West Indies. *Proc. Int. Congr. Entomol.* 10:475–78
140. Simmonds, F. J. 1965. A brief history of the development and a resume of the activities of the Commonwealth Institute of Bio-

logical Control. *Commonw. Agr. Bureaux,* 24 pp. (mimeogr.).
141. Simmonds, F. J., Hughes, I. W. 1963. Biological control of snails exerted by *Englandina rosea* (Ferussac) in Bermuda. *Entomophaga* 8:219–22
142. Smith, H. S. 1919. On some phases of insect control by the biological method. *J. Econ. Entomol.* 12:288–92
143. Smith, H. S. 1929. Multiple parasitism: its relation to the biological control of insect pests. *Bull. Entomol. Res.* 20:141–49
144. Smith, H. S. 1935. The role of biotic factors in the determination of population densities. *J. Econ. Entomol.* 28:873–98
145. Smith, H. S. 1939. Insect populations in relation to biological control. *Ecol. Monogr.* 9:311–20
146. Smith, H. S. 1941. Racial segregation in insect populations and its significance in applied entomology. *J. Econ. Entomol.* 34(1): 1–13
147. Smith, H. S., Armitage, H. M. 1931. The biological control of mealybugs attacking citrus. *Univ. Calif. Agr. Exp. Stat. Bull. 509.* 74 pp.
148. Smith, H. S., Basinger, A. J. 1947. History of biological control in California. *Calif. Cultivator* 94: 720–29
149. Smith, R. F., Allen, W. W. 1954. Insect control and the balance of nature. *Sci. Am.* 190(6): 38–42
150. Souza, H. D. 1942. A broca da cana e seu parasito em Campos, R.J. *Boletim do Instituto de Experimentacao Agricola* 4:22
151. Speyer, E. R. 1927. An important parasite of the greenhouse white-fly (*Trialeurodes vaporariorum,* Westwood). *Bull. Entomol. Res.* 17:301–8
152. Steinhaus, E. A. 1956. Microbial control—The emergence of an idea. A brief history of insect pathology through the nineteenth century. *Hilgardia* 26(2): 107–60
153. Stern, V. M., Smith, R. F., van den Bosch, R., Hagen, K. S. 1959. The integrated control concept. *Hilgardia* 29:81–101
154. Swan, L. A. 1964. *Beneficial Insects.* New York: Harper & Row, 429 pp.
155. Sweetman, H. L. 1958. *The Principles of Biological Control.* Iowa: Wm. C. Brown, 560 pp.
156. Taylor, T. H. C. 1935. The campaign against *Aspidiotus destructor,* Sign. in Fiji. *Bull. Entomol. Res.* 26:1–102
157. Taylor, T. H. C. 1937. *The biological control of an insect in Fiji. An account of the coconut leaf-mining beetle and its parasite complex.* London: Imp. Inst. Entomol., 239 pp.
158. Telenga, N. A. 1958. Biological method of pest control in crops and forest plants in the USSR. *Int. Conf. Quarantine Plant Protection Against Pests and Diseases, Moscow, 9th, 1958.* 15 pp.
159. Thewke, S. E., Puttler, B. 1970. Aerosol application of lepidopterous eggs and their susceptibility to parasitism by *Trichogramma. J. Econ. Entomol.* 63: 1033–34
160. Thompson, W. R. 1930. The biological control of insect and plant pests. A report on the organization and progress of the work of Farnham House Laboratory. *Imp. Inst. Entomol., Farnham House, Empire Marketing Bd. Publ. 29.* 124 pp.
161. Thompson, W. R. 1952. The work of the Commonwealth Institute of Biological Control. *Brit. Agr. Bull. No. 19,* 53–57
162. Thompson, W. R. 1956. The fundamental theory of natural and biological control. *Ann. Rev. Entomol.* 1:379–402
163. Timberlake, P. H. 1927. Biological control of insect pests in the Hawaiian Island. *Proc. Haw. Entomol. Soc.* 6(3):529–56
164. Tooke, F. G. C. 1955. The eucalyptus snout-beetle, *Gonipterus scuttellatus* Gyll. A study of its ecology and control by biological means. *Un. S. Africa, Entomol. Mem.* 3:282
165. Tothill, J. D., Taylor, T. H. C., Paine, R. W. 1930. *The Coconut Moth in Fiji. A History of its Control by Means of Parasites.* London: Imp. Bur. Entomol., 269 pp.
166. Trotter, A. 1908. Due precursori

nell'applicazione degli insetti carnivori a difesa della piante coltivate. *Redia, Firenze* 5:126–32

167. Trouvelot, B. 1925. Directives à suivre dans l'importation pour les besoins de l'agriculture d'insectes entomophages etrangers. *Rev. Zool. Agr.* 24:125–48
168. Tryon, H. 1910. The "wild cochineal insect," with reference to its injurious action on prickly pear (*Opuntia* spp.) in India, etc. and its availability for the subjugation of this plant in Queensland and elsewhere. *Queensland Agr. S.* 25:188–97
169. Turnbull, A. L., Chant, D. A. 1961. The practice and theory of biological control of insects in Canada. *Can. J. Zool.* 39: 697–753
170. Tusnadi, G. 1954. Défense biologique par des rapaces contre les alifères nuisibles des nos étangs a poisons. *Allattan Kozlemények* 44:239–42 (Hung.)
171. Usinger, R. L. 1964. The role of Linnaeus in the advancement of entomology. *Ann. Rev. Entomol.* 9:1–16
172. van den Bosch, R. 1971. Biological control of insects. *Ann. Rev. Ecol. Syst.* 2:45–66
173. van den Bosch, R., Messenger, P. S. 1972. *Biological Control,* New York: Intext. In press
174. Villa, A. 1845. Degli insetti carnivori adoperati a distruggere le species dannose all'agricoltura. *Spettatore* 3(19):35
175. Watanabe, C. 1958. Review of biological control of insect pests in Japan. *Proc. Int. Congr. Entomol., 10th.* 4:515–17
176. Whitten, M. J. 1970. Genetics of pests in their management. In *Concepts of Pest Management,* ed. R. L. Rabb, F. E. Guthrie, 119–135. Raleigh: N. Carolina St. Univ., 242 pp.
177. Wille, J. E. 1958. El control biológico de los insectos agrícolas enel Perú *Proc. Int. Congr. Entomol. 10th.* 4:519–23
178. Wilson, F. 1960. A review of the biological control of insects and weeds in Australia and Australian New Guinea. *Tech. Comm. Commonw. Inst. Biol. Control, Ottawa, Canada.* 1:102
179. Wolcott, G. N. 1958. The evanescence of perfect biological control. *Proc. Int. Congr. Entomol. 10th.* 4:511–13
180. Wood, B. J. 1968. *Pests of oil palms in Malaysia and their control.* Malaysia: Inc. Soc. Planters, 204 pp.
181. Woodworth, C. W. 1908. The theory of the parasitic control of insect pests. *Science* 28:227–30
182. Yasumatsu, K. 1964. The possible control of rice stem borers by use of natural enemies. In Symposium on major insect pests of rice. *The International Rice Institute. Los Banos, Laguna, Philippines* Paper 2–1, 20 pp.
183. Yasumatsu, K. 1967. Distribution and bionomics of natural enemies of rice stem borers. (Research on the natural enemies of rice stem borers.) *Mushi* 39(suppl.):33–44

183a. Yasumatsu, K., Torii, T. 1968. Impact of parasites, predators and diseases on rice pests. *Ann. Rev. Entomol.* 13:295–324

184. Zaslavskii, V. A., Sugonyaev, E. S. 1967. Biological control of pests as a problem of modern ecology. *Zool. Zh.* 46:1536–51. In Russian with English summary

THE HISTORICAL DEVELOPMENT OF NINETEENTH AND TWENTIETH CENTURY STUDIES ON THE BEHAVIOR OF INSECTS

G. Richard
Faculté des Sciences, Université de Rennes
Rennes-Beaulieu 35, France

None of the branches of Biology is so suffused with the ideas of their time as is the study of behavior. This is why we shall try to emphasize here the relationship between research themes, authors' interpretations, and the general context in which these studies take place.

The Sources

At the end of the eighteenth century, during which time the spirit of philosophical enquiry was spreading throughout all classes, there developed among the more cultivated members of society a strong interest in nature; and all of European society was becoming "scientific".

In everyone's consciousness were notions of Cartesian rationalism, Leibnitzian determinism, Newtonian empiricism, and Lavoisier's materialism. J. J. Rousseau preached antiscience while, at the same time, opening wide the doors of nature to his disciples; Kant radically separated science from philosophy, and the great movement of ideas and structures released by the French revolution or by its inherent premises encouraged the development of the scientific mind.

Regarding the natural sciences, the eighteenth century saw changes in the explorers' habits of indiscriminately collecting anything that came to hand. Henceforth every important expedition included at least one professional naturalist. Thus, G. Banks, who accompanied Captain Cook, showed the way to future discoverers such as Bonpland, Humboldt, and Darwin. At the same time, Linne's method of classification and Lamarck's transformism were gradually gaining ground.

In the field of entomology, Réaumur and his disciples (Bonnet, de Geer, and Trembley) showed the way in precise observation, detailed experimentation, and accurate recording of phenomena. The *Mémoires pour Servir à l'Histoire Naturelle des Insectes* dominated the eighteenth century and served as a model for many nineteenth century authors. Réaumur's scientific discipline ("the more facts deviate from the rule, the more they have to be proved. . . ." a letter to Ch. Bonnet), as well as his new approaches (the existence of fringes of intelligence in animals) are much superior in terms of accuracy to the often flimsy ideas of Buffon, who nevertheless, with his unconvincing transformism, introduced a wedge of science into the doctrine of the Church.

Starting off from this painstakingly acquired basis, the nineteenth and twentieth centuries progressively led—via controversies between creationists and transformists, vitalists and mechanicists—from the simple description of animal mores, from the elementary action of environment on the living creature, to research attempts to find definition and purpose for a neuroethology in which insects played a prominent role.

Moreover, the increasing pace of research over the last 50 years of this century has been such that it is now utterly impossible to quote all sources—at least, within the limits of the present review. I have thought it wise to divide this survey into five periods: 1800–1850 when methods and ideas scarcely manage to emerge from the history of habits; 1850–1900 when the objective analysis of the relations between the insect and its environment, as perceived through its behavior, becomes more precise; 1900–1930 when mechanism begins to outpace vitalism, and transformism triumphs; 1930–1950 when new techniques are developed, and when the problems of complex behavior are formulated in terms of causality; 1950 to the present day when Schneirla's interpretation of behavior, Lorenz's objective ethology, Skinner's behaviorism, and Wiener's cybernetics are leading to neuroethology (Figures 1 and 2).

1800–1850: Continuation of the History of Habits

The first half of the nineteenth century, during which F. E. Melsheimer created the Entomological Society of Pennsylvania (1842), saw the foundation of the most important European entomological societies (France 1832, Great Britain 1833, Germany and Holland 1857); but it was not as fruitful in the field with which we are concerned as in the other branches of biology: morphology, embryology, or taxonomy.

In America, entomologists collected and described new species, but in Europe, François Huber's genius, kindled by Réaumur, enabled him, a blind man endowed with a strong sense of scientific investigation, to produce exemplary studies on bees and ants. His son and successor, Jean Pierre Huber, extended his work on ants and began to observe bumble bees.

Both father and son were interested in all aspects of biology, in the social equilibrium and in the individual behavior of the insects under examination; it was a time when general observation was more important than specialization. Thus, if François Huber demonstrated, among other phenomena, the existence and the function of ventilation in the hive, if he described mating flights, and if he detailed with more precision several of the facts adduced by Réaumur, Jean Pierre Huber, using a strict experimental method, discovered the mode of dealation of the founder female ant, and related its dealation to separation from the colony, to fertilization, and to its aptitude to found a new colony. Using the same method he ascertained, among the social Hymenoptera, the insect's ability to orient itself in its own environment and the existence of communication between individual members of the same species. The same authors proved beyond any shadow of doubt the existence

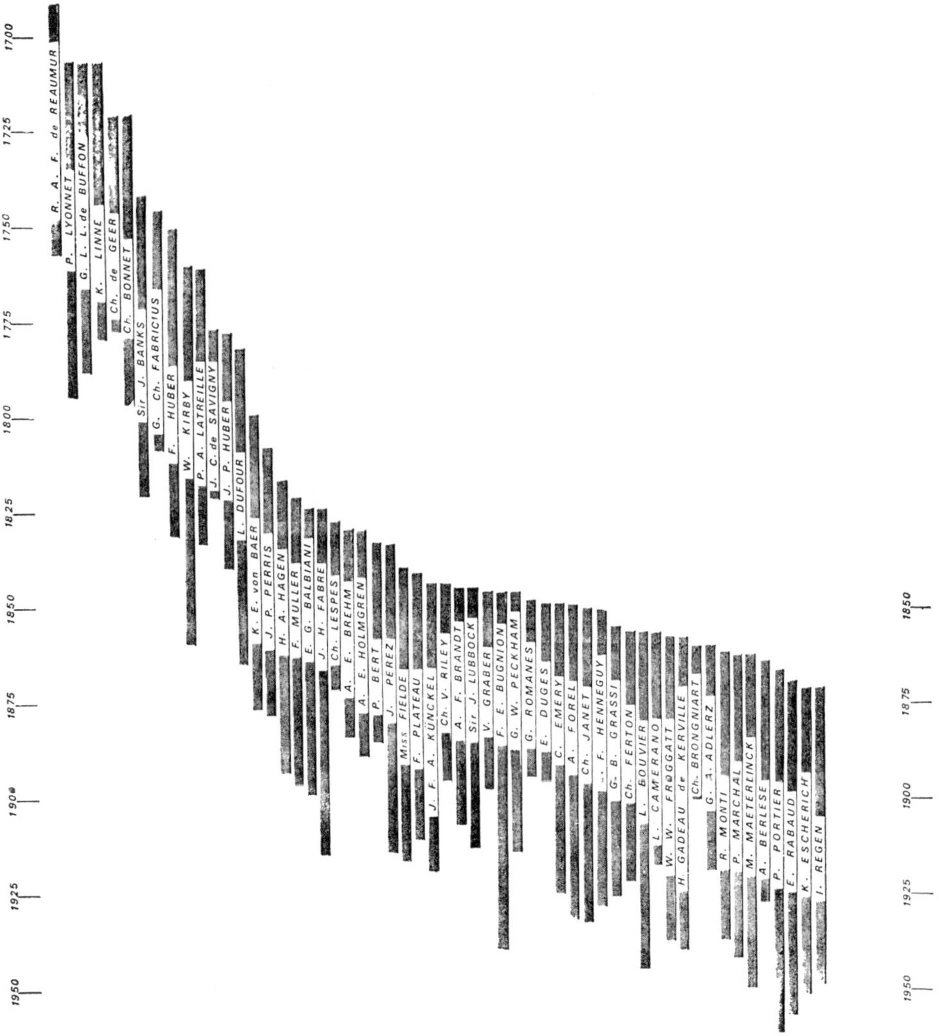

FIGURE 1. Life periods of 56 entomologists.

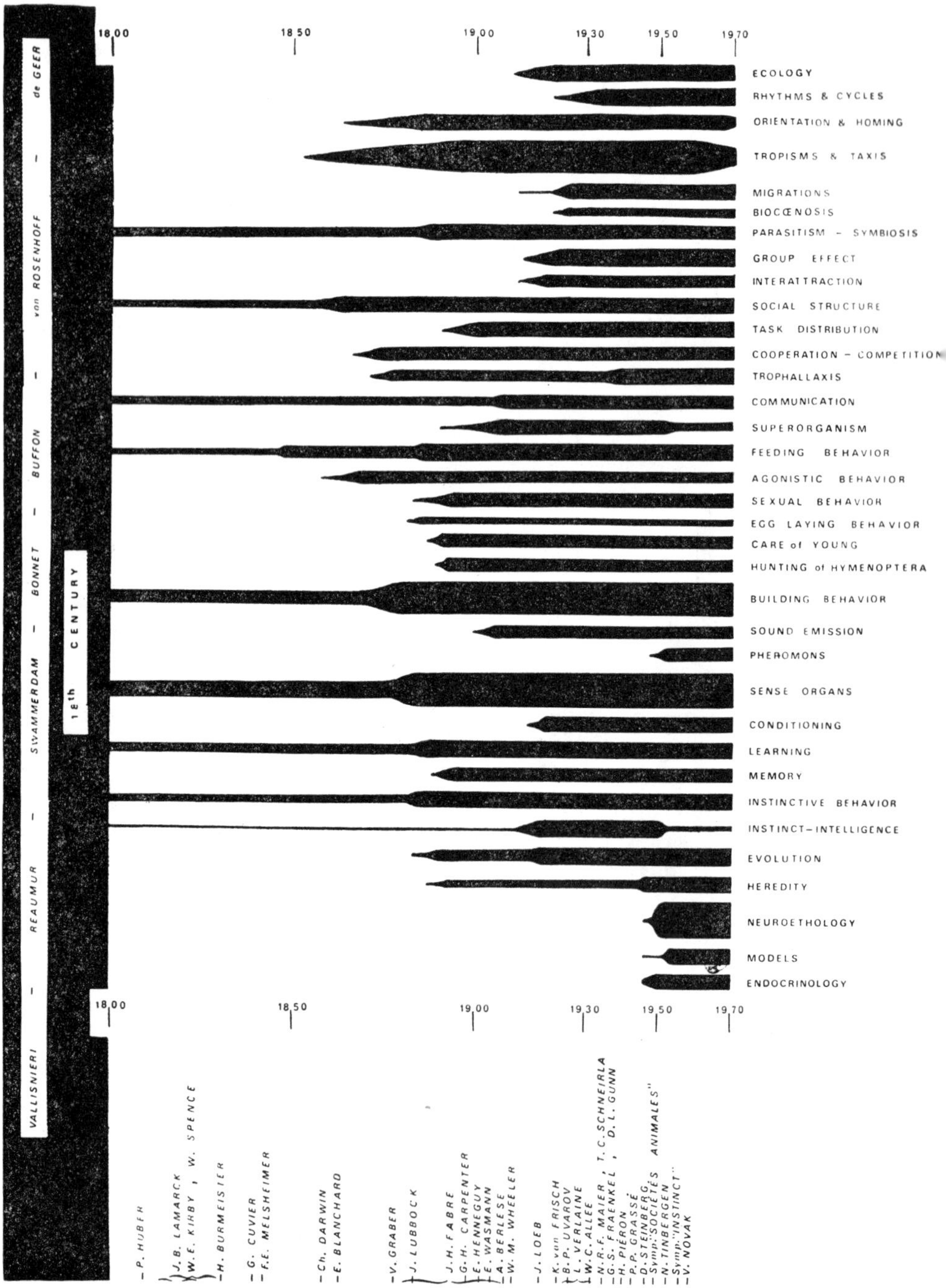

FIGURE 2. Schematic representation of the origin and importance of major behavioral research during the nineteenth and twentieth centuries. Names of some authors are mentioned facing their first fundamental publication.

of esclavagism and brood transport among ants. However, the fact remains that very often enthusiasm and admiration affected the precision of observations and that interpretation of the facts went beyond pure objectivity to embrace a particular philosophical creed.

J. P. Huber was certainly a precursor of future taxonomists of behavior when, for example, he maintained that "each species has its own habits, each individual has its own constitution"; but he went beyond the essential limits of objective observation when he perceived, among slave ants whose world is ruled by "order and harmony", the "Creator writing out with his almighty hand the laws of a republic free of all corruption which could serve as a model for those more complex societies where the common good necessitates slavery".

Observers of the period were obviously most interested in insect behavior, which led them to pose the problems of instinct in the terms in which they were to be discussed for the next hundred years. Comparisons were made between the set behavior of animals and man's freedom: "Instinct in animals is the analogue of the soul in man; a soul differing only from the human soul by the necessity with which it does everything, whereas the human soul is independent of necessity, and freely resolves upon its actions" (H. Burmeister). However, there were also considerable possibilities of regulation exhibited by at least some insects (Darwin's *Sphex,* Lund's ants) "which appears to be the result of a free and rational consideration and of a certain degree of reflection . . . the reflection of insects consisting in the choice of the best means to attain the object" (Burmeister). Above all, an ability to acquire a certain amount of experience and to combine previous behaviors was credited to the mnemonic powers of insects: "the experience gained must remain in constant recollection if it be to yield a constant advantage, and it is made so by a quality of the soul which we call memory" (Burmeister); observations which seemed to support this were made by C. K. Sprengel on the effraction of flowers by bumble bees, by Kirby on the location of swarms from the same hive and by Huber on the behavior of ants on their trails.

Despite a tendency to subjective interpretation, the facts accumulated over these 50 years; Audouin detected the preliminaries to mating in the *Cantharis;* Rathke, Burmeister, Dufour, and von Siebold noted precisely the modes of pairing in the Odonata; Goureau was the first to describe those of mantids. Oviposition behavior also attracted observers' attention; von Siebold described the endophytic laying behavior of the *Lestes* and Kirby reexamined and clarified de Geer's observations concerning the projection of the *Tipula*'s eggs onto the ground.

Dufour described most carefully the more complex behavior of predatory Hymenoptera, while other observers began to make out some aspects of caste biology in insect societies and the behavioral links between myrmecophilous insects and ants. Sykes and Savage directed attention to ants' provisioning problems, Westwood thought it correct to relate the numerical regulation of members in the different castes to the action of worker ants on the population, while Smeathman fully described the role of larvae in lower termite societies. Huber, Muller, Grimm, and Markel analyzed the relationships between ants and other insects, which either live with them in their nests or which they use (e.g. aphids).

But it is perhaps only when one considers the problem of the insects' relationships with one another and with their environment that one fully appreciates the discrepancy between the growing amount of descriptive material and the absence of definitive interpretation. Perris (on the ichneumonids), Lefebvre (on bees), Duges and Dumeril (on Muscidae), Hauser and Erichson (on various other insects) located the sense of smell in the antennae. F. and J. P. Huber proved that ants recognize one another through smells. The Hubers, once again, then Kirby, Spence, and Goureau ascertained the existence of communication among the social Hymenoptera, but the means of this communication, of its connection with chemical signals, remained undiscovered.

The treatises and manuals on entomology written by Kirby and Spence, Burmeister, and Blanchard reflect this total lack of generalization.

1850–1900: Objective Analysis

The second half of the nineteenth century saw the beginning of a new expansion both in the amount and in the specialization of work undertaken. However, the general naturalist tendency inherited from the previous century did not completely disappear. Descriptions were given of agglomeration phenomena among insects, especially among Coleoptera (Kirby and Spence, Camerano, Fabre, Targioni, Osten, etc). Observations were reported on short range migration (processionaries studied by Berlese or Fabre) or long range migration (locusts observed in America by Riley; butterflies, aphids, and dragonflies by a great many authors).

The behavior of social insects became the subject of numerous studies along the lines of the methods used by predecessors. Réaumur had been the first to observe and interpret correctly the mating flight of the *Lasius* in 1731; J. P. Huber had analyzed the process of dealation of the fertilized female; but Lubbock, Muller, Potts, and Blochman described colony foundation in divers species of ants. Lubbock proved the longevity of the queen and tackled the problem of cooperation and communication. Dujardin provided conclusive proof of communication among bees: he placed a cup of honey between two hives in a place little frequented by the bees, then he watched for some days as workers bees from one hive used this food supply, while those from the second one ignored it. However, Lubbock's experiments with wasps certainly showed that even if such distribution of information was possible in a society, the use made of this capacity was not automatic.

The foraging behavior of ants (Bates and Lubbock), parasitical or symbiotic relationships among social insects (Lespes, Hagens, and Lubbock), and the social structure (Schenk, Espinas, Bates, Lubbock, McCook, and Nasonov) were studied in depth. Particular mention must be made of a type of behavior first noticed by Forel: mouth to mouth exchange of food between two individuals of a society of ants. At the time, this phenomenon (later called trophallaxis) seemed relatively episodic; its true significance was only realized at the beginning of the twentieth century.

Following the example of P. Bert, one of the founders of psychophysics, many authors became interested in the sensorial basis of behavior. They were particularly interested in vision. Murray compared blind insects to those having normal eyes. Lubbock, Perez, and the Peckhams demonstrated the existence of color vision in bees. Lubbock showed the different powers of attraction between pieces of red and blue cardboard and Perez induced bees to visit a *Pelargonium* flower by putting honey on its corolla; but Lubbock emphasized the specific characteristics of the visual spectrum after demonstrating bees' sensitivity to ultraviolet light and then posing the color problem that this kind of radiation may represent for the insect. In a complementary study adding to the observations of Graber and Merykowsky, Plateau demonstrated that colors are distinguished not by wave-length but by the intensity of radiation to which they correspond. Taking much the same line, Pierantoni emphasized the stimulating value for behavior of shapes (the attraction of the chrysanthemum flower to insects) and Forel experimented with *Bombus* and artificial flowers; thereby indicating the probable existence of other senses in addition to vision as factors controlling honey-gathering behavior. Plateau and Exner analyzed movement perception and its role in stimulation.

Chemical sensitivity presented an experimental field which, if still fairly mysterious, allowed for plenty of research. Balbiani analyzed most carefully the role of the antennae in mating in *Bombyx mori* and proved both the location in the female of a source of odor, and the existence in the male of sensitive organs. Fabre, in his memorable experiments on the greater and smaller emperor moths, enlarged upon Balbiani's results which have been repeatedly taken up and expanded.

Graber demonstrated the reactivity of different insects to contact with chemical substances; but Plateau and Forel qualified his results by showing that by using only nonirritant substances only the reactions of the antennae would be demonstrated.

In the case of ants, Bethe described a type of aggressive behavior with olfactory stimulation, and Forel demonstrated the existence of a topochemical sense which would explain certain aspects of the structuring of their environment by means of odorous trails on which the insects move about.

The powers of attraction of various physicochemical factors in the environment were first applied to the catching of insects by means of a light source (Berce and Dupuiset). This led to a recognition of the importance of orientation phenomena in the expression of behavior patterns. In this respect those who first insisted on insects' orientation ability (Lubbock, Fielde, Bates,

Bethe, Peckham, Romanes, etc), appeared as pioneers of the great interest later roused by tropisms at the beginning of the next century.

Another characteristic of these 50 years was the general tendency towards precise observation and the description of particular sequences in behavior.

Sexual behavior was the subject of many studies: Brauer described the mating behavior of the *Bittacus,* during which the same prey may be eaten by both partners; Balbiani observed multiple mating by a single male in *Phylloxera;* Olfers, then Reuter and Levander, worked on the behavior of Collembola (description of *Sminthurus* mating, among other species). Beauregard continued the analysis of mating behavior in cantharid beetles which was later completed by Fabre. Kolbe and Ingenitzky completed the description of the mating modes of Odonata. But very often the analysis of sexual behavior was limited to a careful description which was affected by anthropomorphism, as witnessed by numerous references to sexual perversion. Peragallo observed the coupling of telephorid (cantharid) males; Gadeau de Kerville, after Germar, described the same phenomenon in many Coleoptera of different species. Dubois observed *Bombyx* males coupling with the substratum on which a female had rested.

Oviposition behavior was also the object of many studies; Robin, then Balbiani, described that of *Chironomus;* Regimbart the endophytic laying of *Dytiscus;* Adler and Rilec that of cynipids; Taschenberg that of *Panorpa;* Perez that of *Histeropterus* and Berlese that of Homoptera. Cockroaches were the interest of Duchamp, Kadyi, Wheeler; mantids that of Brongniart; and termites that of Grassi. Kunckel d'Herculais' description of the mechanical means used by the female acridian to drive her abdomen into the ground ("as if one were driving in a stake, and not drilling") remained a classic for a long time.

But with the descriptive research of Dufour, Fabre, Janet, Kellog, Marchal, the Peckhams, Perez, Perris, and Williams, among others, on behavior patterns in nest-building, prey-catching, and care of the young by many insects (especially solitary Hymenoptera), fuel was added to the transformist controversy concerning instinct which was to dominate the end of the century and nourish the thought of Bergson.

Behavior patterns were described in great detail and the sequence observable in the sphegids (sphecids) became standard: digging the burrow, hunting the prey, the latter's capture and immobilization, its transfer to the burrow, laying the egg, renewal of provisioning, closing the burrow.

Réaumur's careful and reliable descriptions did not go beyond the bare facts. Dufour, a disciple of Latreille and an enthusiastic recorder, added precision to his predecessor's findings. Fabre's descriptions are a psychological and poetic interpretation of nature. Without ever trying to define instinct, Fabre tried to point out what he considered to be its characteristics, describing sequences of invariable acts inevitably linked together in a necessary order. The insect "is unaware of its marvellous talents, just as the stomach is unaware of its complex chemistry. It builds, weaves, hunts, stabs and paralyses in the same way as it digests, secretes its weapon's venom, its cocoon's silk, its comb's wax: without being aware of the means or the purpose" (*Souvenirs Entomologiques,* 4° Série, Chapter III).

The blind fatalism contained in instinctive phenomena which Fabre emphasized by means of often incomplete observations led him to insist even more than his predecessors on the taxonomic value of behavior. Fabre underlined the fact that it was psychical differences "which mark most dramatically the impassable dividing line between two closely related species".

The rigid routine of the species does not deny some discrimination, however: "the insect can distinguish between dry and wet, sheltered and exposed, solid and fragile; and can choose the most easily established dwelling-place" (loc cit, chapter V). It can even allow for some possible readjustment in privileged species: the *Copris* which adapts its behavior to the collection and preparation of its stercoral building material and *Osmia* which dictates the sex of its progeny; but it is rigid enough to make Fabre adhere to a narrow fixism which in turn obliges him to combat the rising tide of Darwinian transformism. The two men respected one another, but Fabre implacably sought to "prick the enormous transformist bubble in order to reveal its inanity" (V. Legros). The mimicry of the *Volucella* and of some other insects seemed to Fabre a suitable example with which to demonstrate the limitations of Darwin's conceptions. After the latter's death, Perez tried to win over Fabre to "some kind of understanding", but the two men differed too much in their conceptions of entomology, notation of observations and evolution.

It was Ferton who was to point out the gaps in Fabre's work and to complete his observations. At first full of admiration for the work of the "hermit of Sérignan", then later convinced of his limitations, Ferton observed and experimented on precisely determined insects, using a strict method. His criticism gave rise to the mechanicalist reaction in France, which was at first fruitful.

1900–1930: Mechanism

During this period, the fundamental behavior patterns of insects, including alimentary, sexual, interindividual, and building behavior, were more rigorously analyzed (Boldyrev, Fedorov, Fabre, Haber, Hamm, Marshall, Ondeman, Poulten, etc). The role of olfaction in behavior attracted a great deal of attention (Fabre and Noel continued Balbiani's experiments and clearly demonstrated the role of the antenna in sex approach in saturniid Lepidoptera); Haase, then Illig, recognized the presence on the wings of some butterflies of odorous organs which play a fundamental role in sexual behavior. However, anthropomorphism often still concealed reality: Fabre asserted that he could not smell anything on a female emperor moth, and was astonished by the sight of a male going to the female in a room smelling strongly of arum. For similar reasons, Noel referred to the antennae as receiving electric waves. The analysis of the trigger effect of sounds on behavior led the same authors to the same misinterpretations, despite the excellent work of Froggatt and Kunckel d'Herculais on the song of the acridians, and despite Regen's irrefutable demonstration of the attraction of the female cricket to the chirping male and on the very real dialogues carried on by *Pholidoptera.*

The reaction against vitalist and finalist ideas took place as a corollary to the extension given to the reflexological theory in the field of physiology. It also happened as a result of the new interest in both the animal's environment and in its analysis by means of the sensory organs whose capacities could now be precisely described by psychophysics.

Eltringham, Mast, Biest, and Brun, as physiologists, analyzed the functioning of insects' sensory organs, but authors tackled the problem mainly through behavior reactions. Following Loeb, many of them had started work on tropisms and especially on phototropism. If Mast and Patten, like Loeb, both insisted on the role of light intensity in triggering off reaction or the reversal of its sense, they differed violently about the problem of forced movements. Loeb synthesized the theories of the botanists Sachs and de Candolle and having studied the aimless circular movements of a blinded insect, he considered that tropistic reactions depend on a combination of two factors: the direction of light flux and the symmetry of reception. Others, e.g. Dolley, Garrey, Minich, rediscovered and completed Loeb's findings. Many others, e.g. Baldus, von Buddenbrock, Clark, Holmes, Mast, Radl, and Rabaud, denied these results and described a readjustment of this circular movement provided it is allowed to go on long enough.

Mast championed the idea of a connection between visual stimulation and muscular tonus: according to his theory asymmetric lighting gave rise to a tonic asymmetry which alone explained the circular movement of the insect. Rabaud, based on his own work and that of Pictet, extended this idea of tonic response to cover stimuli other than visual (the butterfly's tilting reaction to heat) and he attempted to classify insects from this point of view in a continuous series according to criteria related to eye-size and to relative importance of locomotory muscles. The possibility of sensory substitution or re-education was frequently referred to then.

However, if the explanation of tropistic orientation by tonic variations seemed to be proved, the fact had to be accounted for that this orientation seldom happens on its own; and that it is often combined with locomotion. Patten distinguished between photokinesis and movement orientated under the influence of light and von Buddenbrock analyzed the variations in the orientated response during the time of movement. He discovered a limit to Loeb's rule of parallelogram of forces, since he observed that between two light sources there is a point of decision for most insects.

Many authors investigated the direction of insects' tropistic reactions. Some believed in the existence of negative tropisms having the same characteristics as positive ones; it could not be said that the criticizable experiment with the locust confined in a tube and crawling backwards towards the light source represented conclusive proof concerning the problem of reversal preceding orientated locomotory expression. Undoubtedly, the newest experiments, both in form and essence, were those which tried to link one particular type of reaction of the insect with internal factors: Mast showed that by manipulating a butterfly, its positive response could be made negative; Woodsedalek showed that the *Dermestes* is photonegative in the larval state and photopositive after mating and egg-laying. Bethe, in 1897, had already shown that the *Hydrophilus*' photonegativity could be eliminated by effecting a median section of its brain; Matula experimented on *Aeschna* larvae by removing different parts of their cerebroid ganglions; this kind of research was carried on by Baldi (on Coleoptera) and Alverdes (on *Cloeon*), but their results were often at variance with the theory of tropisms.

As the authors of the beginning of the twentieth century developed a reflexological theory of insects' orientation in their environment, they differed more and more from Loeb, perhaps influenced by those who, like Jennings, could see finality in orientated behavior. But even more, they formed the habit of interpreting behavior as the result of a whole series of complex actions to which both the external and internal environment of the insect contributed. Unfortunately, their experiments were too numerous and their procedures were not sufficiently comparable, and Kuhn's classification of tropisms did not provide the required guideline. Thus, Manquat accounted for the butterfly's reactions in front of a flame by the conflict between mechanism and self-interest and Manquat, Patijaud, and Pictet believed that behavior is dictated by the law of self-interest, by certain painful or pleasant sensations, sometimes even by feelings of a higher mental order.

Those who adhered to a more precise nervous determination sought after a more integrated interpretation of complex behavior patterns. If Poiret described the excitation of the male *Mantis* as he is being decapitated by the female, Rabaud now put forward the idea that the reflexes of the last abdominal ganglia depended on the inhibitory influences of the cephalic ganglia. McCracken's admirable experiments doubtless laid the foundations of subsequent research which developed considerably after 1950: the suppression of the thorax does not prevent egg-laying provided one stimulates the abdomen directly with the fingers, but the eggs are disordered; the cutting of the abdominal ganglions produces a lack of coordination.

The insect's ability to orientate itself in accordance with its environment, by physiological mechanisms behavioral scientists tried to understand, was more totally analyzed in integrated behavior such as the return to its nest of the solitary Hymenoptera (Bouvier, Malyshev, and Rabaud), or ants following their tracks. Concerning the latter, Brun demonstrated the chemical polarization of tracks leading to aphids; he emphasized the importance of visual landmarks all along the tracks; following on from Santschi who proved that insects can take bearings on the sun, he introduced the idea of the compass-eye in 1916.

As at the end of the nineteenth century, solitary nest building Hymenoptera continued to provide the basis for the controversy about instinct. Ferton's work, despite its often anthropomorphic vocabulary, stated more surely than that of any of his predecessors that instinct and anatomical dispositions should be considered of equal value by systematicians. Along with Adlerz, Nielsen, Picard, and Saunders, he brought forward a great number of precise observations on the paralyzing sting of the Hymenoptera, and he believed instinctive behavior to be a reflex movement. The Peckhams and Picard insisted that the sting is dependent on the prey's reactions, and that it is never admirably precise; Marchal emphasized that the predatory insect licks the liquids oozing from the prey as a function of its self-interest (alimentary) and not, as Fabre explained, in the interest of its progeny.

With Rabaud, one got away from the vitalist finality concept of an aim known to the animal before any expression of complex behavior. After Ferton, Janvier, Marchal, Minckiewicz, and Picard, he insisted:

> . . . any animal movement is the resultant of two elements of equal value both of which have definite properties which do not appear simultaneously. Just as the external environment varies, the animal's physiological state may be normal or accidental, temporary or permanent: of necessity therefore, the general interpretation of behavior must equally take into account these two groups of variables which merge, ineluctably into one complex whole: the organism x environment complex.

All this was very far from Fabre, and von Uexkull's ideas pervaded behavioral science: in a definable *umwelt,* each species of insect possesses a *merkwelt* whose limits must be found. Instinctive behavior was no longer rigid and self-sufficient; it became a complex series of sequential elements each of which is an automatic response (devoid of finality) on the part of the animal to the external and internal situation produced by the previous phase.

At this time, one particular environment began to attract the interest of research workers: the one constituted by the congeners in a group. Bohn had just shown that in ship-worm (*Convoluta*) the density of the animals in a determined environment modified the physiology of these animals. And Rabaud directed attention to grouping, distinguishing between chance gatherings and more coordinated ones which lead to the formation of societies. But migratory locusts were to become the center of interest when Uvarov proved that their shape and activity are sensitive to the density of their populations, thus bringing up phase phenomena and greatly simplifying these insects' taxonomy.

Much research was conducted into environmental actions affecting more than the insect's actual behavior: von Frisch's bees learned visual discrimination based on form, color, and contrast in substrata; Woodsedalek's dragonflies learned a detour and flew to a stone which provided shade and shelter; Forel's ants learned to avoid honey containing strychnine; Fielde's ants learned to recognize the colony's specific smell; Beling's bees were trained to collect honey at times differing from the normal 24 hr rhythm; Brown's whirligig beetles accustomed themselves to their environment.

But it was undoubtedly Verlaine and Schneirla who led the way in precise research experiments concerning the learning capacity of insects in the laboratory. Ants' ability to solve maze problems was then analyzed and continued to be the subject of much work in the following periods. All this pertained to insects' psychic life as it was conceived of by Bouvier, Hungerford, Peckham, and Williams and provided a basis for hereditary adaptation (Pictet and Schröder) which was in accordance with evolutionist conceptions of behavior as represented by Ferton, Janvier, and Minkiewicz.

More and more research workers became interested in insect societies, and apart from Fielde, Beebe, Brun, Bugnion, Cornetz, Descy, Doflein, Eidman, Emerson, Escherich, and Roubaud, who must be mentioned for their work on Hymenoptera or termites, some authors stood out. Among them were Rösch who discovered in the bee behavioral aging and task specialization according to physiological aging; Wheeler, whose immense studies led among other things to the study of polyethism in the ants; Forel who, with Emery, Fielde, Nager, Roubaud, and Wheeler, proved the general importance of a fundamental phenomenon in social insects: trophallaxis which soon headed all research on the interattraction of individuals and on social regulation; Wassman who specialized in analyzing the relationship of ants with their commensals; Lameere who, following Spencer and Emery among others, started the discussion on the evolutive origin of insect societies; but especially von Frisch who discovered in bees the existence of intercommunication by means of dances and who noted precisely the sensory summation of movement and odor, thus opening an immense field of experiment for many future research workers.

1930–1950: New Techniques and Theories of Behavior Analysis

This period, despite political and economic upheaval, was fruitful for the development of behavior research. New techniques and methods appeared. Hans Berger had just produced the first encephalogram in 1929. In America, an interdisciplinary group founded cybernetics, the value of which was revealed in Europe by Wiener's book in 1942. The ethological method using models opened up a great research field. More important was the high quality of thought of a number of original minds (not all entomologists) who were later to become, in the next period, the leaders of the different groups and schools. Allee, Bey Bienko, von Buddenbrock, Butler, Emerson, von Frisch, Gösswald, Grassé, Lorenz, Pavlov, Schneirla, Tinbergen, and Uvarov were among the most eminent and their renown quickly spread internationally.

The animal's relationship with its environment preoccupied research workers, as also did the forms and means of structuration in insect groups and societies. The analysis of causality was facilitated by the increased precision of techniques.

The study of sensory reception was one of the first diciplines to benefit from the new techniques: e.g. Autrum's work on sensitivity to vibrations and on visual sensitivity. At the same time, research of the same type as used in the preceding period was carried on; and Faber on Orthoptera and Malek on *Notonecta* brought precision to the reactions to mechanical stimuli arising from the environment. Vision remained a favorite research subject: authors were interested in the general properties of light (Bertholf showed that most insects are totally blind to the color red; Ilse and Opfinger defined the influence of various colored radiations of the spectrum); also in environment structures (Hertz and Hundertmark showed the effect of contours on visual selection and compared them to the effects of surfaces); as well as in the structure of receptors (Ludtke described the role of different parts of the eye in the *Notonecta*'s orientation reactions).

A new method was used in the analysis of the relative movement of the environment in relation to the insect: the optomotor reflex method (von Buddenbrock and Gaffron).

All this analysis was immediately integrated into tropistic and taxic theory, and the tendency developed to closely relate stimulus reception and the positioning of the insect (von Buddenbrock defined, among other things, the light compass reaction). But what most characterized the new research on taxis was the study of the combined effects of the summation of divers factors to the orientated behavior. Bohn insisted upon the phenomenon of differential sensitivity which modified the rigidity of Loeb's theory; Brandt on *Ephestia* and Weyrauch on *Notophilus* combined photo- with scototaxis.

These studies, of course, referred back to complex reaction in the natural environment (Bodenheimer); and just as Kalmus analyzed geotaxis, Buck demonstrated the response of *Photinus* to intermittent flashes of light emitted by its congeners, Strelnikov analyzed various insects' photokinesis after sunset, Fraenkel analyzed the combination between tarsal contact with the ground and the insect's taking off and alighting, and Bierens de Haan analyzed the diverse phases in the orientation of the *Geotrupes* during its olfactory alimentary taxis.

Building behavior was analyzed by Dembowsky and Frankenhauser on Trichoptera; Baerends, Carpenter, Malyshev, Minkiewicz, and the Japanese teams of Iwata, Matsuda, Tsuneki, Uchida, and Yasumatsu on Hymenoptera; and Chen, Grassé, and Weyrauch on social insects. Alimentary behavior was analyzed by Lukyanowitch who showed the specialization of phytophagous insects; Sweetman who attempted to establish a continuous series from parasitism to predation; and Schneirla and many others who described harvesting in ants.

Using as a starting point most older studies and those of the present period, Berland forcefully put forward the idea of a behavior specificity. Similarly, Fulton, working on the cricket's chirping, Kozhanchikov working on alimentary behavior, and Verlaine studying aggressiveness in the bee experimentally, proposed the idea of a possible heredity of certain behavior patterns.

But the influence of Lorenz and Tinbergen's objective ethological methods gave rise to two studies of importance: that of Tinbergen, Meeuse, Varossiau, and Boerema (1942) on the sexual behavior of *Eumenis,* which for more than ten years was one of the few examples of the analysis of insect behavior using the models method; and that of Baerends on the nest-building and parental behavior of the *Ammophila* (1942), which gave a solid interpretation of sequential phenomena in the simultaneous maintenance of several nests.

A great number of research workers, doubtless influenced by the predilection of white rat psychologists for mazes, but also influenced by von Frisch's stimulating theories, became very interested in the study of temporal variations in behavior: memory, conditioning, learning. Brandt on *Ephestia,* Gates on cockroaches, Thorpe in *Nemeritis,* Hoagland and Schneirla on ants, von Frisch, Hörmann, Hunter, Rau, and Verlaine on bees, all contributed to demonstrate that retention phenomena are linked to the physical environment (Hunter on the effect of cold), that they may be helped by the grouping of the animals (Gates and von Frisch), and that certain types at least are hereditary (Verlaine). In this direction, as in many others, Schneirla's work was extremely important; it aimed at defining the role of efficient motivation in learning, and of the elements depending on stereotyped insect behavior. Thus the objective possibility of comparative characterization in animal-environment interaction in invertebrates and higher vertebrates was arrived at. This led in turn to a very new presentation of older concepts of instinct and intelligence.

Something which was also an important tendency in the 1930–1950 period was that numerous research workers tried to adapt pollinating insects' learning capacity to agriculture (Gubin) and to use predator-prey or parasite-host relationships in what was later to be known as biological control.

Nevertheless the tendencies of the period revealed themselves most clearly in connection with social insects. Studies were made of social structure and its premises in different types of groupings, individuals' sensory or alimentary relationships, and division of labor.

Following Uvarov's example, research workers became involved in a study of sensory inputs upon which depends the phasic effect. This involvement marked the beginning of a great number of entomological studies on group effect, a phenomenon which Grassé first tried to define in a series of lectures given in 1942, and then later in numerous publications. Many insects respond, either in their shape, physiology, or behavior, to the density of the specific grouping in which they live, and consequently to the intensity of their visual, tactile, and olfactory interactions (Chauvin, Chen, Chopard, Landowski, Badonnel, Bonnemaison, and Waloff, supplied the first proof of this). But what are the factors which determine grouping? Picard attempted a classification of biocoenosis, Rabaud discussed the levels of interrelations between individuals, Allee created a socio-biology; but it was the school of Grassé (Chauvin on Orthoptera and Ledoux on cockroaches) which first demonstrated the existence of interattraction among some insects. This fruitful concept was to go on gaining ground and in 1942, with his masterful causal analysis of the swarming of termites, Grassé opened up the way to a renewal of the French ethological school. Not only did he confirm the existence of interattraction among termites but he also for the first time integrated this social factor with physiological events inside the colony (sexual maturation of swarming insects) and to ecological development (seasonal rhythms: of swarming and of the relative numerical composition of castes in the society).

For his part, Schneirla contributed a great deal with his field study of *Eciton* by which he proved the integration of ecology, physiology, and behavior in doryline societies, since the alternation of this ant's nomadic and stationary phases depend on a complex summation of climatic factors, alimentation of queen, workers and larvae, and brood comb maturation.

As concerns interindividual exchanges within a society, trophallaxis appears to be a privileged link. Van Boven, Clark, von Frisch, Gösswald, Grassé, Haskins, Le Masne, Noirot, Raignier, Schneirla, and Wheeler emphasized its diverse forms. Moreover, part of these exchanges concerned the relationships of social insects with their symbionts, as analyzed by Hingston, Le Masne, Reichensperger, and Wasmann.

The idea gradually took shape that this nutritive exchange might constitute one of the elements of behavioral cohesion in the social group, and the work of von Frisch's school integrated this idea more and more into the postural communication characteristic of the language of the bees so diverse in its modalities (Lindauer).

This represented another way of achieving synthesis (by the study of behavior) between ecological response (summation of the climatic and astronomic space and time factors of the environment) and physiological response (variation of maturation expressed by the succession of tasks carried out by the bees, general excitation level related to food, and sensory analysis of events external and internal to the insect).

Emerson's ideas concerning the superorganism derived directly from the global idea contained in this synthesis; they were challenged by Schneirla, among others, who tried to show that a comparison of an insect society and a metazoan organism resulted from an over-schematic interpretation of the facts.

Apart from this original research, other more descriptive work was carried out on the division of labor (Chen, Ehrhardt, Goetsch, Grassé, Kiel, Steiner, Verlaine), on mutual aid (Stager), on building behavior, and in particular on the meaning of diverse specific structures built by termites (nest building of the *Apicotermes* described by Desneux, mushroom combs interpreted hesitantly and even completely erroneously by Ghidini, Kemmer, etc).

Anthropomorphic interpretations were of course still current, but they were fortunately more often to be found in literature than in science: one example of this was Maeterlinck's so-called vulgarization on bees, ants, and termites, which was full of errors and warped by finalism.

As a logical consequence of all this work on insect societies, an international assessment was necessary. Grassé took the initiative in this and, in 1950 in Paris, a CNRS Conference was held on animal societies, at which the most eminent researchers compared their points of view, often trying to establish the characterization of insect society opposed to that of vertebrate society.

1950–1970: Toward Modern Synthesis

In this last period, it has become completely impossible to mention more than a minute fraction of the names of those doing field or laboratory work; but all of them are tackling the problems of insect behavior in one of six principal ways: 1. Research into the homeostasis achieved by means of behavior between the animal and its environment; 2. Research into the immediate and more remote causal relationships existing between stimuli and responses; 3. Research into the physiological mechanisms underlying behavior; 4. Study of the ontogenesis of behavior; 5. Study of the evolution of behavior; 6. Research into models of behavior.

At the same time as research has been continuing in these diverse ways, the vocabulary has also been modified, and by borrowing from operational vocabulary, objective descriptive terms are gradually replacing anthropomorphic ones.

Insects are becoming a widely used laboratory material; there is no doubt that in this connection *Shistocerca gregaria* and *Locusta migratoria,* with diverse cockroaches close behind, are the most studied from the point of view of behavior, about as much as the *Drosophila* is studied in genetics.

All the new approaches are descended from the ethological approach of the preceding period, according to which, movement, a motor act, with its more or less complex sequential combinations, made for an objective definition of behavior.

The ecological environment imposes either short- or long-term rhythms (Bodenheimer, Cloudsley-Thompson, Corbett, Danilievskii, Ghilarov, and Lobashev) whose nervous and endocrine regulation is analyzed by very many authors (Harker, Pittendrigh). Some of these rhythmic locomotor movements correspond to migration studies of the acridians carried out by the team of researchers working directly or indirectly for Anti-locust Research (Uvarov, Haskell, Bey Bienko, Kennedy, Le Berre, Michel, Rainey, and Waloff) and of other terrestrial or aquatic insects (Macan, Popham, Richard, and Southwood).

Work done on taxis has a completely new approach, both in their description and their functional interpretation. If the orientation of response still seems important, the variations of its direction in respect of the physiological state of the insect is studied a great deal (Schöne, Medioni, Rabe, Richard, de Ruiter, de Wilde, and Viaud). But what is most important is that they are gradually being integrated either with ecological space (Dufay, Ellis,

Kennedy, and Verheijen) or with that space which is more significant for more complex behavior (Baerends, Chmurzynski, Papi, Queinnec, Sakagami, Tinbergen, Thorsteinson, Tsuneki, etc). Research is being done into the total value of the cyclic summation of taxis (Pajunen and Perttunen) and most often the problem is being tackled with a practical point of view (Brown, Carmichael, Parker, Roth, etc on mosquitoes; many authors succeed in destroying insects harmful to crops by sexual trapping). The nomenclature of Fraenkel and Gunn is falling into disuse because of the efforts of Busnel, Viaud, and many others; and as a result of Huber, Roeder, and Vowles' experiments, the idea is taking root that central nervous correlations play a role in the complex regulation of immediate or successive orientating influences exercised by the environment on the insect taxic centers.

Research on sensory organs and their role in behavior is following the same lines and, if morphological, phsyiological or electrophysiological analysis is progressing at a great speed (Autrum, Burtt, Fynlayson, Lopatina, Mazokhin Porshnyakov, Roeder, Schneider, Slifer, Voskressenskaya, etc); it is functional theory that is becoming the rule [unicist or dualistic theory of hygroreception (Begg, Bursell, Dodds, Perttunen, Syrjamaki, etc); theory of visual, auditory, olfactory, and chemical codings (Dethier, Frings, Hartline, Haskell, Popov, Schneider, Wilson, etc)]. The role of sensory organs in apprehending the environment and the methods derived from optomotor behavior are leading to models suggested by von Holst's studies of the vertebrates (Hassenstein, Mittlestaedt, Reichardt, Schöne).

Ethological description is also making progress and precise information is being accumulated concerning the displays of Mantidae (Chopard and Crane), Gryllidae (Alexander, Huber, and Leroy), acridians (Jacobs and Zolessi), cockroaches (Barth, Roth, and Willis), Diptera (Bastock, Manning, Petit, Spieth, etc), Odonata (Buccholtz, Pajunen, Richard, etc), Lepidoptera (Roeder and Tinbergen).

Four types of behavior are becoming of particular interest: 1. Alimentary behavior, which allows penetration to the level of eco-ethological liaison: phagostimulants, inhibitory factors (Brian, Dethier, Fraenkel, Free, Jermy, Matsumoto, Sugiyama, Thorsteinson, etc). 2. Building behavior, which demonstrates one precise interpretation of stereotypes in insect behavior ("thème cyclique temporel fondamental" described by Deleurance on *Polistes*) and a particular connection between the animal and its environment through the work it performs in this environment [Dembowsky and Denis on Trichoptera; Desneux, Emerson, and Grassé (theory of stigmergie) on termites; Darchen, Vuillaume on bees and wasps]. 3. Stinging behavior of the solitary wasps which allows greater precision in establishing the relationship between temporal theme and external environment as represented by the prey (Baerends, Evans, Iwata, Maruyama, Steiner, and Tsuneki). 4. Exploratory behavior which poses a completely new problem in terms close to those concerning vigilance phenomena in vertebrates (Darchen, Goustard, and Chauvin).

As in the past, the problem of the animal's ability to learn continues to preoccupy research workers; but the problem is posed either in unitary terms of regulation for the individual (Chesnokova, Hoyle, Le Masne, Lopatina, Montagner), or in terms of global behavior modification which lead to a definition of psychological or motivational levels (Akre, Brown, Dobrzánski, Grosslight, Hunsperger, Schneirla), or in terms of the possibilities for ecological regulation in biocoenosis (countershading analyzed by de Ruiter).

All this led to the necessity of confronting instinct theories and it was once again in France, on the initiative of Grassé, that an international congress was held, which was both animated and extremely interesting (1956).

Social insect behavior underwent a renewal of interest which led in 1952 to the foundation of the International Union for the Study of Social Insects; and thus also to the discovery of very important phenomena. General ethology benefitted from this as new behavior patterns in ants, termites, wasps, and bees were described for the first time. The psychophysiological division of labor is everywhere established, but very precise analyses were also made of building (stigmergie of Grassé), cooperation or mutual aid (Chauvin, Dobrzánski and Dobrzánska, and Sudd, etc), relationship with commensals (Le Masne, Rettenmeyer, etc). The role of chemical substances secreted by insects and deposited on their trails or in their nests was demonstrated (Brian, Luscher, Schneirla, Stumper, Wilson, etc.); and that of chemical substances transferred from mouth to mouth by licking or trophallaxis (pheromones of the queen: Butler and Pain); the speed of trophallactic communication was measured by means of radioisotopes (Gösswald, Chauvin, Lecomte, Le Masne, Pavan, etc).

New concepts emerged from the work of Emerson, Gösswald, Schneirla, and especially from that of Grassé; the strength of social regulation may dominate all the other reactions of the individual (sociotomy during which termite imaginals move out of the nest into the marching column). Termite mushroom combs are a permanent food source built up at the base, destroyed at the top, transformed by the mycelium; the climatic atmosphere of the nest is regulated by the society; many species have an undoubted agrological role.

As in the work of the previous period, the physiological aspect is always integrated (Joly, Lebrun, Novák, Steinberg, de Wilde, Williams, etc) with the morphological, ecological, and ethological aspects.

The work of behavior geneticists, especially of population geneticists, opened the way toward an interpretation of speciation and evolution (Bösiger, Dobzhansky, Hirsch, Manning, Medioni, etc). Moreover the role of behavior was fundamental in the interpretation of the speciation of social insects, in which Noirot saw all evolutive potential concentrated on the level of ethology: "In its ultimate stage society tends to free itself from ecological servitude: instead of submitting itself to the environment, it moulds the latter to its own end. . . ."

In this respect, the study of groups like melipons (Araujo, Kerr, Michener,

Neto, Sakagami, and Darchen) can help to clarify the comprehension of the evolution of aphids, just as the analysis of the structure of *Halictus* populations (Plateaux-Quenu) allows an enlarged concept of society.

Today, insects have become a privileged model for the study of the response of organisms to stimuli coming from the environment. Their behavior lends itself particularly to the analysis of causality; their small size, the characteristics of their nervous system, the accessibility of most of them, the relative ease with which they can be raised in laboratories, their numerous presence among populations most studied, all make them suitable for different new studies. And if one were to compare the present day with the beginning of the nineteenth century, one might say, like Hoyle, that behavior research leading to neuroethology has progressed at such a speed over the last few years that its results have outpaced those of morphology and anatomy. Henceforth, it is to be desired that, while carrying on with the development of methods and results of behavior study, scientists will use the very precise new information becoming available, thanks to such important technical tools as the electron microscope and electrophysiology, in forming their interpretations.

ACKNOWLEDGMENTS

I am very grateful to Mrs. Kerbrat for the time and consideration she put into her perfect translation of my manuscript.

I wish to thank all colleagues who kindly helped me to collect information for this essay: and particularly Drs. d'Aguilar, Bey Bienko, Chmurzynski, Erhan, and Tsuneki.

LITERATURE CITED

Akre, R. D. 1964. *Behavior of insects associated with army ants with emphasis on chemical trails*. PhD thesis. Kansas St. Univ.

Akre, R. D., Rettenmeyer, C. W. 1968. Trail following by guests of army ants. *J. Kans. Entomol. Soc.* 41:165–74

Alexander, R. D. 1956. *A comparative study of sound production in insects with special reference to the singing* Orthoptera *and* Cicadidae *of the Eastern United States*. PhD thesis. Ohio St. Univ.

Alexander, R. D. 1961. Aggressiveness, territoriality and sexual behaviour in field crickets (Orthoptera, Gryllidae). *Behaviour* 17:2–3, 130–223

Alexander, R. D. 1962. Evolutionary change in cricket acoustical communication. *Evolution* 16:4, 433–67

Allee, W. C. 1931. *Animal Aggregations: A Study of General Sociology*. Chicago: Univ. Chicago Press

Alverdes, F. 1927. *Social Life in the Animal World*. New York: Norton

Autrum, H. Ueber Lautaeusserungen und Schallwahrnehmung bei Arthropoden. *Z. Vergl. Physiol.* 23:332–73

Autrum, H. 1950. Die Belichtungspotentiale und den Gehen des Insekten. *Z. Vergl. Physiol.* 32:176–227

Baerends, G. P. 1941. Fortpflanzungsverhalten und Orientierung der Grabwespe *Anmophila campestris*. *Tijdschr. Sch. Entomol.* 84:68–275

Baerends, G. P. Ethological studies of insect behavior. *Ann. Rev. Entomol.* 4:207–34

Baldus, K. 1927. Untersuchungen zur Analyse der Zwangsbewegungen der Insekten. *Z. Vergl. Physiol.* 6:99–150

Barth, R. 1959. Considerações gerais sôbre os órgãos odoriferos sexuais dos machos des Lepidopteros. *Mem. Inst. Oswaldo Cruz* 51:187–220

Beebe, D. 1952. Introduction to the ecology of the Arima valley—Trinidad. *Zoologica* 37:158–84

Berland, L. 1935. Quelques traits

du comportement des hyménoptères Sphégiens. *Ann. Sci. Nat.,* 10 sér., 18:53–66

Berland, L. 1939. *Les Guêpes.* Paris: Stock

Berlese, A. 1909. *Gli Insetti.* Milan: Libraria. 1004 pp.

Bertholf, K. 1932. The extent of the spectrum for Drosophila and the distribution of stimulative efficiency in it. *Z. Vergl. Physiol.* 18:32–64

Bethe, A. 1898. Dürfen wir den Ameisen und Bienen psychisches Qualitäten zuschreiben? *Pflüger's Archiv.* 70: 15–99

Bey Bienko, G. 1936. *Insectes Dermaptères Fauna U.R.S.S.,* No. 5:1–240. In Russian

Bey Bienko, G. 1950. *Nasekomye tarakanovye* (Insecta, Blattoptera) *Fauna SSSR,* No. 40:1–344. In Russian

Bey Bienko, G. 1961. On some regularities in changes of the virgin steppes invertebrate fauna under cultivation. *Entomol. Obozr. Leningrad* 40: No. 4, 763–75. In Russian

Bey Bienko, G. 1969. *Ectobius duskei* as a characteristic inhabitant of steppes in the U.S.S.R. *Mem. Soc. Entomol. Ital.* Vol. 48

Bodenheimer, F. S. 1937. Population problems of social insects. *Biol. Rev.* 12:393–430

Bohn, G. 1906. Les tropismes. les réflexes et l'intelligence. *Ann. Psychol.* 12:137–56

Bohn, G. 1911. *La Nouvelle Psychologie Animale.* Paris: Gauthier-Villars

Bohn, G. 1940. *Actions Directrices de la Lumière.* Paris

Boldyrev, B. 1912. Das Liebeswerben und die Spermatophoren bei einigen Locustodeen und Gryllodeen. *Horae Soc. Entomol. Rossicae* 40:1–54

Bonnier, G. 1907. Le socialisme chez les Abeilles. *Bull. Inst. Gén. Psychol.* 7:397–426

Bouvier, L. 1922. *The Psychic Life of Insects,* transl. L. O. Howard. New York: Appleton. 377 pp.

Bouvier, L. 1924. La mémoire chez les insectes. *Scientia* 18:103–16

Brandt, A. 1868. Die Entwicklung der Eier von Agrion Virgo. *Tagebl. Vers. Deut. Naturf.* pp. 42–104

Brandt, A. 1874. Über die Eiröhren der Blatta orientalis. *Mem. Acad. Sci. St. Petersburg* Vol. 21

Brian, A. D. 1954. The foraging of bumble bees. *Bee World* 35:61–67, 81–89

Brown, J. L., Hunsperger, R. W. 1963. Neuroethology and the motivation of agonistic behaviour. *Anim. Behav.* 11:439–49

Brun, R. 1914. *Die Raumorientierung der Ameisen.* Jena: Fischer. 234 pp.

Brun, R. 1916. Le problème de l'orientation lointaine chez les Fourmis. *Rev. Suisse Zool.* 24:355–88

Brun, R. 1932. *Psychologische Forschungen an Ameisen* Abderhalden's Handbuch 6D

Bruytendijk, F. J. 1928. *Psychologie des Animaux.* Paris

Buddenbrock, W. von 1928. *Grundriss der vergleichende Physiologie.* Berlin

Buddenbrock, W. von 1931. Beiträge zur Lichtkompassorientierung (menotaxis) der Arthropoden. *Z. Vergl. Physiol.* 15:597–612

Bugnion, F. Z. 1873. Moeurs de *Haemonia equisete. Arch. Sci. Phys. Nat.*

Bugnion, F. Z. 1882. Ruckenrüben Schäden durch die Larvae von *Silpha reticulata. Mitt. Schweiz. Entomol. Ges.* Vol. 6

Bugnion, F. Z. 1886. Moeurs de l'*Hylesinus oliependa* et les galeries du *Phloetribus oleae. Mitt. Schweiz. Entomol. Ges.* 7:218, 225

Bugnion, F. Z. 1923. La Fourmi fileuse de Ceylan. *Riv. Sci.* 10:17–30

Burmeister, H. 1836. *A Manual of Entomology,* transl. W. E. Shuckard. London: Edward Churton

Burtt, E. T., Catton, W. T. 1956. Electrical responses to visual stimulation in the optic lobes of the locust and certain other insects. *J. Physiol.* 133: 68–88

Busnel, R. G., Dumortier, B. 1955. La trémulation du corps et la transmission aux supports des vibrations en résultant comme moyen d'information à courte portée chez Ephippigères. *Bull. Soc. Zool.* 80:18

Busnel, R. G., Loher, W. 1956. Etude des caractères physiques réactogènes de signaux acoustiques artificiels déclencheurs de phonotropismes chez les Acrididae. *Bull. Soc. Entomol. Fr.*

Busnel, R. G. 1954. Observation sur le comportement acoustico sexuel de la femelle d'*Ephippiger bitterensis. C. R. Soc. Biol.* 148:1589

Busnel, R. G. 1963. On certain aspects of animal acoustic signals. *Acoustic Behavior of Animals,* pp. 69–111. Amsterdam: Elsevier

Butler, C. G. 1945. The behaviour of bees when foraging. *J. Roy. Soc. Arts,* 93:No. 4699, 501–24

Butler, C. G. 1958. Queen supersedure and swarming. *Cent. Assoc. Bee Keepers,* pp. 1–12
Butler, C. G. 1962. Some recent work on bee behaviour. *Proc. Roy. Soc. Med.* 55:545–48
Butler, C. G. 1962. *The World of the Honeybee.* London: Collins
Butler, C. G. 1967. Insect pheromones. *Biol. Rev. G. B.* 42:42–87
Chauvin, R. 1956. *Vie et Moeurs des Insectes.* Paris: Payot. 223 pp.
Chen, S. G. 1937. Social modification of the activity of ants in nest building. *Physiol. Zool.* 10:420–36
Chen, S. G. 1937. The leaders and followers among the ants in nest building. *Physiol. Zool.* 10:437–55
Chmurzynski, J. A. 1966. Research on animal behavior at the Nencki Institute of Experimental Biology. *Acta Biol. Exp., Warsaw* 26:79–94
Chmurzynski, J. A. 1969. Orientation of blowflies towards white light of various intensities. *Bull. Acad. Pol. Sci.* Cl. 2, 17, No. 3
Chopard, L. 1938. La biologie des Orthoptères. *Encycl. Entomol. Paris,* sér. A 20:541 pp.
Coaton, W. G. H. 1947. The Pienaars River complex of wood-eating termites. *J. Entomol. Soc. S. Afr.* 9:No. 2, 130–77
Cornetz, V. 1910. Trajets de fourmis et retours au nid. *Inst. Gen. Psychol. Mém.* No. 2:167
Cornetz, V. 1914. *Les Explorations et les Voyages des Fourmis.* Paris: Flammarion
Cott, H. B. 1940. *Adaptive Coloration in Animals.* Oxford: Oxford Univ. Press
Crozier, W. J. 1928. Tropisms. *J. Gen. Psychol.* 1:213–38
Crozier, W. J. 1929. The study of living organisms. In *Foundations of Experimental Psychology,* ed. Murchison. Clark Univ. Press.
Cuenot, L. 1925. *L'Adaptation.* Paris
Danilievski, A. S. 1961. *Photoperiodism and Seasonal Development of Insects.* Leningrad, Russian ed.; English ed., 1965, Oliver & Boyd
Darchen, R. 1968. La construction dans la ruche. *Traite de Biologie de l'Abeille.* Paris: Masson
Darwin, Ch. 1874. Recent researches on termites and stingless honey bees. *Am. Nat.* Vol. 7
Deleurance, E. Ph. 1957. Contribution à l'étude biologique des Polistes. I—L'activité de construction. *Ann. Sci. Nat. Zool.* 19:91–222
Desneux, J. 1923. Quelques points particuliers de la Société des Termites. *Bull. Soc. Entomol. Belgique* 4:No. 13, 135–42
Desneux, J. 1923. Fondation de colonies nouvelles par des adultes ailés. *Bull. Soc. Entomol. Belgique* 5:No. 1–2, 17–24
Desneux, J. 1952. Notes sur la phylogénie des nidifications des Apicotermes. *Ann. Sci. Nat. Zool.* 14:393–445
Dethier, V. G. 1957. Communication by insects: physiology of dancing. *Science* 125, 3243:331–36
Dethier, V. G. 1967. Feeding and drinking behavior of invertebrates. *Handb. Physiol. Am. Physiol. Soc.* Vol. 1
Doflein, F. 1905. Beobachtungen an den Weberameisen. *Biol. Centralblatt* 25:497–507
Doflein, F. 1916. *Der Ameisenlöwe.* Jena: Fischer. 138 pp.
Dujardin, F. 1852. Observations sur les Abeilles. *Ann. Sci. Nat.* Vol. 15
Dumortier, B. 1963. Ethological and physiological study of sound emissions in arthropods. *Acoustic Behavior of Animals,* pp. 583–654. Amsterdam: Elsevier
Eidmann, H. 1927. Die Sprache der Ameisen. *Rev. Zool. Russe* 7:37–47
Ellis, P. E. 1953. Social aggregation and gregarious behaviour in hoppers of *Locusta migratoria. Behaviour* 5: 225–60
Ellis, P. E. 1956. Learning and social aggregation in two species of locust. *Nature* 178:1007
Ellis, P. E. 1959. Learning and social aggregation in locust hoppers. *Anim. Behav.* 7:91–106
Emerson, A. E. 1929. Communication among termites. *Trans. Int. Entomol. Congr., 10th, Ithaca* 2:722–27
Emerson, A. E. 1929. The social life of termites. *Rep. Symp. Termite Invest. Comm., Univ. California* 13:24–30
Emerson, A. E. 1937. The termite problem. *Nat. Hist. NY* 39:No. 4, 249–54
Emerson, A. E. 1937. Termite nests. A study of the phylogeny of behavior. *Science* 85:No. 2193, 56
Emerson, A. E. 1938. Termite nests. A study of the phylogeny of behavior. *Ecol. Monogr.* 8:No. 2, 247–84
Emerson, A. E. 1952. The supraorganismic aspects of the society. In France; Cent. Nat. Rech. Sci. 34 *Structure et physiologie des societes animales*—Paris, Service des Publi-

cations, 13:333–54
Emerson, A. E. 1958. The evolution of behavior among social insects. *Behav. Evol.*, pp. 311–35
Emerson, A. E. 1962. The impact of Darwin on biology. *Acta Biotheoretica* 15:175–216
Escherich, K. L. 1906. *Die Ameise.* Braunschveig. Berlin
Escherich, K. L. 1912. Von der Beukunst der Termiten. *Biolog. Centralblatt* 32:211
Ewing, A., Manning, A. The evolution and genetics of insect behavior. *Ann. Rev. Entomol.* 12:471–94
Fabre, J. H. 1879–1908. *Souvenirs Entomologiques.* 10 volumes. Paris: Delagrave
Ferton, Ch. 1923. *La Vie des Abeilles et des Guêpes.* Paris: E. Chiron
Fielde, A. 1900. Portable ant nests. *Biol. Bull. Boston* Vol. 2
Fielde, A. 1904. The power of recognition among ants. *Biol. Bull.* 7:227–50
Forel, A. 1886–1888. Experience et remarques critiques sur les sensations des insectes. *Recueil Zool. Suisse* 2: 1–76
Forel, A. 1908. *The Senses of Insects,* transl. Yearsley, p. 324. London: Methuen
Forel, A. 1910. *Das Simmelsleben der Insekten.* Munchen:
Forel, A. 1921–1923. *Le Monde Social des Fourmis,* 5 volumes. Geneve: Kundig
Fraenkel, G. S., Gunn, D. L. 1940. *Orientation in Animals; Kineses, Taxes and Compass Reactions.* Oxford: Oxford Univ. Press
Free, J. B. 1962. The foraging behaviour of honeybees in relation to pollination. *Rep. Cent. Assoc. Brit. Beekeepers Assoc.*, pp. 1–10
Frings, H., Frings, M. 1958. Uses of sounds by insects. *Ann. Rev. Entomol.* 3:87–106
Frings, H., Frings, M. 1960. *Sound Production and Sound Perception by Insects,* pp. 1–108. Pennsylvania St. Univ. Press
Frisch, K. von 1919. Über den Geruschsinn der Bienen und seine blutendbiologische Bedeutung. *Zool. Jahrb. Zool. Physiol.* 37:1–238
Goetsch, H. 1930. Beiträge zur Biologie Körnersammeln der Ameisen. *Morphol. Oekol. Tiere* 16:371–452
Gösswald, K. 1942. Die rote Waldameise und ihre Umwelt. *Der Biologe* 11:180–91
Gösswald, K. 1951. Die rote Waldameise im Dienste der Waldhygiene. 160S., 50 Abb. (Metta-Kinau-Verlay)
Gösswald, K. 1954. Der Termitenstaat. *Kosmos / Handweisen Naturfreunde* 50:59–65
Graber, V. 1877–1879. *Die Insekten* 2 bd. Munchen: Oldenbourg
Grassé, P. P. 1939. La reconstruction du nid et le travail collectif chez les Termites supérieurs. *J. Psychol.* 36: 370–96
Grassé, P. P. 1939. Comportement et particularités des soldats de Termites. *Bull. Soc. Zool.* 44:251–62
Grassé, P. P. 1942. L'essaimage des termites. Essai d'analyse causale d'un complexe instinctif. *Bull. Biol. Fr. Belg.* 76:No. 4, 347–82
Grassé, P. P. 1952. L'effet de groupe chez les insectes. *Bull. (Sect. Franç) Union Int. Etude Insectes Sociaux* 1: No. 1, 32–43
Grassé, P. P. 1952. Le fait social: ses critères biologiques, ses limites. *Cent. Nat. Rech. Sci. Structure et physiologie des Societes animales, 34*
Grassé, P. P., Noirot, C. 1948. La "climatisation' de la termitière par ses habitants et le transport de l'eau. *C. R. Acad. Sci.* 227:869–71
Grassé, P. P., Noirot, C. 1951. La sociotomie: Migration et fragmentation de la termitière chez les Anoplotermes et les Trinervitermes. *Behaviour* 3:No. 2, 146–66
Grassi, B. 1887. Nuove ricerche sulle termiti. Nota preliminare. *Bull. Soc. Entomol. Ital.* 79:75–80
Grassi, B. 1888. Re e regine di sostituzione nel regno delle termiti. *Bull. Soc. Entomol. Ital.* 20:139–47
Grassi, B. 1888. Esatzpaar bei dem Termiten. *Zool. Anz.* 11:63
Grassi, B. 1888. Weitere Mittheilungen über die Ersatz-Könige und Königinnen im Reiche der Termiten. *Zool. Anz.* 11:615–18
Hagen, H. A. 1858. *Monographie der Termiten. Limaea Entomol.* 12
Haskell, P. T. 1957. Stridulation and associated behaviour in certain Orthoptera. Analysis of the stridulation of, and behaviour between, males. *Brit. J. Anim. Behav.* 5:139–48
Haskell, P. T. 1958. Stridulation and associated behaviour in certain Orthoptera. II—Stridulation of females and their behaviour with males. *Brit. J. Anim. Behav.* 6:27–42
Haskell, P. T. 1961. *Insect Sounds.* Chicago: Univ. Press
Haskell, P. T. 1966. Insect Behaviour.

Symp. Roy. Entomol. Soc. London, No. 3

Haskins, C. P., Haskins, E. P. 1950. Note on the method of colony foundation of the Ponerine ant, *Brachyponera lutea. Psyche* 57:1–9

Haskins, C. P., Haskins, E. P. 1950. Note on the biology and social behavior of the archaic Ponerine ants of the genera Myrmecia and Promyrmecia. *Ann. Entomol. Soc. Am.* 43:461–91

Hassenstein, B. 1968. Über die Wahrnehmung der Bewegung von Figuren und unregelmässigen Helligkeitsmustern. *Z. Vergl. Physiol.* 40:556–92

Highman, K. C. 1964. Hormones and behaviour in insects. *Viewpoints Biol.* 3:219–55

Hingston, R. W. G. 1931. *Problèmes de l'instinct et de l'intelligence chez les Insectes.* Paris

Hodgson, E. S. 1962. Neurosecretion and behaviour in arthropods. *Gen. Compar. Endocrinol.* 2:Suppl. 1,180–85

Holmes, S. J. 1911. *The Evolution of Animal Intelligence.* New York: Macmillan

Holmgren, N. 1909–1912. Termitenstudien. Uppsala and Stockholm: Almqvist et Wiksells

Hoyle, G. 1970. Cellular mechanisms underlying behavior-Neuroethology. In *Advances in Insect Physiology,* Vol. 7. New York: Academic

Huber, F. 1963. The role of the central nervous system in Orthoptera during the coordination and control of stridulation. *Acoustic Behaviour of Animals,* pp. 440–88. Amsterdam: Elsevier

Huber, J. P. 1810. *Recherches sur les Moeurs des Fourmis Indigènes.* Paris and Geneve: J. J. Paschoud

Ilse, D. 1932. Zur "Formwarhrnehmung" der Tagfalter. Spontane Bevorzugung von Formmerkmalen durch Vanessen. *Z. Vergl. Physiol.* 17:537–56

Imms, A. D. 1925. *A General Textbook of Entomology.* London: Methuen

Imms, A. D. 1931. *Social Behaviour in Insects.* London: Methuen. 117 pp.

Imms, A. D. 1951. *Insect Natural History,* pp. 33, 291–92, 298. London: Blakiston

Janet, C. 1893–1920. Etudes sur les Fourmis. *Notes Ann. Soc. Entomol. France Mem. Soc. Zool. France*

Janvier, Pere 1933. Etude biologique de quelques Hyménoptères du Chili. *Ann. Sci. Nat.* 16, 2:209–356

Jeannel, R. 1944. Les Fossiles Vivants des Cavernes. Paris: Gallimard. 322 pp.

Jones, M. D. R. 1966. The acoustic behaviour of the bush cricket *Pholidoptera griseoaptera.* I—Alternation synchronism, and rivalry between males. *J. Exp. Biol.* 45:1, 15–30

Jucci, C. 1926. Il problema della caste nei termiti. *Riv. Biol.* 8:No. 3, 404–9

Jucci, C. 1952. Symbiosis and phylogenesis in the Isoptera. *Nature* 169: 837

Jucci, C. 1960. Genetics, evolution and systematics in some Insects. *Quaderno No. 47 Evoluzione e genetica. Colloquio Internationale.* Roma 8–11 aprile 1959, pp. 105–27. In Italian

Kalmus, H. 1937. Photohorotaxis eine neue Reaktionsart, gefunden an den Eilarven von Dixippus. *Z. Vergl. Physiol.* 24:644–55

Kennedy, J. S. 1939. The behaviour of the Desert Locust (*S. gregaria*) in an outbreak centre. *Trans. Roy Entomol. Soc. London* 89:385–542

Kennedy, J. S. 1945. Observations on the mass migration of Desert locust hoppers. *Trans. Roy. Entomol. Soc. London* 95:247–62

Kirby, W., Spence, W. 1815–1828. *Introduction to Entomology.* 4 volumes. London

Kuhn, A. 1919. *Die Orientierung der Tiere im Raum.* Jena: Fischer. 71 pp.

Kuhn, A. 1927. Ueber den Farbensinn der Bienen. *Z. Vergl. Physiol.* 5:762–800

Kuhn, A. 1929. Phototropisms und Phototaxis der Tiere. In Bethe's *Handbuch der Norm. Pathol. Physiol.* 11:17–35

Lameere, A. 1915. L'origine des sociétés d'insectes. *Rev. Gén. Sci. Pures Appl.* 26:459–64

Lameere, A. 1920. The origin of insect societies. *Ann. Rep. Smithson. Inst.,* pp. 511–21. Translated

Lecomte, J. 1961. Le comportement agressif des ouvrières d'*A. mellifica. Ann. Abeille* 4:165–270

Leroy, Y. 1966. Signaux acoustiques, comportement et systématique de quelques espèces de Gryllides (Orthoptères, Ensifères). *Bull. Biol. Fr. Belg.* 1:3–134

Loeb, J. 1918. *Forced Movements, Tropisms and Animal Conduct.* Philadelphia and London

Lopatina, N. G., Chesnokova, E. G. 1969. The significance of the individual experience of bee on distance information to honey bees family. *Trudi* Vol. 53

Lubbock, J. 1883. *Fourmis, Abeilles et Guêpes.* Paris

Luscher, M. 1953. The termite and the cell. *Sci. Am.* 188:No. 5, 74–76, 78

Luscher, M. 1960. Hormonal control of caste differentiation in termites. *Ann. NY Acad. Sci.* 89:549–63

Luscher, M. 1961. Demonstration of a trail pheromone in termites. *Symp. Gen. Biol. Ital.* 11:190–92

Maier, N. R. F., Schneirla, T. C. 1964. *Principles of Animal Psychology.* New York: Dover

Malyshev, S. I. 1936. The nesting habits of solitary bees. *Eos* 11:201–309

Malyshev, S. I. 1959. *Hymenoptera, Their Origin and Evolution.* Moscow. In Russian

Mast, S. O. 1911. *Light and the Behavior of Lower Organisms.* New York: Wiley

Mazokhin Porshnyakov, G. A. 1969. The colour vision of the honey bee and the problem of the adaptable significance of flower colours. *Trudi* Vol. 53

Michener, C. D. 1953. Problems in the development of social behavior and communications among insects. *Trans. Kans. Acad. Sci.* 56:1–15

Michener, C. D., Michener, M. H. 1951. *American Social Insects,* pp. 191–227. New York: Van Nostrand

Miller, E. M. 1940. Chemical integrative mechanisms in insect societies. *Proc. Florida Acad. Sci.* 5:136–47

Morgan, C. Lloyd 1912. *Instinct and Experience.* London

Muller, Fritz 1873. Beiträge zur Kenntniss der Termiten. *Jenaische Z. Nat.* pp. 75–87

Muller, Fritz 1874. Recent researches on termites and honey bees. *Nat. Febr.* 19, B. 9

Nasonov, N. V. 1893. Ob orghanizatzii Termitov v svyazis postroikoi imi ghnyezd. *Entomol. Izslyedovaniya* pp. 1–30

Novak, V. 1958. *Neznámý svět Hmyzn.* Prague: Orbis

Noyes, F. K. 1912. Termites the first civilized people. *Volta Rev.* 13:No. 10, 561–69

Packard, A. S. 1898. *Textbook of Entomology.* New York: Macmillan

Pain, J. 1961. Sur la phéromone des reines d'Abeilles et ses effets physiologiques. *Ann. Abeille* 4:73–152

Peckham, G. W., Peckham, E. G. 1898. *On the Instincts and Habits of Solitary Wasps.* Boston: Houghton-Mifflin

Picard, F. 1933. *Les phénomènes Sociaux chez les Animaux.* Paris: A. Colin

Pieron, H. 1941. *Psychologie Zoologique* Paris: P. U. F.

Popov, A. V. 1969. Comparative investigation of sound signals of communication and some principles of hearing system organization in cicadas and orthopteras. *Trudi* Vol. 53

Regen, J. 1912. Experimentelle Untersuchungen über das Gehör von *Liogryllus campestrus. Zool. Anz.* 40: 305–16

Regen, I. 1914. Untersuchungen Über die Stridulation und das Gehör von *Thamnotrizon apterus Fab.* (Männchen). *Anz. Akad. Wiss.; Wien* 123: 853–92

Regen, I. 1923. Über die Orientierung des Weibchens von *Lyogryllus campestris* L. nach dem Stridulationsschall des Männchens. *S. B. Akad. Wiss; Wien* 132:81–88

Regen, I. 1926. Über die Beeinflussung der Stridulation von *Thamnotrizon apterus* Feb. durch kunstlich erzeugte Tone verschiedenartige Geräusche. *S. B. Akad. Wiss; Wien* 135:329–68

Richard, G. 1949. Les réactions phototaxiques des termites. *Ann. Sci. Nat. Zool. Biol. Anim.* 10:67–73

Richard, G. 1951. Le phototropisme du termite à cou jaune (*Calotermes flavicollis*) et ses bases sensorielles. *Ann. Sci. Nat. Zool. Biol. Anim.* 12:485–603

Roeder, K. D. 1935. An experimental analysis of the sexual behavior of the praying mantis. *Biol. Bull.* 69: 203–20

Roeder, K. D. 1955. Spontaneous activity and behavior. *Sci. Mon.* 80: 362–70

Roeder, K. D. 1963. *Nerve Cells and Insect Behavior.* Cambridge: Harvard Univ. Press

Roth, L. M. 1962. Hypersexual activity induced in females of the cockroach *Nauphoeta cinerea. Science* 138: 1267–69

Snyder, T. E. 1920. "White ants" as pests in the United States and methods of preventing their damage. *Calif. St. Dept. Agr. Mon. Bull.* 9: No. 1–2, 7–20

Snyder, T. E. 1948. *Our Enemy the Termite.* Ithaca: Comstock. 261 pp.

Steiner, A. 1962. Etude du comportement prédateur d'un Hyménoptère Sphégien: *Liris nigra. Ann. Sci. Nat. Zool.* 4:1–126

Svidersky, V. 1969. Problèmes actuels de la structure et de la fonction du Système nerveux des Insectes. *Horae Soc. Entomol. Unionis Soveticae* Vol. 53

Tinbergen, N. 1963. On aims and methods of ethology. *Z. Tierpsychol.* 20: 410–33

Tinbergen, N., Meeuse, B. J. D., Boerema, L. K., Varossieau, W. W. 1942. Die balz des sentfalters, *Eumenis* (=Satyrus) *semele* L. *Z. Tierpsychol.* 5, h. 1:182–225

Tshumakova, B. M., Gorjunova, Z. S. 1963. Development of males of *Prospaltella perniciosi* parasite of San Jose Scale. *Rev. Entomol. U.R.S.S.* 42:320–28

Uexkull, J. von 1909. *Umwelt und Innenwelt der Tiere.* Berlin

Uvarov, B. P. 1928. *Locusts and Grasshoppers.* London: Imp. Bur. Entomol.

Verlaine, L. 1924–1934. L'instinct et l'intelligence chez les Hyménoptères. *Mem. Ann. Bull. Acad. Roy. Sci., Soc. Roy. Zool., Soc. Entomol. Belg. Soc. Roy. Sci. Liege*

Viaud, G. 1950. *Les Tropismes.* Paris: P. U. F.

Voskresenskaya, A. K. 1969. The regulating function of insect nervous systems. *Trudi* Vol. 53

Vowles, D. M. 1961. Neural mechanism in Insect behaviour. In *Current Problems in Animal Behaviour,* ed. W. H. Thorpe, O. L. Zangwill. Cambridge: Cambridge Univ. Press.

Wasmann, E. 1905. *Instinkt und Intelligence im Tierreich.* Fribourg

Wheeler, W. M. 1911. The ant colony as an organism. *J. Morphol.* 22:307–25

Wheeler, W. M. 1920. *Ants.* New York: Harcourt, Brace

Wheeler, W. M. 1920. The termitodesca or biology and society. *Sci. Mon.* 10:No. 2, 113–24

Wheeler, W. M. 1923. *Social Life Among the Insects.* New York: Harcourt, Brace. 375 pp.

Wheeler, W. M. 1928. *The Social Insects.* New York: Harcourt, Brace. 378 pp.

Wheeler, W. M. 1934. Animal societies. *Sci. Mon.* 39:289–301

Wilson, E. O. 1965. Chemical communication in the social insects. *Science* 149, No. 3688:1064–71

Zolessi, Lucrecia Covelo de. 1958. Bioecologia y ontogenesis de *Scotussa cliens* (Stäl) Lieb. (Acridoidea, Catantopidae) en el Uruguay. *Univ. Rep.,* Montevideo

Zolessi, Lucrecia Covelo de. 1963. Reproduccion y ontogenia de *Coryacris angustipennis* (Bruner) 1900 (Acrididae, Romaleinae). *Univ. Fac. Agron.* Bol. 67, Montevideo

1966. Insect behaviour. *Symp. Roy. Entomol. Soc. London* No. 3

1962. Biological receptor mechanisms. *Symp. Soc. Exp. Biol.* Vol. 16

1961. Entomological studies from a high tower in Mpsanga Forest, Uganda. *Trans. Roy. Entomol. Soc. London* 113

SUBJECT INDEX

I